T0239098

Lecture Notes in Civil Engineering

Volume 213

Lecture Notes in Civil Engineering (LNCE) publishes the latest developments in Civil Engineering - quickly, informally and in top quality. Though original research reported in proceedings and post-proceedings represents the core of LNCE, edited volumes of exceptionally high quality and interest may also be considered for publication. Volumes published in LNCE embrace all aspects and subfields of, as well as new challenges in, Civil Engineering. Topics in the series include:

- Construction and Structural Mechanics
- Building Materials
- Concrete, Steel and Timber Structures
- Geotechnical Engineering
- Earthquake Engineering
- Coastal Engineering
- Ocean and Offshore Engineering; Ships and Floating Structures
- Hydraulics, Hydrology and Water Resources Engineering
- Environmental Engineering and Sustainability
- Structural Health and Monitoring
- Surveying and Geographical Information Systems
- Indoor Environments
- Transportation and Traffic
- Risk Analysis
- Safety and Security

To submit a proposal or request further information, please contact the appropriate Springer Editor:

- Pierpaolo Riva at pierpaolo.riva@springer.com (Europe and Americas);
- Swati Meherishi at swati.meherishi@springer.com (Asia - except China, and Australia, New Zealand);
- Wayne Hu at wayne.hu@springer.com (China).

All books in the series now indexed by Scopus and EI Compendex database!

More information about this series at https://link.springer.com/bookseries/15087

Guangliang Feng
Editor

Proceedings of the 8th International Conference on Civil Engineering

 Springer

Editor
Guangliang Feng
Institute of Rock and Soil Mechanics
China Academy of Sciences
Wuhan, China

ISSN 2366-2557 ISSN 2366-2565 (electronic)
Lecture Notes in Civil Engineering
ISBN 978-981-19-1262-7 ISBN 978-981-19-1260-3 (eBook)
https://doi.org/10.1007/978-981-19-1260-3

This Springer imprint is published by the registered company Springer Nature Singapore Pte Ltd.
The registered company address is: 152 Beach Road, #21-01/04 Gateway East, Singapore 189721, Singapore

Preface

Civil engineering is closely related to people's life, and the quality of civil engineering directly affects people's life and personal safety and affects the development of society to a great extent. Therefore, it is very necessary to analyze the development status of civil engineering, to improve the problems existing in the development of civil engineering.

This book contains the proceedings of the 8th International Conference on Civil Engineering (ICCE 2021) which was held on December 5, 2021, as a hybrid conference (both physically and online via Zoom) at Nanchang Institute of Technology in Nanchang, China. The conference is hosted by Nanchang Institute of Technology and Civil Engineering Academy of Jiangxi Province, co-organized by Journal of Rock and Soil Mechanics, Key Laboratory for Safety of Water Conservancy and Civil Engineering Infrastructure in Jiangxi Province, Key Laboratory of Sichuan Province for Road Engineering, Southwest Jiaotong University. More than 150 participants were able to exchange knowledge and discuss the latest developments at the conference. The book contains 139 peer-reviewed papers, selected from more than 300 submissions and ranging from the theoretical and conceptual to strongly pragmatic and addressing industrial best practice.

The book shares practical experiences and enlightening ideas from civil engineering and will be of interest to researchers and practitioners of civil engineering everywhere.

Organizations

Conference Committee

Chairs

Bin Xu	Nanchang Institute of Technology
Chengxiang Xu	Wuhan University of Science and Technology
Kaihua Zeng	Nanchang Institute of Technology
Pizhong Qiao	Shanghai Jiaotong University
QingQuan Liang	AASCCS

Members (Alphabet Sequence)

Fang Xu	China University of Geosciences (Wuhan)
Guannan Liu	China University of mining and Technology
Hongtao Zhang	North China University of Technology
Jiawen Li	North China University of Technology
Jie Sun	Wuhan University of Science and Technology
Lei Zeng	Changjiang University
Libiao Bai	Chang'an University
Qiuxin Liu	Wuhan University of Science and Technology
Tiequan Ni	Huanggang Normal University
Xiao Yu	Wuhan University of Science and Technology
Yifeng Zhong	Chongqing University
Yongjun Ni	Beijing Jiaotong University
Zongwu Chen	China University of Geosciences (Wuhan)

Scientific Committee

Chair

Pizhong Qiao	Shanghai Jiaotong University

Members (Alphabet Sequence)

Abdulqader Said Ali Al-Najmi	The University of Jordan
Abdul Rahman M. Sam	Universiti Teknologi Malaysia
Ali Kaveh	Iran University of Science and Technology
Antonio Panico	Pegaso Online University
Awrejcewicz Jan	Lodz University of Technology
Belén González-Fonteboa	University of A Coruña
Bingxiang Yuan	Guangdong University of Technology
Chikhotkin Victor	Russian Academy of Natural Sciences
Chung Bang Yun	Zhejiang University
Claudia Vitone	Politecnico di Bari
Eduardo Rojas	Universidad Autónoma de Querétaro
Ehsan Seyedi Hosseininia	Ferdowsi University of Mashhad
Evangelos J. Sapountzakis	National Technical University of Athens
Francesco Pellicano	University of Modena and Reggio Emilia
Francisco Agüera-Vega	Universidad de Almería
Guofeng Du	Yangtze University
Harvinder Singh	Guru Nanak Dev Engineering College
Jamal Khatib	University of Wolverhampton
Jiming Xie	Zhejiang University
Krzysztof Schabowicz	Wroclaw University of Technology
Lei Wang	Changsha University of Science and Technology
Muhammad Hadi	University of Wollongong
Mohammad Arif Kamal	Aligarh Muslim University
Mohamed Elgawady	Missouri University of Science and Technology
Paritosh Srivastava	Galgotias University Greater Noida
Pawel K. Zarzycki	Koszalin University of Technology
Qian Wang	Manhattan College
Shu-Lung Kuo	Open University of Kaohsiung
Yi Bao	Stevens Institute of Technology
Zengfu Wang	Northwestern Polytechnical University

Publication Committee

Chair

Guang-Liang Feng	Institute of Rock and Soil Mechanics, Chinese Academy of Sciences

Members (Alphabet Sequence)

Ali Zaidi	University of Sherbrooke
Junqing Lei	Beijing Jiaotong University
Xiaohua Li	Chongqing University
Yuncheng He	Guangzhou University

Contents

Particle Morphology Effect on the Soil Pore Structure

M. Ali Maroof[1], Danial Rezazadeh Eidgahee[2], and Ahmad Mahboubi[3(✉)]

[1] Shahid Beheshti University, Tehran, Iran
[2] Faculty of Civil Engineering, Semnan University, Semnan, Iran
[3] Faculty of Civil, Water and Environmental Engineering, Shahid Beheshti University, Tehran, Iran
mahboubi@sbu.ac.ir

Abstract. The soil fabric can be expressed as a network model. Granular media voids connectivity and constriction size distribution may lead to movement of air, fluids, and solids in the soil, and therefore affect the chemical, physical and mechanical properties of soils. Understanding the soil voids areas and their inter-connection might be helpful in understanding different phenomena such as transport in porous media, water retention, fluid flow in the soil, soil contamination, internal erosion, suffusion, and filtration. In addition, specifying the soil voids interconnectivity can help researchers and practical engineers to provide the best rehabilitation and remediation approaches. The pore network was investigated in the current study, assuming the soil particles to be similar to discrete spheres and particles with different shapes. Also, based on the modelling techniques, the profiles of pore connectivity and constriction size distribution were assessed.

Keywords: Soil structure · Soil fabric · Pore network model · Constriction size distribution · Particle shape

1 Introduction

The porous medium can be considered as the interconnected networks of bonds and nodes. The nodes represent the pores, and bonds show constriction between these pores. Fatt [1] pioneered to use of two-dimensional square and regular networks, including cross-sectional pipes with various radii, to describe the stone pore structure [1]. The physics of air and fluid flow and soil mass transport can be simulated using the network model [2]. Network modelling is one of the methods that can be recruited to estimate the transport in porous media [1–5], fluid flow in porous media and permeability [6, 7] and the amount and size of eroded particles from soil skeleton [7–11]. A pore network should specify the geometric characteristics and topology of porous media. In order to conduct an accurate simulation of the porous media, the nature of the convergence and divergence of pore distribution, multiple connectedness of pore space, and pore size distribution should be considered [12]. In practice, a network model can be created by average connectedness (coordination number), pore size distribution, and pore length

© The Author(s) 2022
G. Feng (Ed.): ICCE 2021, LNCE 213, pp. 1–10, 2022.
https://doi.org/10.1007/978-981-19-1260-3_1

distribution [3]. The regular networks, such as those comprised of square shapes or honeycomb forms, cannot describe irregular soils and rock structures due to those real amorphous and tortuous structures. By changing the length or removing some of the constrictions, a network similar to the actual soil structure voids can be created. Voronoi and Kylie lattice are two examples of these networks. Based on the network configuration and the coordination number, different networks can be generated. Figure 1 illustrates some examples of two-dimensional networks of soil and rock pore structures, bonds, and nodes which are equivalent to the soil pores and the paths of their connectedness in these networks, respectively. The coordination number (Cn) can describe this relationship, which represents the bonds connected to each node [2].

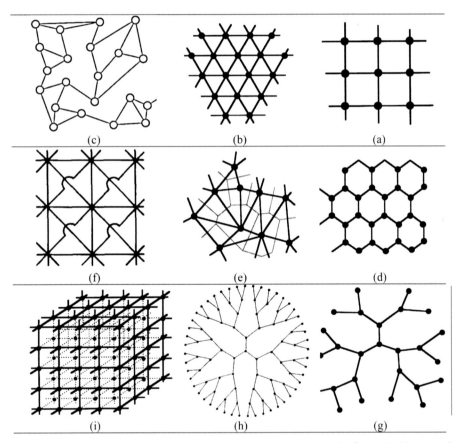

Fig. 1. Examples of the bi-dimensional networks, (a) square network, Cn = 4, (b) rectangle network, Cn = 6, (c) irregular network, Cn = 3, (d) hexahedral network, Cn = 3, (e) Voronoi lattice, Cn = 4, (f) intercept square network, Cn = 8, (g) Kylie lattice (simple graph), Cn = 3, (h) Bethe lattice, Cn = 3, (g) cubic network, Cn = 6. [3, 13–16]

The shape of the pores can be spherical, cylindrical, incomplete conical, crossed, bubble-shaped, similar to concrete pores, or flat such as mica or clay. However, pores mostly have irregular shapes [17]. Pore geometry and its tortuosity and topology play a significant role in the structure of porous media [18]. The coordination number can be considered as the main feature of network topology. A network model should determine the mean of the coordination number, the topology of pore connectivity on the microscopic scale, and its distribution [14, 19]. The coordination number of the different proposed networks ranges from 1 to 26. Raoof and Hassanizadeh [19] proposed a three-dimensional cube network, in which each pore has 13 outputs in different directions, and its coordination number is 26. The network models can be divided into quasi-static and dynamic displacement types [20]. Quasi-static models create a capillary pressure in the network and calculate the final static position of fluid-liquid interfaces. An inflow rate is usually applied in dynamic models, which relates the transient pressure and interface position.

In order to describe each network element, some parameters such as radius, shape factor, volume, area, and length should be determined [21, 22]. It is difficult to measure and calculate these parameters due to the complexity and amorphous of the porous medium; however, these parameters are calculated for mono-spheres [23, 24]. Regular and irregular models of the proposed pore network for simulating soil structure are shown in Table 1. Some of these models were suggested to mimic fluid flow in porous media; the model of the proposed network tube of Fatt [1], the ball and tube network model of Chandler et al. [25], Raoof and Hassanizadeh [19], Indraratna & Vafai [8], Kovacs [7], conical pore and tube Network Model of Toledo et al. [2].

Primary constrictions for fluid flow are through the pore throats, where pores interconnect with each other. In sediments, the dimension of the two interconnected pores might specify the constriction size or the pore thought size (d_{throat}) is typically half of the pore dimension (d_{pore}). The following geometric relations apply to a simple cubic packing of monosized spheres: $d_{pore} = 0.73d_{grain}$, $d_{throat} = 0.41d_{grain}$ and $d_{throat} = 0.56d_{pore}$. So the correlation between k-and-d_{pore} can be extended to k-and-d_{throat}, applying an appropriate correction factor [30]. In the diagenetically altered, and sedimentary structures, the pore and constriction size ratio may deviate apparently from $d_{throat} \approx 0.5d_{pore}$ [31]; thus, considerations must be taken into account for the constriction size. For example, the hydraulic conductivity of carbonates can be associated with the square of the maximum constriction size ($\max[d_{throat}]^2$) and maintain the quadratic relation in Hagen-Pouseuille [30]. The soil porosity relies on different parameters, including packing density, the extent of the particle size distribution (polydisperse cases versus monodisperse ones), the particles shape, and cementation conditions. Three-dimensional pore space networks of the particles were obtained through direct imaging, usually through computed tomography, micro-CT, use of a probabilistic method to produce a synthetic three-dimensional structure that obtains its properties from two-dimensional thin sections, and the simulation of particle packing by using geological processes such as sedimentation and compaction [32]. In the next section, a review of the previous studies including sedimentation method, analytical method, numerical method and experimental methods are presented. Additionally, CT scan image and replica technique are implemented in the current research on the pore soil structure characterization.

Table 1. Pore structure network model

Schematic of pore network	Pore network model	Reference	Schematic of pore network	Pore network model	Reference
	tri-dimensional - parallel tubes	[7, 8]		One-dimensional - parallel tubes	-
	bi-dimensional – ball and tube	[25]		tri-dimensional – tubular	[1]
	tri-dimensional – ball and tube	[29]		One-dimensional - parallel tubes	[27, 28]
	tri-dimensional – ball and tube	[19]		bi-dimensional – conical pore and tube	[30]

2 Methodologies

2.1 Sedimentation Method

Sherard et al. [33] estimated the diameter of the flow channels by collecting the suspended particles in the water passing through the samples. Kenney *et al.* [9] also determined the diameter of the particles passing through the filter by the dry-vibration method. This method is effective for sand filters. Soria et al. [34] evaluated the performance of a filter with the same materials and variable thickness. In this method, base materials were suspended in the filter with a constant head, and passed particles were collected and graded [10, 11]. The diameter of the passing particles might indicate the distance of the particles moving within the filter, based on which Soria et al. [34] recommended a method for specifying the constriction size distribution (CSD).

2.2 Analytical Method

In the analytical method, CSD can be calculated by using PSD. A more realistic structure of the pores can be estimated by considering the relative density and using a probabilistic pattern for particles arrangement [35]. In this case, soil particles are often assumed as spheres. Spherical particles can be considered due to the simplicity and applicability of the calculations and modelling of a large number of particles. Silveira [36] provided a model of spherical particle placement in the densest and loosest state and estimated the size of the eroded particles from the filter. Soil porosity depends on several factors, including density, particle size distribution, shape, and cementation [37]. By simulating the soil with equal size spherical and geometric calculations, soil porosity, n, range between 0.26 and 0.48 based on the arrangement of the particles.

2.3 Numerical Method

Uniform spheres packing has been commonly studied in conditions of random loose and dense packing. However, due to a significant increase in the computation power recently, modelling of more complicated and, simultaneously, realistic particles have become widespread. Thus, large packs of irregular particles can be modelled and analyzed. Utilizing the discrete element method (DEM), the samples with different porosity can be constructed, and the pore network and soil structure are examined [38]. However, creating samples with various shapes in the DEM method is rather difficult. Using the clustering and clumped forms, it would be possible to create various complicated particle shapes by putting spherical particles into a group.

2.4 Experimental and Imaging Techniques

Mercury porosimetry is one of the most common methods for measuring the pore structure of porous materials. This method only characterizes the pore size distribution connected to the mercury reservoir and does not measure the closed pore. This method is usually used to measure pore sizes of 3 nm to 100 μm [3]. The gas absorption method is another approach, which works based on surface absorption and pore surface measurements. This method measures lesser pore size ranges compared with mercury porosimetry [39]. Larger pores of porous media can be measured using 3D imaging or replica technique, by which the results are transferable to finer soils because of the similarity of structure and shape of grains [40]. The samples' three-dimensional imaging can be implemented using a 3D laser scan or micro CT scan (μCT). μCT is a non-destructive method that can determine pore space topology and the skeleton using computed tomography [5]. In addition, nuclear magnetic resonance can be used to quantify the pore structure [41]. Also, three-dimensional images of pores can be created by using statistical reconstruction of two-dimensional images of the thin section of the sample. Several 2D view images of cross-sections from a 3D image can also be produced (as shown in Fig. 2). These images show pore space topology, pore-connectivity, and porosity of samples.

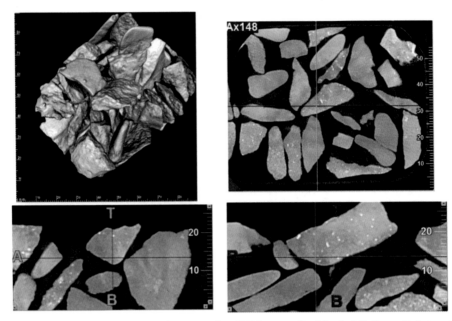

Fig. 2. CT scanning 3D image and cross-sectional 2D images, (R: right, L: left, T: top, B: bottom)

3   Experiments Using Replica Technique

In order to make a model of the coarse-grained pore structure, the liquid resin with high elastic properties enters into the pore until it is saturated. When the resin is hardened, the particles can be removed from the elastic structure, and thus a trace remains from the pore structure. It can also be submerged with resin, slurry, or ceramic powder, deposited in the fluid or immersed in the paste [40]. Sherard et al. [33] filled gravel pores with molten wax and cut the sample after the wax was hardened to examine the channel pores of the gravel.

Following the mentioned pore network modelling using replica technique, an imprint of the pore space for four uniformly graded gravel includes Spherical, Angular, Subangular, and flaky particles, replicated as shown in Fig. 3. Patterns or imprints of the pore network and individual pores show pore geometry, pore number per volume, pore size distribution, and constrictions number per pore (coordination number).

4   Result and Discussion

The acquired results indicated that the particles sphericity or roundness reduction might lead to mean pore length decrease, the tortuosity increase, and constriction sizes reduction [10]. In the case of dealing with flaky and elongated particle samples, the particle arrangement had a higher contribution to pore size distribution than that of specimens with rounded or crushed grains. In these samples, grains' flatness and orientation affected the pores network, whether the grains are placed on their largest, medium, and smallest face or oriented.

Individual pores	Imprint of the pore network	Particle image	Particle Shape
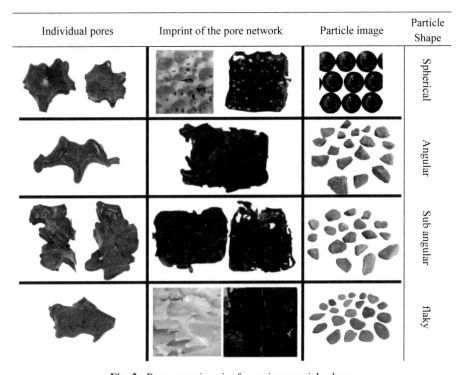			

Fig. 3. Pore space imprint for various particle shape

The average coordination number, pores connection and constriction size distribution, pore length, and the tortuosity coefficient should be determined for soil structure description. These parameters are measured using three methods: computed tomography, modeling by discrete element method, and laboratory tests. The shape of most soil particles can be simulated with spherical, pyramidal, cubical, and flaky particles. The soil structure was evaluated by assuming that particles were spherical, rounded, angular (cubical, pyramidal), and flaky [42]. Analytical methods examined the soil structure for spherical particles of different sizes and densities. Nevertheless, the structure of soils with hetero-shaped particles is complex. Determining pore size distribution becomes impossible by increasing the coefficient of uniformity of the soil (Cu) and density changes. However, by probabilistic methods, an equivalent soil structure can be estimated.

For monospherical particles, the coordination number is between 6 and 12, and the centres of the layers vary between 1.41 and 2.0 times the radius of the particles. The length of the path is a function of the diameter of the particles. In addition, porosity is 25.55 and 47.64 for the densest and loose conditions, respectively. Regarding spherical particles with two-size, three-size, and four-size, both the sample density and the arrangement of particles are essential. The pore network of these materials is similar to spherical particles, although the probability of the placement of the particles near each other makes it impossible to calculate the soil structure accurately. As the angularity of the particles increases, the irregularity of the soil structure and the tortuosity path increase. In addition, the probability of such phenomena as particle bridging and particle breakage in high-pressure increases. Specific surface area (SSA) and shape coefficients differences in flaky specimens can be associated with variations in particle thickness, where thickness increase leads grains SSA to be reduced.

On the other hand, gravity plays another role in packing flaky particles than elongated particles, mainly due to the centre of mass finding lower energy states more easily [43]. The arrangement of the particles in these types of soils plays the most significant role in the pore structure. The placement of the particles on the largest, medium, smaller side or its inclined placement greatly influences the pore network.

5 Conclusion

An imprint of pore space of four uniformly graded gravel includes spherical, angular, subangular, and flaky particle produced, and pore geometry has been evaluated. The analytical method usually can be used for spherical particles, for natural soils, analytical methods should consider and qualify the angularity elongation and roughness of grains.

In the current study pore structure of granular soils and methods to visualize, analyze, simulate and describe the structure of porous media by assuming the soil particle similar to the discrete sphere and particles with different shapes have been presented and discussed. Further, pore network and constriction size distribution were studied using experimental replica technique, and CT-Scan imaging.

An imprint of pore space of four uniformly graded gravel includes spherical, angular, subangular, and flaky particle produced, and pore geometry has been evaluated. Imprints of the pore network and individual pores show pore geometry, pore number per volume, pore size distribution, and constrictions number per pore (coordination number). As the sphericity of the particles decrease, the irregularity of the soil pore structure and the tortuosity path increase.

References

1. Fatt, I.: The network model of porous media, III. Dynamic properties of networks with tube radius distribution. Trans. Am. Inst. Min. Met. Pet. Eng. **207**, 164–177 (1956)
2. Berkowitz, B., Ewing, R.P.: Percolation theory and network modeling applications in soil physics. Surv. Geophys. **19**, 23–72 (1998)
3. Sahimi, M.: Flow and Transport in Porous Media and Fractured Rocks. (2011)
4. Blunt, M.J.: Flow in porous media pore-network models and multiphase flow. Curr. Opin. Colloid Interface Sci. **6**, 197–207 (2001)
5. Mostaghimi, P.: Transport phenomena modelled on pore-space images. PhD thesis, Imp. Coll. London (2012)
6. Witt, K., Brauns, J.: Permeability-anisotropy due to particle shape. J. Geotech. Eng. **109**, 1181–1187 (1983)
7. Kovács, G.: Seepage Hydraulics. Elsevier Scientific Pub. Co., New York (1981)
8. Indraratna, B., Vafai, F.: Analytical model for particle migration within base soil-filter system. J. Geotech. Geoenviron. Eng. **123**, 100–109 (1997)
9. Kenney, T.C., Chahal, R., Chiu, E., Ofoegbu, G.I., Omange, G.N., Ume, C.A.: Controlling constriction sizes of granular filters. Can. Geotech. J. **22**, 32–43 (1985)

10. Maroof, M.A., Mahboubi, A., Noorzad, A.: Effects of grain morphology on suffusion susceptibility of cohesionless soils. Granular Matter **23**(1), 1–20 (2021). https://doi.org/10.1007/s10035-020-01075-1

11. Maroof, M.A., Mahboubi, A., Noorzad, A.: Particle shape effect on internal instability of cohesionless soils. In: Rice, J., Liu, X., McIlroy, M., Sasanakul, I., and Xiao, M. (Eds) Proceedings of the 10th International Conference on Scour and Erosion (ICSE-10), Arlington, Virginia, USA, 18–21 October 2021, pp. 1–12 (2021)

12. Thauvin, F., Mohanty, K.: Network modeling of non-Darcy flow through porous media. Transp. Porous Media. **31**, 19–37 (1998)

13. Berkowitz, B., Ewing, R.P.: Percolation Theory and Network Modeling, pp. 23–72 (1998)

14. Chatzis, I., Dullien, F.A.L.: Modelling pore structure by 2-D and 3-D networks with applicationto sandstones. J. Can. Pet. Technol. **16**, 97–108 (1977)

15. Joekar-Niasar, V., Hassanizadeh, S.M.: Analysis of fundamentals of two-phase flow in porous media using dynamic pore-network models: A review. Crit. Rev. Environ. Sci. Technol. **42**, 1895–1976 (2012)

16. Pozrikidis, C.: Creeping flow in two-dimensional channels. J. Fluid Mech. **180**, 495 (1987)

17. Dullien, F.A.L., Batra, V.K.: Determination of the structure of porous media. Ind. Eng. Chem. **62**, 25–53 (1970)

18. Ghanbarian, B., Hunt, A.G., Ewing, R.P., Sahimi, M.: Tortuosity in porous media: A critical review. Soil Sci. Soc. Am. J. **77**, 1461–1477 (2013)

19. Raoof, A., Hassanizadeh, M.: A new method for generating pore-network models of porous media. Transp. Porous Media. **81**, 391–407 (2010)

20. Celia, M., Reeves, P., Ferrand, L.: Recent advances in pore scale models for multiphase flow in porous media. Rev. Geophys. **33**, 1049–1057 (1995)

21. Joekar-Niasar, V., Prodanović, M., Wildenschild, D., Hassanizadeh, S.M.: Network model investigation of interfacial area, capillary pressure and saturation relationships in granular porous media. Water Resour. Res. **46**, 1–18 (2010)

22. Raeini, A.Q., Bijeljic, B., Blunt, M.J.: Generalized network modeling: Network extraction as a coarse-scale discretization of the void space of porous media. Phys. Rev. E. **96**, 1–17 (2017)

23. Sufian, A., Russell, A.R., Whittle, A.J., Saadatfar, M.: Pore shapes, volume distribution and orientations in monodisperse granular assemblies. Granular Matter **17**(6), 727–742 (2015). https://doi.org/10.1007/s10035-015-0590-0

24. To, H.D., Scheuermann, A., Galindo-Torres, S.A.: Probability of transportation of loose particles in suffusion assessment by self-filtration criteria. J. Geotech. Geoenvironmental Eng. **142** (2016)

25. Chandler, R., Koplik, J., Lerman, K., Willemsen, J.F.: Capillary displacement and percolation in porous media. J. Fluid Mech. **119**, 249–267 (1982)

26. Dullien, F.A.L.: Porous Media: Fluid Transport and Pore Structure. Academic press, Cambridge (1992)

27. Dullien, F.A.L.: New network permeability model of porous media. AIChE J. **21**, 299–307 (1975)

28. Witt, K.J.: Reliability study of granular filters. In: Filters in Geotechnical and Hydraulic Engineering, pp. 35–41. Routledge, Rotterdam (1993)

29. Toledo, P.G., Scriven, L.E., Davis, H.T.: Pore-space statistics and capillary pressure curves from volume-controlled porosimetry. SPE Form. Eval. **9**, 46–54 (1994)

30. Ren, X.W., Santamarina, J.C.: The hydraulic conductivity of sediments: A pore size perspective. Eng. Geol. **233**, 48–54 (2018)

31. Saar, M.O., Manga, M.: Permeability-porosity relationship in vesicular basalts. Geophys. Res. Lett. **26**, 111–114 (1999)

32. Dong, H., Blunt, M.J.: Pore-network extraction from micro-computerized-tomography images. Phys. Rev. E - Stat. Nonlinear, Soft Matter Phys. **80**, 1–11 (2009)

33. Sherard, J.L., Dunnigan, L.P., Talbot, J.R.: Basic properties of sand and gravel filters. J. Geotech. Eng. **110**, 684–700 (1984)
34. Soria, M., Aramaki, R., and Viviani, E.: Experimental determination of void size curves. In: Filters in Geotechnical and Hydraulic Engineering, pp. 43–48. Balkema, Rotterdam (1993)
35. Locke, M., Indraratna, B., Adikari, G.: Time-dependent particle transport through granular filters. J. Geotech. Geoenvironmental Eng. **127**, 521–529 (2001)
36. Silveira, A., de Lorena Peixoto, T., Nogueira, J.: On void size distribution of granular materials. In: 5th Panamerican Conference on Soil Mechanics and Foundation Engineering, Buenos Aires, Argentina, pp. 161–176 (1975)
37. Maroof, M.A., Mahboubi, A., Vincens, E., Noorzad, A.: Effects of particle morphology on the minimum and maximum void ratios of granular materials. Granul. Matter. (2022). https://doi.org/10.1007/s10035-021-01189-0
38. Tahmasebi, P.: Packing of discrete and irregular particles. Comput. Geotech. **100**, 52–61 (2018)
39. Xiong, Q., Baychev, T.G., Andrey, J.: Review of pore network modelling of porous media: Experimental characterisations, network constructions and applications to reactive transport. J. Contam. Hydrol. **192**, 101–117 (2016)
40. Vincens, E., Witt, K.J., Homberg, U.: Approaches to determine the constriction size distribution for understanding filtration phenomena in granular materials. Acta Geotech. **10**(3), 291–303 (2014). https://doi.org/10.1007/s11440-014-0308-1
41. Xiong, Q., Jivkov, A.P., Ahmad, S.M.: Modelling reactive diffusion in clays with two-phase-informed pore networks. Appl. Clay Sci. **119**, 222–228 (2016)
42. Maroof, M.A., Mahboubi, A., Noorzad, A., Safi, Y.: A new approach to particle shape classification of granular materials. Transp. Geotech. **22**, 100296 (2020)
43. Maroof, M.A., Mahboubi, A., Noorzad, A.: A new method to determine specific surface area and shape coefficient of a cohesionless granular medium. Adv. Powder Technol. **31**, 3038–3049 (2020)

Improvement Measures for Structure System Conversions Caused by Utilising SPMTs to Lift Trusses

Zitong Wen[1], Jian Zhou[2], Jing Dong[3], and Chizhi Zhang[4(✉)]

[1] Faculty of Engineering, University of Bristol, Bristol BS8 1TR, UK
[2] Department of Engineering, Durham University, Durham D1 3LE, UK
[3] Engineering Department, University of Cambridge, Cambridge CB2 1PZ, UK
[4] Faculty of Science and Engineering, University of Hull, Hull HU6 7RX, UK
c.z.zhang@hull.ac.uk

Abstract. Since the demand for accelerated construction is increasing these years, much attention has been paid to accelerated bridge construction (ABC) methods. The self-propelled modular transporters (SPMTs) are widely utilised in the ABC method as a versatile transport carrier. However, since the limitation of the SPMTs method, several structural system conversions will happen during truss installation, and tensile stress will potentially appear at the upper chord of the truss. Moreover, it is worth noticing the dynamic effects caused by utilising SPMTs to lift the truss can enlarge the impact of tensile stress. As one type of prestressing, beams prestressed with external tendons can effectively reduce the tensile stress. In order to reduce the impact of cracks caused by tensile stress, the feasibility of adopting temporary external pre-stressing tendons is discussed combined with the simulation results of MIDAS in this research.

Keywords: Spmts · Tensile stress · Accelerated bridge construction · Midas · External pre-stressing

1 Introduction

In recent years, the development of land in urban areas has increased rapidly with the development of the economy [1]. Consequently, the development of urban regions directly leads to an increase in commuter time. For example, the average commute time increased by 10.5% from 2000 to 2015 for the Seoul Metropolitan Area. Therefore, since the traffic pressure is increasing in cities at the present stage, the most significant problem has transformed into minimising the adverse impact of construction on urban traffic.

As a counterpart measure, the accelerated bridge construction (ABC) method has been developed to reduce traffic disruption [2]. ABC method is a novel method, which could reduce traffic and construction period, to expedite bridge construction by utilising new technologies and advanced management methods. In the US, the ABC method has

© The Author(s) 2022
G. Feng (Ed.): ICCE 2021, LNCE 213, pp. 11–17, 2022.
https://doi.org/10.1007/978-981-19-1260-3_2

been utilised as a powerful tool for reducing the social and economic costs of the possible closure to repair, rehabilitate and replace more than 150,000 bridges [3].

The self-propelled modular transporters (SPMTs) method is an important component of ABC methods. SPMTs is computer-controlled multiple platforms that can pivot, lift, and carry large, heavy loads of many types. Currently, SPMTs has already been applied in a series of bridge construction projects. In the past ten years, more than 100 bridges have been moved by SPMT in the US [4]. Furthermore, the State of Utah in the United States has listed the SPMT method as the recommended construction method and formulated relevant regulations with the SPMT as the core part [5].

However, as a lifting installation method, the SPMTs method also has some limitations. Previous studies proved that the trusses could not be lifted from beam-ends. It is not difficult to foresee that the structural system of the truss will convert several times during installation. Moreover, since SPMTs is one kind of crane, dynamic effects should also be considered. Thus, tensile stress is potentially appearing at the upper chord of the truss at this stage considering the compressive pre-stress is often applied at the web members and lower chord of the trusses in general designs. Due to the low tensile strength of concrete, cracks will appear on the surface of concrete structure even when the tensile force is not very large [6]. When the structure cracks, the concrete at the crack section will completely withdraw from the work. Also, suppose cracks are not properly sealed. In that case, it can increase the risk of corrosion because cracks will undermine bridge appearance, increase maintenance and repair costs, and decrease bridge riding quality and smoothness [7].

Researchers from Tongji University have pointed out that the use of external pre-stressing tendons can reduce the tensile stress of the upper flange of the beam and improve the crack resistance of the beam under a negative bending moment [8]. This research will utilise MIDAS to simulate the whole truss installation process to explore the impact of tensile stress. The feasibility and economic efficiency of adopting external pre-stressing tendons will also be discussed according to MIDAS results.

2 Methodology

2.1 Problem Description

Previous studies are utilising finite element simulation to analyse the internal force of bridge structures. In this research, the internal force of both truss and SPMTs will be analysed by using a similar approach. Simultaneously, the study will combine the dynamic effects with the internal force obtained from the analysis results of MIDAS to calculate the tensile stress caused by structure system conversions during installation.

In the research, a simply supported truss will be designed to simulate the whole process of installation. The design process of the truss will follow GB 50010–2010 Code and GB 50017–2017 Code. The length L of the truss will be 32 m, and the height H will be 4 m. The cross-section of the truss member is rectangular, and its area is bh, where b is width and h is height. Simultaneously, the study assumes that the support between the truss and SPMTs is point support, and the body of SPMTs will be regarded as a rigid body. The research will regard truss as a plane structure, and the following assumptions will be used during analysing: 1. All members are connected only at their ends by frictionless hinges; 2. The axis of each bar in the truss is straight, and the axis passes through the centre of the hinge; 3. The external load on the truss acts on the node and is located in the plane of the truss.

For optimisation measures, the research will start from preventing tensile stress to determine the details of the measures and relevant costs. Then, the economic efficiency of the measures will be discussed.

2.2 Design Variables

Table 1. Description of truss members.

Parameter	Value
Length of truss	$L = 32$ m
Height of truss	$H = 4$ m
Cross-section width of the truss members	$b = 400$ mm
Cross-section height of the truss members	$h = 600$ mm
Area of truss members cross-section	$A = 240000$ mm^2
Area of longitudinal tension steel reinforcement	$A_s = 2454$ mm^2
Area of shear reinforcement within spacing s	$A_v = 314$ mm^2

Table 2. Description of design parameters.

Parameter	Value	Rationale
The elasticity modulus of steel	$E_s = 2.06 \times 10^5$ MPa	GB 50017–2017 Code
The elasticity modulus of concrete	$E_c = 3.45 \times 10^4$ MPa	GB50010–2010 Code
The specific mass of steel	$\rho_s = 7850$ kg/m^3	The study of Yeo and Gabbai
The specific mass of concrete	$\rho_c = 2400$ kg/m^3	
Poisson's ratio of steel	$\mu_s = 0.3$	GB 50017–2017 Code
Poisson's ratio of concrete	$\mu_c = 0.2$	GB 50010–2010 Code
Concrete cover (includes a radius of longitudinal tension reinforcement)	$d' = 65$ mm	The study of Yeo and Gabbai
Longitudinal spacing of shear reinforcement	$s = 150$ mm	
External pre-stressing tendons tensile strength	$f_t = 1860$ MPa	JTG 3362–2018 Code

Table 1 lists the description and value of the truss member. In this study, the size of the web member and chord member is consistent. Thus, width b and height h are two fixed parameters. Since MIDAS will be utilised to analyse the internal force of structure system conversions during installation, the parameters of steel and concrete in Table 2 will be regarded as design parameters [9]. Also, the research will not consider the discrete case, where the steel reinforcement positions and the selection of steel and concrete are invariant.

2.3 Methods for Improvement

The external pre-stressing tendons will be adopted to decrease the tensile stress caused by structure system conversions. Preventing from producing tensile stress in truss members will be taken as design principles of external pre-stressing tendons. In this study, the 1 × 7 steel strands whose tensile strength f_t is 1860 MPa will be utilised as temporary external pre-stressing tendons during installation, as shown in Table 2. It will calculate the area of external pre-stressing tendons combined with the analysis results of MIDAS.

$$\sigma_t = \frac{F_t}{A} \tag{1}$$

Equation (1) represents the tensile stress during installation and the required area of external pre-stressing tendons. σ_t is the tensile stress in truss members, and F_t is the maximum tensile force obtained by MIDAS simulation considering dynamic effects. Furthermore, dynamic effects will also be considered in this study. According to GB 3811–2008-T Code, the dynamic effects are taken as the dynamic coefficient multiplying the tensile stress caused by self-weight effect of the concrete structure, as shown in Eq. (2). σ_{dt} is the tensile stress of the upper chord of the truss considering dynamic effects, and φ_1 is a dynamic coefficient that can be taken as 1.5 for hoisting and transportation.

$$\sigma_{dt} = \sigma_t \times \varphi_1 \tag{2}$$

$$A_p = \frac{\sigma_{dt}}{f_t} \tag{3}$$

Finally, the required area of external pre-stressing tendons can be obtained from Eq. (3), in which A_p is the required area of external pre-stressing tendons. And the expected cost of the improvement measures can be considered by Eq. (4). P is the total cost of the improvement measures, m_p is the required weight of steel strands, is the unit price of steel strands, and P_c is the unit price of strand cable pre-stressed anchor.

$$P = m_p \times P_p + 2 \times P_c \tag{4}$$

3 Results

According to Table 1 and Table 2, the study simulates the whole truss installation process by utilising SPMTs. Figure 1 (a), (b) and (c) show the results of internal force analysed by MIDAS. It can be obtained that the structural system has been converted two times during the whole installation. It is worth noting that the maximum tensile force, 294.4

kN, appears at the upper chord of the truss during lifting when the structure is a cantilever truss. Thus, combined with Eq. (1), it can obtain that σ_t is 1.23 MPa.

Since SPMTs is one kind of crane, the dynamic effects need to be considered during lifting. According to Eq. (2), the maximum tensile stress of the upper chord can be regarded as 1.85 MPa. Considering the design value of axial tensile strength of concrete f_{td} is 1.83 MPa in China GB 50017–2017, it is reasonable to believe the crack can appear during lifting and can decrease the effective cross-sectional area.

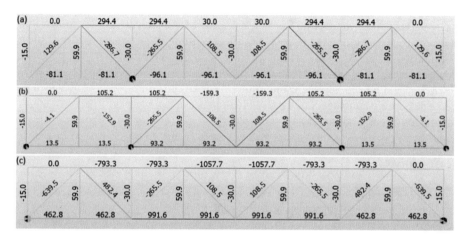

Fig. 1. Internal force of the truss: (a) lifting; (b) installing; (c) completion (kN, negative value represents compression).

In order to reduce the impact, it is necessary to adopt external pre-stressing tendons. Combined with Eq. (3), one 1×7 steel strand is required if it considers that no tensile stress appears at the upper chord of the truss. Moreover, if it ignores the labour cost, according to the current market price, the total cost P can be obtained as 26.46 dollars per truss by Eq. (4).

4   Conclusions

Since the limitation of the SPMTs method, the trusses cannot be lifted from beam-ends. Therefore, two structural system conversions will happen during installation. In this study, a new improvement measure for lifting simply supported trusses by utilising SPMTs is explored.

Combined with Table 1 and Table 2, the research simulates the whole process of installation of a simply supported truss by utilising MIDAS. The simulation results in Fig. 1 illustrate that tensile stress is easy to appear at the upper chord of the trusses during lifting. Furthermore, the calculation shows that the maximum tensile stress of the upper chord is 1.85 MPa considering the dynamic effects. It exceeds the design value of axial tensile strength of concrete, which is 1.83 MPa, in GB 50010–2010 Code. The upper chord of the truss is relatively fragile, considering the compressive pre-stress is often applied at the web members and lower chord of the trusses in general designs. Thus, adopting one 1×7 steel strand as a temporary external pre-stressing tendon is reasonable to reduce tensile stress. Moreover, the materials cost of this improvement measure can be obtained as 26.46 dollars per truss by Eq. (4). Due to the potential damage and maintenance cost caused by cracks, it will be helpful to reduce tensile stress by adopting temporary external pre-stressing tendons in practice.

Nowadays, ABC is becoming more popular, and SPMTs are utilised more often in construction. This research shows the rough idea of the improvement measure during lifting trusses by using SPMTs. It can obtain from the research that the dynamic effects have a considerable impact on the internal force of the structures. For further study, it can consider exploring the impacts of different lifting velocities on dynamic effects. It can also research the influence of offering additional constraints, which potentially could reduce accelerations, on dynamic effects. Furthermore, improving the structure of SPMTs would be another interesting future research direction. It can be considered to add a detachable component to the SPMTs to improve the internal force distribution of the trusses during installation.

References

1. Jun, M.: The effects of polycentric evolution on commute times in a polycentric compact city: A case of the Seoul Metropolitan Area. Cities **98**, 102587 (2020)
2. Salem, O., Salman, B., Ghorai, S.: Accelerating construction of roadway bridges using alternative techniques and procurement methods. Transport **33**(2), 567–579 (2018)
3. Tazarv, M., Saiid, S.M.: Low- for accelerated bridge damage precast columns construction in high seismic zones. J. Bridg. Eng. **21**(3), 04015056 (2016)
4. Culmo, M.P., Halling, M.W., Maguire, M., Mertz, D.: Recommended Guidelines for Prefabricated Bridge Elements and Systems Tolerances and Recommended Guidelines for Dynamic Effects for Bridge Systems. National Academies of Sciences, Engineering and Medicine, Washington DC (2017)
5. Zhang, C., Chen, H.P., Tee, K.F., Liang, D.: Reliability-based lifetime fatigue damage assessment of offshore composite wind turbine blades. J. Aerosp. Eng. **34**(3), 04021019 (2021)
6. Dorafshan, S., Maguire, M., Halling, M.: Dynamic effects caused by SPMT bridge moves. J. Bridg. Eng. **24**(3), 04019002 (2019)
7. Hopper, T., Manafpour, A., Radlińska, A.: Bridge deck cracking: Effects on in-service performance, prevention, and remediation (2015)
8. Chen, S.: Experimental study of pre-stressed steel-concrete composite beams with external tendons for negative moments. J. Constr. Steel Res. **61**(12), 1613–1630 (2005)
9. Yeo, D., Gabbai, R.: Sustainable design of reinforced concrete structures through embodied energy optimisation. Energy Buildings **43**(8), 2028–20331 (2011)

Evaluation the Level of Service of Signalized Intersection: Al-Amreia Intersection as a Case Study

Noor A. Rajab[1], Hamid A. Awad[1(✉)], and Firas Alrawi[2,3]

[1] College of Engineering, University of Anbar, Ramadi, Iraq
{Eng.noor85,hamid.awad}@uoanbar.edu.iq
[2] School of Geographical Sciences and Urban Planning, Arizona State University, Tempe, AZ, USA
Fhamodi@asu.edu
[3] Urban and Regional Planning Center, University of Baghdad, Baghdad, Iraq
dr.firas@uobaghdad.edu.iq

Abstract. One of the main element in the network is the intersection which consider as the critical points because there are many conflict in this element. The capability and quality of operation of an intersection was assessed to provide a better understanding of the network's traffic efficiency. In Baghdad city, the capital of/Iraq the majority of the intersections are operated under the congestion status and with level of service F, therefore theses intersection are consider as high spot point of delay in the network of Baghdad city. In this study we selected Al-Ameria signalized intersection as a case study to represent the delay problem in the intersections in Baghdad. The intersection is located in the west of Bagdad city, this intersection realizes a huge traffic, and there are a lot of tourist attractions near to the study area. The aim of this research is to enhance traffic operations, improve the level of service and decrease the delay in Al-Ameria signalized intersection by examine four suggested alternative. Special teams with a special tools are collected traffic and geometric data for the intersection. HCS 2010 program are used in this study to measure the delay and evaluate the level of service in each approach and for the hall of the intersection. The result of this study show that the intersection is operated under the breakdown condition with level of service F for all approaches. The results highlighted that the fourth alternative is the best suitable suggestion to enhance the level of service for the intersection. The fourth alternative recommended to construct a flyover from the North bound towards the South bound the level of service improve from F to C for the base year and for the target year.

Keywords: Signalized intersection · Delay · Level of service · HCS

1 Introduction

The intersection can be consider as one of the significant part of the network due to the conflict that will be happened at this point and may be leads to many problems such as traffic congestion, traffic accidents. According to Eastern Asia Society for Transportation Studies, there are two types of intersections grade and grade separated. At the same time

G. Feng (Ed.): ICCE 2021, LNCE 213, pp. 18–30, 2022.
https://doi.org/10.1007/978-981-19-1260-3_3

the intersections can be classified as signalized and unsignalized due to the type of control [1]. Traffic capacity, the key items that are used to determine traffic activity in the intersection are the volume to capability ratio, deviation, and quality of service [2]. The capacity of intersection can be defined as the maximum number vehicles that can move through a given section during one hour under the dominant conditions, the capacity for the intersection is measured for the lane group (for all lanes in the approach) [3]. The volume to capacity ratio refer to the degree of saturation of the intersection which refer to the ability of the intersection to operate the traffic demand under best condition. Any intersection with volume to capacity ratio less than 0.85 is consider as an ideal condition for the intersection and the traffic volume for this intersection will not expected any congestion and delay. On the other hand, any intersection with volume to capacity ratio more than 1.0, the intersection will operate with unstable condition and the traffic volume in the intersection will expected more delay and there will be queuing in all approaches [4]. Delay can be consider as an adequate indicator to assess the traffic operation for the intersection [5]. Based on the HCM 2010 the delay can be define as the "the additional travel time experienced by a driver, passenger, or pedestrian". The delay of the intersection can be classified into: uniform delay, incremental delay and initial queue delay [2].

The average control delay can be calculated according to the Eq. 1:

$$d = d_1 + d_2 + d_3 \qquad (1)$$

Where:
d_1: is uniform control delay,
d_2: is incremental delay, and.
d_3: is initial queue delay.
The uniform control delay is:

$$d_1 = \frac{0.5C(1 - g/C)^2}{1 - [min(1, X)g/C]} \qquad (2)$$

Where
C: is cycle length in seconds and g is the lane group's efficient green period (second).
X: represents the lane group's v/c ratio.
The incremental latency is as follows:

$$d_2 = 900T\left[(X - 1) + \sqrt{(X - 1)^2 + \frac{8KIX}{cT}} \right] \qquad (3)$$

Where:
T: analysis period duration (hour),
K: signal controller mode-dependent delay adjustment factor,
I: upstream filtering/metering adjustment factor.
c: lane group potential (veh/hr),
X: lane group v/c ratio.

The initial queue delay is:

$$d_3 = \frac{3600}{vT}\left(t_A \frac{Q_b+Q_e-Q_{eo}}{2} + \frac{Q_e^2+Q_{eo}^2}{2c_A} - \frac{Q_b^2}{2c_A}\right)$$

$$Q_e = Q_b + t_A(v - c_A)$$

$$\text{If } v \geq c_A \text{ then: } Q_{eo} = T(v - c_A) \qquad (4)$$

$$t_A = T$$

$$\text{If } v < c_A \text{ then: } \begin{matrix} Q_{e0} = 0.0 \cdot veh \\ t_A = Q_b/(c_A - v) \leq T \end{matrix}$$

Where:

T: is the analysis period's time in hours,

v: the request flow rate in vehicles per hour, and.

tA: is the adjustment period for unmet demand during the analysis period (hour),

cA: average potential of lane category (veh/h),

Qb: denotes the initial queue at the start of the analysis period (veh),

Qe: denotes the initial queue at the end of the analysis period (veh),

Qeo: denotes the initial queue at the end of the analysis period when cA > cA and Qb = 0.0 (veh).

To evaluate the quality of the any part in the transportation network the traffic engineers used the term level of service (LOS) which is represented the delay that occur in the traffic stream that used this part of the network [6]. According to HCM 2010 the LOS can be defined as "a quantitative stratification of a performance measure or measures that represent the quality of service". The LOS of intersection can be classified into six level centered on the normal intersection delay from A to F The LOS for a signalized intersection is shown in Table 1.

Table 1. LOS criteria for signalized intersection [2]

LOS	Average delay sec/veh
A	≤10
B	10–20
C	20–35
D	35–55
E	55–80
F	More than 80

2 Previous Studies

The HCS software has been used in several studies to evaluate and improve the LOS for signalized intersections in various Iraqi cities. HCS 2000 was used to assess the

Al-Thawra signalized intersection in Al-Hilla district, Iraq. The intersection operates with a 263.7 s/veh F LOS delay. The analysis proposed building a flyover to boost the LOS; as a result, the LOS would be C with a latency of 22.8 s/veh [7]. Karim used HCS 2000 in 2011 to assess the Al-Quds signalized intersection in Baghdad, Iraq. With an average delay of 328.7 s per vehicle, the intersection was found to fit with LOS F. By inserting one lane for each approach, the intersection's LOS improves to C, with an average delay of 34.6 s per vehicle [8]. Another research used HCS 2000 to assess the AL-Mustansiriyah Intersection in Baghdad, Iraq. The best idea for improving the LOS in this intersection was to construct a flyover between Al-Mustansiriyah University Street and Al-Talibia Street, according to the study [9]. In addition, the LOS for the AL-Kafa'at signalized intersection in AL-kut district, Iraq, is evaluated using HCS 2000. The current LOS for this intersection is F, with an average delay of 102.8 s per vehicle; but, according to this report, adding more lanes for the right turn would increase the LOS to D, with an average delay of 38.1 s per vehicle [10]. HCS 2010 was used to test the LOS for the Al-Furqan intersection in Al-Fallujah district, Iraq. The operational review for this intersection indicates that the intersection operates at LOS F with an average delay of 105.2 s per vehicle. This study recommended that traffic from the west bound be avoided in order to increase the LOS from C, which has an average pause of 34.5 s per vehicle [11].

3   Objectives of the Study

The main objectives in this study are:

- Establish the peak hour for Al-Amreia intersection, which is consider as the highest traffic volume in all approaches.
- Evaluate the current LOS at the Al-Amreia intersection with both approaches.
- Suggestion different proposals to improve the LOS at Al-Amreia intersection.
- Evaluation the LOS for all suggested proposals for all approaches at Al-Amreia intersection.
- Selections the best propsal to improve the LOS at Al-Amreia intersection for the base and target year

4   Study Area

Baghdad is the capital of Iraq; it considers as one of the congested cities in the world because the huge number of vehicles that using the network in this city especially after 2003. All intersections in Baghdad city are operated under breakdown condition with LOS F. For this reason, one of the congested intersection are selected in this study. Al-Amreia intersection is selected as a case study for these reasons:

- Al-Amreia intersection connects the traffic volume that are coming from the West provinces to Baghdad city.
- This intersection has high traffic volumes in all approaches

- There are many attraction locations (residential, educational and commercial) close to the study area.

Figure 1 represented Al-Amreia intersection and the boundary area of the selected intersection study.

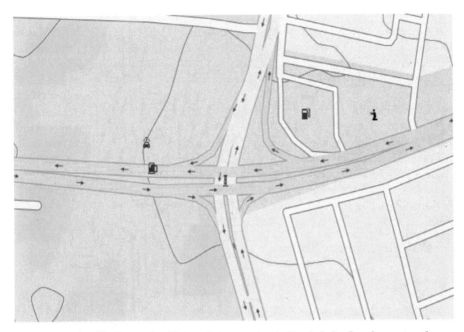

Fig. 1. Satellite image for Al-Amreia intersection in Bagdad city, Iraq [open street]

5   Methodology

To obtain the LOS for Al-Amreia intersection this study will follow the methodology that describe in HCM 2010. Figure 2 shows the main steps that must be follow to obtain the LOS which is the primary output.

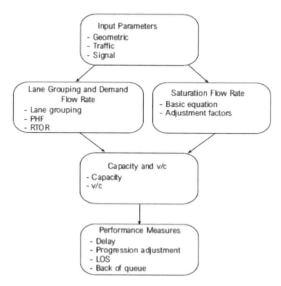

Fig. 2. Signalized intersection methodology

6 Data Collection

To evaluate the traffic operation at Al-Amreia intersection in terms of LOS a field data survey is made by special teams these data including traffic and geometric data. The measurements of these data are made manually on working days (Monday to Thursday) at January 2021 to spot the peak hours.

6.1 Traffic Volumes

Traffic data survey is made for Al-Amreia intersection at the workdays from (6:00 am to 7:00 pm) during the 2nd week on January 2021, the traffic volume is counted in each approach for the three movement (left, through and right) and the highest number of traffic volume during the survey time was highlighted as peak hour. The traffic volume flow is classified into two categories:

- Passenger vehicles: Any vehicle contains four tires only.
- Heavy vehicles: Any vehicle contains more than four tires.

Table 2 shows the traffic volume at Al-Amreia intersection for each approach according to their movement form (6:00 am to 7:00 pm).
While Table 3 shows the Heavy vehicles percentage at Al-Amreia intersection.

Table 2. One-hour traffic level at the Al-Amreia intersection for both approaches

Time (hr)	EB (Al-Khadra)			WB (Abu Ghareeb)			NB (Al-Ghazalia)			SB (Al-Amreia)		
	RT	TH	LT	RT	TH	LT	RT	TH	LT	RT	TH	LT
6–7 am	150	250	120	50	240	55	30	220	400	110	200	50
7–8 am	250	520	442	32	320	92	30	230	490	170	320	120
8–9 am	240	540	480	36	280	180	25	232	480	240	320	240
9-10 am	260	400	616	28	320	184	24	344	300	160	260	60
10–11 am	260	432	640	12	160	100	28	268	420	140	172	80
11–12 pm	256	380	800	28	408	200	20	400	548	184	304	56
12–1 pm	560	360	624	80	620	304	32	356	524	152	428	120
1–2 pm	360	320	480	84	440	320	40	488	500	180	400	152
2–3 pm	600	316	908	100	584	300	60	660	664	240	756	120
3–4 pm	465	250	682	82	456	250	74	582	574	272	290	78
4–5 pm	240	200	660	60	360	160	60	400	528	240	240	48
5–6 pm	120	256	408	70	278	160	32	288	273	220	400	55
6–7 pm	88	200	256	65	220	130	25	320	389	250	534	53

Table 3. Heavy vehicles percentage at Al-Amreia intersection.

Approach	% Heavy vehicles
EB (Al-Khadra)	7
WB (Abu Ghareeb)	12
NB (Al-Ghazalia)	9
SB (Al-Amreia)	6

6.2 Saturation Flow Rate

One of the main effective parameter on the capacity of intersection is the saturation flow rate. To calculate this parameter for Al-Amreia intersection. The software HCS 2010 is employed. The calculated saturation flow for each approach at Al-Amreia intersection is shown in Table 4.

6.3 Existing Geometric Design

It is important to determine the number of lanes and the direction of each movement when evaluating the quality of operation (LOS) at the Al-Amreia intersection. Figure 3 illustrates the intersection's current geometric layout.

Table 4. Saturation flow rate calculated at Al-Amreia intersection.

Approach	Movement	Saturation flow rate (vphg)
EB (Al-Khadra)	RT	1615
	TH	5187
	LT	1805
WB (Abu Ghareeb)	RT	1733
	TH	5033
	LT	1723
NB (Al- Ghazalia)	RT	1499
	TH	5123
	LT	1902
SB (Al-Amreia)	RT	1644
	TH	4944
	LT	1899

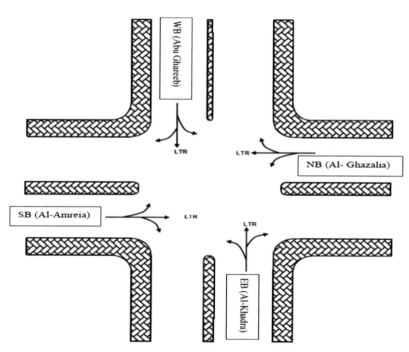

Fig. 3. Existing geometric design of Al-Amreia intersection

7 Analysis and Results

7.1 Peak Hour Volumes

The following findings were drawn from the site inspection and traffic analysis:

- The peak hour at the Al-Amreia intersection is between 2:00 and 3:00 p.m. At this hour, the overall traffic volume at the Al-Amreia intersection was 5038 pc/h (see Fig. 4).

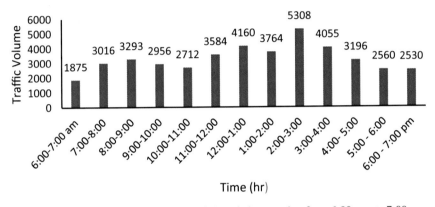

Fig. 4. Distribution of traffic volume at Al-Amreia intersection from 6:00 a.m to 7:00 p.m.

7.2 Peak Hour Factor (PHF)

According to HCM 2010, the PHF can be described as the ratio of total volume to the hour's maximum 15-min rate of flow. The following table summarizes the PHF values for both routes to the Al-Amreia intersection (Table 5).

7.3 Existing LOS

To assess the current LOS HSC 2010 software, it is implemented. The LOS at the base year was determined to be LOS (F), as seen in Table 6.

Table 5. Peak hour factor at Al-Amreia intersection.

Approach	Movement	PHF
EB	RT	0.88
	TH	0.91
	LT	0.89
WB	RT	0.95
	TH	0.94
	LT	0.87
NB	RT	0.88
	TH	0.95
	LT	0.87
SB	RT	0.90
	TH	0.93
	LT	0.91

Table 6. Existing LOS at Al-Amreia intersection

Approach	Average delay sec/veh	LOS
EB (Al-Khadra)	216.3	F
WB (Abu Ghareeb)	102.4	F
NB (Al-Ghazalia)	128.6	F
SB (Al-Amreia)	107.3	F
Intersection	138.7	F

8 Proposals Design Alternative

8.1 First Proposal: Change the Cycle Length and Green Time for All Approaches

The first proposal that will be adopted it to change the cycle length form 90 s to 120 s. In addition, it will increase the green time for the congested direction approaches (EB (Al-Khadra) and SB (Al-Amreia)) in the intersection, it is found from the results shown in Table 7 that the change the cycle length and green time for all, the intersection became operational as a result of these methods (LOS F). As a result, this plan is not recommended for operational improvement, and another one must be adopted.

8.2 Second Proposal: Increase Number of Lanes

The second plan to enhance the intersection's LOS proposed increasing the amount of lanes on both approaches by removing parking in the approach lanes. It is found from the

Table 7. LOS at Al-Amreia intersection within first proposal

Approach	Average delay sec/veh	LOS
EB (Al-Khadra)	172.2	F
WB (Abu Ghareeb)	92.4	F
NB (Al-Ghazalia)	89.3	F
SB (Al-Amreia)	91.8	F
Intersection	111.4	F

results shown in Table 8 that the increase the number of operation lanes at all approaches caused the intersection to work (LOS F). As a result, this plan is not recommended for operational improvement, and another one must be adopted.

Table 8. LOS at Al-Amreia intersection within second proposal

Approach	Average Delay sec/veh	LOS
EB (Al-Khadra)	145.7	F
WB (Abu Ghareeb)	83.7	F
NB (Al- Ghazalia)	75.6	E
SB (Al-Amreia)	81.8	F
Intersection	96.7	F

8.3   Third Proposal: Underground from East Bound Towards West Bound

The third proposal to improve the LOS for the intersection suggested to increase the number of lanes in all approaches by eliminating the parking in the execution of underground along EB (Al-Khadra) towards WB (Abu Ghareeb). It is found from the results shown in Table 8 that the increase the number of operation lanes at all approaches made the intersection operate on (LOS D). Therefore, this proposal is not recommended to improve the operation and it is necessary to adopt another proposal (Table 9).

8.4   Fourth Proposal: Fly Over from North Bound Towards South Bound

The fourth proposal is to execute a flyover that connects NB (Al- Ghazalia) towards SB (Al-Amreia), while the intersection is kept operating with four legs. It is clear from the results that were shown in Table 10 that the LOS was (C). Also the execution of this proposal will not make any improvement on the LOS, therefore; the fourth proposal was adopted.

Table 9. LOS at Al-Amreia intersection within third proposal

Approach	Average delay sec/veh	LOS
EB (Al-Khadra)	30.7	C
WB (Abu Ghareeb)	29.1	C
NB (Al-Ghazalia)	53.7	D
SB (Al-Amreia)	49.7	D
Intersection	40.8	D

Table 10. LOS at Al-Amreia intersection within third proposal

Approach	Average delay sec/veh	LOS
EB (Al-Khadra)	34.3	C
WB (Abu Ghareeb)	27.5	C
NB (Al-Ghazalia)	20.5	C
SB (Al-Amreia)	21.2	C
Intersection	25.9	C

9 Analysis of Forecasted Traffic Data

The HCS software is used to analyze forecasted data (after 20 years at a 2% annual growth rate) through power, pause, and LOS calculations for all approaches and the entire intersection. For the intended year. According to the data gathered, the LOS in the target year would be LOS (C), as seen in Table 11.

Table 11. LOS at Al-Amreia intersection within third proposal

Approach	Average delay sec/veh	LOS
EB (Al-Khadra)	46.0	D
WB (Abu Ghareeb)	30.5	C
NB (Al-Ghazalia)	22.3	C
SB (Al-Amreia)	26.3	C
Intersection	31.3	C

10 Conclusions

From the results that obtained from the analysis for Al-Amreia intersection it can be concluded that the existing LOS F with average delay 138.7 s/veh. The study suggested four proposals to improve the LOS for the intersection, it is concluded that the fourth proposal which is construct a flyover from NB (Al-Ghazalia) towards SB (Al-Amreia), the proposal reflects the best solution to improve the LOS for the intersection on base and target year. The intersection will operate at C LOS with average delay 25.9 s/veh for base year, while for target year the intersection will operate at C LOS with average delay 31.3 s/veh.

References

1. Kidwai, F.A., Karim, M.R., Ibrahim, M.R.: Traffic flow analysis of digital count down signalized urban intersection. In: Proceedings of the Eastern Asia Society for Transportation Studies, vol. 5, pp. 1301–1308 (2005)
2. Manual, Highway Capacity: HCM2010, Transportation Research Board, National Research Council, Washington, DC (2010)
3. Khisty, C.J., Kent Lall, B.: Transportation engineering. Pearson Education India (2016)
4. Garber, N.J., Hoel, L.A.: Traffic and highway engineering. Cengage Learning (2014)
5. Zheng, Y., Hua, X., Wang, W., Xiao, J., Li, D.: Analysis of a signalized intersection with dynamic use of the left-turn lane for opposite through traffic. Sustainability **12**(18), 7530 (2020)
6. Tüydeş-Yaman, H.: Estimation of Level of Service at Signalized Intersections around the Proposed Health Campus in Etlik Ankara (2014)
7. Al-Ubaidy, A.M., Al-Azzawi, Z.T., Dawood, N.: Evaluation the performance of Al-Thawra at-grade intersection using the HCS2000 computer package. Eng. Tech. J. **28**, 15 (2010)
8. Karim, Q.S: Evaluation and improving of Al-Quds intersection in Baghdad city. Eng. Dev. J. **15**(2) (2011)
9. Jasim, I.F.: Improvement of traffic capacity for AL-Mustainsiriyah intersection in Baghdad City. Al-Qadisiyah J. Eng. Sci. **5**(3) (2012)
10. Khalaf, A.Z., Hamodi, H., Riyadh, M.: Analysis of traffic operation for AL-Kafa'at signalized intersection in Al-Kut City. J. Eng. Dev. Eng. Tech. J. **33** (2015)
11. Alwani, K.H.M.: Improving capacity and traffic operation for Al-Furqan intersection at Al-Fallujah City (Iraq). In: Scientific and Technical Conference Transport Systems Theory and Practice, pp. 27–38. Springer, Cham (2019).

An Algorithm to Obtain Moment-Curvature Diagram for Reinforced Concrete Sections

Morteza Mohemmi[(✉)]

Department of Civil Engineering, University of Tehran, Tehran, Iran
m.mohemmi@ut.ac.ir

Abstract. Reinforced concrete is one of the most common materials in construction and knowledge of the behavior of reinforced concrete with lateral reinforcement has been one of the important topics in researches. Due to the high cost of laboratory studies, researchers have tried to develop numerical methods with the help of relationships obtained from analytical researches as well as various programs including finite element codes. In most studies, moment-curvature diagrams in reinforced concrete with confined cores have been significant in studying the behavior and performance of reinforced concrete sections. In this research, using analytical relations and applying MATLAB program, a suitable algorithm for calculating the moment-curvature in reinforced concrete columns with lateral reinforcement is presented. In this algorithm, the famous Mander and Kent and Park models are used and a method for eliminating the non-confined concrete cover during the analysis is proposed. Finally, the moment-curvature was obtained for 4 sections of reinforced concrete and compared with the KSU-RC program, which show a good agreement.

Keywords: Reinforced concrete · Moment curvature · Numerical analysis

1 Introduction

With the development of the industrial community, structures are becoming larger and more complex, and safety and serviceability assessment requires the development of accurate and reliable methods for their analysis. Laboratory methods that may be subjected to static and dynamic loads are introduced to ensure the strength of structures against earthquakes. But it is important to note that laboratory methods require high costs and may provide little information. In concrete structures, due to the brittle behavior of concrete and its early rupture in loading, the analysis of its behavior is more complicated than steel. And it is usually a little difficult to analytically study the composite behavior of two completely different materials such as concrete and steel and the time-dependent changes of the two materials and the effects between the two materials, and researchers are still working on these topics. Due to the development of computers and numerical methods and a clear explanation of the properties of materials, numerical methods are widely used among researchers today. Reinforced concrete is one of the most widely used structural materials today. One of the suitable methods for the safety

© The Author(s) 2022
G. Feng (Ed.): ICCE 2021, LNCE 213, pp. 31–45, 2022.
https://doi.org/10.1007/978-981-19-1260-3_4

of structures or members of reinforced concrete against seismic loads is to increase their ductility. In addition to ductility, other very important parameters such as strength, stiffness, are obtained from the moment-curvature curve and the relationships between them. This curve examines the behavior of reinforced concrete structures under the effect of bending. In the moment-curvature curve, there are three important points related to the cracking of the reinforced concrete section, the yield of tensile rebars and finally the failure of the compressive concrete. This area of the curve indicates the ductile behavior of a flexural reinforced concrete beam, and the lower the cross-sectional tensile rebars leads to the greater the ductility in the beam behavior as well as the moment-curvature curve.

So far, various methods have been proposed to analyze the ductile behavior of reinforced concrete beams. It is important to consider the tensile effects of concrete and steel correctly. In this regard, a simple algorithm and formula to calculate the relationship between tensile reinforcement and flexibility of reinforced concrete beams Presented [1, 2].

Also, the amount of tensile reinforcement for a beam with the ratio of compressive to tensile reinforcement and the desired ductility is obtained. In an experimental study, 15 laboratory beams have been studied experimentally and theoretically [3] The experiments are divided into 5 groups, each group to study a factor in the behavior of the beam, including strength, maximum deformation and type of failure.

Another group of researchers presented experimental relationships to determine the capacity for elastic and final deformation. [4] These relationships are based on parameters such as final strength, ductility of steel and shear slenderness. In another numerical study, Mohemmi and Broujerdian [5], proposed an indirect method for considering the bond-slip interaction between rebars and concrete in analysis of reinforced concrete frames. In this study in order to accurately evaluate the nonlinear behavior of RC frames, a reduction factor has been considered and demonstrated well agreement between numerical analysis and experimental data.

In this research, after presenting Mander et al. [6] and Kent and Park [7] strain stress models, the relationships used to calculate the forces, which are generated from rebars and concrete in the tensile and compressive areas are presented and then an algorithm for calculating the moment-curvature graph is presented. Also, in the process of obtaining the moment-curvature graph, the cover of sections has been gradually removed due to cracking of unreinforced concrete. In this research 4 rectangular reinforced concrete sections are used to obtain moment-curvature graph, for this purpose MATLAB program has been used and the analysis of reinforced concrete sections has been done gradually by increasing the strain in concrete compressive fiber. Finally, for validation, the results of numerical analysis are compared with the KSU-RC program and there is a good compatibility between the results of this research and the output obtained from the program.

2   Stress-Strain Models for Confined Concrete

2.1   Mander, Priestly, Park 1988

In this section, the relations developed by for the stress strain curve of confined concrete are presented. Figure 1 shows the strain stress curve for confined and non-confined concrete [6], as can be seen the use of stirrup, hoop and spiral increase the compressive stress and ultimate compressive strain as well. Clearly it is important to have accurate information concerning the complete stress-strain curve of confined concrete in order to conduct reliable moment-curvature analysis to assess the ductility available from columns with various arrangements of transverse reinforcement.

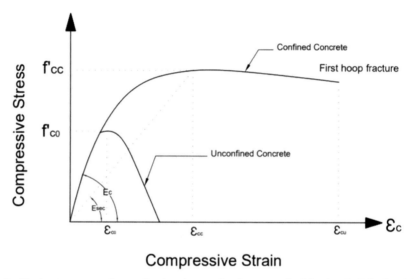

Fig. 1. Strain stress curve for confined and unconfined concrete by Mander, J., Priestley, M. & Park, R., 1988 [6]

Equations 1 to 7 show the relationships proposed by Mander et al. [6]. To calculate the strain stress curve for confined concrete. Equations 8 to 11 are used to calculate the confined stress for circular and 12 to 16 for rectangular sections. As well as Fig. 2 shows the confined core for circular and also Fig. 3 shows the confined core related to rectangular sections, these figures are taken from Mander et al., study [6].

$$f_c = \frac{f'_{cc} \, x \, r}{r - 1 + x^r} \tag{1}$$

Where $f'cc$ is compressive strength of confined concrete

$$x = \frac{\varepsilon_c}{\varepsilon_{cc}} \tag{2}$$

$$\varepsilon_{cc} = \varepsilon_{c0} \left[1 + 5 \left(\frac{f'_{cc}}{f'_{c0}} - 1 \right) \right] \tag{3}$$

$$r = \frac{E_c}{E_c - E_{sec}} \tag{4}$$

$$E_c = 5000\sqrt{f'_{c0}} \tag{5}$$

$$E_{sec} = \frac{f'_{cc}}{\varepsilon_{cc}} \tag{6}$$

where ε_c, is longitudinal compressive concrete strain that is assumed, $\varepsilon_{c0} = 0.002$

$$f'_{cc} = f'_{c0}\left(-1.254 + 2.254\sqrt{1 + \frac{7.94f'_1}{f'_{c0}}} - 2\frac{f'_1}{f'_{c0}}\right) \tag{7}$$

f'$_1$ = lateral confining stresses

$$f'_1 = \frac{1}{2}k_e\rho_s f_{yh} \tag{8}$$

$$k_e = \frac{\left(1 - \frac{s'}{2d_s}\right)^2}{1 - \rho_{cc}} \quad \text{for circular hoops} \tag{9}$$

$$k_e = \frac{\left(1 - \frac{s'}{2d_s}\right)}{1 - \rho_{cc}} \quad \text{for circular spirals} \tag{10}$$

$$\rho_s = \frac{A_{sp}\pi d_s}{\frac{\pi}{4}d_s^2 s} = \frac{4A_{sp}}{d_s s} \tag{11}$$

ps = ratio of the volume of transverse confining steel to the volume of confined concrete core.

$$f'_{1x} = k_e\rho_{sx} f_{yh} \tag{12}$$

$$\rho_{sx} = \frac{A_{sx}}{sd_c} \tag{13}$$

$$f'_{1y} = k_e\rho_{sy} f_{yh} \tag{14}$$

$$\rho_{sy} = \frac{A_{sy}}{sb_c} \tag{15}$$

$$f'_1 = \sqrt{(f'_{1x})^2 + (f'_{1y})^2} \tag{16}$$

where A_{sx} and A_{sy} = the total area of transverse bars running in the x and. y directions and $k_e = 0.75$

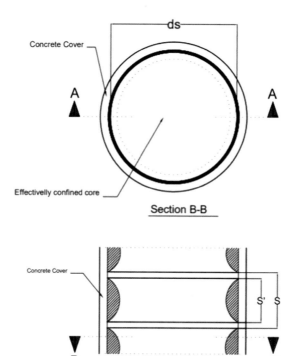

Fig. 2. Confined core for circular sections by Mander, J., Priestley, M. & Park, R., 1988 [6]

2.2 Kent and Park Model

In this section, the Kent and Park model [7] is presented, as seen in Fig. 4 in this model confined concrete sustain more ultimate strain than unconfined concrete based on Kent, D. C, and Park, R. (1971) model [7], but according to the relationships of this compressive stress model, there is no change in its compressive stress.

In this model the ascending branch is represented by Eq. (17).

$$\sigma_c = f_c \left[\frac{2\varepsilon_c}{0.002} - \left(\frac{\varepsilon_c}{0.002} \right)^2 \right] \tag{17}$$

The post-peak branch was assumed to be a straight line whose slope was defined primarily as a function of concrete strength, Eqs. (18–20)

$$\sigma_c = f_c[1 - Z(\varepsilon_c - 0.002)] \tag{18}$$

$$\varepsilon_{50u} = \frac{3 + 0.0285 f_c}{14.2 f_c - 1000} \tag{19}$$

$$\varepsilon_{50h} = \frac{3}{4} \rho_s \sqrt{\frac{b''}{s}} \tag{20}$$

where; σ = Concrete stress; b'' = big size of the core concrete (area inside the stirrup), s = stirrup spacing, ρs = stirrup percent density.

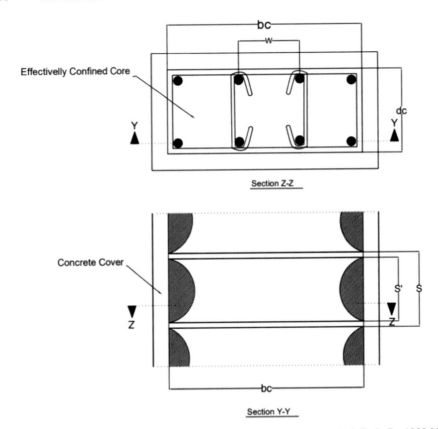

Fig. 3. Confined core for rectangular sections by Mander, J., Priestley, M. & Park, R., 1988 [6].

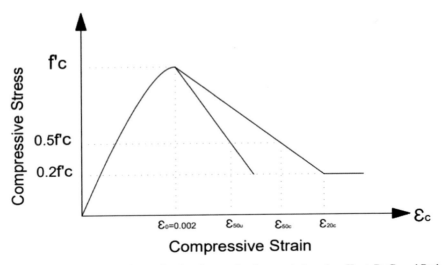

Fig. 4. Stress strain curved for confined and unconfined concrete based on Kent, D. C, and Park, R. (1971) model [7]

3   Calculation of Moment and Curvature

To calculate the moment-curvature curve, we need to calculate the forces in the section that these forces must be calculated step by step with increasing strain and after establishing the tensile and compressive balance in the section, the total moment and curvature are obtained in each step.

It should be noted that the concrete used in the cover is crushed due to not being reinforced at a strain of 0.003 and must be removed from the calculation stage. The subject is included in the analysis. Equations 21 to 29 show how to calculate the forces related to concrete and reinforcements in section and finally the moment and the corresponding curvature as well. Figure 5 shows the forces for concrete and reinforcements for a section with tensile and compressive rebars.

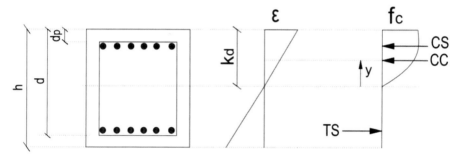

Fig. 5. Concrete and reinforcements forces for a section with compressive and tensile reinforcement

$$\varepsilon_c = \frac{\varepsilon_{ca}}{k_d} \times y \tag{21}$$

$$A_c = \int_0^{k_d} f_c(y)\, dy \tag{22}$$

$$C = \frac{\int_0^{k_d} f_c(y) \times y\, dy}{\int_0^{k_d} f_c(y)\, dy} \tag{23}$$

$$CC = A_c \times b \tag{24}$$

$$TS = As \times E \times \varepsilon_s \tag{25}$$

$$CS = As' \times E \times \varepsilon'_s \tag{26}$$

$$P = CC + CS - TS \tag{27}$$

$$M = CC \times (d - k_d + C) + CS \times (d - dp) \tag{28}$$

$$\phi = \frac{\varepsilon_{ca}}{k_d} \tag{29}$$

$\varepsilon_{ca} = Concrete\ Strain\ that\ is\ assumed$

$k_d = Depth\ of\ Narural\ Axis\ must\ be\ assumed$

$b = width\ of\ sec\ tion$

$A_c = Stress\ Area\ of\ Concrete$

$C = Centroid\ of\ concrete\ force$

$CC = Concrete\ force$

$CS = Compressive\ rebars force$

$TS = Tensile\ rebars force$

$P = Equilibrium\ of\ forces\ (must\ be\ zero)$

$M = Moment$

$\phi = Curvature$

Figure 1 illustrates the calculation of moment-curvature algorithm. In this diagram, first the input data including the compressive strength of unconfined concrete, section depth, section width, area of reinforcement, etc. are given. Then fc and the final strain of the confined concrete are calculated according to the Kent Park and Mander models. And then the balance of forces in the cross section must be checked. If there is balance between tensile and compressive forces then the moment and curvature are calculated. This process should be continued with a gradual increase in compressive strain until the section fails. It should be noted that due to the unconfined concrete cover, this part of the section should be gradually removed from the section after reaching the fracture strain of concrete. These steps are given in full in the algorithm.

As it is noted in previous sections and shown in Fig. 6 algorithm, due to non-confinement of concrete cover it must be eliminated during the analysis. Concrete cover is eliminated when strain in concrete compressive fiber reaches approximately to 0.003. Figure 7 depicts the elimination of concrete cover that leads to depth of section decrease due to experiencing strain upper than 0.003.

4   KSU-RC Software

KSU_RC is a program for analyzing the behavior of reinforced concrete columns, the first version of which has been created and updated by Esmaeily as USC_RC (2002) [8] and then by Esmaeily as KSU_RC (2007) [9]. The current version has the ability of performance analysis under changes in lateral and vertical load patterns but can only be used for rectangular, circular and hollow sections. Shir Mohammadi [10] updated the ksu version in 2015, which was part of his doctoral dissertation.

Rohleder [11] had a research also led to the current version of KSU-RC, which underwent many changes compared to the original version. The program is based on

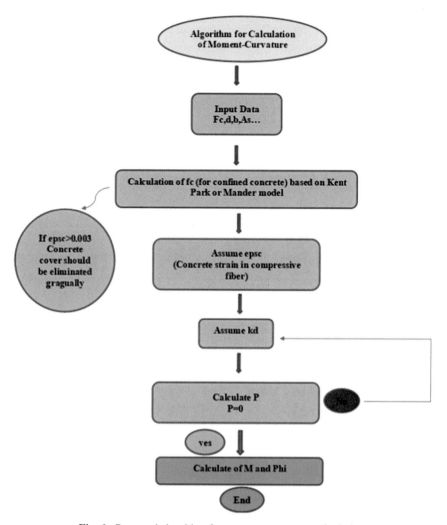

Fig. 6. Proposed algorithm for moment-curvature calculation

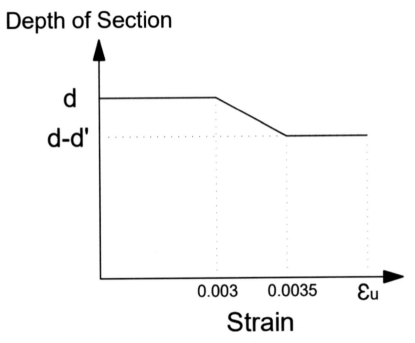

Fig. 7. Gradual elimination of non-confined concrete cover

analytical models of material behavior under monotonic and cyclic loading, all sections can be triangulated by each sectional element following the behavioral model of strain stress based on its deformation. Elements can be defined as both confined and unconfined.

5   Result and Discussion

Figure 8 demonstrates rectangular reinforced concrete sections used in this research in order to calculate moment-curvature diagram. It should be noted that compressive strength of unconfined concrete and the reinforcement yielding strength are assumed; Fc = 22 N/mm^2 assumed; Fy = 400 N/mm^2 respectively.

In this section the result of Moment-Curvature for Mander and Kent-Park models which were created by MATLAB are compared with KSU-RC software.

As can be seen in Fig. 1, 2, 3 and 4, the moment-curvature diagram for the 4 sections are shown. First, the relationship between moment and curvature is linear, and with increasing curvature, the moment also increases. And then after cracking the concrete in the tensile zone, the moment-curvature relationship gradually becomes non-linear to reach the peak point and maximum moment. Then, by increasing the amount of curvature and strain in the concrete compressive fiber, the concrete cover that is not confined is gradually cracked and removed from the section. And we will have a gradual decrease in strength and bending moment. Then, with increasing strain in the compressive farthest wire, the curvature also increases and the tensile rebars reach the yielding point and experience a large deformation without increasing the strength (due to the elastoplastic

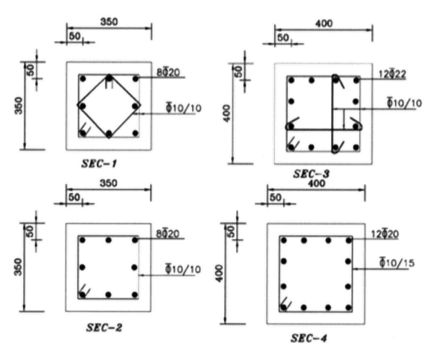

Fig. 8. Reinforced concrete sections used in this study

behavior used in this research for reinforcement) and this causes a large increase in curvature and Keep the amount of moment in section constant. The increase in curvature continues until the concrete reaches its ultimate strain and the section fails.

Figure 9, 10, 11 and 12 show the moment-curvature diagrams for the Kent-Park and Mander models developed in the MATLAB program and the Mander model obtained from the KSU-RC program. This comparison shows that the algorithm proposed in this research to calculate the moment-curvature diagram in confined reinforced concrete sections has a suitable agreement with the KSU-RC reinforced concrete sections analysis program and the part related to concrete cover is removal during analysis with increasing strain well considered in concrete.

Using this algorithm can be used to analyze larger reinforced concrete structures, while similar KSU-RC programs are usually used to analyze a section.

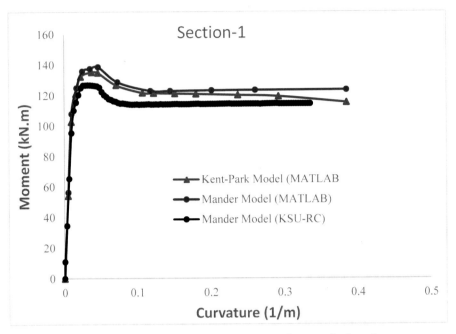

Fig. 9. Comparison of Moment-Curvature diagrams for Sect. 1

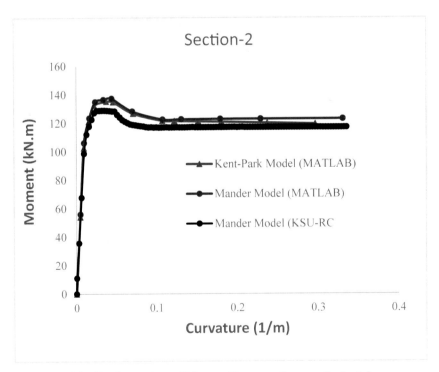

Fig. 10. Comparison of Moment-Curvature diagrams for Sect. 2

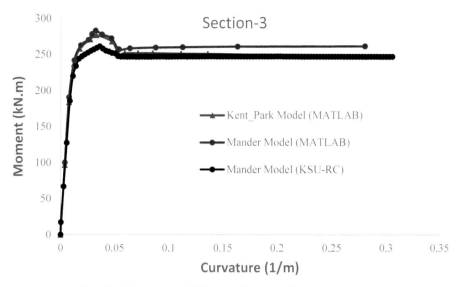

Fig. 11. Comparison of Moment-Curvature diagrams for Sect. 3

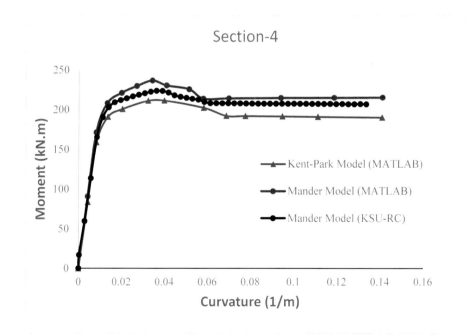

Fig. 12. Comparison of Moment-Curvature diagrams for Sect. 4

6 Conclusion

Experimental tests are the most accurate methods for analyzing reinforced concrete structures in linear and non-linear zones, and reinforced concrete structures have a complex behavior due to nonlinear behavior and cracking of concrete in tension, and researchers still study on the methods of analysis of these structures. But doing laboratory research requires a lot of time and money, and also it is not possible to test a 3D full-scale structure in a model laboratory. For this reason, the use of numerical methods has been developed due to high accuracy and reasonable cost reduction. In this research, an algorithm is proposed to analyze and obtain the moment-curvature diagram in reinforced concrete sections. In this algorithm, Mander and Kent-Park models are used to calculate the bending moment, and step by step, by increasing the strain in the section compressive fiber, the moment and the corresponding curvature are obtained. The important point about this algorithm is that due to the lack of confinement in concrete cover, this part is gradually eliminated from the section at a strain above 0.003, which is very important in gradually reducing the RC sectional strength. Finally, the moment-curvature diagram obtained for four reinforced concrete sections numerically, is compared with the moment-curvature diagram obtained from non-linear analysis software of reinforced concrete structures KSU-RC and shows high accuracy. It is also possible to extend this algorithm to the structure and analyze larger frames and structures.

References

1. Lee, T.-K., Austin, D.E.: Estimating the relationship between tension reinforcement and ductility of reinforced concrete beam sections. J. Eng. Struct. **25**, 1057–1067 (2003)
2. Chandrasekaran, S., Nunziante, L., Serino, G., Carannante, F.: Curvature ductility of RC sections based on eurocode: Analytical procedure. KSCE J. Civ. Eng. **15**(1), 131–144 (2011)
3. Lopes, A.V., Lopes, S.M.R., do Carmo, R.N.F.: Effects of the compressive reinforcement buckling on the ductility of RC beams in bending. J. Eng. Struct. **37**, 14–23 (2012)
4. Raffaele, D., Uva, G., Porco, F., Fiore, A.: A Parametrical analysis for the rotational ductility of reinforced concrete beams. Open Civ. Eng. J. **7**, 242–253 (2013)
5. Mohemmi, M., Broujerdian, V., Rajaeian, P.: An equivalent method for bar slip simulation in reinforced concrete frames. Int. J. Civ. Eng. **18**(8), 851–863 (2020). https://doi.org/10.1007/s40999-020-00507-6
6. Mander, J., Priestley, M., Park, R.: Theoretical stress-strain model for confined concrete. J. Struct. Eng. **114**(8), 1804–1826 (1988)
7. Kent, D.C, Park, R.: Flexural members with confined concrete. J. Struct. Div., ASCE, **97**(7), 1969–1990 (1971)
8. Esmaeily, A., Xiao, Y.: Seismic behavior of bridge columns subjected to various loading patterns. PEER J. **15**, 321 (2002)
9. Esmaeily, A., Peterman: Performance analysis tool for reinforced concrete members. Comput. Concrete **4**(5), 331–346 (2007)
10. Shirmohammadi, F., Esmaeily, A.: Performance of reinforced concrete columns under bi-axial lateral force/displacement and axial load. Eng. Struct. **99**, 63–77 (2015)
11. Rohleder, S.: Performance analysis software for reinforced concrete beam-columns under various load and displacement patterns, Department of Civil Engineering College of Engineering KANSAS State University, B.S., Kansas State University, A thesis for Master of Science degree (2014)

Assessment of the Influence of TH Port's Breakwater on the Hydrodynamic Regime in Cua Lo and Cua Hoi Estuaries, Nghe an Province, Vietnam

Viet Thanh Nguyen[1]([envelope]) and Chi Zhang[2]

[1] Faculty of Civil Engineering, University of Transport and Communications, Hanoi, Vietnam
vietthanh@utc.edu.vn

[2] State Key Laboratory of Hydrology-Water Resources and Hydraulic Engineering, Nanjing 210098, China

Abstract. TH Port is in offshore of Cua Lo estuary, Nghe An Province, Vietnam. In master plan from 2021 to 2030 and vision to 2050 years, an offshore breakwater will be built to protect the harbour basin of the port. This paper will be investigated the influence of the offshore breakwater on the hydrodynamics changes by a couple numerical model. The results indicated that the hydrodynamic regime control by the presence of offshore breakwater in monsoon and storm conditions and the offshore breakwater plays an important role in protection of TH port in NE monsoon and storm waves conditions and the presence of breakwater induced circulations in front of two estuaries and the mid area of Cua Lo beach.

Keywords: Hydrodynamic · Wave · Breakwater · Numerical model · TH port

1 Introduction

Estuary plays in an important role in the human life. There are 22 cities within the 32 largest cities in the world are in estuaries. Many estuaries are important centres of transportation and international commerce. Because of their great commercial and recreational importance, estuaries are often utilized excessively by a burgeoning coastal population. Approximately 60% of the world population now resides near the estuaries.

The regulation work and dredge or the combination of these two can be useful in stabilizing the navigation channel position, diverting water flow, blocking sediment, and minimizing the back-siltation. These methods have been widely used in the estuarine navigation channels over the world. In case of high intensity of back siltation, offshore port will be considered. In 1884, France began regulation work in the Seine Estuary. Two jetties are built along the bank having 60 km length. The distance between two jetties ranges from 300 to 500 m. This work continued until 1980 with a 9 km submerged breakwater constructed in the old North jetty and a dredging volume of 2.5 million cubic meters. The 10.6 m channel depth is designed for vessels 35,000 DWT to reach Rouen. In 1863, for the regulation of Rhine River for Rotterdam Port in Netherland,

G. Feng (Ed.): ICCE 2021, LNCE 213, pp. 46–56, 2022.
https://doi.org/10.1007/978-981-19-1260-3_5

two jetties were constructed in two banks of the Rhine River. Other examples are the Columbia Estuary, southwest channel of Mississippi Estuary in America and Mersey Estuary in United Kingdom [1].

Lacroix et al., used numerical model to simulate investigated the effect of geotextile submerged breakwater on hydrodynamics in La Capte beach, France. The results shown that the presence of the breakwaters has changed the current at La Capte beach in the positive direction, and the significant wave height and current speed have been reduced to acceptable levels [2].

Many estuaries have been investigated by numerical and physical models, as it is very common to calibrate and validate these two methods by each other. It does not save cost or time, but may increase the effect and development speed of the navigation channel research [1]. Nguyen and Zheng proposed four regulation schemes for Dinh An estuary and used numerical model to investigated and analysed with respect to the behaviour of each scheme to improve the water depth in the navigation channel [3].

In this study, the influence of the offshore breakwater in TH port of Cua Lo estuary on hydrodynamics and wave transformation will be investigated by a couple numerical model.

2 Study Area and Methodology

2.1 Description of Study Area

TH port is in offshore of Cua Lo estuary, Nghe An Province, Vietnam. It is the most important estuary in Nghe An province. Cua Lo estuary will become a modern key project of special significance in economic development not only for Nghe An but also for Laos and north-eastern of Thailand. The existing port includes six berths for mooring vessel of 10,000 to 20,000 DWT. Based on the master plan period from 2021 to 2030 and vision 2050, there are three navigation channels to three ports such as ① Cua Lo port, ② DKC port and ③ TH port, and an offshore breakwater will be constructed in the northern of Cua Lo estuary (Fig. 1).

There are two ports of Vissai Nghe An in the north of Cua Lo estuary. It is a specialized mooring area with a general port and a container in which the general ports and containers for vessel of 30,000–50,000 DWT and up to 100,000 DWT, and international passenger vessel of between 3,000 and 5,000 seats, when conditions permit); Specialized wharves for export of cement, clinker, coal for vessel of 70,000- 100,000 DWT of international routes, specialized wharves for coal importation, additives, cement, clinker, vessel of up to 30,000 DWT in domestic and international; Specialized ports specialized in importing liquid petrol, oil, asphalt and petroleum products for vessels of up to 50,000 DWT and wharfs for 5,000 DWT [4, 5].

The TH port (number 3 in Fig. 2) is a priority project for construction for mooring vessel of 100,000 DWT and an offshore breakwater will be built to protect wave impact on the port and ensure the safety operation. The offshore breakwater was proposed by TEDIPORT with 2520 m length and crest level of 6.5 m (Chart Datum).

Wind data was collected from Hon Ngu Island station from 1962 to 2020 with interval time of 3 h. The analysis result shows that offshore wind included two major seasons

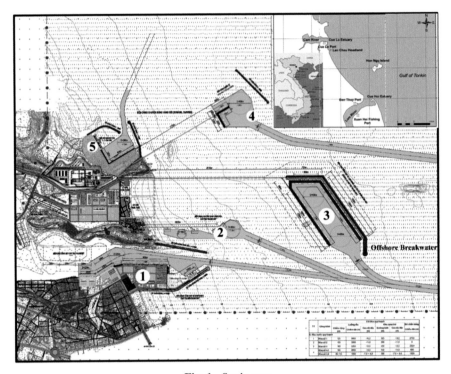

Fig. 1. Study area

are NE monsoon and SE monsoon. The maximum speed of 56 m/s occurred in NNE direction on October 1st, 1964.Wind rose is plotted in Fig. 2, and indicated that the main wind direction for N direction with frequency of 25.85%, for NE direction with frequency of 7.51% in the winter and changes to be S direction with frequency of 8.75%, SE direction with frequency of 9.25% in the summer [4, 6].

Tidal in coastal of Cua Lo estuary is semi-diurnal with tidal ranges from 1.0 to 3.59 m [7]. Before Nghi Quang Dam was built, Cam River flow has significant in the formation and existence of tidal creeks in front of Cua Lo port, now with the presence of Nghi Quang Dam the effect of river flow can be ignored. The observation data from TEDIPORRT in December 2016 show that nearshore current speed of 0.65 m/s with NW direction in the NE monsoon [4, 6].

Nearshore wave depends on the offshore wind, the domination wave in NE monsoon are NE & N waves with the average signification wave height ranges from 0.7 to 1.0 m. In SE monsoon, wave prevailing in SE and SW directions. During storm NANCY (May 18, 1982) significant wave height was recorded of 6.0 m. The frequency of waves occurs as following NE waves of 18.4%, N wave of 15.42%, SE wave of 7.59% and SW wave of 5.16%, respectively [4]. Wave rose plotted in Fig. 4 is result of offshore wave collected from NOAA at the location of 1900N, 106015' E from 1997 to 2016. The maximum wave height observer reached to 4.5 m in NE direction [4, 6].

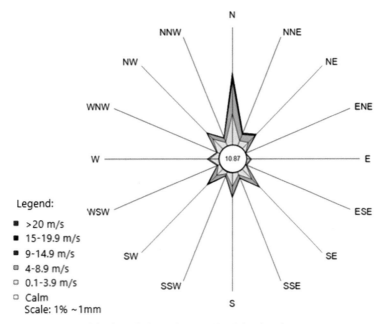

Fig. 2. Wind rose in Hon Ngu Island station

2.2 Methodology

To investigate the influence of offshore breakwater of TH port in the hydrodynamic of Cua Lo estuary based on the two dimensions couple numerical model in a larger coastal area between Cua Lo and Cua Hoi estuaries. The couple model consists of Mike 21HD and Mike 21 SW models, these models were calibrated and verified of by Nguyen et al. (2021).

Mike 21 HD model used a study domain showed in Fig. 3. Study domain with two river boundaries are Cam and Lam rivers. Cam River boundary is far from the Cua Lo estuary about 2 km and Lam River boundary is far from Cua Hoi about 3.5 km. Two river boundaries are near to estuary and subjected to tidal regime. Therefore, tidal levels were used in these boundaries.

Mike 21 SW model used fully spectral formulation with stationary mode. A spectral-form empirical formula was used in initial conditions type. A wave fetch of 160 km and the maximum peak frequency was 0.4 Hz were setup for the JONSWAP fetch growth expression. Other JONSWAP's parameters used the default values. Offshore boundaries were applied timeseries wave data obtained from NOOA, and storm wave with return period of 50 years collected from TEDIPORT (2016) was used in simulation of storm scenario.

Six scenarios of simulation will be carried out in which three scenarios simulate the natural condition and three scenarios simulate with the presence of offshore breakwater in climates condition such as NE monsoon, SE monsoon and storm with return period of 50 years. The results will be used to analysis the hydrodynamics and wave features in the Cua Lo estuary.

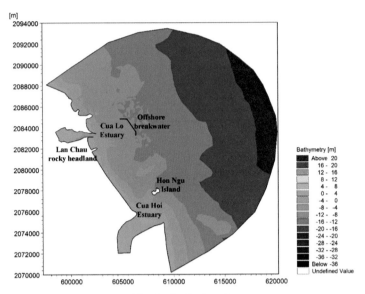

Fig. 3. Study domain

3 Results and Discussion

3.1 Hydrodynamic Features

In NE monsoon, during the flood tidal the current in the nearshore flow from north to south. However, in offshore due to the presence of offshore breakwater two circulations appeared, one occurred in the southwest of the breakwater and one in mid of coastal zone between two estuaries. The current behind the breakwater very small with the maximum speed reached to 0.08 m/s only. In Cua Lo estuary, the maximum flood current speed of 0.15 m/s is smaller than in Cua Hoi estuary of 0.35 m/s. The longshore current in the northern of Cua Lo is larger than in the estuary and Cua Lo beach. The current speed in Cua Hoi estuary reaches to 0.35 m/s (Fig. 4).

During the ebb tidal, the ebb current from Cua Lo estuary flow to the north while in Cua Hoi estuary still existing of the flood current. This phenomena is similarity with the result in the natural condition without offshore breakwater [6]. The current speed in the harbour basin of TH port is a bit smaller than other water areas. Longshore current from Lan Chau rocky headland to the south jetty in Cua Lo estuary and from the north jetty of Cua Lo estuary to the north very small only ranges from 0.02 to 0.05 m/s. The result also indicated that Hon Ngu Island take an important role for distribution of current field in the Cua Lo coastal and also Cua Lo estuary, the presence of Hon Ngu Island induced decreasing of offshore current during ebb tidal (Fig. 5).

In SE monsoon during the flood tidal, the currents in the nearshore flow from the south to the north. The maximum ebb current speed in Cua Lo estuary is 0.41 m/s larger than in Cua Hoi estuary of 0.33 m/s. Longshore current along the coastal of Cua Lo beach ranges from 0.05 to 0.1 m/s (Fig. 6).

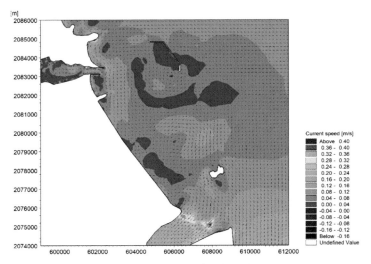

Fig. 4. Flood tidal current field in NE monsoon

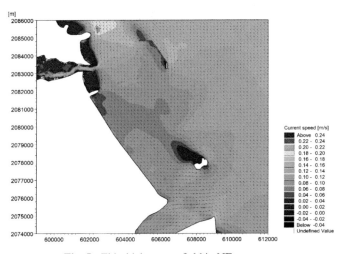

Fig. 5. Ebb tidal current field in NE monsoon

During the ebb tidal, the nearshore currents also flow from the north to the south. A circulation appeared behind the offshore breakwater. Longshore current a bit higher than in flood tidal. The maximum ebb current speed in Cua Lo estuary is 0.25 m/s larger than in Cua Hoi estuary of 0.12 m/s (Fig. 7).

In the storm with return period of 50 years, during the flood tidal, the currents in the nearshore flow from the south to the north. The maximum ebb current speed in Cua Lo estuary is 0.41 m/s larger than in Cua Hoi estuary of 0.33 m/s (Fig. 8). During the ebb tidal, the nearshore currents also flow from the north to the south. The maximum ebb current speed in Cua Lo estuary is 0.25 m/s larger than in Cua Hoi estuary of 0.12 m/s

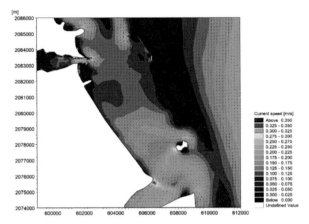

Fig. 6. Flood tidal current field in SE monsoon

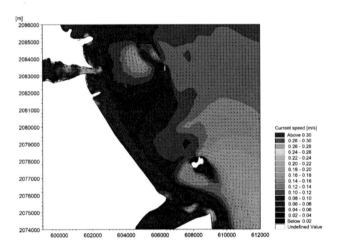

Fig. 7. Ebb tidal current field in SE monsoon

(Fig. 9). In both flood and ebb tidal regime the circulation occurred behind the offshore breakwater, in mid of coastal between two estuaries and in front of Cua Hoi estuary.

The simulation results indicated that the new navigation channel and offshore breakwater induced a complicated coastal current in the study area with many coastal circulations. These circulations will be induced unpredictable sediment transportation and can affect the stability of the navigation channels in the study area.

3.2 Wave Field

During NE monsoon, the NE wave was blocked by offshore breakwater and wave height in front of the gap between two jetties of Cua Lo estuary ranges from 1.2 to 1.4 m and reduced in the south of estuary and Lan Chau rocky headland. The coastal to Cua Lo estuary the wave height ranges from 0.8 to 1.2 m and increasing in the southern part.

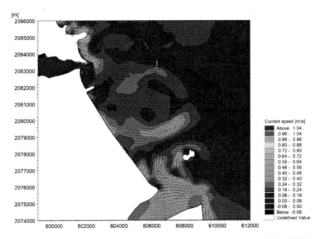

Fig. 8. Flood tidal current field in the storm with return period of 50 years

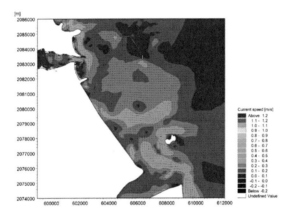

Fig. 9. Ebb tidal current field in the storm with return period of 50 years

This result also indicated that the offshore breakwater has not taken significant in wave height distribution in Cua Hoi estuary in NE monsoon (Fig. 10).

During SE monsoon, the NE wave was blocked by both of Hon Ngu Island and offshore breakwater. However, the present of the offshore breakwater has not taken significance in wave height distribution in Cua Lo estuary while Hon Ngu Island plays an important role in reducing wave height in Cua Lo beach and estuary during the SE monsoon. The wave height in front of the gap between two jetties of Cua Lo estuary ranges from 1.2 to 1.4 m and reduced in the south of estuary and Lan Chau rocky headland. The coastal to Cua Lo estuary the wave height ranges from 0.4 to 0.8 m, and in Cua Hoi ranges from 0.8 to 1.0 m (Fig. 11).

In the storm condition, the present of the offshore breakwater take an important part in reducing the significant wave height in Cua Lo estuary and a half part of coastal area between two estuaries. The Hon Ngu Island also plays an important role in reducing

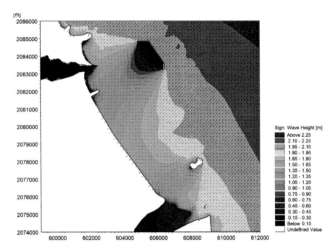

Fig. 10. Wave heigh in NE monsoon

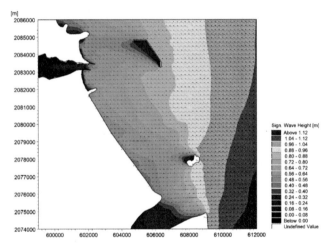

Fig. 11. Wave heigh in SE monsoon

wave height in Cua Hoi estuary. The wave height in front of the gap between two jetties of Cua Lo estuary ranges from 2.0 to 2.4 m and reduced in the south of estuary and Lan Chau rocky headland. The coastal to Cua Lo estuary the wave height ranges from 1.0 to 1.3 m, and in Cua Hoi ranges from 1.4 to 1.8 m (Fig. 12).

Table 1 presented the comparison wave height before (N-BR) and after (W-BR) construction of the offshore breakwater in four locations as show in Fig. 12. The results showed that in the Cua Lo estuary (location t1) the wave height reduced from 22.46 to 32.47%; the Cua Lo beach (location t2) the wave height decreased from 2.26 to 16.22%; the wave height behind the offshore breakwater reduced from 77.36 to 95.74% and in the Cua Hoi estuary from 4.48 to 10.95%.

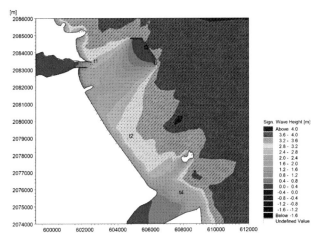

Fig. 12. Wave heigh in the storm with return period of 50 years

Table 1. Comparison wave height before (N-BR) and after (W-BR) construction of the offshore breakwater

Locations	Maximum wave height (m)								
	NE monsoon			SE monsoon			Storm		
	N-BR	W-BR	Difference (%)	N-BR	W-BR	Difference (%)	N-BR	W-BR	Difference (%)
t1	1.87	1.38	26.20	1.54	1.04	32.47	3.23	2.44	24.46
t2	1.78	1.58	11.24	1.33	1.30	2.26	3.76	3.15	16.22
t3	2.07	0.27	86.96	1.59	0.36	77.36	3.76	0.16	95.74
t4	1.65	1.54	6.67	1.23	1.17	4.88	2.1	1.87	10.95

4   Conclusions

1. Offshore breakwater plays an important role in protecting the TH port in both of northeast monsoon and storm conditions.
2. Circulation occurs in both of Cua Lo and Cua Hoi estuaries and in the mid coastal area during storm.
3. The presence of offshore breakwater has good protecting of the TH port from wave impact. However, it may induce sedimentation in existing navigation channels to Cua Lo port due to the complex of hydrodynamic regime induce by the breakwater.

Acknowledgement. This work was financially supported by The Belt and Road Special Foundation of the State Key Laboratory of Hydrology-Water Resources and Hydraulic Engineering (Grant No. 2018490811).

References

1. Thanh, N.V.: Morphological evolution and back siltation of navigation channel in Dinh An Estuary, Mekong River Delta: Understanding, modelling and soluting. 2012, PhD Thesis. Hohai University, Nanjing, China (2012)
2. Lacroix, Y., Vu, T.M., Than, V.V., Thanh, N.V.: Modelling the effect of geotextile submerged breakwater on hydrodynamics in La Capte beach. In: Vietnam-Japan Workshop on Estuaries, Coasts and Rivers 2015. Construction Publishing House, Hoi An (2015)
3. Nguyen, V.T., Zheng, J.H.: Preliminary study of regulation schemes for navigation channel in Dinh An Estuary Vietnam. Appl. Mech. Mater. **212–213**, 117–122 (2012)
4. TEDIPORT: Report on improvement project-Investment project to upgrade the navigation channel for vessel of 10,000 DWT entering Cua Lo port. TEDIPORT: Hanoi, Vietnam (2016)
5. CMB-TEDIPORT-TDSI: Master plan of Vietnam sea port development period from 2021–2030, vision 2050 2021: Hanoi, Vietnam. p. 499
6. Nguyen, V.T., Vu, M.T., Zhang, C.: Numerical investigation of hydrodynamics and cohesive sediment transport in Cua Lo and Cua Hoi Estuaries, Vietnam. J. Mar. Sci. Eng. **9**(11), 1258 (2021)
7. Nguyen, V.T., Do, M.D., Zhang, C.: Effectiveness of maintenance dredging in the navigation channel of Cua Lo Port, Vietnam. In: The 3rd International Conference on Sustainability in Civil Engineering (ICSCE 2020). Transport Publishing House, Hanoi (2020)

Effects of Hole Locations on the Elastic Global Buckling Loads of Cold-Formed Steel Channel Members with Perforations Under Compression or Bending

Ngoc Hieu Pham[(✉)]

Faculty of Civil Engineering, Hanoi Architectural University, Hanoi, Vietnam
hieupn@hau.edu.vn

Abstract. Cold-formed steel members with perforations have been commonly applied to meet the demands for technical installations. The design of the perforated members was regulated in Specification AISI S100-16 using the Direct Strength Method (DSM). This method is based on elastic buckling analyses to predict the capacities of cold-formed steel members. The determination of elastic buckling loads is compulsory for the application of the DSM method in the design and has been presented in the Specification. The specification regulations are only applied for symmetrical and evenly spaced holes. The paper, therefore, investigates the effects of unsymmetrical, unevenly spaced and eccentric holes on the elastic global buckling loads of perforated channel members using finite element analyses. The effect of symmetrical and evenly web holes on the elastic global buckling loads of cold-formed steel channel members in comparison with those of gross section members is also investigated.

Keywords: Hole locations · Elastic global buckling loads · Cold-formed steel channel members · Perforations

1 Introduction

Web holes are found in a variety of cold-formed steel members to allow technical services to pass through. These holes are commonly pre-punched in the webs of channel or Zed sections that have significant impacts on the stability and capacity of this type of structural member. The design of the cold-formed steel members with holes was also included in the Specification AISI S100-16 [1] using the Direct Strength Method (DSM). The DSM method can provide the sectional and member capacities based on the elastic buckling loads. The determination of the elastic buckling loads, therefore, is mandatory for the design of perforated members.

A large number of studies on the cold-formed steel members with perforations have been available in literature. An experimental program presented in [2] investigated the strength of stub columns with circular holes to illustrate the strength decreasing with the increasing of hole diameter. The influences of circular, slotted and rectangular web holes

© The Author(s) 2022
G. Feng (Ed.): ICCE 2021, LNCE 213, pp. 57–69, 2022.
https://doi.org/10.1007/978-981-19-1260-3_6

on the strengths of stub columns were also investigated by many studies [3–5]. A variety of studies carried out by Moen and Schafer [6–13] to study the stability and capacity of cold-formed steel members with holes. These studies were the base to propose a simple method to determine the elastic sectional and global buckling loads of perforated members and the DSM design formulae for the design of perforated members. These proposals were then included in the Specification AISI S100-16 [1]. In order to support the elastic buckling analyses of perforated sections, a simple hole module was developed by the American Iron and Steel Institute [14] to perform the elastic buckling analysis and obtain the elastic buckling loads of perforated sections that can be used for the DSM design. Moen and Yu [15, 16] studied the elastic buckling analysis of cold-formed members with edge-stiffened holes. These previous studies were aimed to investigate the stability and the capacities of perforated members with symmetrical and evenly spaced holes whereas studies on asymmetric and unevenly web holes remain scarce.

The influences of hole locations on the capacities of stub columns were reported in [17–19] and the effects of hole lengths on the behaviors of compression elements were studied in [20]. Ortiz-Colberg [2] carried out 25 intermediate column tests and pointed out that the column strengths were not impacted by a single hole along the length of the column tests. Moen and Schafer [6] investigate the effects of the locations of a single hole on the elastic buckling of an intermediate column length. Their research demonstrated that the locations of the single hole have a minimal impact on the elastic buckling loads of the investigated columns. The influence of unsymmetrical and unevenly web holes of the elastic buckling loads of long perforated members are not reported. This paper, therefore, investigates the effects of unevenly spaced and/or unsymmetrical web holes or eccentric web holes on the elastic global buckling loads of long perforated members under compression or bending using finite element analyses. The finite element models used for this investigation were verified against test results as presented in Pham *et al.* [21, 22]. Also, the influence of symmetrical and evenly web holes on the elastic global buckling loads of cold-formed steel channel members in comparison with those of gross section members according to the specification AISI S100-16 [1] is included.

2   Finite Element Models for Buckling Analyses

Material properties are used for the investigation including Young's modulus $E = 203400$ MPa and Poisson's constant $\mu = 0.3$. The finite element models for buckling analyses were developed and validated against the test results as fully reported in Pham *et al.* [21, 22] including compression and bending models.

In terms of compression models, two model configurations were constructed with the variations of end boundary conditions to obtain different global buckling modes. The first configuration allows the specimens to freely rotating about the minor axis to achieve the flexural buckling mode whereas the free rotation about the major axis was used for the second configuration to obtain the flexural-torsional buckling mode. Warping displacements were prevented at two ends for both two configurations. These configuration models are illustrated in Fig. 1 and Fig. 2. The effective lengths, therefore, were taken as $L_x = L_z = 0.5$ L; $L_y = L$ and $L_y = L_z = 0.5L$; $L_x = L$ for the first and second configurations respectively, where L is the member length. The lateral load was applied at the centroid of one end section.

In terms of the bending model, the web of each end section was contacted with three points through the shear centre of the section. The model configuration was developed with simple supports and free warping displacements, as illustrated in Fig. 3. The effective lengths were taken as $L_x = L_y = L_z = L$, where L is the member length. Vertical loads were applied at the shear centres (see loading points in Fig. 3) of the section.

Fig. 1. The flexural buckling model under compression

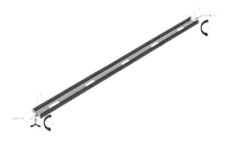

Fig. 2. The flexural-torsional buckling model under compression

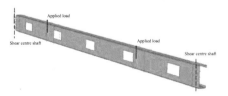

Fig. 3. The flexural-torsional buckling model under bending

3   Effects of Symmetrical and Evenly Spaced Web Holes on the Elastic Global Buckling Loads of Cold-Formed Steel Channel Members

The global buckling loads of perforated members can be determined using the "weighted average" method as regulated in AISI S100-16 [1]. The elastic global buckling loads of a column include the flexural or flexural-torsional buckling loads as presented in Eqs. (1) and (2), whereas the global buckling moment of a flexural member is determined as in Eq. (3).

The elastic flexural buckling load of a compressive member:

$$P_{cre} = \frac{\pi^2 EI_{y,avg}}{(K_y L)^2} \tag{1}$$

The elastic flexural-torsional buckling load of a compressive member:

$$P_{cre} = \frac{1}{2\beta}\left[(P_{ex} + P_t) - \sqrt{(P_{ex} + P_t)^2 - 4\beta P_{ex}P_t}\right] \tag{2}$$

The elastic flexural-torsional buckling moment of a flexural member:

$$M_{cre} = \frac{\pi}{K_y L}\sqrt{EI_{y,avg}\left(GJ_{avg} + \frac{\pi^2 EC_{w,net}}{(K_t L)^2}\right)} \tag{3}$$

where

$$P_{ex} = \frac{\pi^2 EI_{x,avg}}{(K_x L)^2}; \ P_t = \frac{1}{r_{o,avg}^2}\left(GJ_{avg} + \frac{\pi^2 EC_{w,net}}{(K_t L)^2}\right)$$

$I_{x,avg}, I_{y,avg}, J_{avg}, r_{o,avg}$ are the section properties determined using the "weighted average" method as regulated in AISI S100-16, as follows:

$$I_{avg} = \frac{I_g L_g + I_{net} L_{net}}{L}; \ J_{avg} = \frac{J_g L_g + J_{net} L_{net}}{L}; \ r_{o,avg} = \sqrt{x_{o,avg}^2 + y_{o,avg}^2 + \frac{I_{x,avg} + I_{y,avg}}{A_{avg}}}$$

$$A_{avg} = \frac{A_g L_g + A_{net} L_{net}}{L}; \ x_{o,avg} = \frac{x_{o,g} L_g + x_{o,net} L_{net}}{L}; \ y_{o,avg} = \frac{y_{o,g} L_g + y_{o,net} L_{net}}{L}$$

$(I_g, J_g, A_g, L_g, x_{o,g}, y_{o,g})$ and $(I_{net}, J_{net}, A_{net}, L_{net}, x_{o,net}, y_{o,net})$ are properties of the gross section and the net section respectively.

$C_{w,net}$ is the net warping constant of an assumed section with the assumed depth of the hole (h_{hole*}) determined in Eq. (4), where h_{hole} is the actual depth of the hole and D is the depth of the section.

$$h_{hole*} = h_{hole} + \frac{1}{2}(D - h_{hole})\left(\frac{h_{hole}}{D}\right)^{0.2} \tag{4}$$

The C20015 section is selected for this investigation with the nominal dimensions including $D = 203$ mm, $B = 76$ mm, $L = 19.5$ mm, $t = 1.5$ mm, as demonstrated in

Fig. 4. The investigated member has a length of 2.5 m and 05 symmetrical and evenly spaced web holes with the hole depth h_{hole} varying from 0.2 to 0.8 times of the section depth and the hole length varying from 0.5 to 2 times of the section depth. The spacing between these web holes are illustrated in Fig. 5. The loads and boundary conditions have been presented in Sect. 2. The elastic global buckling loads of the perforated members are determined using the "weight average" method and plotted into the graphs in Figs. 6, 7 and 8 in comparison with those of gross section members.

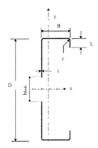

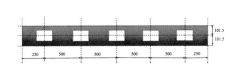

Fig. 4. Nomenclature and nominal dimensions for C20015 section

Fig. 5. The locations of symmetrical and evenly spaced web holes

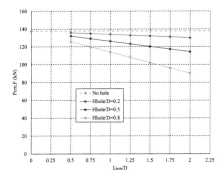

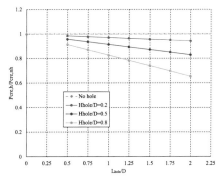

Fig. 6. Flexural buckling loads under compression ($P_{cre,F}$ is the flexural buckling load; $P_{cre,h}$ and $P_{cre,nh}$ are elastic global buckling loads of the gross section and the net section.)

The investigated results show that the elastic global buckling loads decrease as the hole sizes increase, relative to the web depth. The relationship between buckling loads and the hole lengths is linear, interpolation then can be used to determine the buckling loads of the intermediate points between the specific points of the hole lengths.

The web holes were found to have the most significant impacts on the flexural-torsional buckling loads under compression with the reduction of nearly 50% compared to those of the gross section, whereas these reduction values are about 30% for both the flexural mode due to compression and the flexural-torsional buckling mode under bending.

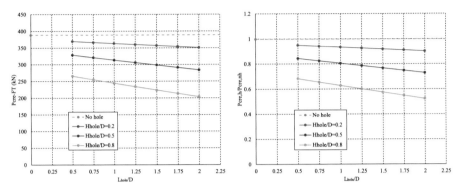

Fig. 7. Flexural-torsional buckling loads under compression ($P_{cre,FT}$ is the flexural-torsional buckling load; $P_{cre,h}$ and $P_{cre,nh}$ are elastic global buckling loads of the gross section and the net section.)

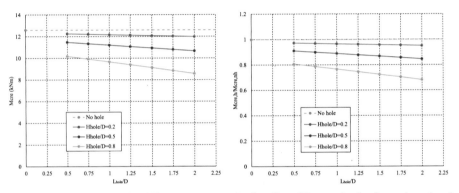

Fig. 8. Flexural-torsional buckling moments under bending ($M_{cre,FT}$ is the flexural-torsional buckling moment; $M_{cre,h}$ and $M_{cre,nh}$ are elastic global buckling loads of the gross section and the net section.)

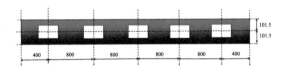

Fig. 9. The original model - Model 0

The variation of hole lengths has insignificant impacts on the elastic global buckling loads if the hole depths are small (see $h_{hole}/D = 0.2$ in the graphs) with the small slopes of the relationship lines. In terms of the large hole depths (see $h_{hole}/D = 0.8$ in the graphs), these slopes significantly increase to demonstrate the noticeable influences of the hole lengths on the elastic global buckling loads.

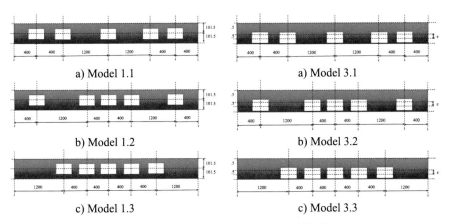

a) Model 1.1

a) Model 3.1

b) Model 1.2

b) Model 3.2

c) Model 1.3

c) Model 3.3

Fig. 10. Model 1 – Symmetrical holes

Fig. 11. Model 3 – Symmetrical and eccentric holes

4   Effects of Web Hole Locations on the Elastic Global Buckling Loads of Perforated Channel Members

The investigated section is the C20030 section with the nominal dimensions including $D = 203$; $B = 76$; $L = 19.5$; $t = 3$, and the nomenclature of this section is illustrated in Fig. 4. The member length is taken as 4000 for investigation. The thickness and the length of the investigated member are selected to obtain the global buckling modes without the interactions with other sectional modes. The finite element models used in this investigation are presented in Sect. 2.

There are 05 rectangular holes in the web of the section with the ratio of hole length and section depth (L_{hole}/D) of 1.0 and the ratios of hole depth and section depth (h_{hole}/D) varying from 0.2 to 0.8. The spacings and locations of these holes are initially symmetrical and even (see Fig. 9), are then rearranged as illustrated in Figs. 10, 11, 12 and 13 to investigate the effects of unsymmetrical, unevenly spaced and eccentric holes on the elastic global buckling loads of channel members under compression or bending. The eccentric values (e) are $0.25D$, $0.125D$ and $0.05D$ for the ratios of h_{hole}/D varying from 0.2, 0.5 to 0.8 respectively. Failure modes are obtained as illustrated in Figs. 14, 15 and 16, and the deviation of buckling loads (in percentage) between investigated models (Model 1 to Model 4) and the original model (Model 0) are illustrated in Fig. 17.

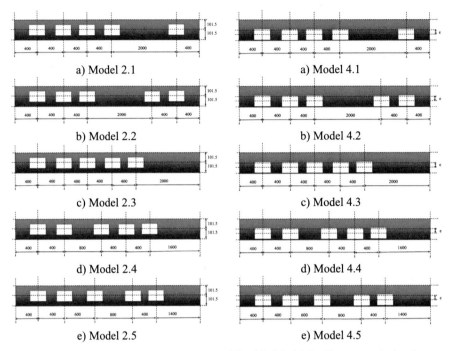

a) Model 2.1

b) Model 2.2

c) Model 2.3

d) Model 2.4

e) Model 2.5

a) Model 4.1

b) Model 4.2

c) Model 4.3

d) Model 4.4

e) Model 4.5

Fig. 12. Model 2 – Unsymmetrical holes

Fig. 13. Model 4 – Unsymmetrical and eccentric holes

Fig. 14. Flexural buckling mode under compression

Fig. 15. Flexural-torsional buckling mode under compression

Fig. 16. Flexural-torsional buckling mode under bending

The investigated results illustrate that the web hole locations have a minimal influence on the elastic global buckling of perforated members for the small hole depth (h_{hole}/D = 0.2), but it becomes significant impacts for the large hole (h_{hole}/D = 0.8).

In terms of compression, the deviation is less than 3% for the flexural-torsional buckling mode regarding the hole depths whereas it reaches 12% for the flexural buckling mode with the large hole (h_{hole}/D = 0.8). The simple method for the determination of elastic global buckling loads for symmetrical and evenly spaced holes, therefore, can be applied to this flexural-torsional buckling mode.

The results show that the elastic global buckling loads significantly decrease as the web holes are arranged closely to the mid-length. This conclusion can be seen in the results of Models 1.3 and 3.3 for both compression and bending.

The buckling loads are noticeable increase with the absence of the web hole at the mid-length as seen in the results of Models 2.2 and 4.2. This illustrates the great impact of the web hole at the mid-length on the elastic global buckling of perforated members.

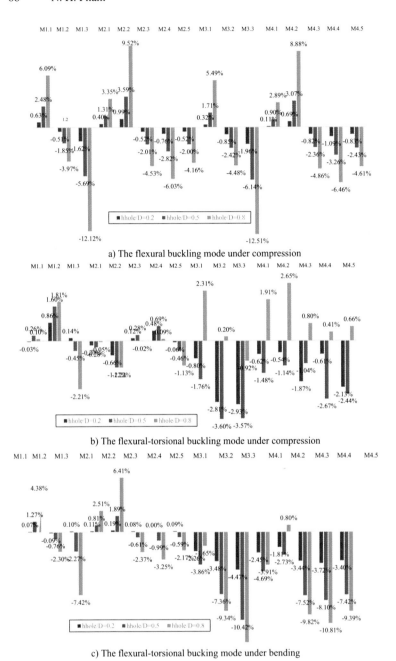

a) The flexural buckling mode under compression

b) The flexural-torsional buckling mode under compression

c) The flexural-torsional bucking mode under bending

Fig. 17. The deviation of buckling loads between the Model 1 to Model 4 and the Model 0

5 Conclusion

The paper investigated the effects of web holes on the elastic global buckling loads of cold-formed steel channel members under compression or bending. The effects of symmetrical and evenly spaced web holes on the elastic global buckling of cold-formed steel channel members were carried out using the simple method according to the Specification AISI S100-16 with the following conclusions: as the sizes of holes increased, the elastic buckling loads of perforated members decreased. In terms of the high ratio of the hole depth and the section depth, the elastic buckling loads were observed to be significantly reduced as the hole length increased.

The influences of hole locations on the elastic global buckling loads were investigated by using finite element analyses. The finite element models were validated in previous studies of Pham *et al.* [21, 22], and they were used for elastic buckling analyses in this investigation. Based on the investigated results, the following conclusions can be drawn:

– The "weighted average" method in the AISI S100 still can be used to determine the elastic flexural-torsional buckling loads under compression due to the minimal impacts of hole locations.
– The presence of the web hole at the mid-length has great impacts on the elastic global buckling loads of perforated channel members under compression or bending.

These remarks are beneficial for the designers in selecting the sizes and locations of the web holes in the design.

References

1. American Iron and Steel Institute: North American Specification for the Design of Cold-formed Steel Structural Members. American Iron and Steel Institute, Washington DC (2016)
2. Ortiz-Colberg, R.A.: The Load Carrying Capacity of Perforated Cold-Formed Steel Columns. Cornell University, Ithaca (1981)
3. Sivakumanran, K.S.: Load capacity of uniformly compressed cold-formed steel section with punched web. Can. J. Civil Eng. **14**(4), 8 (1987). https://doi.org/10.1139/l87-080
4. Banwait, A.S.: Axial Load Behaviour of Thin-Walled Steel Sections with Openings. McMaster, University Hamilton, Ontario (1987)
5. Abdel-Rahman, N.: Cold-Formed Steel Compression Members with Perforations. McMaster, University Hamilton, Ontario (1997)
6. Moen, C.D.: Direct Strength Design for Cold-Formed Steel Members with Perforations. Johns Hopkins University, Baltimore (2008)
7. Moen, C.D., Schafer, B.W.: Experiments on cold-formed steel columns with holes. Thin-Wall Struct. **46**(10), 1164–1182 (2008). https://doi.org/10.1016/j.tws.2008.01.021
8. Moen, C.D., Schafer, B.W.: Elastic buckling of cold-formed steel columns and beams with holes. Eng. Struct. **31**, 2812–2824 (2009). https://doi.org/10.1016/j.engstruct.2009.07.007
9. Moen, C.D., Schafer, B.W.: Impact of holes on the elastic buckling of cold- formed steel columns. In: International Specialty Conference on Cold-Formed Steel Structures, pp. 269–283 (2006)

10. Moen, C.D., Schafer, B.W.: Extending direct strength design to cold-formed steel beams with holes. In: The 20th International Specialty Conference on Cold-Formed Steel Structures - Recent Research and Developments in Cold-Formed Steel Design and Construction, pp. 171–183 (2010)

11. Cai, J., Moen, C.D.: Elastic buckling analysis of thin-walled structural members with rectangular holes using generalized beam theory. Thin-Walled Struct. **107**, 274–286 (2016). https://doi.org/10.1016/j.tws.2016.06.014

12. Moen, C.D., Schafer, B.W.: Direct Strength Design of Cold-Formed Steel Members with Perforations. Johns Hopkins University, Baltimore (2009)

13. Moen, C.D., Schafer, B.W.: Elastic buckling of thin plates with holes in compression or bending. Thin-Walled Struct. **47**(12), 1597–1607 (2009). https://doi.org/10.1016/j.tws.2009.05.001

14. American Iron and Steel Institute: Development of CUFSM Hole Module and Design Tables for the Cold-formed Steel Cross-sections with Typical Web Holes in AISI D100. Research Report, American Iron and Steel Institute (2021)

15. Moen, C.D., Yu, C.: Elastic buckling of thin-walled structural components with edge-stiffened holes. In: Collection of Technical Papers - AIAA/ASME/ASCE/AHS/ASC Structures, Structural Dynamics and Materials Conference, pp. 1–10 (2010)

16. Grey, C.N., Moen, C.D.: Elastic buckling simplified methods for cold-formed columns and beams with edge-stiffened holes. In: Structural Stability Research Council Annual Stability Conference, pp. 92–103 (2011)

17. Rhodes, J., Schneider, F.D.: The compressional behaviour of perforated elements. In: Twelfth International Specialty Conference on Cold-Formed Steel Structures, pp. 11–28 (1994)

18. Loov, R.: Local buckling capacity of C-shaped cold-formed steel sections with punched webs. Can. J. Civil Eng. **11**(1), 1–7 (1984). https://doi.org/10.1139/l84-001

19. Pu, Y., Godley, M.H.R., Beale, R.G., Lau, H.H.: Prediction of ultimate capacity of perforated lipped channels. J. Struct. Eng. **125**(5), 4 (1999). https://doi.org/10.1061/(ASCE)0733-9445(1999)125:5(510)

20. Rhodes, J., Macdonald, M.: The effects of perforation length on the behaviour of perforated elements in compression. In: Thirteenth International Specialty Conference on Cold-Formed Steel Structures, pp. 91–101 (1996)

21. Pham, N.H., Pham, C.H., Rasmussen, K.J.R.: Finite element simulation of member buckling of cold-rolled aluminium alloy 5052 channel columns. In: Ha-Minh, C., et al. (eds.) CIGOS 2019, Innovation for Sustainable Infrastructure: Proceedings of the 5th International Conference on Geotechnics, Civil Engineering Works and Structures, pp. 263–268. Springer Singapore, Singapore (2020). https://doi.org/10.1007/978-981-15-0802-8_39

22. Pham, N.H., Pham, C.H., Rasmussen, K.J.R.: Numerical investigation of the member buckling of cold-rolled aluminium alloy channel beams. In: The 9th International Conference on Steel and Aluminium Structures, p. 11 (2019)

Effect of Bridge Skew on the Analytical and Experimental Responses of a Steel Girder Highway Bridge

Renxiang Lu[1](✉) and Johnn Judd[2]

[1] University of Wyoming, Laramie, USA
`rlu@uwyo.edu`
[2] Brigham Young University, Provo, USA
`johnn@byu.edu`

Abstract. This study examines the effect of bridge skew on the load rating and natural frequencies of a steel girder skewed highway bridge. The analytical load rating was determined based on a line-girder model and the AASHTO bridge design specification. The experimental load rating was determined based on a series of calibrated-weight truck runs. The analytical natural frequency was determined based on correlating the single span response to a continuous span response. The experimental natural frequency was obtained based on the free vibration response from the calibrated-weight truck. The frequency associated with the first spike of the frequency domain plot was identified using a Fast Fourier Transformation. The results show that the analytical load ratings and natural frequencies differed from the experimental values primarily due to effect of bridge skew, which caused the actual load path to be significantly shorter than the bridge span length that was used in the analytical calculations.

Keywords: Load rating · Natural frequency · Skewed bridges · Field testing

1 Introduction

Theoretical equations mathematically developed through summarized empirical studies or finite element simulations provide a convenient and simplified approach to estimate the complex behavior of highway bridges [1]. However, for bridges with parameters not considered in design and with atypical features, the estimations obtained using these equations can produce inaccurate results. For example, the AASHTO specification live load distribution factor equations neglect the influence of parapets which therefore reduce the predicted stresses induced in exterior and interior girders. In one study this simplification produced a 36% and 13% underestimation for exterior and interior girders, respectively [2]. Several other factors also affect the accuracy of analytical methods. Friction at bridge bearings due to accumulation of debris has been shown to increase the load rating capacity estimated in the AASHTO specification by 4% for a straight steel girder bridge [3]. The actual normal stress impact factor for spans larger than 49.5 m (162

© The Author(s) 2022
G. Feng (Ed.): ICCE 2021, LNCE 213, pp. 70–81, 2022.
https://doi.org/10.1007/978-981-19-1260-3_7

ft) in a curved steel box-girder bridge can also be significantly smaller (approximately 60%) than the values predicted in the AASHTO specification [4].

A particularly important feature that is not commonly addressed in equations is the effect of bridge skew. Bridge skew occurs when the direction of the bridge span is not orthogonal to the bridge supports. The degree of bridge skew is generally dictated by space limitations and constraints from anthropogenic or natural obstacles that do not allow the bridge to be built straight [5]. In skewed bridges, the load path is oriented toward the corners of the bridge span with an angle greater than 90°, unlike the straight bridges whose orientation is toward the direction of the span [6]. As a consequence, compared to an equivalent straight bridge a skewed bridge generates larger shear forces at exterior girders [7], higher reactions at corners with an angle greater than 90° [8], and larger transverse moments [9]. As a result, the behavior of a skewed bridge differs significantly from a straight bridge. In particular, the effect of bridge skew on the bridge load rating and on the natural frequencies of the bridge can be significant. However, prior to this study these aspects have not been examined jointly in a single experimental study.

This study investigates the effect of skew by examining the responses of a four-span skewed highway bridge. Analytical load ratings at the critical moment locations were calculated using a line-girder model and the AASHTO Standard Specifications for Highway Bridges [10], and the experimental response was obtained through a field test. In the field test, after instrumenting the critical moment locations for every girder with strain gages, a truck with a calibrated weight was driven at crawl speed across the entire bridge at distinct transverse locations in a series of truck runs. Critical responses were used to complete the experimental load rating. The analytical natural frequency was estimated based on a simple span bridge using the Ontario Highway Bridge Design Code (OHBDC) [11] and then corrected for a continuous span bridge based on the recommendations given in Barth and Wu [12]. The experimental natural frequency was determined based on responses due to the calibrated truck driven at high speed. A Fast Fourier Transform (FFT) was used to transform to the portion of the response in free vibration condition (time domain response) into a frequency domain plot where the first spike was determined to be the experimental natural frequency. Although the study focuses on skewed bridges with steel girders, the methodology could also be applied to straight bridges and to bridges made of different materials (reinforced concrete, prestressed concrete, wood, etc.) and cross-sections (box-section, plate girder section, rectangular section, and so forth).

2 Bridge Description

The four-span skewed highway bridge examined in this study (Fig. 1) is located on the eastbound direction of Interstate 80 over the Bear River in Evanston, Wyoming, USA. The bridge is symmetric with continuous steel girders supporting a nominally 191 mm-thick (7.5 in.-thick) reinforced concrete deck. The total length of the bridge is 124.4 m (408 ft), with outer spans of 25.6 m (84 ft) each and inner spans of 36.6 m (120 ft) each. The outer eastern pier is supported by a pin bearing. Rocker bearings are used at the remaining four supports. The bridge has an angle of skew of 47° relative to the abutments. The bridge has a width of approximately 12.2 m (40 ft) and has a shoulder on both ends. The bridge was originally built with four nominally identical I-shaped steel girders that were designed to act non-compositely with the deck. The bridge was later widened and a fifth steel girder was added. The fifth girder was designed to act compositely with the deck. The bridge cross-section is shown in Fig. 2. The girders are haunched at the pier locations. The height of the girder web increases parabolically and the web and flange are thicker at the pier locations. The dimensions of the girders at selected cross-sectional locations are described in Lu et al. [13]. The minimum specified yield strength of the structural steel is 250 MPa (36 ksi). The steel girders were relatively free of corrosion. The specified minimum concrete compressive strengths of the original and added concrete deck are 22.4 MPa (3.25 ksi) and 27.6 MPa (4 ksi), respectively.

Fig. 1. Bridge examined in this study.

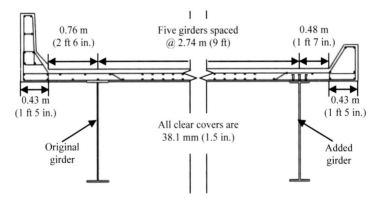

Fig. 2. Cross-section.

3 Load Rating

The load rating capacity of the bridge was obtained based on the rating factor, *RF*, of each individual girder. In the AASHTO rating method, the girder with the lowest *RF* controls the load rating of the bridge [14]. This rating method was applied. Among the different limits used to load rate bridges, the strength inventory rating factor is generally the most critical, and therefore this limit was adopted here.

Since the bridge was originally designed and evaluated using the Load Factor Design (LFD) method and Load Factor Rating (LFR), respectively, the calculation of *RF* was based on the LFD method. Although the HS20 design truck in the LFD method overestimates the bridge capacity compared to the HL-93 design truck in the Load and Resistance Factor Design (LRFD) method [15], the overestimation cancels out when taking the ratio of the analytical and experimental load ratings.

In this study, *RF* was defined as the ratio between the reserve capacity for live load and the limit state design live load and expressed as the number of standardized HS20 design truck loads:

$$RF = \frac{R_n - 1.3D}{2.17LL(1+I)} \tag{1}$$

where R_n is the capacity of the member, D is the dead load effect on the member, LL is the live load effect of the member, and I is the dynamic impact factor. The value of R_n was calculated as the product of the specified yield strength of the structural steel and the elastic section modulus of the girder. The value of D was calculated as the summation of the contributions due to self-weight of the girder, self-weight of the deck and diaphragms, self-weight of the future wearing surface, distributed to a single girder. When *RF* is greater than 1.0, the bridge is regarded as structurally adequate for the design load. On the other hand, when *RF* is less than 1.0, the bridge is inadequate for the design load.

3.1 Analytical Load Rating

The analytical live load, LL_A, accounting for dynamic impact was calculated as follows:

$$LL_A(1 + I_A) = M_{HS20}DF_A m_A(1 + I_A) \tag{2}$$

where M_{HS20} is the maximum analytical moment due to the HS20 design truck at the location of interest, DF_A is the analytical live load distribution factor, m_A is the analytical live load multi-presence factor, and I_A is the analytical impact factor. DF_A is calculated using the AASHTO Standard Specifications for Highway Bridges in section 3.23.2.3.1.5. The AASHTO equations provide the lateral distribution for only a single wheel-line. Since the design truck has two wheel-lines, the DF_A needs to be divided by two. The value of I_A was calculated using the AASHTO Standard Specifications for Highway Bridges equation in section 3.8.2.1.

It was not feasible to instrument the inner span of the bridge. Therefore, Eq. (1) and Eq. (2) were applied to the critical moment locations of the outer span. At the outer span, the critical positive longitudinal moment was located at 0.4 times the span length from the abutment, and the critical negative longitudinal moment was located at the pier between inner and outer spans. The corresponding values of M_{HS20} were determined using a one-dimensional line-girder bridge model, where the bridge was modeled as if it consisted of only a single girder. The single girder was modeled by attributing the value of the nominal non-composite moment of inertia without the concrete deck. This approach is consistent with values of R_n and D calculated for a single girder. To produce the maximum positive moment, M_{HS20} was modeled using three concentrated loads. To produce the maximum negative moment, M_{HS20} was replaced by the uniform-plus-two-concentrated-load configuration in AASHTO specification section 3.7.6. As defined in AASHTO specification section 3.12.1, the live load multi-presence factor accounts for the probability of coincident maximum loading. The multi-presence factor is 1.0 for single or 2-truck loadings, and 0.9 for 3-truck loadings. Since one truck was placed on the model, m_A was equal to 1.0.

The values of the variables in Eq. (1) and Eq. (2) for the positive and negative critical moment locations are given in Table 1. It was determined that $LL_A(1 + I_A)$ was equal to 1320 kN-m (972 k-ft) and -1820 kN-m (-1340 k-ft) at the positive and negative moment, respectively. The strength inventory RF values are equal to 0.98 and 0.91 at the positive and negative moment, respectively.

Table 1. Values of the variables for Eq. (1) and Eq. (2).

Variables	Positive moment	Negative moment
R_n, kN-m (k-ft)	3680 (2710)	-6170 (-4550)
D, kN-m (k-ft)	672 (496)	-1980 (-1460)
M_{HS20}, kN-m (k-ft)	1300 (956)	-1820 (-1340)
I_A	0.24	0.22
DF_A	0.82	0.82

3.2 Experimental Load Rating

The experimental live load accounting for dynamic impact is calculated as follows:

$$LL_E(1 + I_E) = \left(\frac{M_{HS20}}{M_{TRK\_OS}} \right) M_{girder} \left(\frac{M_{TRK}}{M_{TRK\_OS}} \right) m_E(1 + I_E) \tag{3}$$

where M_{girder} is the non-composite moment in the girder, M_{TRK} is the analytical maximum moment due to the calibrated truck used in the field test, M_{TRK_OS} is the analytical maximum moment due to the calibrated truck used in the field test when the truck is at the outer span, m_E is the live load multi-presence factor according to the loading conditions of the actual experiment, and I_E is the experimental impact factor based on the field test.

The line-girder model used to obtain M_{HS20} was also used to determine the values of M_{TRK} and M_{TRK_OS}. The axle loads and axle spacings of the calibrated truck (Fig. 3) were measured and converted into three concentrated loads that were then applied to the model. M_{TRK_OS} was determined because the bridge was only instrumented on the outer span. For the positive moment, M_{TRK} and M_{TRK_OS} were the same value, equal to 890 kN-m (657 k-ft), because they occurred at the outer span. For the negative moment, M_{TRK} (at the inner span) was equal to −762 kN-m (-554 k-ft), and M_{TRK_OS} was equal to −570 kN-m (−420 k-ft).

The values of M_{girder} were obtained through a field test. In the field test, strain gages were installed at the most critical moment locations along the bridge. At the negative moment, the strain gages were installed with a 2.44-m (8-ft) offset from the support to minimize the influence of the support on the response. A larger offset was not used to reduce errors due to extrapolation [16]. Two strain gages were installed on each girder at each location. One strain gage was mounted on the bottom of the bottom flange. The other strain gage was mounted on the web 102 mm (4 in.) below the bottom of the top flange.

The calibrated truck was driven at crawl speed over the bridge in successive runs that traverse the width of the roadway according to the requirements given in the AASHTO Manual for Bridge Evaluation [17]. Figure 4 shows the run sequence and truck position. A total of fifteen runs were conducted to traverse the entire roadway, and the runs were conducted from left to right relative to the direction of travel. In this paper, the girders are numbered 1 to 5. (Girder 5 is the composite girder that was added when the bridge was widened.) In the first run (Run 1), the center of the left front wheel was positioned 0.91 m (3 ft) from the left curb. The position of each successive run was 0.61 m (2 ft) offset to the right of the previous run. The configuration of the run sequence is nearly symmetric relative to Girder 3. A total of fifteen runs were conducted.

Since the truck load was calibrated to produce a response that was within the linear elastic range, the responses for side-by-side truck loading combinations were determined by superimposing the measured results. The controlling combination was defined as the loading that produced the greatest $LL_E(1 + I_E)$ at the positive or negative moment location. For this bridge, the three side-by-side truck loading combination involving Runs 3, 8 and 13 controlled the positive moment, and the two side-by-side truck loading combination involving 1 and 6 superimposed controlled the negative moment. M_{girder} was equal to 317 kN-m (234 k-ft) and occurred in Girder 3 at the positive moment. M_{girder} was equal to −320 kN-m (-236 k-ft) and occurred in Girder 2 at the negative moment. Field tests were not conducted to determine the actual dynamic impact factor; therefore, it was assumed that I_E was equal to I_A. As a result, $LL_E(1 + I_E)$ was equal to 515 kN-m (380 k-ft) and −1640 kN-m (−1210 k-ft) at the positive and negative moment locations, respectively.

An inspection to determine the *in-situ* dead loads was not practical. Therefore, the experimental values of R_n and D were taken equal to the analytical values in Table 1. The resulting experimental strength inventory RF is equal to 2.51 and 1.01 for positive and negative moment, respectively. Based on the HS20 standardized design truck, the bridge was rated for a HS50.2 truck (20 times 2.51) at the positive moment location and HS20.2 (20 times 1.01) at the negative moment location.

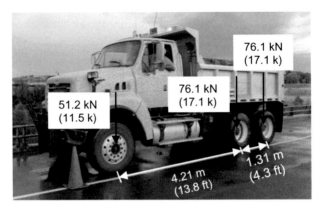

Fig. 3. Calibrated truck.

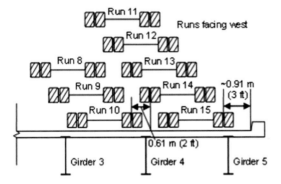

Fig. 4. Run sequence and truck positions.

3.3   Comparison of Analytical and Experimental Load Ratings

The ratio of the experimental load rating to the analytical load rating was used to determine the effect of additional contributors to the bridge actual capacity, including the effect of bridge skew. The ratio of the experimental to the analytical load ratings is equal to 2.56 at the positive moment and equal to 1.11 at the negative moment. The ratio of load ratings can also be taken as the inverse of the ratio of the live load effects (LL_A/LL_E) since R_n and D were taken as constants. This conclusion was expected because the design load rating is generally conservative. It does not consider additional contributors to the bridge capacity (such as the additional stiffness due to curbs and railings, actual longitudinal and lateral distribution factor, unintended composite action, and in special, the effect of the bridge skew). The additional contributions discretized in a skewed bridge are discussed in detail in Lu et al. [13].

4   Natural Frequency

The AASHTO specification limits deflections (e.g., $L/800$ to $L/1000$, where L is the bridge span length) but it does not use natural frequency as a direct measure of serviceability. In contrast, the OHBDC recommends the calculation of natural frequency to control vibration. In the OHBDC approach, a relationship is established between the natural frequency, the live-load deflection, and the pedestrian usage of the bridge. The natural frequency of a simple span bridge (first bending frequency), f_b, is

$$f_b = \frac{\pi}{2L^2}\sqrt{\frac{EI_bg}{w}} \tag{4}$$

where E is the modulus of elasticity of the girder, I_b is the flexural composite moment of inertia of the girder (for levels of serviceability, slippage is not expected to occur even if the bridge is designed to act non-compositely), and w is the weight per unit length of the girder. Equation (4) has been extended for continuous span bridges by Barth and Wu [12] based on the results of parametric studies and regression analysis. In their approach, the value of f_b is corrected (multiplied) by the "natural frequency coefficient" λ^2, calculated as follows:

$$\lambda^2 = a\frac{I_b^c}{L_{max}^b} \tag{5}$$

where L_{max} is the maximum bridge span length. In Eq. (5), when SI units are used for L_{max} (m) and I_b (m^4), a is equal to 1.49, b is equal to -0.033, and c is equal to 0.033 for bridges with three or more spans.

4.1   Analytical Natural Frequency

Equation (4) was applied to the bridge, and the resulting values are given in Table 2. In this study, L was equal to 36.6 m (120 ft), and I_b was calculated based on a weighted average of each girder and each span relative to its corresponding length. For the haunched portions of the bridge, it was opted to simplify the calculation so that the averages of the greatest and smallest moments of inertia were taken. Here w was determined as a summation of the self-weight of the girder (a weighted average was determined for the computation of this component), self-weight of the deck and diaphragms, composite dead load which consists of the weight of the curbs and railings, and the weight of the future wearing surface, all distributed to a single girder. The resultant value of f_b is equal to 2.05 Hz. Applying Eq. (5), the value of λ^2 is equal to 1.49. Thus, the analytical corrected natural frequency for the bridge is equal to 3.07 Hz.

4.2   Experimental Natural Frequency

After the field test, the natural frequency of the bridge was determined experimentally by measuring the acceleration caused by the calibrated truck crossing the bridge at high speed while the bridge was closed to ambient traffic. The truck velocity was set to the

Table 2. Variables for natural frequency per Eq. (4).

Variables	Values
L, m (in.)	36.6 (1440)
E, GPa (ksi)	200 (2.90 × 10^4)
I_b, mm^4 (in.^{4})	3.21 × 10^{10} (7.71 × 10^4)
w, kN/m (k/in.)	20.6(0.118)

speed limit for the bridge, 121 km/h (75 mph). An accelerometer with a range between $-2g$ and $+2g$ was mounted on the top of the bottom flange of Girder 2. Three tests were conducted. The acceleration data was collected using 1000 samples per second. A representative acceleration history is shown in Fig. 5. Small and regular vibrations before the truck enters the bridge are evident in the history. At approximately 6 s, the accelerations increase, reaching a peak of 0.15 g at approximately 14 s (at this instant, the truck is immediately above the accelerometer). As the truck leaves the bridge, a steady decline of the vibration is observed, and the bridge enters the free vibration condition at approximately 16 s. The bridge was reopened for ambient traffic after each test. Therefore, small intermittent increases can be observed starting at approximately 22 s. As a result, the free vibration condition of the bridge between approximately 16 and 22 s was used to determine the natural frequency.

A Fast Fourier Transform (FFT) was conducted in which the acceleration history (time domain plot) of the free vibration condition (approximately 6 s) was transformed into a frequency domain plot (Fig. 6) and used to determine the principal frequencies of the bridge. The data for the free vibration condition was condensed to 4096 (2^{12}) data points. The frequency domain plot displays the magnitude of the Fourier Coefficient in function of frequency with noticeable spikes. Although the coefficients do not have any physical meaning for the method, the frequencies corresponding to the respective spikes are values of the different modes of vibration. The first spike represents the natural frequency, the second the second harmonic frequency, the third the third harmonic frequency, and so forth. For this bridge, the experimental natural frequency was equal to 3.42 Hz.

4.3   Comparison of Analytical and Experimental Natural Frequencies

Comparing the natural frequencies obtained analytically and experimentally, it was observed that the analytical natural frequency underestimated the experimental natural frequency by 11%. The increase of natural frequency with the bridge skew agrees with prior studies [18, 19]. The difference is partially due to the estimation of I_b in the calculation of the analytical natural frequency, but the bridge skew effect is thought to be the main cause. In a skewed bridge, if the span length is significantly greater than the cross-sectional width, flexural behavior tends to govern. The load path tends to be oriented toward the obtuse corners of the bridge, i.e., the "shortest path" across the span. Although the actual load path could vary between the "shortest path" and the longitudinal

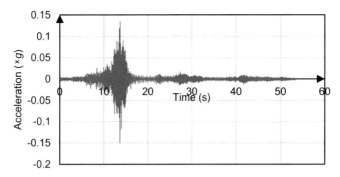

Fig. 5. Representative acceleration history.

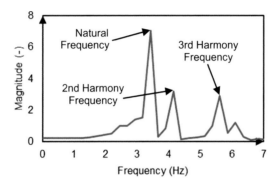

Fig. 6. Frequency domain plot.

distance between abutments (Fig. 7), it is significantly shorter than the bridge longitudinal span length of an equivalent straight bridge. This observation has considerable impact on the variables of L and L_{max} of Eq. (4) and Eq. (5), respectively. In particular, the application of the correction factor λ^2 for the estimating natural frequencies of skewed bridges may exceed the valid range for Eq. (5).

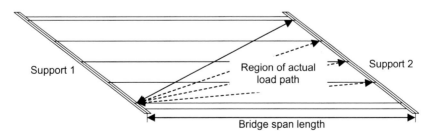

Fig. 7. Shortest path, bridge span length, and actual load path.

5 Conclusions and Future Recommendations

The analytical and experimental load ratings and natural frequencies were determined for a four-span skewed highway bridge. The analytical load rating was calculated using a line-girder model and the AASHTO Standard Specifications for Highway Bridges. The experimental load rating was determined using a field test in which a calibrated-weight truck was driven at crawl speed across the bridge at a range of transverse locations. The results indicated that the analytical load rating was equal to 0.98 and 0.91 at the positive and negative moment, respectively.

The experimental load rating was equal to 2.51 for the positive moment and 1.01 for the negative moment, which were 2.56 and 1.11 times higher than the corresponding positive and negative analytical load ratings. The increase observed was attributed to effects not considered in analytical calculation, namely additional stiffness due to curbs and railings, actual longitudinal and lateral distribution factor, unintended composite action, and, in particular, the effect of bridge skew.

The analytical natural frequency was determined using the OHBDC and an adjustment to account for a continuous span. The experimental natural frequency was determined using a FFT analysis of the free vibration response for truck crossing the bridge at high speed. The results indicate that the analytical natural frequency underestimated the experimental by 11%. The underestimation was attributed to the length of the load path, which is inversely proportional to the natural frequency. In a skewed bridge, the load path length is shorter than in an equivalent straight bridge.

The findings in this study suggest that accuracy of equations for estimating the load rating and the natural frequency of a highway bridge depend on several aspects, including the effect of bridge skew. As a result, additional research is needed to adjust the equations to better estimate the response. It is recommended that the equations account for the main contributors to the bridge load rating, including the effect of skew. Parametric studies are also needed to expand the applicability of the equations for estimating natural frequency to account for the effective length of the load path.

References

1. Nagayama, T., Reksowardojo, A.P., Su, D., Mizutani, T.: Bridge natural frequency estimation by extracting the common vibration component from the responses of two vehicles. Eng. Struct. **150**, 821–829 (2017). https://doi.org/10.1016/j.engstruct.2017.07.040
2. Conner, S., Huo, X.S.: Influence of parapets and aspect ratio on live-load distribution. J. Bridge Eng. **11**, 188–196 (2006). https://doi.org/10.1061/(ASCE)1084-0702(2006)11:2(188)
3. Barker, M.G.: Quantifying field test behavior for rating steel girder bridges. J. Bridge Eng. **6**, 254–261 (2001). https://doi.org/10.1061/(ASCE)1084-0702(2001)6:4(254)
4. Huang, D.: Field test and rating of Arlington curved-steel box-girder bridge: Jackson Florida. Transp. Res. Rec. **1892**, 178–186 (2004). https://doi.org/10.3141/1892-19
5. Coletti, D., Chavel, B., Gatti, W.J.: Challenges of skew in bridges with steel girders. Transp. Res. Rec. **2251**, 47–56 (2011). https://doi.org/10.3141/2251-05
6. Nouri, G., Ahmadi, Z.: Influence of skew angle on continuous composite girder bridge. J. Bridge Eng. **17**, 617–623 (2012). https://doi.org/10.1061/(ASCE)BE.1943-5592.0000273

7. Cross, B., Vaughn, B., Panahshahi, N., Petermeier, D., Siow, Y.S., Domagalski, T.: Analytical and experimental investigation of bridge girder shear distribution factors. J. Bridge Eng. **14**, 154–163 (2009). https://doi.org/10.1061/(ASCE)1084-0702(2009)14:3(154)
8. Badwin, I.Z., Liang, R.Y.: Reaction distribution in highly skewed continuous steel girder bridge. Transp. Res. Rec. **2028**, 163–170 (2007). https://doi.org/10.3141/2028-18
9. Théoret, P., Massicotte, B., Conciatori, D.: Analysis and design of straight and skewed slab bridges. J. Bridge Eng. **17**, 289–301 (2012). https://doi.org/10.1061/(ASCE)BE.1943-5592. 0000249
10. American Association of State Highway and Transportation Officials (AASHTO): AASHTO standard specifications for highway bridges, 17th edn, Washington DC, USA (2002)
11. Ministry of Transportation, Quality and Standards Division: Ontario highway bridge design code/commentary, 3rd edn, Toronto, Ontario Canada (1991)
12. Barth, K.E., Wu, H.: Development of improved natural frequency equations for continuous span steel I-girder bridges. Eng. Struct. **29**, 3432–3442 (2007). https://doi.org/10.1016/j.eng struct.2007.08.025
13. Lu, R., Judd, P.J., Barker, M.G.: Field load rating and grillage analysis method for skewed steel girder highway bridges. J. Bridge Eng. **26**, 05021013 (2021). https://doi.org/10.1061/ (ASCE)BE.1943-5592.0001787
14. Sanayei, M., Reiff, A.J., Brenner, B.R., Imabro, G.R.: Load rating of a fully instrumented bridge: Comparison of LRFR approaches. Perform Constr. Facil. **30**, 04015019 (2016). https:// doi.org/10.1061/(ASCE)CF.1943-5509.0000752
15. American Association of State Highway and Transportation Officials (AASHTO): AASHTO LRFD bridge design specifications, 9th edn, Washington DC, USA (2020)
16. Barker, M.G.: Steel girder bridge field test procedures. Constr. Build. Mater. **13**, 229–239 (1999). https://doi.org/10.1016/S0950-0618(99)00013-6
17. American Association of State Highway and Transportation Officials (AASHTO): AASHTO manual for bridge evaluation, 3rd edn, Washington DC, USA (2018)
18. Maleki, S.: Free vibration of Skewed Bridges. J. Vib. Control **7**, 935–952 (2001). https://doi. org/10.1177/107754630100700701
19. Mohseni, I., Rashid, K.A., Kang, J.: A simplified method to estimate the fundamental frequency of skew continuous multicell box-girder bridges. Lat. Am. J. Solids Struct. **11**, 649–658 (2014). https://doi.org/10.1590/S1679-78252014000400006

Investigation of Sectional Capacities
of Cold-Formed Steel SupaCee Sections

Ngoc Hieu Pham[✉]

Faculty of Civil Engineering, Hanoi Architectural University, Hanoi, Vietnam
hieupn@hau.edu.vn

Abstract. SupaCee section is formed by adding stiffeners in the web of the channel section, and it has been illustrated to be more stable and innovative than the
traditional channel section. The member capacity comparisons between such two
section members investigated in a companion paper showed that SupaCee was
significantly beneficial for small and thin section members, but the reductions of
global buckling strengths of SupaCee section members led to the lower capacities
for long members compared to those of channel section members. These reductions can be prevented by using full bracing systems to avoid global buckling;
this allows member capacities to reach sectional capacities. This paper, therefore,
presents a study on sectional capacities of cold-formed steel SupaCee sections
under compression or bending in comparison with those of channel sections.

Keywords: Cold-formed steel · Supacee sections · Sectional capacities ·
Stiffeners

1 Introduction

Cold-formed channel sections have been commonly used in structural applications for
years [1]. These sections are very sensitive to instability due to their small thickness.
These sections, therefore, have been added stiffeners in the web to increase their stability
that results in significant improvements of strength performances, known as SupaCee
sections. Also, SupaCee's lips have been rounded to become safer while installing on-
site.

Direct Strength Method (DSM) is a new design method developed by Schafer and
Pekoz [2–4] currently regulated in the Australian/New Zealand Standard AS/NZS 4600
[5] or North American Specification AISI S100–16 [6] for the design of cold-formed
steel structures. Elastic buckling analysis is compulsory in this design method that can be
supported by THIN-WALL-2 [7, 8] or CUFSM [9] software programs. The DSM design
method has been demonstrated its advances compared to the traditional design method
[2] and provides insight understandings into the buckling behaviors of cold-formed steel
sections [1]. This method will be used in this investigation.

The presence of stiffeners was found to be the strength improvements of cold-formed
steel sections that have been investigated by a variety of past studies. Experiments on
cold-formed steel storage racks with edge stiffeners have been investigated by Hancock

© The Author(s) 2022
G. Feng (Ed.): ICCE 2021, LNCE 213, pp. 82–94, 2022.
https://doi.org/10.1007/978-981-19-1260-3_8

et al. [10–14] to insightfully understand the distortional buckling behaviors. The behaviors of cold-formed steel channel columns with the edge or intermediate stiffeners were also studied by Wang *et al.*, Yan and Young, Xiang *et al.*, and Manikandan *et al.*, and El-taly *et al.* [15–23]. A variety of stiffeners of cold-formed steel channel sections in bending were investigated by Ye *et al.*, and Chun-gang *et al.* [24, 25] to demonstrate the beneficial effects of these stiffeners on the sectional capacities. Cold-formed steel built-up I sections in the form of back-to-back SupaCee sections were also tested by Manikandan and Arun [26] or Ganeshkumar *et al.* [27] to observe their behaviors under compression or bending. A series of SupaCee section beams were tested by Pham and Hancock [28–31] to provide deep insights into their behaviors under shear or combined bending and shear, followed by the development of numerical studies and design proposals. The past researches presented experiments or numerical studies of cold-formed steel channel sections with stiffeners or SupaCee sections that were then used for design proposals.

The effects of stiffeners on the SupaCee section member capacities under compression or bending were also investigated by Pham and Vu [32] to demonstrate the innovation and the strength improvement of SupaCee section members compared to the traditional channel section members. The shortcoming of these new section members was the reductions of their global buckling strengths that result in the lower capacities for long members. This shortcoming can be solved by using full bracing systems to prevent the occurrence of global buckling. In this case, the member capacities reach the sectional capacities that will be examined in this paper. The sectional capacities of SupaCee sections are determined by using the DSM design equations formulated in AS/NZS 4600 [5], and are then compared to those of channel sections to evaluate the strength improvements of SupaCee sections. The investigated sections are commercial sections from BlueScope Lysaght catalogs [33] with steel grade G450 regulated in AS 1397 [34].

2 Summarize the Investigated Capacities of SupaCee Section Members

SupaCee member capacities were determined in comparison with those of channel members, as fully reported in Pham and Vu [32]. The dimensions of the channel and SupaCee sections are listed in Table 1, and their nomenclatures are illustrated in Fig. 1. The same amount of material is required for channel and SupaCee sections to assess the effectiveness of SupaCee sections. The material properties of investigated steel grade G450 according to AS 1397 [34] include the yield stress $f_y = 450$ (MPa), Young's modulus $E = 200$ (GPa) and the Poisson ratio $\upsilon = 0.3$. The model configurations for compression and bending are shown in Fig. 2 and Fig. 3 with the bracing at the mid-length of the members. The member lengths varied from 2.0 to 8.0 m. The investigated results are illustrated in percentage diagrams, where the horizontal axis is for the capacities of channel members and the vertical axis is for the deviations (in %) between capacities of SupaCee and channel members (see Fig. 4).

The investigated results in Fig. 4 demonstrated that SupaCee member capacities, in general, are higher than those of channel members, but the capacities of several former

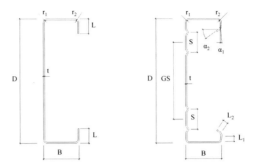

Fig. 1. Nomenclature for channel and SupaCee sections

Table 1. The nominal dimensions of the channel and SupaCee sections[a]

Sections	t	D	B	L_1	L_2	L	GS	S	α_1	α_2
SC/C15012	1.2	152	64	7.5	7.5	14.5	64	42	5	35
SC/C15015	1.5	152	64	7.5	7.5	14.5	64	42	5	35
SC/C15019	1.9	152	64	7.5	7.5	14.5	64	42	5	35
SC/C15024	2.4	152	64	7.5	7.5	14.5	64	42	5	35
SC/C20012	1.2	203	76	10	10	19.5	115	42	5	35
SC/C20015	1.5	203	76	10	10	19.5	115	42	5	35
SC/C20019	1.9	203	76	10	10	19.5	115	42	5	35
SC/C20024	2.4	203	76	10	10	19.5	115	42	5	35
SC/C25015	1.5	254	76	11	11	21.5	166	42	5	35
SC/C25019	1.9	254	76	11	11	21.5	166	42	5	35
SC/C25024	2.4	254	76	11	11	21.5	166	42	5	35
SC/C30019	1.9	300	96	14	14	27.5	212	42	5	35
SC/C30024	2.4	300	96	14	14	27.5	212	42	5	35
SC/C30030	3.0	300	96	14	14	27.5	212	42	5	35
SC/C35019	1.9	350	125	15	15	30.0	262	42	5	35
SC/C35024	2.4	350	125	15	15	30.0	262	42	5	35
SC/C35030	3.0	125	125	15	15	30.0	262	42	5	35
SC/C40019	1.9	400	125	15	15	30.0	312	42	5	35
SC/C40024	2.4	400	125	15	15	30.0	312	42	5	35
SC/C40030	3.0	400	125	15	15	30.0	312	42	5	35

[a]The inner radius $r_1 = r_2 = 5$mm; t, D, B, L1, L2, GS, S (mm); α_1, α_2 $(^0)$

members are still lower than latter member capacities. Pham and Vu [32] pointed out the reason for these lower capacities of SupaCee members was the reductions of global buckling strengths led to the lower strengths for long members with the governing of the global buckling modes or led to the significant decrease of local buckling strengths due to the interaction between global and local buckling modes. The global buckling failure modes can be avoided using the full bracing models that will be presented in Sect. 4 of this paper.

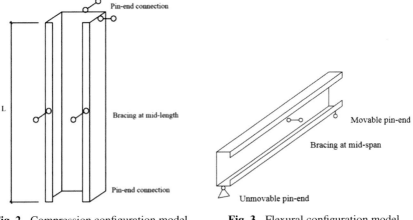

Fig. 2. Compression configuration model   **Fig. 3.** Flexural configuration model

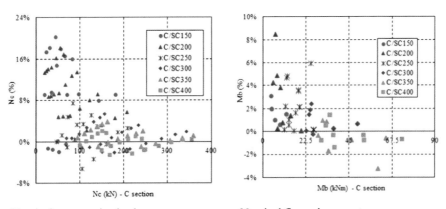

Nominal compressive loads   Nominal flexural moments

Fig. 4. Compression and flexural capacities of the channel and SupaCee cold-formed steel members [32]

3   Determination of Sectional Capacities of Cold-Formed Steel Section Using the Direct Strength Method

Sectional capacities of cold-formed steel sections can be determined using the DSM design based on elastic buckling stresses (f_{ol}, f_{od}) and the yield stress (f_y). Local and distortional buckling strengths can be directly determined by using these stresses as shown in Eqs. (1)–(3) and (4)–(6) for compression and bending respectively.

The nominal capacity of a cold-formed steel section under compression is less of the nominal strengths of local buckling (N_{cl}) and distortional buckling (N_{cd}).

$$N_s = Min\,(N_{cl},\,N_{cd}) \tag{1}$$

$$N_{cl} = \begin{cases} N_y & for \; \lambda_1 \leq 0.776 \\ \left[1 - 0.15\left(\frac{N_{ol}}{N_y}\right)^{0.4}\right]\left(\frac{N_{ol}}{N_y}\right)^{0.4} N_y & for \; \lambda_1 > 0.776 \end{cases} \qquad (2)$$

$$N_{cd} = \begin{cases} N_y & for \; \lambda_1 \leq 0.561 \\ \left[1 - 0.25\left(\frac{N_{od}}{N_y}\right)^{0.6}\right]\left(\frac{N_{od}}{N_y}\right)^{0.6} N_y & for \; \lambda_1 > 0.561 \end{cases} \qquad (3)$$

where $\lambda_l = \sqrt{N_y/N_{ol}}$; $\lambda_d = \sqrt{N_y/N_{od}}$; N_y is the nominal yield capacity of the member in compression; N_{ol} is the elastic local buckling load; N_{od} is the elastic distortional buckling load.

The nominal moment capacity (M_b) is the smaller of the nominal strengths in local buckling (M_{bl}), and distortional buckling (M_{bd}).

$$M_s = Min \, (M_{bl}, \, M_{bd}) \qquad (4)$$

$$M_{bl} = \begin{cases} M_y & for \; \lambda_1 \leq 0.776 \\ \left[1 - 0.15\left(\frac{M_{ol}}{M_y}\right)^{0.4}\right]\left(\frac{M_{ol}}{M_y}\right)^{0.4} M_y & for \; \lambda_1 > 0.776 \end{cases} \qquad (5)$$

$$M_{bd} = \begin{cases} M_y & for \; \lambda_1 \leq 0.673 \\ \left[1 - 0.22\left(\frac{M_{od}}{M_y}\right)^{0.5}\right]\left(\frac{M_{od}}{M_y}\right)^{0.5} M_y & for \; \lambda_1 > 0.673 \end{cases} \qquad (6)$$

where $\lambda_l = \sqrt{M_y/M_{ol}}$; $\lambda_d = \sqrt{M_y/M_{od}}$; M_y is the yield moment; M_{ol} and M_{od} are the elastic local and distortional buckling moments respectively.

Local and distortional buckling stresses can be determined using numerical tools such as THIN-WALL-2 [7, 8] or CUFSM [9] software packages.

4 Comparisons of Sectional Capacities Between the Channel and SupaCee Sections

As discussed in Sect. 2, the capacities of several SupaCee section members were lower than those of the channel section members due to the significant reductions of global buckling strengths. In order to improve the effectiveness of SupaCee sections, the global buckling modes should be prevented using bracing systems. The bracing systems can be made by the installations of numerous restraint points along the member length to reduce the effective length of the member as illustrated in Fig. 5 and Fig. 6. The bracing systems can be used as the bracing members attached to the webs and/or the cladding plates/ sheet plates attached to the flanges of the sections. As the global buckling modes are prevented, the member capacities reach the sectional capacities of the cold-formed steel sections that can be determined as presented in Sect. 3 using the Direct Strength Method. Elastic buckling analysis, therefore, is compulsory to determine in the application of this method. This analysis is carried out using the THIN-WALL-2 software program [7, 8] to obtain the local and distortional buckling stresses. The THIN-WALL-2 software program was developed at The University of Sydney on the basis of the finite strip method. The outcome is a signature curve to demonstrate the relationship between the elastic buckling stress and the half-wavelength of buckling mode for each section. Figure 7 illustrates the signature curve of a cold-formed steel channel section with two minimum points. The first point provides the local buckling stress with the shortest half-wavelength, while the second point gives the distortional buckling stress with the intermediate half-wavelength. The investigated sections are listed in Table 1, and the elastic local and distortional buckling stresses are presented in Tables 2 and 3. These analysis results show the significant increase of the local buckling stresses f_{ol} of SupaCee sections for compression or bending with the appearance of stiffeners in the webs. In terms of distortional buckling stresses of SupaCee sections, a minor reduction is seen for compression by less than 5% whereas they are generally higher for bending compared to those of channel sections with the deviation reaching 10.81% as seen in C25019 section.

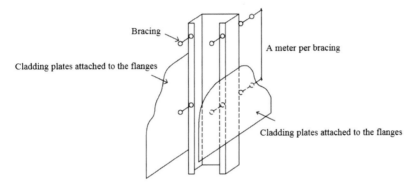

Fig. 5. The compression configuration model with bracing systems

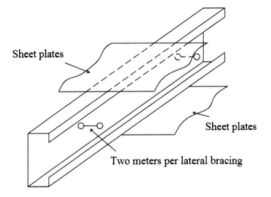

Fig. 6. The flexural configuration model with bracing systems

The sectional capacities of the investigated sections are achieved with two following cases of bracing systems. The investigated results are illustrated in percentage diagrams for the comparisons of sectional capacities between the SupaCee and the channel sections

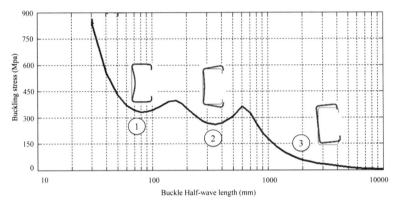

Fig. 7. A signature curve of a cold-formed steel channel section.

Table 2. The sectional buckling stresses of the investigated sections under compression

Sections	f_{ol} (MPa)			Sections	f_{od} (MPa)		
	Channel	SupaCee	$\Delta(\%)^a$		Channel	SupaCee	$\Delta(\%)$
15012	68.57	125.17	82.54%	15012	141.36	134.68	−4.73%
15015	104.25	168.23	61.37%	15015	179.64	172.67	−3.88%
15019	167.31	235.09	40.51%	15019	239.88	231.68	−3.42%
15024	266.47	332.76	24.88%	15024	317.74	308.91	−2.78%
20012	37.25	62.31	67.28%	20012	104.3	100.9	−3.26%
20015	58.23	86.19	48.02%	20015	133.94	130.53	−2.55%
20019	93.59	123	31.42%	20019	176.06	172.95	−1.77%
20024	149.53	179.62	20.12%	20024	231.71	231.29	−0.18%
25015	37.7	50.57	34.14%	25015	88.65	85.33	−3.75%
25019	60.53	74.02	22.29%	25019	118.19	116.12	−1.75%
25024	96.54	110.68	14.65%	25024	160.18	159.82	−0.22%
30019	42.82	50.12	17.05%	30019	105.26	102.22	−2.89%
30024	68.42	75.77	10.74%	30024	138.76	136.3	−1.77%
30030	107.17	114.49	6.83%	30030	185.47	186.13	0.36%
35019	30.68	34.7	13.10%	35019	86.81	82.67	−4.77%
35024	49.55	53.33	7.63%	35024	112.75	109.71	−2.70%
35030	77.57	81.35	4.87%	35030	145.81	142.46	−2.30%
40019	23.96	26.1	8.93%	40019	63.69	60.83	−4.49%
40024	38.24	40.51	5.94%	40024	84.45	82.11	−2.77%
40030	59.77	62.13	3.95%	40030	112.8	111.63	−1.04%

[a] $\Delta\%$ is the buckling stress deviation between SupaCee and channel sections, unit %

as shown in Fig. 8 and Fig. 9, where the horizontal axis is for sectional capacities of the channel sections and the vertical axis is for the sectional capacity deviations (in %) between two types of investigated sections.

Table 3. The sectional buckling stresses of the investigated sections under bending

Sections	f_{ol} (MPa)			Sections	f_{od} (MPa)		
	Channel	SupaCee	Δ (%)[a]		Channel	SupaCee	Δ (%)
15012	323.82	515.68	59.25%	15012	266.93	286.5	7.33%
15015	486.77	700.24	43.85%	15015	333.62	353.76	6.04%
15019	782.32	1018.9	30.24%	15019	447.14	461.25	3.16%
15024	1199.89	1416.16	18.02%	15024	582.9	605.02	3.79%
20012	190.43	315.35	65.60%	20012	224.34	238.94	6.51%
20015	294.28	432.78	47.06%	20015	285.28	307.95	7.95%
20019	464.42	616.87	32.83%	20019	368.48	404.73	9.84%
20024	754.61	871.54	15.50%	20024	498.45	528.98	6.12%
25015	195.83	278.64	42.29%	25015	254.94	279.15	9.50%
25019	315.86	411.48	30.27%	25019	335.19	371.43	10.81%
25024	511.97	579.62	13.21%	25024	448.01	471.02	5.14%
30019	225.21	266.14	18.17%	30019	278.12	285.08	2.50%
30024	368.86	408.86	10.84%	30024	371.27	385.13	3.73%
30030	555.52	585.86	5.46%	30030	463.53	485.92	4.83%
35019	167.52	191.35	14.23%	35019	205.96	202.57	−1.65%
35024	251.77	273.52	8.64%	35024	250.29	251.54	0.50%
35030	406.71	424.77	4.44%	35030	333.42	335.83	0.72%
40019	120.25	134.79	12.09%	40019	167.24	165.08	−1.29%
40024	200.88	214.66	6.86%	40024	226.75	226.7	−0.02%
40030	311.56	323.01	3.68%	40030	291.99	292.62	0.22%

[a] Δ% is the buckling stress deviation between SupaCee and channel sections, unit %

In the first case, only bracing members are used to prevent the global buckling modes (see Figs. 5 and 6). The sectional capacities are determined as Eqs. 1 to 3 and Eqs. 4–6 for compression and bending respectively, as illustrated in Figs. 8(c) and 9(c). The investigated shows that the global buckling failures although are prevented and local buckling strengths of SupaCee sections are generally higher than those of channel sections (see Figs. 8(a) and 9(a)), but the sectional capacities of the former sections do not demonstrate their strength improvements compared to the latter sections with minor strength deviations (see Figs. 8(c) and 9(c)). The reason is the sectional capacities of the investigated sections are governed by distortional buckling modes for almost all sections. Therefore, the innovative capacities of SupaCee sections can be utilized if the distortional buckling failures are prevented as presented in the second case below.

In the second case, both bracing members attached to the webs and cladding plates/ sheet plates attached to the flanges of the sections are used to prevent both the global and distortional buckling modes. The investigated sections, therefore, are only failed due to the combination of yield strengths and local buckling strengths, as illustrated in Eqs. (2) and (5). The sectional capacities are demonstrated in Figs. 8(d) and 9(d). The investigated results show that the sectional capacities of SupaCee sections are significantly higher than those of the channel sections due to the addition of stiffeners in the web. These capacity improvements reach 25% for compression (see Fig. 8(d)) and 15% for bending (see Fig. 9(d)). The investigated results also show that the high effectiveness of stiffeners for local buckling strengths is found in thin and small dimension sections as presented in Pham and Vu [32].

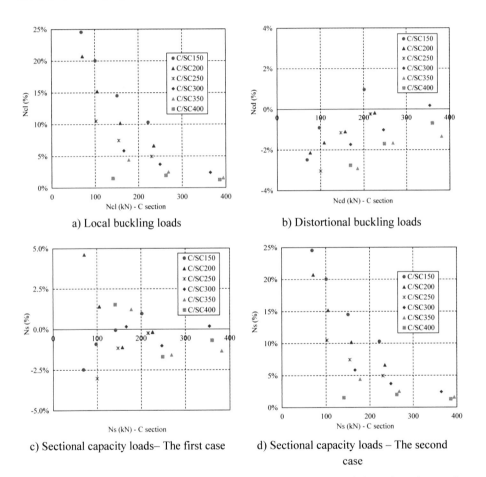

a) Local buckling loads

b) Distortional buckling loads

c) Sectional capacity loads– The first case

d) Sectional capacity loads – The second case

Fig. 8. Comparisons of sectional capacities between the SupaCee and channel sections under compression

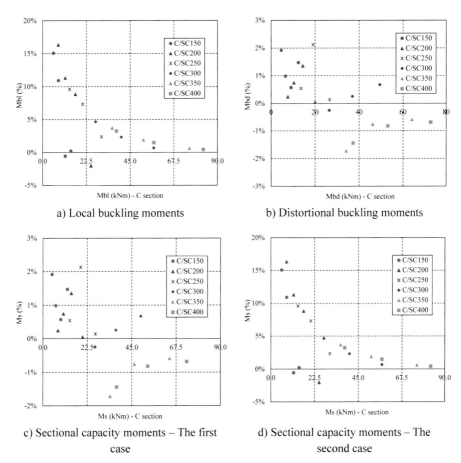

a) Local buckling moments

b) Distortional buckling moments

c) Sectional capacity moments – The first case

d) Sectional capacity moments – The second case

Fig. 9. Comparisons of sectional capacities between the SupaCee and channel sections under bending

5 Conclusion

This paper studies the sectional capacities of SupaCee sections compared to those of channel sections under compression or bending. SupaCee and channel sections were commercial sections from the BlueScope Lysaght catalogs [33]. The sectional capacities are determined using the DSM design regulated in AS/NZS 4600 [5] with the support of the THIN-WALL-2 software program [7, 8]. Based on the investigated results, the remark conclusions are given as follows:

(1) The local buckling strengths of SupaCee sections show their strength improvements compared to those of channel sections, but the sectional capacities of former sections are approximate to those of later sections due to the governing of distortional buckling. Therefore, the innovative capacities of SupaCee sections are only utilized if the distortional buckling is avoided.

(2) Web stiffeners of SupaCee sections show their high effectiveness for thin and small dimension sections.

(3) These remark conclusions provide deep insights into of sectional capacities of the channel and SupaCee sections, this allows the users to use these sections properly in each structural situation.

References

1. Hancock, G.J., Pham, C.H.: New Section Shapes Using High-Strength Steels in Cold-Formed Steel Structures in Australia. Elsevier Ltd (2016)
2. Schafer, B.W., Peköz, T.: Direct strength prediction of cold-formed members using numerical elastic buckling solutions. In: The Fourteenth International Specialty Conference on Cold-Formed Steel Structures (1998)
3. Schafer, B.W.: Local, distortional, and euler buckling of thin-walled columns. J. Struct. Eng. **128**(3), 289–299 (2002). https://doi.org/10.1061/(ASCE)0733-9445(2002)128:3(289)
4. Schafer, B.W.: Review: the direct strength method of cold-formed steel member design. J. Constr. Steel Res. **64**(7–8), 766–778 (2008). https://doi.org/10.1016/j.jcsr.2008.01.022
5. AS/NZS 4600–2018: Australian/New Zealand Standard: Cold-formed steel structures. The Council of Standards Australia (2018)
6. AISI: North American Specification for the Design of Cold-Formed Steel Structural Members. American Iron and Steel Institute (2016)
7. Nguyen, V.V., Hancock, G.J., Pham, C.H.: Devemopment of the thin-wall-2 for buckling analysis of thin-walled sections under generalised loading. In: The 8th International Conference on Advances in Steel Structures (2015)
8. Nguyen, V.V., Hancock, G.J., Pham, C.H.: New developments in the direct strength method (DSM) for the design of cold-formed steel sections under localised loading. Steel Constr. **10**(3), 227–233 (2017). https://doi.org/10.1002/cepa.191
9. Li, Z., Schafer, B.W.: Buckling analysis of cold-formed steel members with general boundary conditions using CUFSM: Conventional and constrained finite strip methods, Saint Louis, Missouri, USA (2010)
10. Hancock, G.J.: Distortional buckling of steel storage rack columns. J. Struct. Eng. ASCE **111**(12), 2770–2783 (1985). https://doi.org/10.1061/(ASCE)0733-9445(1985)111:12(2770)
11. Hancock, G.J., Kwon, Y.B., Stefan, B.E.: Strength design curves for thin-walled sections undergoing distortional buckling. J. Constr. Steel Res. **31**(2–3), 169–186 (1994). https://doi.org/10.1016/0143-974X(94)90009-4
12. Lau, S.C.W.: Distortional Buckling of Thin-Walled Columns. University of Sydney, Sydney (1988)
13. Kwon, Y.B.: Post-Buckling Behaviour of Thin-Walled Channel Sections. University of Sydney, Sydney (1992)
14. Kwon, Y.B., Hancock, G.J.: Tests of cold - formed channels with local and distortional buckling. J. Struct. Eng. ASCE **118**(7), 1786–1803 (1992). https://doi.org/10.1061/(ASCE)0733-9445(1992)118:7(1786)
15. Wang, L., Young, B.: Cold-formed steel channel sections with web stiffeners subjected to local and distortional buckling - Part I: Tests and finite element analysis. In: The 22nd International Specialty Conference on Recent Research and Developments in Cold-Formed Steel Design and Construction, pp. 229–242 (2014)

16. Wang, L., Young, B.: Cold-formed steel channel sections with web stiffeners subjected to local and distortional buckling - Part II: Parametric study and design rule. In: The 22nd International Specialty Conference on Recent Research and Developments in Cold-Formed Steel Design and Construction, pp. 243–257 (2014)

17. Yan, J., Young, B.: Column tests of cold-formed steel channels with complex stiffeners. J. Struct. Eng. **128**(6), 737–745 (2002). https://doi.org/10.1061/(ASCE)0733-9445(2002)128:6(737)

18. Xiang, Y., Xuhong Zhou, Y., Shi, L.X., Yunpeng, X.: Experimental investigation and finite element analysis of cold-formed steel channel columns with complex edge stiffeners. Thin-Wall. Struct. **152**, 106769 (2020). https://doi.org/10.1016/j.tws.2020.106769

19. Wang, C.G., Ma, P., Song, D.J., Yu, X.Y.: Design of cold-formed thin-walled steel fixed-ended channels with complex edge stiffeners under axial compressive load by direct strength method. Appl. Mech. Mater. **226–228**, 1232–1235 (2012)

20. Wang, C., Zhang, Z., Zhao, D., Liu, Q.: Compression tests and numerical analysis of web-stiffened channels with complex edge stiffeners. J. Constr. Steel Res. **116**, 29–39 (2016). https://doi.org/10.1016/j.jcsr.2015.08.013

21. Wang, L., Young, B.: Design of cold-formed steel channels with stiffened webs subjected to bending. Thin-Wall. Struct. **85**, 81–92 (2014). https://doi.org/10.1016/j.tws.2014.08.002

22. Manikandan, P., Sukumar, S., Kannan, K.: Distortional buckling behaviour of intermediate cold-formed steel lipped channel section with various web stiffeners under compression. Int. J. Adv. Struct. Eng. **10**(3), 189–198 (2018). https://doi.org/10.1007/s40091-018-0191-3

23. El-Taly, B.B.A., Fattouh, M.: Optimization of cold-formed steel channel columns. Int. J. Civil Eng. **18**(9), 995–1008 (2020). https://doi.org/10.1007/s40999-020-00514-7

24. Ye, J., Hajirasouliha, I., Becque, J., Pilakoutas, K.: Development of more efficient cold-formed steel channel sections in bending. Thin-Wall. Struct. **101**, 1–13 (2016). https://doi.org/10.1016/j.tws.2015.12.021

25. Wang, C.-G., Zhang, Z.-N., Jia, L.-G., Xin-yong, Y.: Bending tests and finite element analysis of lipped channels with complex edge stiffeners and web stiffeners. J. Central South Univ. **24**(9), 2145–2153 (2017)

26. Manikandan, P., Arun, N.: Behaviour of partially closed stiffened cold-formed steel compression member. Arab. J. Sci. Eng. **41**(10), 3865–3875 (2016). https://doi.org/10.1007/s13369-015-2015-0

27. Ganeshkumar, R., Suresh, B.S., Leema, R.A.: Experimental study on flexural and compressive behavior of cold formed steel back to back supacee section. Int. Res. J. Eng. Technol. **5**, 2850–2854 (2018)

28. Pham, C.H., Hancock, G.J.: Direct strength design of cold-formed C-sections in combined bending and shear. In: 20th International Specialty Conference on Cold-Formed Steel Structures - Recent Research and Developments in Cold-Formed Steel Design and Construction, pp. 221–236 (2010)

29. Pham, C.H., Hancock, G.J.: Finite element analyses of high strength Cold-Formed SupaCee® Sections in Shear. In: Proceedings of SDSS' Rio 2010: International Colloquium Stability and Ductility of Steel Structures, vol. 2, pp. 1025–1032 (2010)

30. Pham, C.H., Hancock, G.J.: Experimental investigation and direct strength design of high-strength, complex C-sections in pure bending. J. Struct. Eng. **139**(11), 1842–1852 (2013). https://doi.org/10.1061/(ASCE)ST.1943-541X.0000736

31. Pham C.H., Hancock G.J.: Direct strength design of cold-formed sections for shear and combined actions. In: Proceedings of SDSS' Rio 2010: International Colloquium Stability and Ductility of Steel Structures, vol. 1, pp. 101–114 (2010)

32. Pham, N.H., Vu, Q.A.: Effects of stiffeners on the capacities of cold-formed steel channel members. Steel Constr. **14**(4), 270–278 (2021). https://doi.org/10.1002/stco.202100003
33. BlueScope Lysaght: Supapurlins Supazeds & Supacees. Blue Scope Lysaghts (2014)
34. AS1397:2011: Continuous Hot-dip Metalic Coated Steel Sheet and Strip - Coating of Zinc and Zinc Alloyed with Aluminium and Magnesium. Standards Australia (2011)

Preliminary Regression Study on Air Quality Inside a Road Tunnel: A Case Study in Vietnam

Thu-Hang Tran[(✉)]

Faculty of Civil Engineering, University of Transport and Communications, Hanoi, Vietnam
tranthuhang@utc.edu.vn

Abstract. Normal air quality that ensures the visibility and brings no harmful impact to the health of the vehicle drivers is essential for all road tunnels. It is affected by various issues. In this paper, the air quality – meteorology – traffic volume correlations were quested in the case study of an opened road tunnel on Vietnam National Highway 1A. The linear regression modelling technics using the least squares method with 95% of confidence was executed. Four representative models of the total suspended particles and airborne lead concentration against the meteorological parameters (temperature, relative humidity, wind velocity) and the vehicle flow density inside tunnel were proposed. The correlations with the volatile organic compounds were also studied but no representative model was proposed. Further studies on a richer source of data were suggested. The study confirmed the role of the in-tunnel vehicle volume and the meteorology on the tunnel's air quality.

Keywords: Air quality · Meteorology · Road tunnel

1 Introduction

The in-tunnel air quality is one of big concerns for not only the owners but also the users of road tunnels during the tunnel's service life. The main source of the air pollution inside an in-service road tunnel is the exhaust gas emitted from vehicles. While the category and quantity of the daily vehicle flows can be listed through the car counting campaigns, the actual exhaust emission is difficult to be figured out since it is affected by various issues: engine's type, car's quality, fuel's type, fuel's quality, driving regime, … The small and closed space inside a tunnel makes it more vulnerable to air pollution than other open-air transport structures (e.g., roads, bridges, viaducts) [1]. The natural and mechanical ventilation systems are equipped for road tunnels to provide the under-controlled air quality following the given design and operation technical regulations. When the tunnel's traffic volume in real conditions surpasses the traffic growth scenarios in the planning and design documents, the tunnel's ventilation effectiveness reduces [2]. As consequence, the in-tunnel air quality becomes vulnerable due to the inadequate ventilation regimes.

Studies on the correlation between the road tunnel's air quality and meteorological parameters have been published. Knowing that the wind takes responsive in the dispersion and dilution of vehicle's exhausted gas in tunnels [3], the actual influences of the

G. Feng (Ed.): ICCE 2021, LNCE 213, pp. 95–102, 2022.
https://doi.org/10.1007/978-981-19-1260-3_9

wind on the air quality of different tunnels have been carried out [1, 4–6]. The temperature inside is normally higher than outside tunnels [2], so it creates the natural ventilation in the tunnel and the impacts to different air parameters inside tunnels [7–9]. Since the variation of humidity also plays a role in the pollution dispersion in road tunnels, the case studies on this meteorological parameter have been executed [7, 9–11].

Aiming at the understanding of the meteorology – air quality correlation in road tunnels, a case study on an open road tunnel on Vietnam National Highway 1A (NH1A) was carried out based on the recorded data of a previous air quality measurement. During two continuous days in November 2017, various parameters of the air quality, and meteorological data were recorded at one measurement station [12]. In this paper, the statistical data treatments were executed and the linear regression models between some typical ambient parameters were quested. Each model's statistical reliability was analysed to find out the representative regression model for the correlation, and the commentaries were presented. The preliminary study showed the multiple impacts on the in-tunnel air quality through the proposed regression model of the total suspended particles, airborne lead, and three volatile organic compounds against the meteorological parameters and the vehicle flow density inside tunnel.

2   Tunnel and Meteorology Description

The tunnel was 6.28 km long, 11.5 m wide, stayed in the middle central area of Vietnam, on the busiest national transport rout (NH1A). The tunnel included of two longitudinal tubes: one main tube for two bi-directional circulation vehicle lanes, and one auxiliary tube for the evacuation. There were fifteen cross passages connecting the two tubes. The tunnel located in the climate boundary of the country. Some typical tunnel's meteorological features at the measurement time were shown in Table 1.

Table 1. Typical tunnel's meteorological features at the measurement time.

Parameter	Out-tunnel [13]		In-tunnel [12]
	Northern part	Southern part	
Season/Type of climate	Winter/Four-seasonal tropical monsoon	Rainy/Two-seasonal tropical monsoon	–
Average temperature (^{0}C)	20.00 (night) 24.00 (day)	22.00 (night) 30.00 (day)	31.10 (night) 32.57 (day)
Average daily relative humidity (%)	90.50	88.00	62.18
Average annual rainfall (mm)	2456.00	2066.00	–
Average wind velocity (m/s)	1.70	3.30	2.35
Wind direction	North-East	North	North-South

3 Linear Regression Study

Three meteorological parameters were analysed: temperature (T, ^{0}C), relative humidity (H, %), wind velocity (W, m/s). Because the wind direction was always North-South, it was not considered in the study. At the same time, several typical air quality parameters were examined: total suspended particles (TSP, $\mu m/m^3$), airborne lead (Pb, $\mu m/m^3$), benzene (C_6H_6, $\mu m/m^3$), toluene (C_7H_8, $\mu m/m^3$), xylene (C_8H_{10}, $\mu m/m^3$). The simple and multiple linear regression modelling technics using the least squares method with 95% of confidence was applied on the data to build the regression models between the parameters. The vehicle flow density (expressed by the number of vehicles per hour (V, car/h)) was also taken into consideration during the statistical treatments.

3.1 Total Suspended Particles

The total suspended particles concentration (TSP, $\mu m/m^3$) in the air were studied and the correlations with the meteorological parameters (T, H, W) and vehicles per hour (V, car/h) were analysed. The simple and multiple linear regressions were done with the regression equations as follows:

$$TSP = a_1 + b_1.T + c_1.H + d_1.W + e_1.V \qquad (1)$$

The detail parameters of the linear regressions were presented in Table 2.

Table 2. Linear regression parameters of total suspended particles against meteorological parameters and vehicle per hour measured inside tunnel.

Model	a_1	b_1	c_1	d_1	e_1	R value	Significance F
1st simple	−4274.24	152.66	0	0	0	0.71	0.11%
2nd simple	3523.14	0	−47.25	0	0	0.49	1.44%
3rd simple	−172.66	0	0	322.55	0	0.57	0.40%
4th simple	190.63	0	0	0	1.07	0.67	0.04%
1st multiple	−1682.89	103.61	−22.58	159.05	0	0.78	0.02%
2nd multiple	−1369.08	93.93	−22.6	136.46	0.14	0.78	0.08%

The measured data and the regression lines were sketched in Fig. 1.

3.2 Airborne Lead

The airborne contaminants were released from the exhaust gas during the circulations of the motor vehicles in the tunnel. Emitted from gasoline vehicles, lead causes the direct and strong dangers to human being's health. The data of particulate lead content of aerosols (Pb, $\mu m/m^3$) in tunnel were examined. The influences of the meteorological

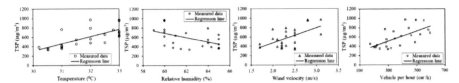

Fig. 1. Measure data and regression lines of total suspended particles against meteorological parameters and vehicle per hour inside tunnel.

parameters (T, H, W) and vehicle per hour (V) on the airborne lead concentration (Pb) were investigated with the proposed regression equations as follows:

$$Pb = a_2 + b_2.T + c_2.H + d_2.W + e_2.V \qquad (2)$$

The detail parameters of the linear regressions were shown in Table 3.

Table 3. Linear regression parameters of airborne lead against meteorological parameters and vehicle per hour measured inside tunnel.

Model	a_2	b_2	c_2	d_2	e_2	R value	Significance F
1st simple	−2.99	0.10	0	0	0	0.84	0.00%
2nd simple	1.57	0	−0.02	0	0	0.43	3.77%
3rd simple	−0.26	0	0	0.18	0	0.59	0.25%
4th simple	−0.11	0	0	0	0.001	0.88	0.04%
1st multiple	−2.28	0.08	−0.01	0.07	0	0.87	0.00%
2nd multiple	−1.02	0.04	−0.01	−0.02	0.001	0.92	10^{-5}%

The measured data and the regression lines were sketched in Fig. 2.

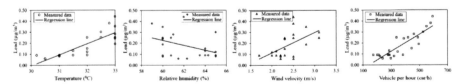

Fig. 2. Measure data and regression lines of airborne lead against meteorological parameters and vehicle per hour inside tunnel.

3.3 Volatile Organic Compounds

Dangerous to human, the volatile organic compounds (VOCs) are top priority chemical pollutants in the air [14]. The concentration of benzene (C_6H_6, $\mu m/m^3$), toluene (C_7H_8,

$\mu m/m^3$), and xylene (C_8H_{10}, $\mu m/m^3$) inside tunnel were studied to find out the correlation with three mention meteorological parameters (T, H, W) and the traffic volume (V). Among the built regression models, the ones whose Significance F greater than 5% were listed in Table 4.

Table 4. Linear regression parameters of volatile organic compounds against the meteorological parameters measured inside tunnel.

Correlation	R	Significance F
C_6H_6 and H	0.31	13.66%
C_6H_6 and W	0.26	21.84%
C_6H_6 and T, H, W	0.52	9.63%
C_7H_8 and H	0.31	13.86%
C_8H_{10} and H	0.20	35.87%
C_8H_{10} and T, H, W	0.55	5.88%

3.4 Discussions

The significance F smaller than 5% was the criterion for the model's acceptance. The correlation coefficient R was utilised to give commentaries on the correlation status (R $= 0$: non correlated, R $= 1$: strong correlated) and to select the representative regression model. The preliminary discussions were obtained.

3.4.1 Case 1

In the linear regressions of TSP against T, H, W, and V, five models were all concluded to be statistical reliable. While the temperature and the wind velocity showed the positive relationships, the relative humidity expressed the negative correlations with the total suspended particles concentration. The positive slope of the (TSP-V) fitted line showed the direct proportion of the TSP to the traffic volume inside tunnel. The multiple models exhibited the stronger correlation coefficients than the simple ones. Two multiple regression models were recommended to be the representatives.

The 1st representative model represented the combined action of the meteorological parameters on the total suspended particles concentration in tunnel:

$$TSP = -1682.89 + 103.61T - 22.58H + 159.05W \tag{3}$$

The 2nd representative model represented the combined action of the meteorological parameters and the traffic volume on the total suspended particles concentration in tunnel:

$$TSP = -1369.08 + 93.93T - 22.6H + 136.46W + 0.14V \tag{4}$$

3.4.2 Case 2

In the linear regressions of Pb against T, H, W, and V, five models were all considered to be statistical reliable. While the temperature and the wind velocity showed the positive relationships, the relative humidity expressed the negative correlations with the airborne lead concentration. The positive slope of the (Pb-V) fitted line expressed the direct proportion of the airborne lead concentration to the traffic volume inside tunnel. The vehicle-related models owned the stronger correlation coefficients than the non-vehicle-related ones. The negative trend of Pb against W in the 2nd multiple regression model was not clearly understood. Further studies should be supplied. In the scoop of this paper, the 4th simple and the 1st multiple regression model were recommended as the representatives.

The 1st representative model represented the correlation of the airborne lead concentration and the traffic volume in tunnel:

$$Pb = -0.11 + 0.001V \tag{5}$$

The 2nd representative model represented the combined action of the meteorological parameters on the airborne lead concentration in tunnel:

$$Pb = -2.28 + 0.08T - 0.01H + 0.07W \tag{6}$$

3.4.3 Case 3

Among the built linear regression models of VOCs against T, H, W, and V, those who were presented in Table 4 were found to be non statistical reliable and were eliminated. It did not mean the non correlation between them. A richer source of measured data to increase the regression observations was suggested. Other statistical reliable regression models were found but they did not show the strong correlations ($R < 0.5$). In the scoop of this paper, no representative regression model for the VOCs – meteorology – traffic volume correlation was proposed.

4   Conclusions

The in-tunnel air quality strongly impacts the operation of a road tunnel. The vehicle's clear visibility is essential for the smooth and safe circulation inside tunnels. Besides, the non harmful effects to the drivers' health during their passage through tunnel are also demanded. A bad air quality potentially provides the unsafeness to the circulations inside road tunnels. Using the statistical treatments, the paper explored the recorded data of a previous air quality measurement in an opened road tunnel on Vietnam National highway 1A. The linear regression models using the least squares method with 95% of confidence were found. Three study cases were attacked and the representative models for two cases were proposed. From the representative models, it was understood that the combination of high temperature, strong wind and low humidity potentially decreased the tunnel's air quality. The negative slope of the regression lines of TSP and Pb against H proved that a wet road surface could probably reduce the pollution in road tunnels as mentioned in [10, 11]. The important role of the in-tunnel traffic volume in the deterioration of the tunnel's air quality was also confirmed. The study in the scoop of this paper was hoped to underline of role of the meteorological parameters in the in-tunnel air quality studies.

Acknowledgements. The original data of the air quality measurement mentioned in the paper is from Environment mission MT163008 executed by Trong-Khang Dinh et al. (Vietnam Institute of Transport Science and Technology), funded by Vietnam Ministry of Transport in 2017. Deep gratitute is expressed to Mr. Trong-Khang Dinh for giving us the permission of access to the data for research purposes.

References

1. Zhou, R., et al.: Study on the traffic air pollution inside and outside a road tunnel in Shanghai, China. PLOS One **9**(11), e112195 (2014)
2. Longley, I., Kelly, F.: Air quality in and around traffic tunnels - Final report 2008. National Health and Medical Research Council, Canberra (2008)
3. NIWA: Guidance for the management of air quality in road tunnels in New Zealand: NIWA research report for NZTA. New Zealand transport agency, Auckland (2010)
4. Staehelin, J. et al.: Emission factors from road traffic from a tunnel study (Gubrist tunnel, Switzerland). Part I: concept and first results. Sci. Total Environ. **169**(1–3), 141–147 (1995)
5. Tao, Y., Dong, J., Xiao, Y., Tu, J.: Numerical analysis of pollutants dispersion in urban roadway tunnels. In: Proceedings of the 20th Australasian Fluid Mechanics Conference, 508, Perth (2016)
6. Solazzo, E., Cai, X., Vardoulakis, S.: Modelling wind flow and vehicle-induced turbulence in urban streets. Atmos. Environ. **42**(20), 4918–4931 (2008)
7. Indrehus, O., Vassbotn, P.: CO and NO_2 pollution in a long two-way traffic road tunnel: investigation of NO_2/NO_x ratio and modelling of NO_2 concentration. J. Environ. Monit. **3**, 220–225 (2001)
8. Chow, W., Chan, M-Y.: Field measurement on transient carbon monoxide levels in vehicular tunnels. Build. Environ. **38**(2), 227–236 (2003)
9. Gokce, H., Arioglu, E., Copty, N., Onay, T., Gun, B.: Exterior air quality monitoring for the Eurasia Tunnel in Istanbul, Turkey. Sci Total Environ. **699**, 134312 (2020)
10. Gustafsson, M., et al.: Particles in road and railroad tunnel air. Sources, properties and abatement measures - VTI rapport 917A. Swedish National Road and Transport Research Institute, Linköping (2016)
11. Davy, P., Trompetter, B., Markwitz, A.: Concentration, composition and sources of particulate matter in the Johnstone's Hill tunnel, Auckland. GNS Science Consultancy Report 2010/296, Auckland (2011)
12. Dinh, T.K., et al.: MT163008 Mission of preparing a standards basic of air quality in road tunnels - Summary report. ITST, Hanoi (2017). (In Vietnamese)
13. JV of NK, NKV and HoangLong: Haivan pass road tunnel expansion investment. Package HV2-XL1: Design report. Deoca investment Jsc., Hanoi (2017)
14. Tsai, J.H., Lu, Y.T., Chung, I.I., Chiang, H.L.: Traffic-related airborne VOC profiles variation on road sites and residential area within a microscale in urban area in Southern Taiwan. Atmosphere **11**(9), 1015 (2020)

Experimental Study of Reinforced Concrete Beams Strengthened with CFRP

Yihong Hong[(✉)]

Shanghai Urban Construction Vocational College, Shanghai 200438, China
hyhzju@126.com

Abstract. To test the strengthening effect of CFRP sheets on reinforced concrete (RC) beams, an experimental study was performed on five RC beams to analyze the influence of the amount of CFRP sheets and the sustained load of RC beams on the strengthening effect. The results showed that CFRP sheets could significantly improve the bending bearing capacity of the beams. However, there was no linear relationship between the amount of CFRP sheets and the strengthening effect, and the load on the beams had a great impact on the rigidity of the beams in the yield stage.

Keywords: Carbon Fiber Reinforced Polymer (Cfrp) · Strengthening · Reinforced-concrete beam · Bending test

1 Introduction

With the development of China's construction industry, the technology of strengthening concrete with CFRP sheets has been widely used in China's structures such as bridges and buildings due to the light weight, high strength, corrosion resistance, easy application and other characteristic of CFRP sheets [1]. Scholars at home and abroad have carried out a number of experimental studies and theoretical analyses on bending resistance of reinforced concrete (RC) beams strengthened with CFRP sheets [2–4], most of which are focused on the intact beams. However, the structures that need to be strengthened in the actual engineering are those of the existing engineering with insufficient bearing capacity and other conditions, so an analysis on the strengthening of RC beams under secondary load is necessary. At present, some achievements have been made in this field [5, 6]. In this study, an experimental analysis will be carried out on five beams on the basis of previous studies to compare the influence law of the amount of CFRP sheets and the sustained load state of RC beams on the failure characteristics of RC beams, concrete strain, rigidity of beams and ultimate bearing capacity.

2 Experimental Design

2.1 Specimen Design and Fabrication

According to the Code for Design of Concrete Structures (GB 50010-2010), five beams of the same size (the sectional size was 120 mm × 250 mm, and the length was 2500 mm)

© The Author(s) 2022
G. Feng (Ed.): ICCE 2021, LNCE 213, pp. 103–108, 2022.
https://doi.org/10.1007/978-981-19-1260-3_10

were designed and fabricated, as shown in Fig. 1. The strength grade of concrete of the test beams was designed to be C20, the tensile main reinforcement was HRB335, the erection reinforcement was HPB300, the stirrup was HPB300, and the reinforcement ratio was 1.52%. The tensile strength of CFRP sheets was 3550 MPa, the elasticity modulus was 235 GPa, and the thickness was 0.111 mm. Parameters such as the amount of carbon fiber and the sustained load point of each specimen are shown in Table 1.

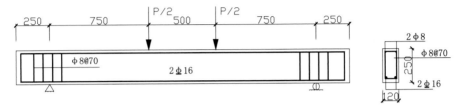

Fig. 1. Diagram of test beam reinforcement

Table 1. Specimen parameters

Specimen No	Ratio of longitudinal reinforcement	Amount of CFRP (Sheets)	Sustained load point	Remark
Beam1	1.52%	0	–	Not strengthened
Beam2	1.52%	1	–	Intact beam strengthened
Beam3	1.52%	2	–	Intact beam strengthened
Beam4	1.52%	1	$0.6P_y$	–
Beam5	1.52%	2	$0.6P_y$	–

2.2 Loading Scheme

The test beam was in the form of simply supported beam, with the test length of 2 000 mm and the length of simple bending segment of 500 mm. The loading diagram is shown in Fig. 1. Hydraulic jack was used for opposite side loading, and the experiment was carried out according to the Standard for Test Method of Concrete Structures [7]; when the load applied was close to the characteristic point load (such as the cracking point of concrete and the yield point of reinforcement), the loading level was reduced to accurately control the characteristic points; after yield of reinforcement, the loading was controlled by deflection until the specimen was destroyed. To keep the load on the beam unchanged when pasting the CFRP, the hydraulic jack was used for self-locking.

The surface of the tensile side of the beam was polished until the coarse aggregate was exposed, and the surface of the polished specimen should be flat; the dust was removed and the surface was washed with acetone. The surface was step-loaded to the pre-defined load grade according to the loading history in Table 1 and remained under this load (for ease of the following analysis, this point was referred to as the "sustained load point"), and then the CFRP was pasted.

3 Experimental Results and Analysis

The failure process of strengthened beam mainly had the following characteristics: the longitudinal bar yielded first, then the CFRP sheets peeled off the concrete in the simple bending segment or the bending-shear zone below the loading point, and finally the concrete in the compression zone was crushed; there were obvious elastic stage, stage of working with cracks and stage of yield of reinforcement during the destruction; there was no failure of over-reinforced beam that the CFRP sheets at the end peeled off the surface of concrete or the concrete in the compression zone was crushed first.

The load, deflection, rigidity and failure modes of the five specimens in each loading stage are shown in Table 2.

Table 2. Main experimental results

Specimens	Load (kN)			Deflection (mm)			Rigidity (N/mm, $\times 10^4$)			Failure modes
	P_{cr}	P_y	P_u	cr	y	u	B_{cr}	B_y	B_u	
Beam1	10.0	65.2	68.7	1.0	7.6	17.2	0.971	0.840	0.037	Bending failure, concrete compression failure
Beam2	10.1	73.0	82.4	0.8	7.9	21.0	1.217	0.887	0.072	Protective layer in the bending segment peeled off
Beam3	9.1	71.6	85.1	1.0	8.6	18.8	0.910	0.819	0.133	Protective layer in the bending segment peeled off, concrete compression failure
Beam4	12.1	80.1	92.1	0.7	7.3	23.7	1.862	1.026	0.073	CFRP 30% tensile failure, concrete compression failure
Beam5	12.6	81.2	96.1	0.8	7.4	17.2	1.658	1.032	0.153	Protective layer in the bending segment peeled off, concrete compression failure

3.1 Characteristic Load Analysis

3.1.1 Cracking Load

The experimental data showed that there was no effect on the cracking load of beams strengthened with CFRP sheets. The cracking loads of Beam1, Beam2 and Beam3 without sustained initial load were basically the same, and the cracking load of RC beams did not change with or without CFRP. The reason is that the tension of the concrete at the bottom of the beam is mainly borne by the concrete in the tensile zone at the bottom of the beam, and when the tensile stress of the concrete in the tensile zone exceeds the tensile strength, the concrete is cracked. Due to the small load at the beginning of loading, the pasted on the tensile surface has not played the role of reinforcement and strengthening.

3.1.2 Yield Load and Ultimate Load

When the concrete was cracked and stopped working, the tension was borne by tensile reinforcement and CFRP together. As the load gradually increased, the original tension on the concrete at the bottom of the beam was borne by the CFRP sheets pasted at the bottom of the beam, which reduced the stress borne by the reinforcement and delayed the yield of reinforcement, thus improving the yield load of the beam to a certain extent. Comparison of the experimental results showed that the yield load was improved by 17% with one CFRP sheet and 18% with two CFRP sheets.

After yield of longitudinal reinforcement, the high strength and high modulus of CFRP were put to good use with the further increase of the load, and the ultimate bearing capacity of the beam strengthened with CFRP sheets was greatly improved, of which it was improved by 27% with one sheet and 32% with two sheets. The experimental results also showed that with the increase of the amount of the CFRP, the resistance to bending was improved, although it did not increase linearly.

3.2 Deflection and Rigidity Analysis

Analysis on the data in Table 2 showed that the deflection of Beam3 and Beam5 pasted with two CFRP sheets was lower than that of Beam2 and Beam4 pasted with one CFRP sheet under the ultimate load, suggesting that the amount of CFRP sheets had great impact on the deflection of the beams. Meanwhile, the amount of CFRP sheets could significantly improve the rigidity of the beams under the ultimate load, of which the rigidity of Beam2 and Beam4 pasted with one CFRP sheet was nearly twice that of Beam1 that was not strengthened under the ultimate load, and the rigidity of Beam3 and Beam5 pasted with two CFRP sheets was nearly twice that of Beam2 and Beam4 pasted with one CFRP sheet under the ultimate load. This suggested that CFRP sheets could effectively improve the rigidity of the beams under the ultimate load, and the degree of the improvement was linearly related to the amount of the CFRP sheets.

4 Conclusion

In this experiment, one control beam and four test beams were used to analyze the impact of the amount of CFRP sheets and the sustained load points on the strengthening effect of CFRP sheets on RC beams, and the following conclusions were drawn. 1) The bearing capacity of the beams strengthened with CFRP sheets was greatly improved. However, the improvement rate was not linearly related to the amount of the CFRP. 2) The ultimate state of the RC beams strengthened with CFRP sheets was delayed under a certain load (such as $0.6 P_y$), and the ultimate bearing capacity was also improved with the increase of the amount of CFRP. 3) The rigidity of the RC beams strengthened with CFRP was significantly improved, compared with the ordinary RC beams, and the CFRP sheets could improve the rigidity and control the deflection of the components mainly resistant to bending. With the increase of the amount of the CFRP, the rigidity of the beams also increased. 4) The load of the strengthened beams had a great impact on the rigidity of the beams in the yield stage.

Acknowledgement. We are grateful for the support from construction of first-class engineering cost major in Shanghai project (Z106-0603-20-001).

References

1. Chen, X.B.: Guide for the Design and Construction of Fiber Reinforced Plastics in Civil Engineering, pp. 1–4. China Building Industry Press, Beijing (2009)
2. You, S.Q., Liu, B., Zhang, S.Y.: Experimental and numerical studies on flexural bearing capacity of reinforced concrete beams strengthened with carbon fiber-reinforced polymer laminates. Build. Sci. **21**(4), 1–5 (2005)
3. Yu, Q.F., Deng, L., Wang, J.: Experimental study on flexural behavior of steel fiber concrete beam strengthened with carbon fiber sheets. Build. Sci. **22**(6), 1–3 (2006)
4. Wang, S.Y., Yang, M.: Experimental study of flexural behavior of high-strength RC beams strengthened with CFRP sheets. Build. Sci. **22**(5), 34–38 (2006)
5. Wang, S.Y., Yang, M.: Experimental study on flexural behavior of damaged high-strength RC beams strengthened with CFRP. Earthq. Resist. Eng. Retrofit. **28**(2), 93–96 (2006)
6. Liu, X.: Experimental study on flexural performance in damage of RC beams strengthened with CFRP sheets. Earthq. Resist. Eng. Retrofit. **38**(4), 114–120 (2016)
7. Standard for Test Method of Concrete Structures GB50152-2012. China Building Industry Press, Beijing (2012)

The Influence of Different Excavation Methods on Deep Foundation Pit and Surrounding Environment

Yongming Yang[✉], Xiwen Li, and Yao Chen

School of Mechanics and Civil Engineering, China University of Minning and Technology, Beijing, China
yangym@cumtb.edu.cn

Abstract. Taking a deep foundation pit in Shijiazhuang, Beijing as the background of the project, combined with the excavation range of the deep foundation pit and the spatial location and geometry of the surrounding engineering bodies, FLAC3D software is used to establish a numerical model of the deep foundation pit excavation containing the deep foundation pit, surrounding buildings and underground tunnels and other engineering bodies. The numerical model truly reflects the geometric size and spatial location relationship between deep foundation pit and surrounding engineering bodies. Based on the numerical model, the numerical simulation of deep foundation pit excavation under different excavation methods was carried out. The analysis of the horizontal displacement of the support structure during the excavation of the deep foundation pit, the investigation of the surrounding buildings and surface settlement and the deformation of the metro tunnel. The influence mechanism of different excavation methods on the safety of deep foundation pit, surrounding buildings and tunnel were discussed. It is concluded that the excavation method has less influence on the deformation and displacement of the deep foundation pit and the surrounding buildings and tunnels under the condition of shallow excavation depth. As the excavation depth increases, the effect of the excavation method on the deformation of the deep foundation pit, buildings, tunnels and surface settlement increases. The layered excavation method is suitable for deep foundation pits with shallow excavation depths, while the layered section excavation method is more suitable for deep foundation pits with deep excavation depths. Deep foundation pit excavation has a smaller effect on the deformation of the tunnel. The surrounding buildings have a greater influence on the deformation during the excavation of the deep foundation pit, but the effect of the buildings on it is less than the limiting effect of the prestressing anchor cables on the deformation of the deep foundation pit.

Keywords: Deep foundation pit · Different excavation methods · Complex engineering environment

1 Introduction

With the acceleration of urbanisation, the scale of deep foundation works is gradually becoming larger, and research into the depth and complexity of the project has become

G. Feng (Ed.): ICCE 2021, LNCE 213, pp. 109–129, 2022.
https://doi.org/10.1007/978-981-19-1260-3_11

a major development objective, while at the same time its safety issues are facing new challenges and problems [1–5]. The safety of deep foundation pit is facing new challenges and problems. In recent years, deep foundation pit stability problems occur from time to time, especially in some deep foundation pit projects with complex environment and difficult support, which has caused great harm to China's economic and social stability [6–10].

Deep foundation pit has obvious spatial effect, which has a serious impact on the deformation and stability of deep foundation pit [11–13]. In recent years, scholars have done extensive research work on the safety of deep foundation pit. For example, Jianhang Liu launched an analysis of deep foundation pit deformation based on the application of spatial effect theory to explore the influence of the soil itself on the control of deep foundation pit deformation, in order to achieve the purpose of controlling deep foundation pit deformation by using the law of soil change [14]. Ruiwu Qi et al. studied the the effect of deformation of the enclosure structure itself and the deformation of different locations around the deep foundation pit under different working conditions of the giant deep foundation pit [15]. Based on the field measured data, Peixin Wang et al. analysed the deformation law, settlement causes and control measures of subgrade and foundation pit [16]. Jiangtao Liu et al. established a three-dimensional deep foundation pit model, studied the deformation law of deep foundation pit excavation based on different constitutive models, and analysed the law of surface settlement and retaining structure deformation [17]. John simulates the displacement and deformation of a square deep foundation pit without enclosure structure based on linear elastic and finite element analysis methods [18]. Mana and Clough analysed different widths of deep foundation pits based on the two-dimensional finite element method to reveal the relationship between the lateral displacement of deep foundation pits and the change of excavation width [19]. Wong and Broms used a two-dimensional finite element method to establish a deep foundation excavation model to explore the effects of various factors on the deformation of the deep foundation pit,including the form and stiffness of the support structure, the depth of entry, and the depth and width of the excavation [20]. By focusing on the spatial effect, other scholars have revealed the influence law of deep foundation pit geometric size, length width ratio, support structure, construction sequence, construction method and stiffness of support on deep foundation pit deformation [21–25].

The above research results are important for people to understand the deformation characteristics of deep foundation pit enclosure structures, the ground settlement pattern and the influence mechanism on the safety of the surrounding environment. At present, the research on the deformation law of deep foundation pit mostly focuses on the support form and excavation sequence of deep foundation pit, while there is little research on the impact of the excavation method on the deformation of the foundation pit itself and the safety for the surrounding structures [26–28]. People have insufficient understanding of the deformation of deep foundation pit under different excavation methods and its impact on the safety of surrounding environment, which needs to be further studied.

This paper takes a deep foundation pit in Shiliuzhuang, Beijing as the engineering background, and uses FLAC3D software to conduct numerical engineering simulation research in a complex environment, and analyzes the deformation law of the deep foundation pit by changing the excavation method to reveal the influence mechanism of different excavation methods on the safety of the deep foundation pit on its surrounding tunnels and buildings.

2 Introduction to Numerical Simulation Method

FLAC (Fast Lagrangian analysis of continua) is an English numerical simulation software based on the finite difference method. The software has a simple interface, easy operation and powerful operation, which greatly saves cost and time compared with traditional experiments. FLAC3D is widely used in the field of geotechnical engineering because it contains a comprehensive material model, flexible solution method, accurate results, and can meet various engineering problems. velocity, plastic state and other internal variables that cannot be obtained from real experiments. In addition, the post-processing function of FLAC3D is also powerful, users can directly get the stress and displacement diagrams of the simulated objects, which can express the engineering problems in a more intuitive and concise way.

3 Project Overview and Geological Conditions

Based on the background of a deep foundation pit to be excavated in Shiliuzhuang, Beijing, the surrounding environment of the deep foundation pit is relatively complex. The west side is close to a 6-storey residential building with a horizontal distance of 4 m; On the east side has a subway tunnel, horizontal distance of 16 m, 12 m in buried deeply. Excavation scope is 56 m long, 40 m wide. Figure 1 is the plane position of deep foundation pit.

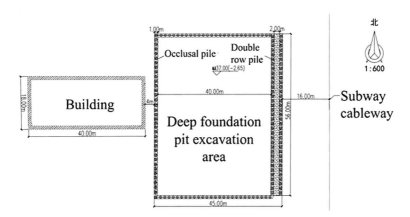

Plan location of deep foundation pit

Fig. 1. Diagram of plane location of deep foundation pit.

According to the survey data provided on site, the deep foundation pit and surrounding buildings and subway tunnel are located from the surface. The stratum can be divided into 7 layers according to different lithology, including Miscellaneous soil layer, Clay silt layer, Silty clay layer-1, Fine medium sand layer-1, Silty clay layer-2, Fine medium sand layer-2 and Pebble layer. The thickness and mechanical parameters of different strata are shown in Table 1.

Table 1. Stratum thickness and mechanical parameters.

Soil layer name	Thickness (m)	Cohesion (KPa)	Density (KN/m^3)	Elastic modulus (MPa)	Poisson's ratio	Tensile strength (KPa)
Impurity soil	1	10	18	30	0.30	5
Clayey silt	1	23	19.7	17	0.29	11.5
Silty clay-1	7	15	19.9	21	0.29	7.5
Medium fine Sand-1	4	0	20	38	0.28	0
Silty clay-2	6	26	20	22	0.29	13
Medium fine Sand-2	5	0	20.4	40	0.28	0
Pebble	14	0	22	60	0.27	0

4   Numerical Model of Deep Foundation Pit Excavation

With the purpose of studying the influence of diverse excavation methods on the safety of deep foundation pit and surrounding environment, a numerical model containing engineering bodies such as deep foundation pits, surrounding buildings and underground tunnels was established using numerical methods based on information such as the actual spatial location distribution and geometric dimensions of the proposed excavation range of deep foundation pits and surrounding engineering bodies, which can truly reflect the deep foundation pits and surrounding engineering.

4.1   Numerical Model Dimensions of Deep Foundation Pit

The size of the numerical model for the pit under study are selected in this paper taking into account the influence of spatial effects. If the size of the established model is too small, it will cause the deformation of itself and the surrounding engineering body to be bounded by the boundary conditions, resulting in size effects; if the model size is too large, it will lead to an infinite increase in computation time, which can be reduced by increasing the number of meshes, but will reduce the computation accuracy. Increasing the number of grids can reduce the calculation time, but the calculation accuracy will be reduced. According to the research, the range of surface settlement caused by general deep foundation pit excavation is about 3 times of the excavation depth of deep foundation pit. In this paper, we not only consider the proposed excavation depth but also the geometric information of the surrounding engineering body, and the established geometric model of deep foundation excavation is 151 m in length, 56 m in width, 38 m in height, and 11.5 m in depth of deep foundation excavation. The geometric model is meshed and optimized, and the number of optimized grid elements is 2.58 million. Figure 2 shows the deep foundation pit excavation calculation model established by us. The figure clearly shows the excavation scope of the proposed deep foundation pit, the stratum of the area and the spatial location of surrounding buildings, subway tunnels and other engineering bodies.

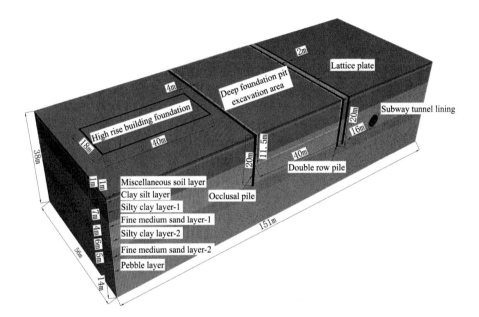

Fig. 2. Numerical model of deep excavation.

4.2 Pile Anchor Support Structure and Surrounding

This paper presents a numerical simulation modelling of a solid unit support system using a deep foundation support method with pile-anchor support. The outer diameter of the underground tunnel lining on the east side of the deep foundation pit is 3 m. On the side of the deep foundation pit adjacent to the tunnel, a double-row pile + anchor cable support system + slurry flower pipe is used, with a double-row pile diameter of 1 m, a pile spacing of 1.5 m, a pile length of 20.0 m and a row spacing of 3.0 m. A lattice plate is connected between the double-row piles, with a lattice plate size of 56 m in length, 2 m in width and 0.5 mm in height. A total of 1 anchor cable and 4 grouting pipes are arranged with a horizontal spacing of 1.5 m and a vertical spacing of 2 m. There are buildings on the west side of the deep foundation pit, 40 m long and 18 mm wide. In the immediate vicinity of the building side using the occlusion pile + anchor cable support system, occlusion pile diameter 1mm, pile spacing 1.5 m, pile length of 20.0 m. A total of 3 anchor cables are arranged, with a transverse spacing of 1.5 m and a longitudinal spacing of 3 m. Figure 3 shows a partial enlarged view of the numerical model of the pile anchor support system, in which the geometric information and spatial location of the pile anchor support system, surrounding buildings and tunnels are marked in detail. It should be noted that the model has been simplified to take into account the complexity of the model by converting the double-row piles and the occluded piles into a ground link wall in accordance with the principle of flexural stiffness equivalence, and the modulus of elasticity of the ground link wall is taken as 70% of the modulus of elasticity of the double-row piles and the occluded piles selected by the piles. Even the wall is 1 m in width, consistent with the diameter of the pile; even the wall depth to 20 m, and is consistent with the length of the pile.

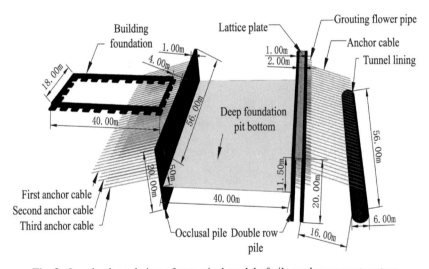

Fig. 3. Local enlarged view of numerical model of pile-anchor support system.

4.3 Basic Assumptions, Boundary Conditions and Material Parameters

The following assumptions have been made in the modelling and calculations with the aim of removing the effects of some minor factors, while ensuring that the results are accurate:

- Assume that the soil is isotropic and homogeneous elastic plastic body.
- It is assumed that building foundation, tunnel lining, lattice plate and diaphragm wall (double row pile and bite pile) are ideal linear elastomers.
- Different types of piles (such as cast-in-place piles, rotary jet grouting biting, etc.) have different soil penetration processes. The soil penetration process of piles is not a completely static process, involving large deformation problems, and the stress state

will change. Since the deep foundation excavation is the focus of this simulation, the process of pile entry is ignored in this simulation.

- When applying boundary conditions, the full displacement constraint is applied at the bottom of the whole model, the top of the model is a free surface, and the normal displacement constraints are applied on the sides around the model. The building on the west side is converted into gravity load and applied to the building foundation, with a total of 108 MN. The effect of gravity is considered in the whole model.
- Mohr Coulomb criterion is adopted as the constitutive model of stratum soil material. See Table 1 for the selection of soil material parameters. Concrete material with strength grade of C25 is selected for building foundation and lattice plate, and concrete material with strength grade of C25 is also selected for diaphragm wall (double row pile and bite pile), but its elastic modulus is 70% of C25 concrete elastic modulus according to the principle of equivalent flexural stiffness, The subway tunnel lining is made of concrete with strength grade of C40. See Table 2 for the mechanical parameters of concrete with each strength grade. See Table 3 for the material parameters of the anchor cable and grouting flower pipe model adjacent to the east of the subway tunnel and the three anchor cables model adjacent to the west of the building.

Table 2. Mechanical parameters of supporting structure materials.

Structure name	Concrete strength	Elastic modulus (GPa)	Poisson's ratio
Diaphragm wall (double row pile, bite pile)	C25	19.6	0.2
Tunnel lining	C40	32.5	0.2
Foundation and lattice plate	C25	28	0.2

4.4 Calculation Conditions

Before the deep foundation pit model is excavated, the initial stress balance calculation is carried out for the overall model, that is, the stress balance of the overall model is calculated under the simultaneous action of the self-weight of the building and the model as a whole, and the displacement is cleared to zero after the calculation. Then the deep foundation pit excavation calculation is carried out, and the excavation is carried out in layers along the vertical direction, 1.5 m per layer, one layer at a time, and divided into four excavation areas along the horizontal direction, as shown in Fig. 4. To study the effects of different excavation methods, this paper adopts two methods for excavation:

Method 1: Layered excavation, that is, the four excavation areas excavate one layer as a whole, and then construct the anchor cable and grouting flower pipe in Zone 1 and Zone 2, and then the anchor cable and grouting flower pipe in Zone 3 and Zone 4. This is repeated until the 11.5 m deep foundation pit is excavated in 8 layers, a total of 8

Table 3. Mechanical parameters of anchor cable and grouting flowered pipe structural materials.

Name		Incident angle	Total length (grouting section) (m)	Pre applied axial force (KN)	Elastic modulus (GPa)	Tensile strength (GPa)	Perimeter of grouting hole(m)
Grouting flower pipe		15°	6(6)	/	206.0		0.471
East anchor cable		15°	22(17)	300	205.0	7.44	0.471
The west side	First way	15°	26(17)	400	205.0	7.44	0.471
	Second way	15°	24(18)	450	205.0	7.44	0.471
	Third way	15°	23(18)	430	205.0	7.44	0.471

Note: The bonding stiffness of the grouting section of anchor cable and grouting flower pipe is 560 MN/m, the bonding strength is 0.15 MP/m, and the bonding friction angle is 25°; The bond strength of the free end of the anchor cable is considered to be 0 MP/m and the adhesive friction angle is 0°

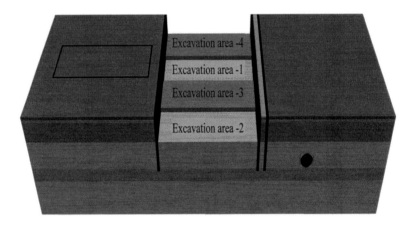

Fig. 4. Diagram of four excavation areas of deep foundation pit.

calculation conditions. Method 2: Excavate in layers and sections, that is, excavate one layer at the same time in Zone 1 and Zone 2, and then construct the anchor cable and grouting flower pipe in Zone 1 and Zone 2; Then excavate one layer in Zone 3 and Zone 4 at the same time, and then construct the anchor cable and grouting flower pipe in Zone 3 and Zone 4. Repeat this until the 11.5 m deep foundation pit is excavated in 8 layers, a total of 16 calculation conditions.

5 Impact of Different Excavation Methods on Deep Foundation Pit and Surrounding Environment

Based on the previously established numerical model, under different excavation methods was carried out to analyse the horizontal displacement of the support structure, the settlement of surrounding buildings and ground surface and the deformation law of subway tunnel in the process of deep foundation pit excavation are analyzed, and the influence mechanism of different excavation methods on the safety of deep foundation pit, surrounding buildings and tunnel is discussed.

In order to fully investigate the settlement displacement and deformation pattern of the deep foundation pit, surrounding buildings and tunnels during excavation, multiple displacement measuring points and lines are set on the model. Firstly, in order to detect the deformation of the deep foundation pit during excavation, a measuring point is arranged in the middle of the top of the support structure on the west side of the deep foundation pit, numbered ZH-1 respectively, and three vertical measuring lines are arranged at an equal interval of 22 m on the support structure, numbered SC-X1, SC-X2 and SC-X3. One measuring point, numbered ZH-2, is arranged in the middle of the top of the support structure on the east side. Two vertical measuring lines, numbered SC-X4 and SC-X5, are arranged on the support structure at an interval of 44 m. In order to detect the settlement and displacement of the building and its surrounding surface, two measuring points are arranged at an interval of 25.2 m on the foundation of the north side of the building, numbered JZ-1 and JZ-2 respectively. On the surrounding surface, two measuring lines are symmetrically arranged on both sides of the building along the east-west direction, 11 m away from the building, numbered DB-X1 and DB-X2. In order to explore the deformation and displacement of the subway tunnel, a survey line, numbered SD-X1, is arranged along the tunnel direction on the side of the tunnel close to the deep foundation pit, and a survey line, numbered SD-X2, is also arranged at the top of the tunnel. Two survey lines, numbered DB-X3 and DB-X4, are arranged along the east-west direction on the surface above the tunnel at an interval of 40 m. Two survey points, numbered DB-1 and DB-2, are arranged on the surface between the two survey lines at an interval of 13 m. The detailed layout is shown in Fig. 5.

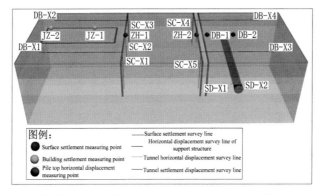

Fig. 5. Layout diagram of measuring points and lines.

5.1 Effect of Deep Foundation Pit Excavation on Deep Foundation Pit Deformation

The horizontal lateral displacement of the supporting structure on both sides of the deep foundation pit can reflect the overall deformation of the deep foundation pit. In order to analyze the deformation of the deep foundation pit under the two excavation methods after all excavation along the depth of 11.5 m.

Figure 6 shows the horizontal displacement diagram of the foundation pit after excavation. Figure 7 shows the horizontal displacement curves of the measured lines of the support structures on the east and west sides of the pit after excavation. Figure 8 shows the measured results of horizontal displacement of supporting structures.

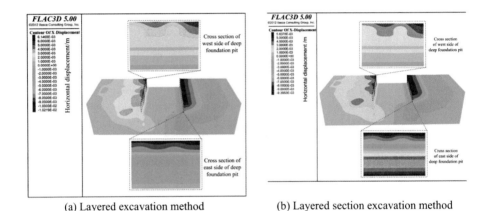

(a) Layered excavation method   (b) Layered section excavation method

Fig. 6. Horizontal displacement of deep foundation pit.

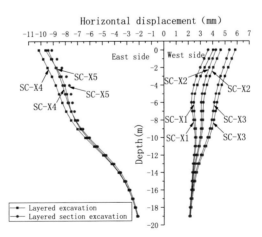

Fig. 7. Numerical results of horizontal displacement of retaining structure.

It can be seen from Fig. 6 and Fig. 7 that when the excavation is carried out in accordance with the depth of 11.5 m, the largest displacement of the support structure on both the west and east sides of the deep foundation pit under the two different excavation methods occurs at the top of the deep foundation pit, and the horizontal displacement on both the east and west sides of the deep foundation pit show a trend of gradual reduction as the depth increases. As can be seen from Fig. 7, the maximum horizontal displacement on the west side is 5.85 mm, and the maximum horizontal displacement on the east side is 10.22 mm, which is significantly greater than that on the west side, with a difference of 4.37 mm. Analysing the reasons for this, three prestressing anchor cables were applied during the support of the western side of the deep foundation pit as the western side was adjacent to the building, while one prestressing anchor cable was applied for the metro tunnel which was farther away from the deep foundation pit. It is obvious from the curves in Fig. 7 that the horizontal displacements on both sides under the layered section excavation method are overall smaller than those under the layered excavation method, so the layered section excavation method has less influence on the deformation of the deep foundation pit. As the calculation in this paper assumes that whichever excavation method is used, the prestressed anchor cables in Zone 1 and Zone 2 are excavated first, and then the prestressed anchor cables in zone 3 and zone 4 are constructed. The first applied anchor cables limit the deformation of the supporting structures in Zone 1 and Zone 2, so the horizontal displacement of the survey lines in Zone 1 and Zone 2 is less than that of the survey lines in Zone 3 and Zone 4. This can be clearly seen in Figs. 6 and 7. It should be pointed out that the west side of deep foundation pit Numbers for SC - X2 line, fall in the area of SC - X2 line located close to the side of the building. Due to the influence of the building's self-weight, the horizontal displacement of the SC-X2 measuring line is greater than that of the SC-X1 measuring line falling in Zone 1. It can be seen that the building still has a great impact on the deformation of the deep foundation pit. The overall horizontal displacement of the measurement line SC-X3 is greater than the horizontal displacement of the measurement line SC-X2, and it can be concluded that the effect of the building on the deformation of the deep foundation pit is less than the limiting effect of the prestressing anchor cable on the deformation of the deep foundation pit.

To verify the validity of the numerical simulation results, the horizontal displacement of the excavation was monitored during excavation, and the measurement lines were arranged in the same way as the numerical simulation. Through comparison, the trend of the horizontal displacement curve of each measurement line in the numerical simulation results matches well with the field monitoring data, thus verifying the accuracy of the numerical simulation.

In order to further analyze the influence of different excavation methods on the defor-mation law of deep foundation pit in the excavation process, Fig. 9 shows the relationship curve of the horizontal displacement of the top measuring points of the support structure on the east and west sides with the excavation depth under different excavation methods. The numerical simulation curve and field measured curve are given in the figure. The field measured data are in good agreement with the numerical simulation results. As can be seen from the diagram, the horizontal displacement at the top position of the deep foundation support structure under both excavation methods gradually increases as the excavation depth increases. At shallow excavation depths there is little difference in the horizontal displacement of the deep foundation pit caused by the two excavation methods. But when the excavation depth increases to a certain value, the horizontal displacement under layered excavation begins to be greater than that under layered section excavation. The critical value in the east is 3 m and that in the west is 7.5 m. It is not difficult to conclude that the reinforcement effect of anchor cables can weaken the effect of excavation methods on the deformation of deep foundation pits. The difference of horizontal displacement between the two excavation methods increases with the increase of excavation depth. The horizontal displacement under layered excavation method is greater than that under layered and segmented start method. In summary, it can be concluded that the layered excavation method is suitable for excavating shallow deep pits, while the layered section excavation method is more suitable for deeper deep pits.

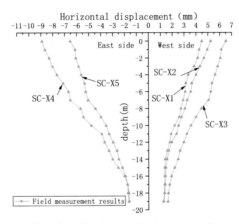

Fig. 8. The measured results of horizontal displacement of supporting structures.

5.2 Impact of Deep Foundation Pit Excavation on Buildings and Surrounding Surface

5.2.1 Impact on Surface Settlement around Buildings

The excavation of deep foundation pit will inevitably cause the settlement of adjacent buildings and surrounding surface. In the process of excavation, when the settlement displacement of buildings and surrounding surface reaches a certain range, it will affect the normal use, and more serious settlement will lead to safety accidents. Analyze Fig. 10, the settlement displacement curve of the ground surface around the building after the excavation of the deep foundation pit, to investigate the effect of different excavation methods on the settlement displacement of the ground surface around the building.

As can be seen from Fig. 10, when the deep foundation pit is fully excavated, whether it is excavated in sections or in layers and sections, the surface settlement displacement decreases as the surface position is far away from the deep foundation pit. At a distance of 2 m from the deep foundation pit, the curve shows an extreme value, indicating a large settlement of the ground surface in the middle position due to the combined effect of the building and the deep foundation pit. The maximum surface settlement displacement

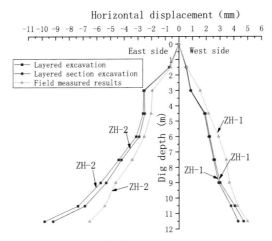

Fig. 9. The relation curve between the horizontal displacement of the top of the supporting structure and the excavation depth.

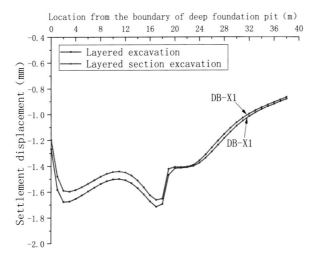

Fig. 10. The surface settlement displacement curve around the building.

occurs at 17 m, with a maximum of 1.71 mm for layered excavation and 1.67 mm under layered section excavation. By comparing the curves under the two excavation methods, it can also be seen that the overall surface settlement displacement under the layered section excavation method is less than that under the layered excavation. The closer it is to the boundary of the deep foundation pit, the more obvious this phenomenon is. It can be seen that different excavation methods have a greater influence on the surface settlement around the deep foundation pit. Compared with the layered excavation method, The layered section excavation method has less impact on the surface settlement and displacement and is safer, which is consistent with the conclusion obtained from the above analysis of deep foundation pit deformation.

5.2.2 Impact on Building Settlement

To analyze the effect of different excavation methods on the settlement of buildings, Fig. 11 shows the settlement displacement diagram of the building after deep foundation pit excavation, and Fig. 12 shows the relationship curve between building settlement displacement and excavation depth.

Whether it is layered excavation or layered section excavation, when the excavation is complete, the closer the building settles to the deep foundation pit, the less it settles, while on the side furthest from the deep foundation pit, the building settles to a maximum. The reason is that there is anchor cable reinforcement in the soil close to the deep foundation pit, and the overall performance of the soil is better. In the absence of anchor cables in the soil away from the deep foundation, the building settles and displaces more under the self-weight of the building. The maximum settlement of buildings under layered excavation is 4.3 mm, and the maximum settlement under layered section excavation is 4.12 mm. It can be seen that layered excavation has a greater impact on building settlement. With the increase of excavation depth, the settlement displacement of buildings also increases. When the excavation depth is shallow, the settlement displacement of buildings under the two excavation methods is almost the same. When the excavation depth increases, the settlement displacement of buildings under layered excavation method is significantly greater than that under layered section excavation method. By analyzing the influence law of building settlement, it can also be concluded that the layered excavation method is suitable for shallow and deep foundation pit, and the layered and section excavation method is more suitable for deeper deep foundation pit.

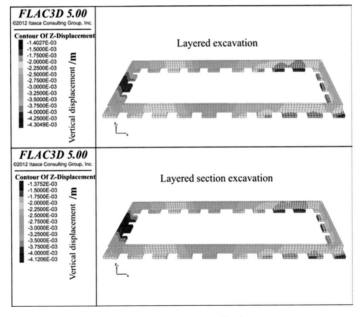

Fig. 11. Building settlement displacement.

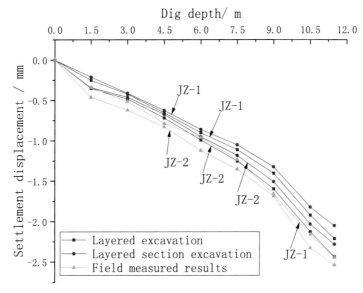

Fig. 12. The relationship between foundation settlement and excavation depth.

5.3 Influence of Deep Foundation Pit Excavation on Tunnel and Surface

5.3.1 Influence on Tunnel Deformation

In order to analyze the effect of deep foundation excavation on the deformation of metro tunnels, Fig. 13 shows the horizontal and vertical displacement diagrams of the tunnel after all the deep foundation pit excavation under two different types of excavation.

As can be seen from Fig. 13, the impact of deep foundation excavation on the overall deformation of the tunnel is not significant for either excavation method. In the vertical direction, the excavation of deep foundation pit causes downward displacement in the top area of the tunnel and upward displacement in the bottom area, which makes the tunnel flatten, but the displacement is relatively small. The upper and lower displacement difference caused by the two excavation methods does not exceed 0.5 mm, that is, the deformation of the tunnel in the vertical direction does not exceed 0.5 mm; In the horizontal direction, the excavation of deep foundation pit has caused the displacement to the West on the left and right sides of the tunnel, and the displacement is basically the same. It can be seen that the tunnel has basically no deformation in the horizontal direction.

5.3.2 Influence on Tunnel Displacement

To analyze the effect of deep foundation excavation on settlement and horizontal displacement in metro tunnels, Fig. 14 and Fig. 15 show the tunnel settlement displacement curve and horizontal displacement curve after all deep foundation pit excavation.

The displacement curve shows that the maximum settlement displacement at the top of the tunnel is 0.262 mm in the layered excavation method and 0.256 mm in the layered section excavation method, while the vertical settlement displacement of the whole

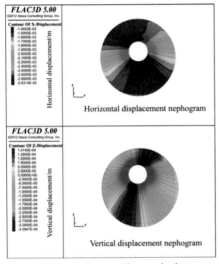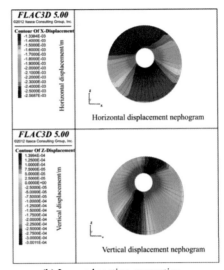

(a) Layered excavation method   (b) Layered section excavation

Fig. 13. Deformation of tunnel under two different excavation methods

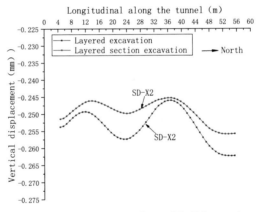

Fig. 14. The settlement displacement curve of tunnel Building settlement displacement.

tunnel is less than 0.3 mm, which means that the deep foundation excavation has a small effect on the settlement of the tunnel; However, deep foundation excavation has a greater impact on the horizontal displacement of the tunnel. As can be seen from the graphs, the horizontal displacement of the tunnel towards the deep foundation pit was generated in both excavation methods, with a maximum horizontal displacement of 2.03 mm in the layered excavation and 2.01 mm in the layered section excavation. Whether it is settlement displacement or horizontal displacement, the displacement under layered excavation is greater than that under layered section excavation, but the difference is small. It can be obtained that the different excavation methods of the deep foundation pit have less influence on the displacement of the tunnel.

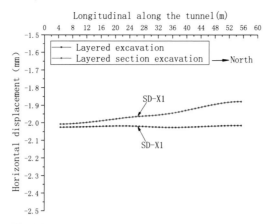

Fig. 15. The horizontal displacement curve of tunnel.

5.3.3 Impact on Surface Settlement Above Tunnel

Settlement displacement curves for the surface survey line on the east side of the deep foundation pit after excavation are shown in Fig. 16, and surface settlement curves with increasing excavation depth are shown in Fig. 17. The analysis shows that the surface settlement displacements near the boundary of the deep foundation pit are the largest, exceeding 3.5 mm for both excavation methods, whether the excavation is carried out in layers or in layers sections. When the surface is far away from the deep foundation pit, the settlement shows a decreasing trend. A peak in settlement displacement occurs at a distance of 16 m from the boundary of the deep foundation pit, directly below the tunnel. It can be concluded that the presence of the tunnel still has some influence on the ground settlement. The surface settlement caused by layered excavation is greater than that caused by layered and section excavation. As the excavation depth increases, the

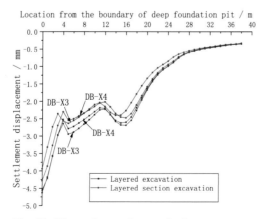

Fig. 16. The surface settlement displacement curve

difference between the two becomes more obvious. Once again, it is verified that the layered section excavation method has less influence on the deformation and displacement of the deep foundation pit and the surrounding buildings and tunnels.

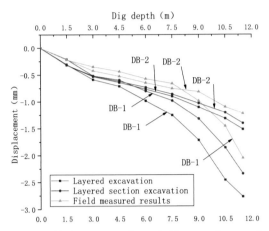

Fig. 17. Relationship between surface subsidence and excavation depth.

6   Conclusion

Taking a deep foundation pit in Shiliuzhuang, Beijing as the engineering background, FLAC3D was used to establish a numerical model of deep foundation pit excavation in a complex environment, and a numerical simulation study of deep foundation pit excavation under the conditions of different excavation methods was carried out to reveal the mechanism of the impact of different excavation methods on the safety of deep foundation pit on its surrounding tunnels and buildings, with the following main conclusions:

- The greater the excavation depth of the deep foundation pit in both excavation methods, the greater the horizontal displacement of the support structure on both sides. When the excavation depth is not large, the excavation method has no obvious influence on the horizontal displacement caused by excavation, but as the depth increases, at a certain value, the horizontal displacement under the layered excavation method starts to be larger than the horizontal displacement under the layered section excavation method. And the difference is more obvious as the depth deepens. It is easy to conclude that the layered excavation method is suitable for deep pits with shallow excavation depths, while layered section excavation method is suitable for deep pits with deep excavation depths. The reinforcement effect of the anchor cables reduces the effect of the excavation method on the deformation of the deep foundation pit. The surrounding buildings have a greater influence on the deformation of the deep foundation pit, but the influence of the buildings on the deformation of the deep foundation pit is less than the effect of the prestressing anchor cable on limiting the deformation of the deep foundation pit layered section excavation.

- The excavation of deep foundation pit causes the settlement of adjacent buildings and surrounding surface. As the distance to the deep foundation pit increases, the settlement displacement of the building gradually increases and the surface settlement displacement gradually decreases. Different excavation methods have a great impact on the settlement of buildings and surrounding surface. The overall settlement displacement of buildings and surface under the layered section excavation method is less than that under the layered excavation. The closer to the deep foundation the more pronounced this phenomenon becomes. Compared with the layered excavation method, the layered section excavation method has less and safer impact on the settlement displacement of buildings and surface.
- The deep foundation excavation has a small effect on the settlement of the tunnel and a large effect on the horizontal displacement of the tunnel. Different excavation methods have little impact on the displacement of the tunnel. Whether it is the settlement displacement or horizontal displacement of the tunnel, the displacement under the layered excavation method is greater than that under the layered section excavation method, but the difference is small. The deep foundation excavation did not have a significant impact on the overall deformation of the tunnel.
- Deep foundation excavation will cause large settlement of the surface above the tunnel. Maximum ground settlement displacement near the boundary of the deep foundation pit. When the surface is far away from the deep foundation pit, the settlement shows a gradual decreasing trend. There is a peak value of increasing settlement displacement directly above the tunnel. The surface settlement caused by layered excavation is greater than that caused by layered section excavation.

Acknowledgments. . The authors are grateful for the financial support from the National Natural Science Foundation of China (Grant numbers. 51874309).

References

1. Ding, K.W., Zhao, F., Zhang, X.Z.: Effect of large deep foundation pit dewatering on the neighbouring buildings. Appl. Mech. Mater. **3307**, 580–583 (2014)
2. Zhang, X.D., Lv, J.W.: Deep foundation pit engineering technology research. Appl. Mech. Mater. **256–259**, 332–335 (2012)
3. Yang, F., Qi, X.L., Wu, Y.Z., Cao, S.Y.: Research on deep foundation pit excavation based on data monitoring. IOP Conf. Ser. Earth Environ. Sci. **525**, 1 (2020)
4. Feng, Z.W., Shen, Y.H.: Discussion on construction technology of deep foundation pit support in construction engineering. IOP Conf. Ser. Mater. Sci. Eng. **787**, 1 (2020)
5. Zhang, X.T., Zhang, Q.Y., Wang, C., Wang, Y.F.: Monitoring analysis and safety evaluation of deep foundation pit under complex conditions. Appl. Mech. Mater. **3307**, 580–583 (2014)
6. Liu, J.H., Hou, X.Y.: Excavation Engineering Handbook. China Architecture & Building, Beijing, pp. 1–20 (1997)

7. Han, G.D., Guan, C.S., Zhou, J.: Environmental impact and hazard control method on deep foundation pit supporting engineering. Adv. Mater. Res. **2195**, 639–640 (2013)
8. Yang, Z.S., Liu, J.X., Wang, Y.R.: Weight analysis of accident factors in deep foundation excavation based on analytic hierarchy process. Appl. Mech. Mater. **711** (2014)
9. Alireza, V., Nordin, Y., Norhazilan, M.N., Jurgita, A., Jolanta, T.: Hybrid SWARA-COPRAS method for risk assessment in deep foundation excavation project: an Iranian case study. J. Civ. Eng. Manag. **23**, 4 (2017)
10. Qian, Q.H., Chen, Z.L.: Development and utilization of underground space in the 21st century. In: Annual Meeting of China Civil Engineering Society, pp. 162–169 (1998)
11. Li, L.J., Liang, R.W.: Research on the spatial effect of double row piles structure system in deep foundation. Adv. Mater. Res. 374–377 (2011)
12. Feng, S.L., Wu, Y.H., Li, J., Li, P.L., Zhang, Z.Y., Wang, D.: The analysis of spatial effect of deep foundation pit in soft soil areas. Procedia Earth Planetary Sci. **5** (2012)
13. Zhu, C.Z., Wang, S.J.: Analysis of pile-anchor supporting structure internal force and building settlement around foundation pit. Int. J. Appl. Math. Stat.™ **51**, 22 (2013)
14. Liu, J.H.: Theory and practice of space time effect in soft soil foundation pit engineering. In: Cross Strait Symposium on Tunnel and Underground Engineering, pp. 7–10 (1999)
15. Qi, R.W., Deng, H.Y., Dai, G.L.: Analysis of giant foundation pit excavation influence on the spatial effect of surrounding environment. J. Water Resour. Architect. Eng. **19**(02), 178–184 (2021)
16. Wang, P.X., Zhou, S.H., Di, H.G., et al.: Impacts of foundation pit excavation on adjacent railway subgrade and control. Rock Soli Mech. **37**(S1), 469–476 (2016)
17. Liu, J.T., Nie, M.J., Li, C.J., Yan, J.P.: Deformation and numerical simulation analysis of deep foundation pit excavation of Nanjing Yangtze River floodplain metro station Earth and Environmental Science. IOP Conf. Ser. Earth Environ. Sci. **719**(3), 1–4 (2021)
18. John, H.D.S.: Field and theoretical studies of the behaviour of ground around deep excavations in London clay. University of Cambridge, Cambridge, pp. 15–26 (1975)
19. Mana, A.I., Clough, G.W.: Prediction of movements for braced cuts in clay. Geotech. Spec. Pub. **107**(118), 1840–1858 (1981)
20. Wong, K.S., Broms, B.B.: Lateral wall deflections of braced excavations in clay. J. Geotech. Eng. **115**(6), 853–870 (1989)
21. Zhou, Y., Zhu, Y.W.: Interaction between pile-anchor supporting structure and soil in deep excavation. Rock Soli Mech. **39**(9), 3246–3252 (2018)
22. Jiang, Y.Q.: Study on spatial effect analysis and environmental symbiosis comprehensive evaluation of deep foundation pit under complex environmental conditions. Chongqing University, pp. 1–20 (2015)
23. Ding, W.S.: Monitoring and research on spatial effect of foundation pit supporting structure. In: National Conference on Structural Engineering, pp. 702–706 (2001)
24. Jasim, M.A.: Evaluation conduct of deep foundations using 3D finite element approach. Am. J. Civ. Eng. **1**, 3 (2013)
25. Cao, H.Y., Jia, D.B., Chen, T.J.: Study on deformation characteristics of deep foundation pit in unsaturated soil. Adv. Mater. Res. 374–377 (2011)
26. Peng, H., Qi, X.X., Dong, Z.Y.: Study on deformation influence of deep foundation pit pile-anchor supporting system nearby metro structure. Appl. Mech. Mater. **3489**, 638–640 (2014)
27. Tan, W.H., Li, C.C., Sun, H.B.: Study on methods for deep foundation pit supporting in Beijing. Adv. Mater. Res. **2195**, 639–640 (2013)
28. Tong, S.H.: Research on the application of deep foundation pit support construction technology in civil engineering. Int. J. Educ. Econ. **3**, 3 (2020)

Study on Critical Shaft Length in Underground Metro Tunnel with One End to Outside

Tingting Wang[1], Pengqiang Geng[2], Miaocheng Weng[1,3,4(✉)], and Fang Liu[1,3,4(✉)]

[1] School of Civil Engineering, Chongqing University, Chongqing, China
1058031901@qq.com
[2] Chongqing Pharmaceutical Design Institute of Sinopharm Group, Chongqing, China
[3] Key Laboratory of New Technology for Construction of Cities in Mountain Area of Ministry of Education, Chongqing University, Chongqing, China
[4] Joint International Research Laboratory of Green Buildings and Built Environments, Chongqing, China

Abstract. A 1:15 small-scale model was built based on a blocked tunnel at one end of Chongqing rail transit, the reliability of FDS numerical simulation was verified by experiments. Adopts the FDS18.0 software for the shaft seal at the end of the tunnel, and discusses the fire location, fire heat release rate on the effects of the shaft of natural smoke extraction mode when fire between seal and shaft, shaft smoke exhaust volume significantly greater than other fire source position, the critical length of shaft is bigger, when fire heat release rate between 2.5 MW to 7.5 MW. The critical shaft length decreases with the increase of shaft height, but has nothing to do with the heat release rate of fire source. A critical shaft length prediction model for tunnel closure at one end is proposed.

Keywords: Block the tunnel at one end · Shaft natural ventilation · FDS numerical simulation · Critical shaft length

1 Introduction

Due to the limitation of ground space, various forms of one-end blocking tunnels have gradually appeared in the planning, design and construction of mountainous cities, which is different from the common open tunnels at both ends. Due to the existence of the blocking end, the flue gas is easier to gather and is not easy to discharge out of the tunnel, and the flue gas flow law of such one-end blocking tunnels in case of fire is less involved. Putting forward a suitable ventilation and smoke exhaust design scheme for one end blocked tunnel has become an important solution to the difficulty of smoke control.

The research shows that compared with the top opening natural ventilation and smoke exhaust mode, the shaft natural ventilation has stronger chimney effect, which can produce greater pressure difference and obtain better smoke exhaust effect. Its smoke exhaust effect has been verified by a large number of experimental data and examples [1–4]. At the same time, the influence of shaft length on flue gas flow under natural

© The Author(s) 2022
G. Feng (Ed.): ICCE 2021, LNCE 213, pp. 130–141, 2022.
https://doi.org/10.1007/978-981-19-1260-3_12

ventilation is carried out, and the theoretical model of critical shaft length based on two open tunnels under different conditions is obtained [5–9]. However, the above research is mainly based on the open tunnel with shaft at both ends. Whether the research conclusion is applicable to the tunnel blocked at one end needs further research.

FDS numerical simulation method is used to study the critical shaft length of one end blocking tunnel when the shaft width is constant, the shaft in the tunnel is rectangular, and the shaft width is equal to the tunnel width. The smoke spread length with different parameters such as fire source location, fire source heat release rate and shaft height is discussed, and then the critical shaft length is analyzed.

2   Critical Shaft Length

The concept of critical shaft length was proposed in the study of the influence of open tunnel shafts on flue gas flow. The so-called critical shaft length usually refers to the minimum shaft length to ensure that the flue gas generated by the fire source and spread to the shaft is completely discharged from the shaft and will not spread to the rear of the shaft. In the blocked tunnel at one end, when the section of the shaft is too small or the height of the shaft is too low, the buoyancy generated by the pressure difference inside and outside the shaft is small, and the chimney effect is limited. A large amount of flue gas will spread through the shaft and continue to gather in the tunnel area behind the shaft, so the flue gas cannot be discharged through the shaft, which is not conducive to personnel evacuation. In order to make the shaft discharge flue gas effectively and reduce the spread distance of flue gas to the greatest extent, the critical shaft length of one end blocking tunnel is studied.

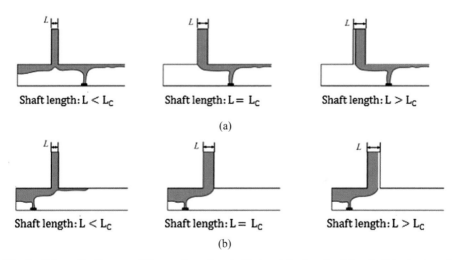

Shaft length: $L < L_C$ Shaft length: $L = L_C$ Shaft length: $L > L_C$

(a)

Shaft length: $L < L_C$ Shaft length: $L = L_C$ Shaft length: $L > L_C$

(b)

Fig. 1. Schematic diagram of the smoke exhaust effect and the length of the shaft in the tunnel with one closed portal (a) $X_f < X_s$ (The fire source is located between the plugging end and the shaft); (b) $X_f > X_s$ (The fire source is located between the shaft and the open end).

Assuming that only one shaft is set in the blocked tunnel at one end and close to the blocked end, L_c is the critical shaft length. The relationship between smoke exhaust effect of tunnel shaft blocked at one end and shaft length is shown in Fig. 1. It can be seen from Fig. 1 (a) that when the fire source is located between the blocking end and the shaft, in order to meet the conditions of full exhaust of flue gas in the critical shaft length, the shaft needs to continuously discharge all flue gas generated by the fire source to ensure that the tunnel area behind the shaft is a smoke-free area. It can be seen from Fig. 1 (b) that when the fire source is located between the open end and the shaft, the appropriate critical shaft length can discharge all the flue gas spreading to the blocking end, avoiding the accumulation of flue gas at the blocking end, so that the flue gas in the tunnel will not settle for a long time and ensure the safety of personnel in the tunnel.

3 FDS Numerical Simulation

3.1 Validation of FDS Numerical Simulation

Before using FDS for numerical simulation, the experimental model is established according to the small-scale experimental conditions, and the accuracy of FDS is demonstrated by comparing the small-scale experimental results with the FDS numerical simulation results. Taking a blocked tunnel at one end of a rail transit in Chongqing as the prototype, a 1:15 small-scale model test-bed was built with Froude similarity model rate [10]. The small-scale test-bed is a tunnel blocked at one end, in which the tunnel is 5.0 m long, 0.32 m wide and 0.48 m high. The model uses K-type armored thermocouples to measure the tunnel temperature. 38 thermocouples are arranged on the longitudinal centerline in the tunnel. The spacing above and near the fire source is 0.1m, and the spacing of the far fire source is 0.2 m. Using ethanol as fuel, oil pans of different sizes are designed and calculated according to the fuel loss rate. The heat release rate of four ethanol oil pools was designed, which was $8 \times 8, 10 \times 10, 12 \times 12, 15 \times 15$. According to the above experimental conditions, the tunnel geometry, fire source fuel and tunnel material settings established by FDS are consistent with the experiment. The comparison between small-scale test and FDS simulation results is shown in Fig. 2.

As can be seen from Fig. 2 (a), for the longitudinal temperature distribution of the tunnel roof, the FDS numerical simulation results are slightly smaller than the experimental results near the fire source, while there is little difference between the experimental results and FDS numerical simulation results in the far fire source area. There are individual experimental data measurement points that fluctuate. In general, FDS numerical simulation can better reflect the experimental results.

3.2 Grid Independence Test

Before FDS numerical simulation, grid independence and reliability verification shall be carried out. The FDS user manual [11] provides a grid division criterion for users. The grid size can be determined by the calculated value of $D^*/\delta x$, δx is the grid size and D^* is the characteristic diameter of fire source. The calculation formula is as follows:

$$D^* = \left(\frac{Q}{\rho_a c_p T_a g^{1/2}} \right)^{\frac{2}{3}} \tag{1}$$

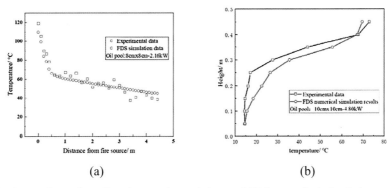

(a)   (b)

Fig. 2. Comparison of small-scale experimental data and FDS numerical simulation result (a) Temperature distribution in longitudinal position of tunnel roof; (b) Temperature distribution of thermocouple tree 0.5 m away from tunnel opening.

At present, it is generally believed that when the calculated value of $D^*/\delta x$ is between 4–16, the numerical simulation will have a better result. The fire source power used in this paper is 2.5 MW, 5 MW and 7.5 mw. Through calculation, it can be obtained that the more reasonable grid size should be between 0.114 m–0.458 m. Four grids with different scales of 0.100 m, 0.125 m, 0.160 m and 0.200 m are selected to calculate the same fire condition. Under different grid sizes, the longitudinal temperature distribution diagram of tunnel ceiling blocked at one end is shown in Fig. 3.

It can be seen from Fig. 3 that the calculation results of the four grid sizes have little difference. The maximum temperature at the fire source location of the grid of 0.2 m is low, while the grid of 0.160 m is close to the maximum temperature of 0.100 m and 0.125 m, and the temperature in the far fire source area of the four grids has little difference. In this simulation, 0.160 m is uniformly used as the simulated grid size.

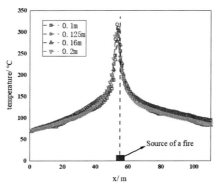

Fig. 3. Grid independence analysis

3.3 FDS Simulation Scheme

The heat release rate of fire source is 2.5 MW, 5.0 mw and 7.5 mw respectively. Four fire source positions are selected. The shaft center is fixed at $X_s = 10$ m. The shaft length is taken according to the shaft height (1.2 m–2.2 m, with an interval of 0.2 m). The specific simulation arrangement is shown in Table 1.

Table 1. Simulation scheme setting

Simulation scheme	Fire source location	HRR/MW	Shaft height/m	Shaft length/m
Case1–3	X = 5 m	5	0	2.2, 2.6, 3.0
Case4–7			5	1.6, 1.8, 2.0, 2.2
Case8–10			10	1.2, 1.4, 1.6
Case11–13			15	1.2, 1.4, 1.6
Case14–16			20	1.0, 1.2, 1.6
Case17–19		2.5	5	1.8, 2.0, 2.2
Case20–22		7.5	5	1.8, 2.0, 2.2
Case23–25	X = 15 m, 55 m, 95 m	5	10	1.2

When the height of the shaft is 5 m, the heat release rate of the fire source is 5 MW and the fire source is located at $X_f = 5$ m, the temperature distribution of flue gas in the roof of one end blocked tunnel under different shaft lengths is shown in Fig. 5. When the difference between the tunnel ceiling temperature and the ambient temperature is less than 5 °C, it can be approximately considered that this area is a smoke-free area. Since the tunnel ceiling temperature measuring points are all at the ambient temperature of 20 °C after 50 m, in order to better compare the smoke propagation length under different shaft lengths, this figure only shows that the tunnel length is 50 m. It can be seen from Fig. 5 that under the working conditions of different shaft lengths, the temperature of each tunnel has a sudden drop at x = 10 m, but the cooling range is different. Obviously, the longer the shaft length, the greater the cooling range. When the temperature of flue gas directly drops to the ambient temperature after crossing the shaft, the shaft length at this time can be considered as the critical shaft length. The change of shaft length will not have a great impact on the tunnel ceiling flue gas temperature between the blocking end and the shaft. From the shaft to the open end, the flue gas propagation length is significantly shortened with the increase of shaft length. When the shaft length L = 1.6 m is increased to 1.8 m and 2.0 m, the smoke spread length is shortened from 25 m to 18 m and 12 m. When the shaft length L = 2.2 m, the smoke spread length is shortened to 6 m, which is approximately equal to the distance from the shaft to the fire source. There is no smoke spread in the tunnel area behind the shaft, and the tunnel behind the shaft is a smoke-free area. Therefore, when the shaft height $H_s = 5$ m, the fire source heat release rate is 5 MW and the fire source is located at $X_f = 0.5$ m, the critical shaft length $L_c = 2.2$ m.

According to the calculation method shown in Fig. 4, the critical shaft length under different fire source heat release rates and different shaft height conditions is counted, as shown in Table 2.

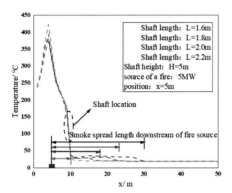

Fig. 4. The tunnel ceiling's smoke temperature distribution under different shaft lengths.

Table 2. The critical shaft length at the fire source location X = 5 m.

Shaft height/m	HRR/MW		
	2.5	5	7.5
0	–	3	–
5	2.2	2.2	2.2
10	–	1.6	–
15	–	1.4	–
20	–	1.2	–

4   Analysis of Simulation Results

4.1   Influence of Fire Source Location

The distribution diagram of flue gas temperature at one end of the blocked tunnel roof at four different fire source locations is shown in Fig. 6, in which the shaft height $H_s = 10$m, the shaft length $L = 1.2$ m, and the heat release rate HRR of the fire source is 5 MW. It can be seen from Fig. 6 that fire source position 1 is located between the blocking end and the shaft, and the maximum temperature of flue gas in the tunnel ceiling blocked at one end is significantly higher than that in the other three fire source positions. The conclusion shows that when the fire source is close to the plugging end and the fire source is located between the plugging end and the shaft, the flue gas generated by the fire source will continue to gather at the plugging end, resulting in a significantly higher maximum temperature. When the fire source is located between the shaft and the open end, the fire source will deflect violently to the plugging end. The continuous consumption of air at the plugging end and the continuous discharge of flue gas from the shaft, while the continuous input of air at the open end leads to the low side pressure at the plugging end of the fire source, resulting in the imbalance of pressure difference at both ends of the fire source and severe deflection. When the fire source is located between the plugging end and the shaft, in order to achieve the effect of full exhaust of shaft flue gas at the critical shaft length, the shaft needs to discharge all flue gas generated by the combustion of the fire source. When the fire source is located between the shaft and the open end, the value of the critical shaft length only needs to discharge about half of the flue gas generated by the combustion of the fire source, then $l_{x_f=5} \geq l_{Otherfiresourcelocations}$. It can be seen from Fig. 5 that the shaft length $L = 1.2$ m, the temperature of the flue gas generated by the combustion of the three fire sources at fire source locations 2, 3 and 4 drops sharply below 25 °C after crossing the shaft, while the flue gas generated by the combustion of fire source location 1 continues to spread to x = 15 m after crossing the shaft, which also proves that the critical shaft length of fire source locations 2, 3 and 4 is less than or equal to 1.2 m. The critical shaft length of fire source position 1 is the largest. Among the four fire source positions, the next section will study the change law of critical shaft length when the heat release rate of fire source changes for fire source position 1 with x = 5m.

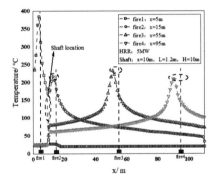

Fig. 5. The tunnel ceiling's smoke temperature distribution under different fire locations

4.2 Effect of Fire Source on Heat Release Rate

The flue gas temperature distribution of one end blocked tunnel ceiling under different fire source heat release rates is shown in Fig. 6. When the parameters of the shaft are the same, there is a large difference in the temperature of flue gas produced by fire sources with different fire source heat release rates before crossing the shaft. The greater the fire source heat release rate, the higher the flue gas temperature of the tunnel ceiling; However, after the flue gas passes through the shaft, the flue gas temperature of the heat release rate of the three fire sources decreases sharply, but the flue gas propagation length of the three fire sources is almost the same. Comparing the three shaft lengths, it is found that with the increase of shaft length, the smoke spread length of the heat release rate of the three fire sources decreases to the same extent, and when the shaft length L = 2.2M, the smoke spread length of the three fire sources decreases to 6 m, which is approximately equal to the distance between the fire source and the shaft. Therefore, it can be considered that under the conditions of FDS simulation in this section, the critical shaft length of the three fire source heat release rates is equal, and the critical shaft length is almost independent of the fire source heat release rate, which is the same as the previous research results [12]. This shows that Heskestad and Yuan's theory is also applicable to blocking the tunnel at one end. Since the fire plume and the smoke exhaust mass flow of the shaft are directly proportional to the 1/3 power of the fire source heat release rate, the value of the critical shaft length has nothing to do with the fire source heat release rate.

4.3   Theoretical Model of Critical Shaft Length

According to the above analysis, the fire source location at $X_f = 5$ m. X_f is a relatively unfavorable situation in the study, and its critical shaft length is the largest, which can meet the conditions for the complete discharge of flue gas from other fire source locations to the shaft along the shaft. The following will mainly analyze and summarize the critical shaft length model at $X_f = 5$ m fire source location. In case of fire at the fire source, the flue gas generated by the fire source will continue to spread to both sides of the tunnel after reaching the tunnel ceiling, and part of the flue gas will be discharged along the shaft ($L < L_c$) and take away the heat in the tunnel. The longer the shaft length, the more flue gas will be discharged along the shaft. At this time, the less flue gas will spread to the tunnel area behind the shaft, the lower the tunnel ceiling temperature, and the shorter the flue gas spread distance, which will provide a safer evacuation environment for the trapped people until the shaft length increases to completely discharge the flue gas generated by fire source combustion (Critical shaft length $L = L_c$). If the length of the shaft is $L > L_c$, there will be a section of the shaft that is completely smoke-free, which will not only increase the investment cost, but also cause a serious suction through phenomenon in the shaft. Therefore, we should try our best to avoid the situation of $L > L_c$.

According to heskestad's theory [13], the plume mass flow (M0) generated by fire source combustion can be expressed by Eq. (2);

$$m_0 \propto \rho_0 g^{1/2} Q^{*1/3} H^{5/2} \tag{2}$$

Where: Q is the heat release rate of fire source, HRR, kW; Q^* is the heat release rate of dimensionless fire source; H is the height of the tunnel, m.

According to Yuan's theory [7], the mass flow at the boundary of the control body (m_1), the mass flow of shaft smoke exhaust (m_s), and the temperature rise of flue gas $\Delta T_1/T_0$ meet the following laws:

$$m_1 = m_{ref}^* \rho_0 g^{1/2} Q^{*1/3} H^{5/2} \tag{3}$$

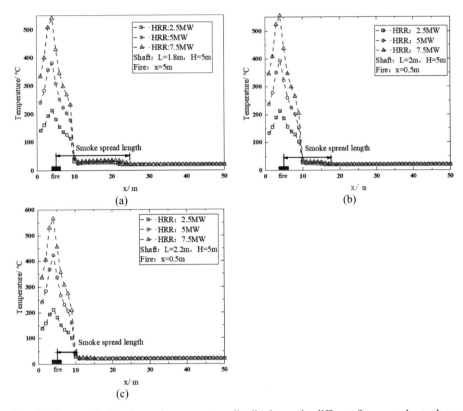

Fig. 6. The tunnel ceiling's smoke temperature distribution under different fire source heat release rates (a) L = 1.8 m; (b) L = 2.0 m; (c) L = 2.2 m.

$$m_s = \left(2m_1 l^2 hw^2 \rho^2 g \varepsilon^{-1} \frac{\Delta T_1}{T_0}\right)^{1/3} \tag{4}$$

$$\Delta T_1 / T_0 = \Delta T_{ref}^* Q^{*2/3} exp\left[-k\left(s - X_{ref}\right)\right] \tag{5}$$

Where: l is the shaft length, h is the shaft height, w is the shaft width, ε is a fixed constant, ref is the reference point position of the fire source section, X_{ref} is the reference point position coordinate, s is the coordinate of the shaft center, and k is the ceiling temperature attenuation coefficient of the fire source section.

For the tunnel blocked at one end, when the flue gas is completely discharged along the shaft, the relationship between m_0, m_1 and m_s can be expressed as the following formula (6):

$$m_0 = m_1 = m_s \tag{6}$$

Combining the above formula, formula (7) can be obtained:

$$h \propto \beta l^{-2} \omega^{-2} H^5 e^{ks} \tag{7}$$

Where: $\beta = \frac{2}{2} o (m^*_{ref})^2 \Delta T^{*-1}_{ref} e^{-kxref}$ and m^*_{ref} (dimensionless flue gas mass flow at the reference position) and $(m^*_{ref})^2 \Delta T^*_{ref}$ (dimensionless temperature rise at the reference position) are constants.

According to the simulation results, the value of k is approximately equal to 0.372, and then fit the results according to formula (7). The results are shown in Fig. 7. The experimental results show that for fire source position 1, the value of $h/\omega^{-2}H^5 e^{ks}$ is approximately linear with the value of l^{-2}, and the following expression is obtained:

$$\frac{h}{\omega^{-2}H^5 e^{ks}} = 0.1678l^{-2} - 0.0144, 0.1m^{-2} < l^{-2} < 0.7m^{-2} \tag{8}$$

Equation 7 can also be rewritten as:

$$h = 0.1678l^{-2}\omega^{-2}H^5 e^{ks} - 0.0144\omega^{-2}H^5 e^{ks}, 1m < l < 3m \tag{9}$$

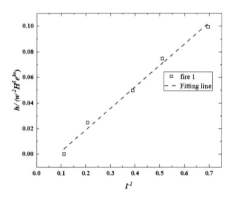

Fig. 7. Fitting graph of the experimental results

5   Conclusion

By comparing the small-scale experiment and FDS simulation, the reliability of FDS for fire smoke simulation of one end blocked tunnel is verified. Taking the planned one end blocked tunnel train inspection depot project as an example, a full-scale FDS numerical simulation calculation model is established. After grid independence verification, the relationship between critical shaft length and fire source location, fire source heat release rate and shaft height is analyzed. According to the numerical simulation results, the following results are obtained:

- Setting appropriate shaft size can effectively discharge all flue gas inside the tunnel and ensure the safety of tunnel structure and personnel evacuation.
- When the fire source is located between the plugging end and the shaft, the natural ventilation and smoke exhaust volume of the shaft is significantly greater than that

of other fire sources, and the critical shaft length is greater. At the same time, the critical shaft length does not change with the change of fire source heat release rate, but decreases with the increase of shaft height.

- When the fire source is located between the blocking end and the shaft, a prediction model of the critical shaft length of the tunnel blocked at one end is proposed.

References

1. Yoon, C.H., Kim, M.S., Kim, J.: The evaluation of natural ventilation pressure in Korean long road tunnels with vertical shafts. Tunn. Undergr. Space Technol. **21**(3), 472 (2006)
2. Tong, Y., Shi, M., Gong, Y., et al.: Full-scale experimental study on smoke flow in natural ventilation road tunnel fires with shafts. Tunn. Undergr. Space Technol. **24**(6), 627–633 (2009)
3. Han, J.Y., Ji, J., Wang, P.Y.: Experimental investigation on influence of shaft cross-sectional area on natural ventilation in urban road tunnel fires. Fire Safety Sci. **22**(1), 36–43 (2013)
4. Yu, L., Wei, Z.: Experimental study of the influence of natural ventilation by shaft on the maximum ceiling temperature of buoyancy plume in tunnel fires. Tunn. Undergr. Space Technol. **108**, 103715 (2001)
5. Yao, Y., Li, Y.Z., Ingason, H., Cheng, X.: Numerical study on overall smoke control using naturally ventilated shafts during fires in a road tunnel. Int. J. Therm. Sci **140**, 491–504 (2009)
6. He, K., Cheng, X.D., Zhang, S.G., Yang, H., Yao, Y.Z., Peng, M.: Critical roof opening longitudinal length for complete smoke exhaustion in subway tunnel fires. Int. J. Therm. Sci. **133**, 55–61 (2018)
7. Yuan, Z., Lei, B., Kashef, A.: Experimental and theoretical study for tunnel fires with natural ventilation. Fire Technol. **51**(3), 691–706 (2013)
8. Xie, B., Han, Y., Huang, H., et al.: Numerical study of natural ventilation in Urban shallow tunnels: Impact of shaft cross section. Sustain. Cities Soc. **42**, 521–537 (2018)
9. Chen, T.: Study on natural ventilation characteristics and critical shaft length of subway tunnel fire. Southwest Jiaotong University (2019)
10. Weng, M.C., Yu, L.X., Liu, F.: Smoke backlayering length and critical wind velocity in a metro tunnel. J. South China Univ. Technol. Nat. Sci. Ed. **42**(6), 121–128 (2014)
11. Weng, M.C.: Study on fire smoke flow and smoke control in subway tunnel. Chongqing University (2014)
12. Zhao, P., Chen, T., Yuan, Z., et al.: Critical shaft height for complete smoke exhaustion during fire at the worst longitudinal fire location in tunnels with natural ventilation. Fire Saf. J. **116**, 103207 (2020)
13. Heskestad, G.: Fire plumes, flame height and air entrainment. In: SFPE Handbook of Fire Protection Engineering. National Fire Protection Association, Quincy (2002)

Research on the Influence of Transverse Limit of Support on Construction Monitoring of Special-Shaped Bridge

Yi Chao and Dong Jun[✉]

Beijing Higher Institution Engineering Research Center of Structural Engineering and New Materials, Beijing University of Civil Engineering and Architecture, Beijing 102616, China
jdongcg@bucea.edu.cn

Abstract. In order to understand whether the alignment, internal force and cable force of the special-shaped cable-stayed bridge can still meet the design and code requirements due to the influence of the transverse displacement of the support during the construction process, the transverse displacement of the girder arch system is restricted before the removal of the support. The difference between the measured value and the designed value at each stage of the boom is within ±5%, which meets the design and specification requirements. In this paper, the pedestrian bridge in Anyi County, Nanchang City, Jiangxi Province is taken as an example. MADIS/Civil software is used for finite element simulation analysis in the construction stage, and the linear shape, internal force and cable force of the superstructure are monitored. The results show that the measured deformation values of arch ribs and beams meet the design and specification requirements during the whole construction process. The difference between the measured value and the designed value at each stage of the boom is within ±5%, which meets the design and specification requirements.

Keywords: Special shaped cable stayed arch bridge · Beam arch composite system bridge · Finite element simulation · Suspender tension · Construction monitoring

1 Introduction

The special-shaped inclined tower cable-stayed arch bridge is a composite bridge with beam-arch system. The structure of the bridge is highly statically indeterminate, and the stress is extremely complex [1]. The most significant characteristic is that the internal force distribution of the structure can be changed by adjusting the hanger cable force, so as to optimize the overall structure stress. In the process of construction, the system transformation of the bridge is complicated due to the phased removal of the arch rib and the beam support, and the change of the linear shape and internal force is uncertain, which brings hidden trouble to the construction [2]. At present, there are few researches on construction monitoring of special-shaped cable-stayed Bridges [3]. Taking the pedestrian landscape bridge in Anyi County, Nanchang City, Jiangxi Province as an example, this paper introduces the construction monitoring of such Bridges.

G. Feng (Ed.): ICCE 2021, LNCE 213, pp. 142–150, 2022.
https://doi.org/10.1007/978-981-19-1260-3_13

2 Project Summary

The length of the bridge is 194.449 m, the full width of the deck is 6.0 m, the clear width is 5.4 m, and the span is arranged according to 15 + 2 × 75 + 15. The main arch ring is pentagonal concrete filled steel tube structure, the main span calculated span 65.0 m, lateral rotation front vector height 24.377, vector span ratio 1/2.666, the arch axis is circular curve. The transverse bridge of the arch rib itself tilts outward 22.088°, and the stability of the arch rib itself is guaranteed by the tension of the inclined hanger rod. Overlooking the main beam is s-shaped, steel box girder is used in design, steel box girder material is Q345qC, steel box girder is 1.5 m high, 6.0 m wide, 1.5 m wide cantilever is set on both sides. The thickness of the roof and bottom plate is 20 mm, the thickness of the web is 16 mm, and the diaphragm is set every 2 m or so. A 1 m wide beam is arranged at the cable tension position of the steel box girder. There are 16 suspender rods in the whole bridge. The upper end of the suspender is projected to be 7 m along the bridge design axis, and the lower end is projected to be 6 m along the bridge design axis. The steel box girder of the bridge deck is rigidly connected by diagonal braces and arch feet. The beam and arch combined stress system of two arch rings and main beam configuration special-shaped double inclined tower cable-stayed arch bridge is shown in the Fig. 1.

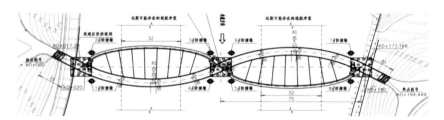

Fig. 1. Layout of bridge structure

3 Finite Element Analysis Model

The large-scale finite element analysis software MADIS/Civil2019 was used to establish the analysis model of the construction stage of the bridge according to the actual construction process [4], so as to simulate and analyze the whole construction process. In the model, arch ribs and steel box beams are simulated by beam element. The suspender is simulated by tension truss element, and the arch rib support and main beam support are simulated by compression only elastic connection. The Finite element analysis model is shown in the Fig. 2.

4 Construction Monitoring Content

The purpose of construction monitoring is to monitor and control the main beam installation, arch rib installation, hanger tension, bracket removal and other stages in the

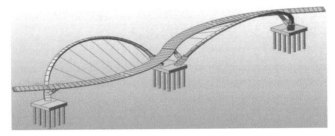

Fig. 2. Finite element analysis model

construction process, so as to ensure that the construction process and its structure are in absolute safety control. According to the actual state of the structure, the linear and internal force control data of each construction stage is given, which can be used to guide and control the construction, prevent the accumulation of errors in the construction, and ensure that the linear and internal force after the bridge meets the design requirements [5].

4.1   The Linear Monitoring

The alignment monitoring of the main beam adopts level elevation monitoring. The alignment monitoring point of the main beam selects the truncation point of the beam and the contact point of the boom and the main beam. The Layout of main beam alignment monitoring point is shown in the Fig. 3 and Fig. 4.

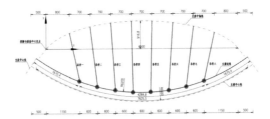

Fig. 3. Layout of main beam alignment monitoring point I

Fig. 4. Layout of main beam alignment monitoring point II

The arch rib alignment monitoring adopts total station displacement monitoring, and the arch rib alignment monitoring point is selected as the cutting point of the arch rib and the contact point between the boom and the arch rib. The Layout of arch rib linear monitoring point is shown in the Fig. 5 and Fig. 6.

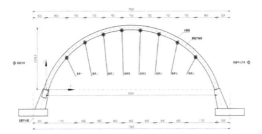

Fig. 5. Layout of arch rib linear monitoring point I

Fig. 6. Layout of arch rib linear monitoring point II

4.2 Stress Monitoring

For the stress monitoring of the superstructure, the pre-embedded and pre-attached strain gauge method is adopted. The stress and strain monitoring points are arranged for the key parts and the stress concentration places, which are the adjacent points of tension on the arch and the adjacent points of tension on the beam. The stress monitoring points are shown in the Fig. 7 and Fig. 8.

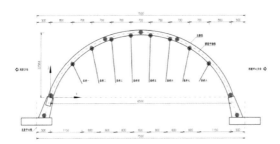

Fig. 7. Layout of stress monitoring points of arch ribs

4.3 Cable Force Monitoring

The vibration signal of the cable under the vibration excitation is picked up by the precision vibration collector, and then the natural vibration frequency of the cable is determined according to the spectrum diagram after filtering and amplification and spectrum

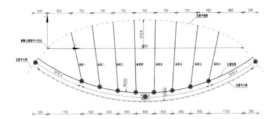

Fig. 8. Layout of stress monitoring points of main beams

analysis, and then the cable force is determined according to the relationship between the natural vibration frequency and the cable force.

5   Construction Monitoring Results

5.1   Linear Monitoring Results

Before the removal of the steel box girder support, the measured lateral displacement value is slightly less than the theoretical displacement value. The transverse displacement and vertical displacement of the arch rib are slightly larger than the theoretical value. After the bracket is removed, the linear state of the bridge is relatively consistent with the theoretical value, and the difference is within a reasonable range. The displacement of main girder is shown in the Fig. 9 and Fig. 10. And the displacement of arch rib is shown in the Fig. 11 and Fig. 12.

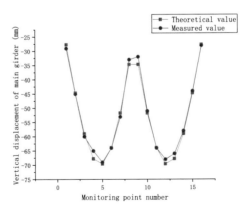

Fig. 9. Vertical displacement of main girder

5.2   Stress Monitoring Results

Before the removal of steel box girder support, the measured stress value of arch rib is slightly greater than the theoretical stress value, considering that the spot welding

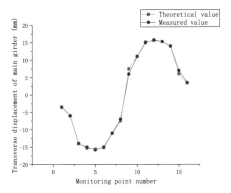

Fig. 10. Transverse displacement of main girder

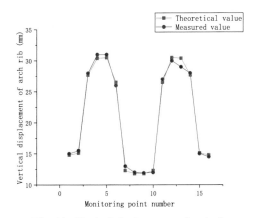

Fig. 11. Vertical displacement of arch rib

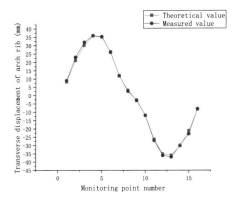

Fig. 12. Transverse displacement of arch rib

anchorage between the bottom of the box girder and the support has the effect of transverse limit, and the theoretical value does not consider the transverse limit. After the removal of the support, the transverse resistance of the support is eliminated, and the arch rib is stretched. The stress state of the bridge is relatively consistent with the theoretical value, and the difference is within a reasonable range. The stress of bridge girder is shown in the Fig. 13 and the stress of arch rib of bridge is shown in the Fig. 14.

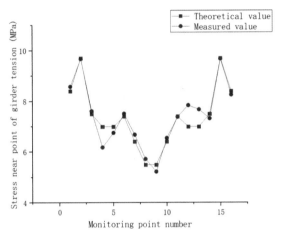

Fig. 13. Stress of bridge girder

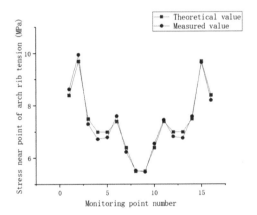

Fig. 14. Stress of arch rib of bridge

5.3   Cable Force Monitoring Results

The suspender is divided into the initial and final tension. The initial tension is tensioned when the arch rib support is removed and the main beam support is not removed. The final tensioning is completed in the load construction of the main girder deck and is tensioned before the main girder is removed [6]. Finally, the main beam support was removed, and the structure formed a combination system of beam and arch. The cable force measured at this time is the final cable force of the bridge [7]. The comparison between finished cable force and design cable force is shown in the Fig. 15.

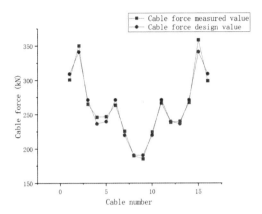

Fig. 15. Comparison between finished cable force and design cable force

6 Conclusion

To Anyi landscape bridge in Nanchang, Jiangxi Province as an example, this paper discusses the special cable stayed arch bridge construction monitoring, the following conclusions:

(1) the demolition of bracket horizontal displacement value of the measured before the overall displacement value is less than the theoretical value, considering the reason is spot welding at the bottom of the box girder with transverse displacement between stent anchoring and restrictions, The theoretical value does not consider the lateral displacement, and the horizontal displacement is within the allowable deviation range [8]. The measured vertical displacement value is slightly higher than the theoretical value, which is considered to be due to the slight lift caused by the vertical component of the boom, and the deviation is within the allowable range. The transverse displacement and vertical displacement of the arch ribs are slightly larger than the theoretical value because the transverse displacement of the box girder caused by the welding between the bottom of the box girder and the bracket is insufficient [9]. The linear state of the bridge after removing the support is consistent with the theoretical value, and the difference is within a reasonable range.

(2) In the construction process of the bridge, the pressure arch ribs and beam tension of adjacent points are close to the theoretical value as a whole, and the key points and beams on the average section of the arch ribs are close to the theoretical value [10].

(3) The cable stress monitoring, after four times of fine adjustment, meets the design value, the error meets the monitoring requirements, the cable stress distribution is reasonable. The cable force is consistent with the design value, and the difference is controlled within $\pm 5\%$

References

1. Xiao, R.C.: Bridge Structure System. People's Communications Press, Beijing (2013)
2. Du, Y.: Design and construction of concrete filled steel tube tied arch bridge. Bridge Constr. **1**, 36–38 (2005)
3. Jin, C.L.: Prestressed Concrete Beam Arch Composite Bridge–Design Research and Practice. People's Communications Press, Beijing (2001)
4. Pan, G.M.: Realization of influence matrix in cable force adjustment in Midis/civil. Sci. Technol. Eng. **37**(8), 6–11 (2003)
5. Han, S.C.: Research on construction monitoring of half through CFST cantilever arch bridge. Chin. Foreign Highw. **30**(3), 172–175 (2010)
6. Chen, G., Ren, W.X.: Identification of cable fundamental frequency of cable stayed bridge based on ambient vibration. Earthq. Eng. Eng. Vib. **23**(3), 100–106 (2003)
7. Ren, W.X., Chen, G.: A practical formula for calculating cable tension from fundamental frequency. J. Civ. Eng. **38**(11), 26–31 (2005)
8. Yao, W. T.: Construction control and analysis of cable-stayed bridge. Master Thesis of Tianjin University (2004)
9. Xiang, Z.F.: Bridge Construction Control Technology and Analysis. People's Communications Press, Beijing (2000)
10. Wang, S.X.: Linear control and spatial simulation analysis of curved continuous rigid frame construction. Master Thesis of Chang'an University (2008)

Probabilistic Analysis of a Braced Excavation Considering Soil Spatial Variability

Shirui Ding[1], Haoqing Yang[2(✉)], and Jiabao Xu[2]

[1] School of Civil Engineering and Architecture, Wuhan University of Technology, 122 Luoshi Road, Wuhan 430070, Hubei, China
[2] School of Naval Architecture, Ocean and Civil Engineering, Shanghai Jiao Tong University, 800 Dongchuan Road, Shanghai 200240, China
yanghaoqing@sjtu.edu.cn

Abstract. Deep braced excavations are generally known to be associated with risks from various sources. The inherent uncertainty of soil strength properties is one of the primary factors that influence the deformation of the retaining wall and the ground settlement. In this study, the numerical model of a braced excavation is firstly established by an elastic-plastic model with Drucker-Prager failure criterion in COMSOL Multiphysics. Random field theory is used to simulate the spatial variability of Young's modulus. The uncertainty of braced excavation on ground settlement and deflection of retaining wall by stages are studied by Monte Carlo simulation based on 500 random fields. The struts can lessen the uncertainty of wall deflection during excavation but have a limited impact on settlement. The deterministic result may underestimate the settlement of braced excavation. The uncertainty of wall deflection is significantly reduced after the first strut. The uncertainty of wall deflection above the depth of struts is well-controlled at the final stage of excavation.

Keywords: Braced excavation · Drucker-Prager · Random field · Uncertainty · Spatial variability

1 Introduction

Deep braced excavations are generally known to be associated with risks from various sources such as retaining systems, subsurface soil conditions, and construction techniques. The inherent uncertainty of the strength properties of soil is one of the primary factors that influence the deformation of the retaining wall and the ground settlement. It is of great importance to study the uncertainty of braced excavations in face with the heterogeneous soil strength properties.

Many scholars used the random field to represent the soil spatial variability and study the uncertainty of braced excavations by probabilistic methods. Luo et al. [1] presented a simplified approach to consider the effect of spatial variability in a two-dimensional random field for reliability analysis of basal heave in a braced excavation in clay. Wu et al. [2]

© The Author(s) 2022
G. Feng (Ed.): ICCE 2021, LNCE 213, pp. 151–159, 2022.
https://doi.org/10.1007/978-981-19-1260-3_14

proposed a novel method of updating the probability distribution of the maximum wall displacement based on the measurements at earlier stages. Qi and Zhou [3] presented an efficient Bayesian back-analysis procedure for braced excavations using wall deflection data at multiple points. Lo and Leung [4] introduced an approach using field measurements to update the parameters characterizing spatial variability of soil properties for subsequent construction stages. Gholampour and Johari [5] presented a practical approach for reliability analysis of braced excavation in spatially varied unsaturated soils. Most of the researchers used PLAXIS software with the Mohr-Coulomb failure criterion to build the models of braced excavations. They found that it is necessary to consider soil spatial variability during braced excavations. Seldom studies discussed the impact of struts on the uncertainty reduction of braced excavations.

In this study, the numerical model of a braced excavation with struts is firstly established by an elastic-plastic model with Drucker-Prager failure criterion in COMSOL Multiphysics. Random field theory is used to simulate the spatial variability of Young's modulus. The uncertainty of ground settlement and deflection of retaining wall controlled by struts are studied by Monte Carlo simulation based on 500 random fields.

2 Methods

2.1 Numerical Model of Braced Excavation with Struts

The numerical model of braced excavation is modeled by COMSOL Multiphysics. Drucker-Prager criterion is used to simulate the elastic-plastic behavior of excavation. Drucker-Prager criterion is the linear combination of the first invariant of stress tensor I_1 and the square root of the second invariant of the deviatoric stress tensor J_2:

$$F = \sqrt{J_2} + \alpha I_1 - k \tag{1}$$

where α and k are two material constants and can be calculated by the following equations:

$$\alpha = \frac{\tan \phi}{\sqrt{9 + 12 \tan^2 \phi}} \tag{2}$$

$$k = \frac{3c}{\sqrt{9 + 12 \tan^2 \phi}} \tag{3}$$

where c and ϕ are the cohesion and the angle of internal friction, respectively. Note that the hardening or softening of soil behavior is not considered, so the dilatation angle is not included.

The example of braced excavation is taken from Schweiger [6]. As shown in Fig. 1, the model is 90 m in width and 60 m in height. The depth of the excavation is 26 m with a retaining wall. Three struts are applied to minimize wall movements. The process of excavation is divided into four stages and the depth of each excavation is 4 m. Three struts placed at 4.8 m, 9.3 m, and 14.35 m are installed during excavation. The soil type

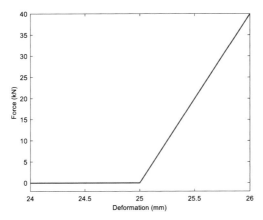

Fig. 1. Force of struts with deformation.

is sand. The Poisson's ratio, cohesion, angle of internal friction, and density of soil are 0.3, 0 Pa, 35°, and 1900 kg/m³, respectively. The stiffness of struts is chosen as 10^4 kN/m. The Young's modulus, Poisson's ratio, and density of retaining wall are 30 GPa, 0.15, and 2400 kg/m³, respectively. These parameters are summarized in Table 1.

Table 1. Parameters of the numerical model and random field.

	Parameters	Definitions	Values
Soil	μ	Mean value of soil Young's modulus	20 MPa
	σ	Standard deviation soil Young's modulus	10 MPa
	l_x	Horizontal correlation length	9 m
	l_z	Vertical correlation length	6 m
	ν_s	Poisson's ratio of soil	0.3
	c	Cohesion	0 Pa
	φ	Angle of internal friction	35°
	ρ_s	Density of soil	1900 kg/m³
Struts	s	Stiffness	10^4 kN/m
Retaining wall	E_w	Young's modulus of retaining wall	30 GPa
	ν_w	Poisson's ratio of retaining wall	0.15
	ρ_w	Density of retaining wall	2400 kg/m³

To simulate the struts, a 1 m-line is defined in geometry for each strut. The struts are active for two conditions. First, the depth of the excavation exceeds the location of a strut. Second, the maximum horizontal deflection is reached. For the second condition, it can be achieved by defining a function as shown in Fig. 1. It means that when the deformation of the wall deflection exceeds 25 mm, the struts start to work. The force provided by the strut is linearly increased by the deformation and the gradient is the stiffness of the strut.

According to the geostatic equilibrium, the in-situ stress in the horizontal and vertical directions are 24 kPa and 35 kPa, respectively. To achieve the second method mentioned before, the in-situ stress is applied by external stress in the whole soil domain and corresponding boundary AB. The boundary load on the retaining wall of boundary AB is 24 kPa and it is 35 kPa for EB and AH. The bedrock is below FG and a fixed constraint is applied on FG. According to the symmetry of the structure, only the right half of the domain is built, and symmetry boundary condition is used on EF. Extrusion operators are applied on the boundaries of BD and CD ensuring that the normal displacement between the retaining wall and the soil stays in contact, with the tangential displacement being unconstrained at the same time. The geometry and boundary conditions of the numerical model are shown in Fig. 2.

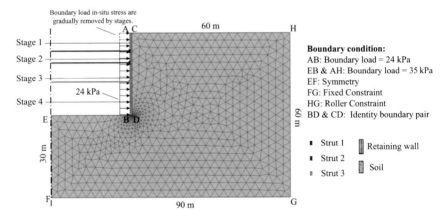

Fig. 2. Geometry and boundary conditions of the numerical model.

2.2 Random Field

Young's modulus E of soil is chosen to consider the spatial variability of strength properties. According to the previous studies [7, 8], E is generally the lognormal distributed. Therefore, $\ln E$ (Natural logarithm of E) is subject to normal distribution. The mean value μ_{ln} and the standard deviation σ_{ln} of $\ln E$ are calculated as:

$$\mu_{ln} = \ln \mu - \frac{1}{2}\sigma^2 \tag{4}$$

$$\sigma_{\ln}^2 = \ln(1 + \frac{\sigma^2}{\mu^2}) \tag{5}$$

where μ and σ are the mean value and standard deviation of E. It is reported that the coefficient of variation (COV) of E ranges from 2%–60%. Therefore, for sandy soil, μ and σ are assumed to be 20 MPa and 10 MPa, respectively (Table 1).

The exponential covariance function $C(\mathbf{x})$ [8–10] is adopted to generate random fields of $\ln E$:

$$C(\mathbf{x}) = \sigma_{\ln}^2 \exp\left\{-\left[\frac{(x_1 - x_2)^2}{l_x^2} + \frac{(z_1 - z_2)^2}{l_z^2}\right]^{1/2}\right\} \tag{6}$$

where l_x and l_z are the horizontal and vertical correlation lengths of $\ln E$, respectively; $\mathbf{x} = [(x_1, z_1), (x_2, z_2)]$ denotes the coordinates of the two points in the domain. The correlation length of $\ln E$ ranges from 1–100 m varies from site to site [11]. In this study, the horizontal and vertical correlation lengths are assumed to be 9 m and 6 m, respectively.

3 Results and Discussions

3.1 Deterministic Results

The deterministic results of braced excavation are shown in Fig. 3. The total displacement at the final stage is shown in Fig. 3(a). The deformation is magnified by 238 times for better visualization. The final maximum displacement is around 25 mm and in the interface between retaining wall and soil. Figure 3(b) is the plastic region of the soil at the final stage. The plastic region is adjacent to the retaining wall indicating the high-risk zone of braced excavation. Figure 3(c) shows the wall deflection in each stage. Note that the struts will be installed if the wall deflection exceeds 25 mm. It can be inferred that the 4.8 m-strut is activated at the second stage and the other two struts are valid at the fourth stage. Figure 3(d) shows surface settlement over stages. The struts can also control the settlement of braced excavation. At the fourth stage, the settlement is reduced significantly because of the installation of two struts.

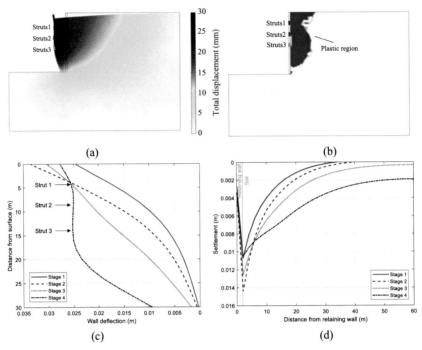

Fig. 3. Deterministic results: (a) final total displacement; (b) final plastic region; (c) wall deflection and (d) settlement.

3.2 Uncertainty of Braced Excavation by Stages

Figure 4 displays the uncertainty of wall deflection during excavation. The blue shadows are the confidence intervals of each percentage. At the first stage, the uncertainty of wall deflection is noteworthy. It gradually decreases with depth. At the second stage, the uncertainty is considerably reduced because of the installation of the first strut. At the third stage, the uncertainty away from the strut is improved but it is still very small compared to the first stage. At the final stage, another two struts are active, and the uncertainty of wall deflection is trivial. Only the lower part of the retaining wall suffers from uncertainty. To conclude, the uncertainty of wall deflection is significantly reduced after the first strut. The uncertainty of wall deflection above the depth of struts is well-controlled at the final stage of excavation.

The uncertainty of settlement during excavation is shown in Fig. 5. As the excavation goes on by stages, the uncertainty of surface settlement increases. It implies that the effectiveness of struts on the reduction of uncertainty for settlement is limited. In addition, the mean results of the settlement are larger than the deterministic results. It gradually closed to the deterministic results. The two results almost overlapped at the final stage. It illustrates that the deterministic results may underestimate the settlement of braced excavation. Controlling the settlement during excavation based on the deterministic result may have potential risks. It is necessary to consider spatial variability of soil strength properties for braced excavation.

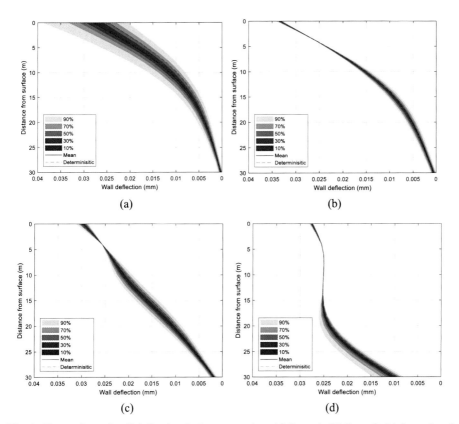

Fig. 4. Uncertainty of wall deflection during excavation: (a) Stage 1; (b) Stage 2; (c) Stage 3 and (d) Stage 4.

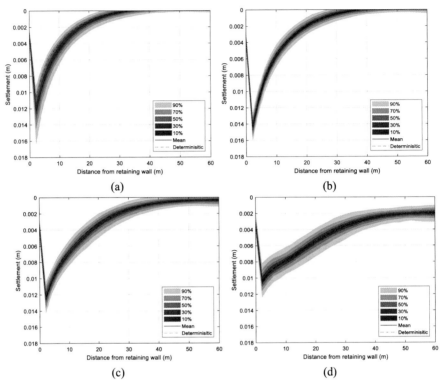

Fig. 5. Uncertainty of settlement during excavation: (a) Stage 1; (b) Stage 2; (c) Stage 3 and (d) Stage 4.

4   Conclusions

In this study, the numerical model of a braced excavation is firstly established by an elastic-plastic model with Drucker-Prager failure criterion in COMSOL Multiphysics. The staged excavation with struts is considered. Random field theory is used to simulate the spatial variability of Young's modulus. The uncertainty of braced excavation on ground settlement and deflection of retaining wall by stages are studied by Monte Carlo simulation based on 500 random fields. Major conclusions are summarized below:

(1) The uncertainty of wall deflection during excavation is significantly reduced after the first strut. The uncertainty of wall deflection above the depth of struts is well-controlled at the final stage of excavation.
(2) The deterministic result may underestimate the settlement of braced excavation. Controlling the settlement during excavation based on the deterministic result may have potential risks. It is necessary to consider spatial variability of soil strength properties for braced excavation.

References

1. Luo, Z., Atamturktur, S., Juang, C.H.: Bootstrapping for characterizing the effect of uncertainty in sample statistics for braced excavations. J. Geotech. Geoenviron. Eng. **139**(1), 13–23 (2013)
2. Wu, S.H., Ching, J., Ou, C.Y.: Probabilistic observational method for estimating wall displacements in excavations. Can. Geotech. J. **51**(10), 1111–1122 (2014)
3. Qi, X.H., Zhou, W.H.: An efficient probabilistic back-analysis method for braced excavations using wall deflection data at multiple points. Comput. Geotech. **85**, 186–198 (2017)
4. Lo, M.K., Leung, Y.F.: Bayesian updating of subsurface spatial variability for improved prediction of braced excavation response. Can. Geotech. J. **56**(8), 1169–1183 (2019)
5. Gholampour, A., Johari, A.: Reliability-based analysis of braced excavation in unsaturated soils considering conditional spatial variability. Comput. Geotech. **115**, 103163 (2019)
6. Schweiger, H.F.: Benchmarking in Geotechnics 1 Computational Geotechnics Group (CGG IR006), 25 (2002)
7. Griffiths, D.V., Fenton, G.A.: Probabilistic settlement analysis by stochastic and random finite-element methods. J. Geotech. Geoenviron. Eng. **135**(11), 1629–1637 (2009)
8. Yang, H.-Q., Zhang, L., Pan, Q., Phoon, K.-K., Shen, Z.: Bayesian estimation of spatially varying soil parameters with spatiotemporal monitoring data. Acta Geotech. **16**(1), 263–278 (2020)
9. Yang, H.Q., Zhang, L.L., Xue, J.F., Zhang, J., Li, X.: Unsaturated soil slope characterization with Karhunen-Loève and polynomial chaos via Bayesian approach. Eng. Comput. **35**(1), 337–350 (2019)
10. Yang, H.Q., Chen, X., Zhang, L.L., Zhang, J., Wei, X., Tang, C.: Conditions of hydraulic heterogeneity under which Bayesian estimation is more reliable. Water **12**(1), 160 (2020)
11. Zhang, L.L., Li, J.H., Li, X., Zhang, J., Zhu, H.: Rainfall-induced Soil Slope Failure: Stability Analysis and Probabilistic Assessment, vol. 280. CRC Press (2016)

Influence of Excavation of Horizontal Adit Tunnel on Main Tunnel and Safety Analysis

Liang Fu[✉]

CCCC Fourth Harbor Engnieering Co., Ltd., Guangzhou Window Headquarters Building, No. 368, Lijiao Road, Haizhu District, Guangzhou, China
52622840@qq.com

Abstract. For the influence of horizontal adit tunnel on the main tunnel, taking a tunnel as an example, MIDAS finite element software is used to analyzes the displacement and stress state of the primary support and secondary lining structure of the main tunnel before and after excavation, gives the local structural safety factor at the junction of the two, and analyzes the safety of the lining structure. The results show that: (1) After the main tunnel lining is removed, the lining has a certain displacement to the free face, the maximum is 55.3 mm, and the displacement changes little after the lining of the horizontal adit tunnel; (2) The safety factor calculation of the main tunnel lining at the junction, except the inverted arch area of primary support, all other parts meet the specification requirements.

Keywords: Horizontal adit tunnel · Main tunnel · Lining structure · Force analysis · Safety factor

1 Introduction

In order to ensure the operation safety of large tunnels and timely evacuation and rescue in case of accidents, it is generally necessary to set up vehicle cross passage and pedestrian cross passage in separate tunnels. The spacing of vehicle cross passage should be 750 m, not more than 1000 m; The setting spacing of pedestrian cross passage should be 250 m, not more than 500 m [1]. The cross traffic passage is excavated in the main tunnel of the existing tunnel, with large free surface at the junction and complex stress. If the construction process is improper and the monitoring is not in place, the lining of the main tunnel of the existing tunnel may be deformed and cracked, and the construction risk is high. At present, the research on construction mechanics of vehicle cross passage and main tunnel is mainly numerical simulation. Shi Yanwen et al. [2] used three-dimensional numerical simulation method to conduct elastic-plastic analysis of the main tunnel and vehicle cross passage; Liu Shanhong et al. [3] used finite element software to conduct numerical analysis on the tunnel crossing section of juyun mountain in Fuling, Chongqing and concluded that in-situ stress and structural stress concentration are two key factors leading to local cracking and affecting the stability of the intersection; Sun Zhijie [4] and others combined the engineering field test with finite

G. Feng (Ed.): ICCE 2021, LNCE 213, pp. 160–170, 2022.
https://doi.org/10.1007/978-981-19-1260-3_15

element simulation to study the deformation law of the main tunnel during the construction of vehicle cross passage; Luo Yanbin [5] et al. carried out three-dimensional finite element numerical simulation of the tunnel of Tianheng mountain in Harbin and studied the influence of cross passage construction on the main tunnel structure of the tunnel; Liu Xiaoliang [6] conducted three-dimensional numerical simulation analysis of tunnel in Danan mountain and studied the influence of vehicle cross passage excavation on the surrounding rock displacement at the intersection of the main tunnel of long-span tunnel.The above scholars have made some progress in the research on the mechanical characteristics and deformation mechanism of the intersection between the main tunnel and the cross passage, but they are limited to the role between the normal main tunnel and a single vehicle cross passage, and rarely mention the role between the main tunnel and the vehicle cross passage when the main tunnel has been weakened. Based on an example of a highway tunnel, using the method of three-dimensional numerical simulation and theoretical calculation, this paper analyzes the deformation and stress law of the lining of the main tunnel when the cross passage is built nearby and the cross passage is excavated, and checks the safety of the lining structure at the interface between the main tunnel and the cross passage after the cross passage is completed, the results can provide reference for similar projects.

2 Project Overview

2.1 Geological Conditions

The tunnel site belongs to the tectonic denudation low mountain landform area, with large topographic fluctuation and change, the relative height difference is about 335 m, the inlet side slope is about 25–35°, and the outlet side slope is about 15–20°. The slope is covered with Deluvial cohesive soil strongly weathered siltstone. The drilling reveals that the upper part of the tunnel site is quaternary eluvial diluvium; The underlying bedrock is Permian Wenbishan formation siltstone and weathered layer, occurrence 118°∠20°; Siltstone and its weathered layer of Carboniferous woodland formation, occurrence 30°∠51°; Quartz sandstone and weathered layer of woodland formation are locally developed. The surrounding rock of the tunnel is a weak permeable layer, and various indexes measured by the groundwater and surface water in the site are slightly corrosive to the concrete and the reinforcement in the concrete.

2.2 Tunnel Overview

The total length of the tunnel is 1022m, and the left and right tunnels are arranged separately. The starting and ending mileage of the tunnel is ZK19+179-ZK20+191 for the left tunnel and YK19+183-YK20+215 for the right tunnel. The left and right tunnels are 1012 m and 1032 m long respectively. The main tunnel of the tunnel adopts Z7-3 lining structure, and the support parameters are: shotcrete C25, thickness 30 cm; 20B I-steel support, spacing 70 cm; C30 waterproof concrete for secondary lining, 50 cm thick; Hollow grouting anchor rod L = 4 m. The entrance and exit of the tunnel are located within the plane curve range. The entrance and exit curve radii of the left line are 1200 m and 1100 m respectively, and the entrance and exit curve radii of the right line are 1200 m and 1000 m respectively. Design elevation of inlet: 349.840 m for left tunnel and 349.796 m for right tunnel; Design elevation of exit: 346.80 m for left tunnel and 337.035 m for right tunnel.

The stake number of the left tunnel of the vehicle cross passage is ZK19+440, 261 m away from the tunnel entrance; The chainage of the right tunnel is YK19+449, 266 m away from the entrance. Z8-2 lining structure is adopted for the vehicle cross passage, and the support parameters are: shotcrete C25, thickness 20cm; Grating steel support, spacing d = 1 m; C30 waterproof concrete for secondary lining, 35 cm thick; Hollow grouting anchor rod L = 2.5 m. The main tunnel and pedestrian cross passage of the tunnel have been completed, and the pedestrian cross passage has been backfilled. The outer contour of the vehicular cross passage is only 3 m away from the outer contour of the pedestrian cross passage. The construction of vehicle cross passage needs to destroy the lining of the original main tunnel, and the surrounding rock will inevitably be disturbed during the construction process. The positional relationship between the main tunnel and the cross passage of the tunnel is shown in Fig. 1.

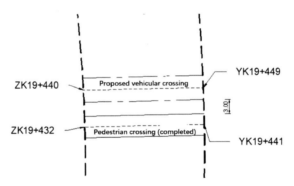

Fig. 1. Position relationship between main tunnel and cross passage of tunnel.

3   Numerical Simulation

3.1   Model Establishment

The three-dimensional model is established by Midas finite element software. The model takes the pedestrian cross passage as the core area, X direction as the cross passage direction and Y direction as the main tunnel direction. Generally, when the boundary range of 3 times the hole distance is taken, the analysis results meet the accuracy requirements [7]. The span of the main tunnel of the project is 14.9 m, the outer contour interval of the main tunnel is 15 m, and the span of the vehicle cross passage is 7.5 m. The model is taken as 110 m in X direction, 70 m in Y direction and 110 m in Z direction. The three-dimensional finite element mesh is shown in Fig. 2.

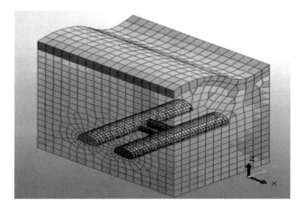

Fig. 2. 3D mesh generation.

3.2 Model Calculation Parameters

The solid element is selected for the surrounding rock, and the Drucker Prager (DP criterion) is adopted for the constitutive model. The plate element is used for the initial support and secondary lining of the main tunnel and vehicle cross passage of the tunnel, and the material calculation parameters are shown in Table 1.

Table 1. Main physical and mechanical parameters.

Structure name	Severe /kN/m^3	Poisson's ratio/υ	Deformation modulus/GPa	Cohesion/MPa	Internal friction angle /°
Deluvial silty clay with gravel and structural fracture zone	18.5	0.43	0.010	0.03	202
Fragmentary strongly weathered siltstone	20	0.38	0.1	0.1	26
Primary support	22	0.2	23	-	–
Secondary lining	25	0.2	30	-	–

4   Analysis of Calculation Results

4.1   Analysis of Surrounding Rock Deformation

When the left tunnel lining is removed, the displacement of the supporting structure of the main tunnel is shown in Fig. 3. Due to the poor self stability of class V surrounding rock, when the lining at the junction is removed, the surrounding rock deforms into the tunnel, with the maximum value of 55.3 mm. The horizontal deformation of the arch waist lining at the junction of the main tunnel and the vehicle cross passage is 12.1 mm, and the horizontal deformation of the lining between the vehicle cross passage and the pedestrian cross passage is 3.7 mm. After the excavation of rock and soil mass, the surrounding rock pressure borne by the lining is released, the vault at the junction of tunnel and vehicle cross passage sinks by 23.3 mm, and the arch bottom bulges by 19.4 mm.

When designed according to the bearing capacity, the deformation of the initial support of the composite lining shall not exceed the design reserved deformation [8]. The deformation of the cross passage is much less than the reserved deformation of 10 cm. However, due to water gushing in the tunnel and poor surrounding rock, the left tunnel shall be reinforced with steel support in advance to reduce the exposure time [9–12].

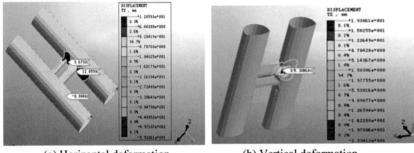

(a) Horizontal deformation.   (b) Vertical deformation.

Fig. 3. Deformation of support structure of main tunnel of tunnel after lining removal at junction.

After the construction of the vehicle cross passage is completed, the displacement of the supporting structure of the main tunnel is shown in Fig. 4. During the excavation of the vehicle cross passage, the excavation and support have little impact on the surrounding rock and left and right tunnels. The maximum displacement in the horizontal direction of the lining is 13.4 mm, which is only 1.3 mm higher than the previous construction step. The arch bottom bulges upward by 22.6 mm, an increase of only 3.2 mm compared with the previous construction step. As the excavation unloading causes the surrounding rock on both sides to squeeze into the vehicle cross passage, resulting in the upward uplift of the arch bottom, the vehicle cross passage lining shall be constructed in time to reduce the deformation of surrounding rock.

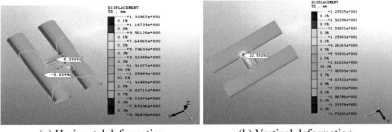

(a) Horizontal deformation.   (b) Vertical deformation.

Fig. 4. Deformation of support structure of main tunnel after completion of vehicle cross passage.

4.2 Stress Analysis of Lining

The axial force of tunnel lining after the removal of left tunnel lining is shown in Fig. 5. When the lining of the main tunnel is removed and the cross passage is excavated, the original arching effect around the main tunnel is destroyed, and the stress concentration occurs at the same time. The maximum axial force of the lining along the X-axis direction is −15244 kN, which is located at the arch waist at the junction of the vehicle cross passage and the tunnel. The maximum axial force along the Y-direction is also located at the arch waist at the excavation edge, which is −10242 kN, which is less than the axial force in the X-axis direction. (Positive values in the figure indicate tension and negative values indicate compression.)

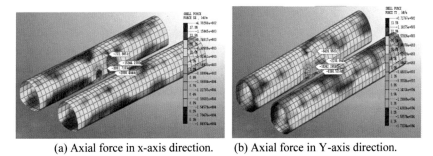

(a) Axial force in x-axis direction.   (b) Axial force in Y-axis direction.

Fig. 5. Axial force of lining after removal of left tunnel.

After the construction of the vehicle cross passage, the axial force of the tunnel lining is shown in Fig. 6. The maximum axial force of the lining along the X-axis direction is located at the arch waist of the boundary edge between the tunnel and the vehicle cross passage, with the maximum value of −9072 kN, and the maximum axial force of the lining along the Y-axis direction is −6421 kN. The stress concentration is obvious here. After the construction of vehicle cross passage lining, the axial force of the main tunnel lining is still large, so the monitoring and support should be strengthened.

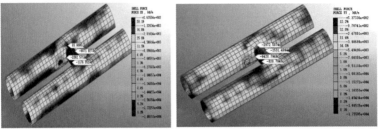

(a) Axial force in X-axis direction.   (b) Axial force in Y-axis direction.

Fig. 6. Axial force of lining after completion of vehicle cross passage.

4.3 Calculation of Safety Factor

4.3.1 Calculation of Safety Factor of Secondary Lining

The secondary lining of the main tunnel is an eccentrically compressed reinforced concrete member. After the cross passage is completed, the stress concentration at the junction of the main tunnel and the cross passage is the highest. The secondary lining structure at the junction is taken for safety checking calculation. The bending moment and axial force are shown in Fig. 7–8 [12–17].

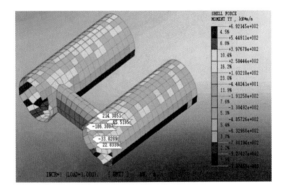

Fig. 7. Bending moment diagram of secondary lining.

The secondary lining of the tunnel is an eccentric compression rectangular member, and the safety factor is calculated according to the comprehensive safety factor method according to article 10.4.25 of *Guidelines for Design of Highway Tunnel* [1] (JTG/TD70-2010). The results are shown in Table 2. It can be seen that the safety factors of representative positions meet the specification requirements.

4.3.2 Calculation of Safety Factor of Initial Support

The initial support of the main tunnel belongs to eccentric compression or tension concrete structure, and the compressive strength or tensile strength of the structure should

Table 2. Internal force value statistics and safety factor of structural control points.

Position	Bending moment valuem (kN.m)	Axial force valuem (kN)	Reinforcement area (mm^2)	Safety factor	Allowable valum of safety factor
Vault	65.51	511.29	1570	18.19	1.53
Spandrel	214.38	6055.47	1570	2.18	1.53
Arch waist	186.38	3382.59	1570	3.57	1.53
Wall bottom	33.82	4021.06	1570	3.71	1.53
Inverted arch	22.83	1452.81	1570	9.90	1.53

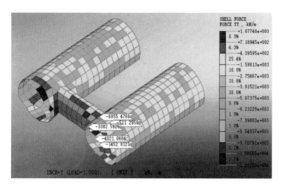

Fig. 8. Axial diagram of secondary lining.

be checked. Similarly, the initial support structure at the junction is selected for safety checking calculation. The bending moment and axial force are shown in Fig. 9–10.

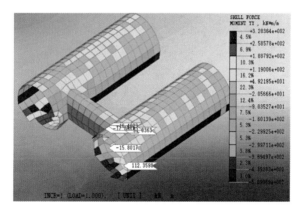

Fig. 9. Bending moment diagram of initial support.

The initial support of the tunnel belongs to eccentric compression or tension rectangular structure, and the safety factor is calculated according to the comprehensive safety factor method according to 10.4.16 or 10.4.17 of *Guidelines for Design of Highway Tunnel* [1] (JTG/Td70-2010), as shown in Table 3. It can be seen that the safety factors of representative positions, except the inverted arch, meet the specification requirements, and the results are relatively safe as a whole.

Table 3. Internal force value statistics and safety factor of structural control points.

Position	Bending moment valuem (kN.m)	Axial force valuem (kN)	Safety factor	Control	Allowable valum of safety factor
Vault	31.03	378.44	4.33	Tensile	2.7
Spandrel	16.4	374	13.50	Compression	1.8
Arch waist	17.05	576.9	9.42	Compression	1.8
Wall bottom	15.80	2976.23	1.91	Compression	1.8
Inverted arch	374	112.35	0.52	Tensile	2.7

According to the calculation results in Table 2 and Table 3, except that the safety factor of the inverted arch area of the initial support of the main tunnel is low, the checking calculation of the safety factor of other parts meets the requirements of *Code for Design of Road Tunnel* [8], and the overall structure is relatively safe. The inverted arch and wall bottom shall be strengthened in combination with the site conditions.

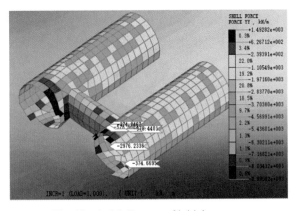

Fig. 10. Axial diagram of initial support.

5 Conclusion

Through the combination of numerical simulation and theoretical checking calculation, the influence of vehicle cross passage construction on the main tunnel is comprehensively analyzed, and the following conclusions are drawn:

(1) Class V surrounding rock has poor self stability. After removing the lining at the junction of the main tunnel and the vehicle cross passage, the surrounding rock will displace into the tunnel, and the surrounding rock at the junction shall be reinforced in advance. After the tunnel lining is excavated, the surrounding rock pressure is released, resulting in a large displacement of the lining at the arch bottom and arch crown of the main tunnel to the free face, with a maximum of 55.3 mm. After the lining of the vehicle cross passage is constructed, the displacement change is small.

(2) After removing the lining of the main tunnel at the junction, the axial force at the arch waist of the interface is large, which is −15244 kn. After the lining of the vehicle cross passage is applied, it is −10242 kn, which is weak compared with other positions.

(3) After the construction of the vehicle cross passage is completed, the safety checking calculation shall be carried out for the initial support and secondary lining of the main tunnel at the junction. Most of the safety factors meet the requirements of the code for design of highway tunnels, and certain strengthening measures shall be taken for weak links.

References

1. Guidelines for design of highway tunnel: JTG/TD70-2010 (2010)
2. Shi, Y.W., Cao, X.Y., Han, C.L.: Numerical simulation analysis of large section highway tunnel main cave and vehicle dedicated cross intersection. Chin. Foreign Highway **29**(4), 405–409 (2009)
3. Liu, S.H., Chen, Y.C., Li, G.: Numerical analysis of local stability at intersection parts of tunnel. J. Chongqing Jiaotong Univ. (Nat. Sci.) **30**(02), 217–220 (2011)
4. Sun, Z.J., Dong, L.S.: Effect of transverse trafic tube to tunnel construction deformation of the main hole. J. Shijazhuang Rail. Univ. **25**(2), 32–36 (2012)
5. Luo, Y.B., Chen, J.X., Wang, M.-S.: Study of influence of skew horizontal adit tunnel construction on main tunnel lining structure. Chin. J. Rock Mech. Eng. **29**(S2), 3792–3798 (2010)
6. Liu, X.L.: The analysis of the influence of excavation of lateral hole on the main tunnel for Nanshan tunnel. Shanxi Sci. Technol. Commun. (05), 43–44+56 (2013)
7. Su, X.K.: Research on the choosing boundary range of surrounding rocks in numerical simulation of tunnel excavation. J. Rail. Eng. Soc. **29**(3), 64–68 (2012)
8. Code for design of road tunnel: JTG3370.1-2018 (2018)
9. Song, Z.B., Deng, J.S.: Analysis of mechanical charac-teristics of long-span tunnel car adit excavation. Road Motor **5**, 179–181 (2010)
10. Zhu, Y.Q., Feng, W.X., Chen, H.X.: The finite element analysis of cross duct in Xidan metro station#2. J. Shijiazhuang Rail. Inst. **6**(3), 39–43 (1993)

11. Zhang, Z.Q., He, B.G., He, C.: Analysis on mechanics of intersection of adit and main tunnel. J. Chin. Rail. Soc. **32**(1), 128–132 (2010)
12. Zhang, Z.Q., Xu, J., Wan, X.Y.: Study of tunnel construction mechanics at intersection of horizontal adit and major tunnel in highway. Rock Soil Mech. **28**(2), 247–252 (2007)
13. Rao, J.Y., Xie, C.J., Zhao, X.: Theoretical stability calculation of surrounding rocks in divergence of deep tunnel. J. Central South Univ. **50**(08), 1949–1959 (2019)
14. Xie, X., Lv, B., Wang, L.N.: Data fusion based surrounding rock stability assessment on a deep burden tunnel. Chin. J. Underground Space Eng. **16**(04), 1108–1115 (2020)
15. Wang, K., Yu, H.Q., Wang, Z.: Analysis on impact of transverse tube construction on the lining structure of main tube for soft rock tunnel. Highway **62**(11), 301–305 (2017)
16. Zhang, J.Y., Shao, G., Xie, J.A.: Research on the construction mechanics of cross section of an interval tunnel and construction channel of Chongqing rail transit. Technol. Highway Transp. **34**(05), 87–91 (2018)
17. Li, X.B.: Research on tunnel transverse passageway arrangement mode based on mechanical effect. Guangdong Highway Commun. **45**(01), 52–56+61 (2019)

Application of Composite Foundation with Long-Short CFG Piles in Engineering Accident Treatment

Qiufeng Tang[1,2], Jianxing Tong[1,2(✉)], Ning Jia[1,2], Xinhui Yang[1,2], Xunhai Sun[1,2], and Mingli Yan[1,2]

[1] China Academy of Building Research Foundation Institute, No. 30 North Third Ring East Road, Beijing, China
tjxcabr@sina.com
[2] State Key Laboratory of Building Safety and Built Environment, No. 30 North Third Ring East Road, Beijing, China

Abstract. Based on the case of composite foundation with long-short CFG piles in handling the engineering accident, this paper analyzes the reasons for the insufficient bearing capacity of existing piles, proposes the reinforcement ideas and design calculation methods of composite foundation with long and short CFG piles, and gives the key construction techniques. The test and monitoring results show that the reinforcement ideas, design calculation methods and key construction techniques of composite foundation with long-short CFG piles are successful in this project, which can provide references for the design of composite foundation with long-short CFG piles and the handling of similar engineering accidents.

Keywords: Ground treatment · Composite foundation with long-short CFG piles · Engineering quality accident treatment · Design of composite foundation with multi-type-piles

1 Introduction

The composite foundation formed by reinforcements of different materials, lengths or diameters is called a multi-pile composite foundation. Multi-pile composite foundation [1–4] is usually used to treat special soil foundation. One type of reinforcement is used to treat the special soil to reduce or eliminate its engineering hazards, and then another reinforcement is used to meet the bearing capacity and deformation requirements. Besides, when the foundation soil of some sites has two good bearing layers, if all the pile tips fall on the shallow layer, the bearing capacity or deformation of the composite foundation cannot meet the design requirements. However, if all on the deep layer, the bearing capacity is too high, which leads to waste. Therefore, it is possible to consider placing pile tips on both the shallow layer and the deep layer to form a composite foundation with long-short piles. It not only meets the design requirements but also saves the cost. In addition, in some actual projects, when the existing single-piled composite foundation fails to pass the bearing capacity test, or the design requirements are not met due to the

G. Feng (Ed.): ICCE 2021, LNCE 213, pp. 171–179, 2022.
https://doi.org/10.1007/978-981-19-1260-3_16

adjustment of the superstructure, the piles need to be reinforced. When the pile material, pile length or pile diameter of the newly added piles is different from the existing piles, the composite foundation with multi-type piles is formed.

The data represented by literature [2–13] has conducted a lot of researches on the stress and deformation behaviors, design calculation methods and engineering applications of composite foundation with long-short rigid piles by means of model tests, field tests and numerical simulations. However, there are few literatures on the application of composite foundation with long-short CFG piles in engineering accident treatment. This article discusses and analyzes its application based on a case of engineering accident treatment.

2 Project Overview

Building 8# of a residential project in Zhuozhou City, Hebei Province, China, has 27 floors above ground and 2 floors underground. The building has a shear wall structure with a raft foundation. The ±0.00 elevation is 32.30 m.

Within the scope of the survey depth, the distribution of the foundation soil layer and its parameters are shown in Table 1. The depth of stable water level is 2.50–9.30 m.

Table 1. The distribution and parameters of each soil layer of the foundation.

Soil layer number	Soil layer	Layer bottom elevation (m)	Average thickness (m)	Unit weight γ (kNm^{-3})	Compression modulus E_s (MPa)	characteristic value of shaft resistance (kPa)	characteristic value of tip resistance (kPa)
①	Fill	29.29		18.0	/	/	/
②	Silt	23.74	5.55	19.5	14.56	23	/
③	Silty clay	22.14	1.60	19.4	11.18	20	/
④	Silt	17.44	4.70	19.3	14.16	20	/
⑤	Silty clay	11.24	6.20	19.3	10.11	20	/
⑥	Fine sand	7.84	3.40	20.0	15.00	23	/
⑥2	Silt	1.74	6.10	19.5	12.24	20	800
⑦	Fine sand	−7.86	9.60	20.0	25.00	25	900
⑦1	Silty clay	−14.16	6.30	19.1	12.60	/	/
⑧	Fine sand	−28.46	14.30	20.0	25.00	/	/

It is required that the characteristic value of the foundation bearing capacity after treatment is no less than 450 kPa, and the final settlement is less than 50 mm. The base soil layer is layer ②, and the characteristic value of the foundation bearing capacity is 120 kPa, which cannot meet the requirement and needs foundation treatment.

The original foundation treatment adopts CFG pile composite foundation. The long auger is used to form 646 piles with 1.2 m pile spacing, 400 mm diameter and 17.3 m effective pile length. The bearing layer is layer ⑥ and layer ⑥$_2$. The designed characteristic value of the single pile bearing capacity is no less than 620 kN. The thickness of the cushion layer is 200 mm.

After construction, 4 CFG piles were randomly selected for the single pile static load test. The results are shown in Fig. 1. The characteristic values of the single pile bearing capacity are 558 kN, 496 kN, 372 kN and 434 kN, which do not meet the design requirements.

In addition, 474 CFG piles were randomly selected for low strain test to verify the integrity of the pile body. The test results show that there are 146 piles of type I, accounting for 30.8%; 100 piles of type II, accounting for 21.1% and 228 piles of type III, with obvious defects in the pile body, accounting for 48.1%.

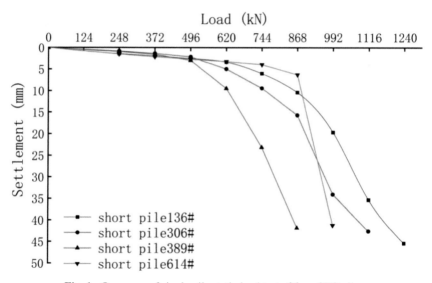

Fig. 1. Q-s curve of single pile static load test of four CFG piles.

Based on the above test results, a reinforcement treatment should be carried out.

3 Reinforcement Scheme Design

The design scheme is to add CFG piles (hereinafter referred to as long piles) on the layer ⑦ of fine sand as the bearing layer to form a composite foundation with long-short CFG piles, with existing CFG piles (hereinafter referred to as short piles).

Based on the results of the single pile static load test of the short piles, the character-istic value of the single pile bearing capacity is taken as 360 kN. The effective pile length of the long piles is 24.0 m, the pile diameter is 500 mm, and the calculated characteristic value of the bearing capacity of a single pile is 956 kN [14], taken as 820 kN. According to Eq. (1) in the "Technical Code for Ground Treatment of Buildings" [1], combined with the plan layout of the short piles, the long piles layout plan is determined and shown in Fig. 2. The calculated characteristic value of the bearing capacity of the composite foundation with long-short CFG piles is 453 kPa, which meets the design requirements.

$$f_{spk} = m_1 \frac{\lambda_1 R_{a1}}{A_{p1}} + m_2 \frac{\lambda_2 R_{a2}}{A_{p2}} + \beta(1 - m_1 - m_2)f_{sk} \tag{1}$$

In Eq. (1), the subscripts of the long piles and short piles are 1 and 2 respectively; λ_1 and λ_2 take the larger value of 0.9 because the characteristic value of the single pile bearing capacity is conservative; β takes the median value of 0.95; f_{sk} takes the characteristic value of natural foundation bearing capacity of 120 kPa due to the use of the long auger to form piles.

According to Eq. (2) and Eq. (3) [1], the soil compressive modulus improvement coef-ficients ζ_1 in the long-short-pile composite reinforcement area and ζ_2 only in the long-pile reinforcement area are calculated respectively, which are 3.77 and 1.98, respectively.

$$\zeta_1 = \frac{f_{spk}}{f_{ak}} \tag{2}$$

$$\zeta_2 = \frac{f_{spk1}}{f_{ak}} \tag{3}$$

In the Eqs. (2) and (3), f_{ak} is the characteristic value of the bearing capacity of the natural foundation under the foundation (kPa); f_{spk1} is the characteristic value of the bearing capacity of the composite foundation only reinforced by long piles (kPa).

According to the deformation calculation theory of Code for Design of Building Foundations [15], the deformation of the composite foundation with long-short piles is estimated according to Eq. (4):

$$s = \psi_s \left[\sum_{i=1}^{n_1} \frac{p_0}{\zeta_1 E_{si}} (z_i \overline{\alpha}_i - z_{i-1} \overline{\alpha}_{i-1}) + \sum_{i=n_1+1}^{n_2} \frac{p_0}{\zeta_2 E_{si}} (z_i \overline{\alpha}_i - z_{i-1} \overline{\alpha}_{i-1}) \right.$$
$$\left. + \sum_{i=n_2+1}^{n_3} \frac{p_0}{E_{si}} (z_i \overline{\alpha}_i - z_{i-1} \overline{\alpha}_{i-1}) \right] \tag{4}$$

The settlement calculation experience coefficient ψ_s is 0.2, and the calculated deformation is 29.94 mm, which meets the design requirements.

The design and calculation parameters of the composite foundation are shown in Table 2.

Table 2. Design parameters of composite foundation with long-short CFG piles.

Pile type	Effective pile length (m)	Pile diameter (mm)	Area replacement rate (%)	Concrete strength grade	Cushion thickness (mm)	Number of piles	Characteristic value of the single pile bearing capacity (kN)	Calculated deformation (mm)
Short piles	17.3	400	8.7	C25	200	646	360	29.94
Long piles	24.0	500	3.4	C25	200	157	820	

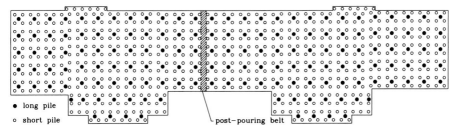

Fig. 2. Layout plan of long and short CFG piles of building 8#.

4   Composite Foundation Construction

The construction of the long piles also adopts the long auger central pressure grouting technique and the ready-mixed concrete.

Here are three main difficulties in the pile construction:

(1)   The groundwater level of the site is high, and there is a thick layer of saturated sand and saturated silt in the range of pile length. The interruption of pouring in this range may cause the concrete to segregate, shrink the diameter of the pile or even break the pile;

(2)   The bearing layer of the long and short piles is the main water-bearing layer, which has strong water permeability and high water pressure. It may cause the drill door to fail to open, and need to redrill many times, resulting in collapsed holes, channeling holes or concrete segregation;

(3)   The distance between short piles is only 1.2 m, which may lead to hole channeling during construction, and then may cause pile tops sinking, piles body mixed with mud, pile diameter reduction and even pile breakage.

In view of Difficulty (1), the supply of the mixture should be ensured, and the pumping speed of the mixture should be controlled to match the lifting speed of the drill pipe, especially in the saturated sand or saturated silt layer. Aiming at Difficulty (2), the drilling rig should use a downward-opening drill bit. For comprehensive Difficulties (1),

(2) and (3), it is advisable to adopt jumping pile driving [2]. During the construction of the long pile, the above requirements were strictly implemented.

The bearing capacity of short piles does not meet the design requirements, and the proportion of type III piles with obvious defects in the pile body is as high as 48.1%. After investigation and analysis, it is believed that the main reason is that jumping pile driving was not adopted and the pump had been stopped many times in the saturated silt layer.

Before the construction of the long pile, the short pile had been excavated to the effective pile top elevation. Therefore, the site is backfilled and compacted with plain soil first, with a thickness of about 0.5 m, which satisfies the safe walking of the long auger drill. The long piles are constructed after meeting the requirements for the thickness of the protective soil layer. After the construction is completed, a small excavator with manual cooperation was used to clear and transport the pile-driving spoil and the protective soil between the piles. The shallow defect piles within a depth of 1m were connected with the original piles.

5   Reinforcement Effect and Evaluation of Composite Foundation

5.1   Composite Foundation Inspection

After the completion of the CFG pile construction, the inspection unit conducted three static load tests on the long piles and three composite foundation with long-short piles static load tests.

The test area contains 1 length and 4 short piles, using a square steel bearing plate with a side length of 2.4 m × 2.4 m, as shown in Fig. 3. A 200 mm-thick crushed stone cushion is laid under the pressure-bearing board.

The static load test results show that the characteristic value of the bearing capacity of the composite foundation is not less than 450 kPa, and the characteristic value of the bearing capacity of the single long pile is not less than 820 kPa, both of which meet the design requirements. The static load test curve is shown in Fig. 4.

Meanwhile, 130 long piles were randomly selected for low-strain testing. The results showed that there were 121 piles of type I, 9 piles of type II, and no piles of type III and IV.

5.2   Building Settlement Monitoring

From the beginning of the construction of the structure, the monitoring unit conducted settlement observations on Building #8, and the settlement-time curve of the building is shown in Fig. 5. It can be seen that the maximum settlement of the building is less than 25 mm. The east and west sides of the post-pouring belt have settled evenly. According to estimates, the final settlement will be less than 30 mm, which meets the design requirements, indicating that the project is successful in using existing CFG piles and newly added CFG piles to form a composite foundation with long-short CFG piles.

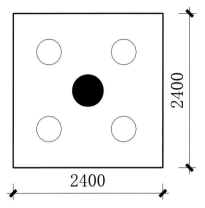

Fig. 3. Schematic diagram of bearing plate for static load test of composite foundation with long-short piles of Building 8#.

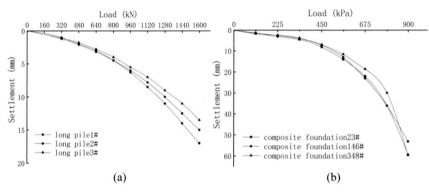

Fig. 4. Static load curve: (a) is the Q-s curve of a single pile, (b) is the p-s curve of the composite foundation.

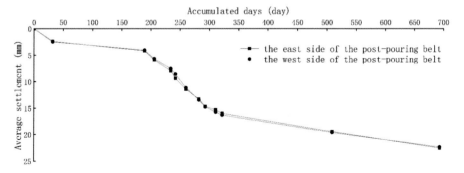

Fig. 5. The settlement-time curve of building 8#.

6 Conclusion

In this project, because the bearing capacity of the existing CFG pile composite foundation does not meet the design requirements, long piles with a larger diameter are used for reinforcement to form a composite foundation with long-short piles with the existing CFG piles. The following conclusions can be drawn from this project:

(1) When designing a composite foundation with long-short CFG piles, a relatively good soil layer should be selected for the long pile as the pile end bearing layer, which can effectively improve the bearing capacity and control the deformation of the foundation.
(2) When using long auger center pressure grouting to form piles, if the groundwater level of the site is high and there is saturated sand or saturated silt layer, the drill pipe should be continuously lifted, and the pump should not be stopped to wait for mixture; if the pile spacing is small, jumping pile driving should be adopted.
(3) The results of building settlement monitoring show that the use of composite foundation with long-short CFG piles has a good effect, which can provide a reference for similar engineering accident treatment.

References

1. Technical Code for Fround Treatment of Buildings (JGJ79-2012), p. 69. China Architecture & Building Press, Beijing (2012)
2. Ji, M., Zhang, D.G., Zhang, Z., Yan, M.L.: Design of the composite foundation with long-short piles. Geotech. Eng. Tech. J. **2**, 86–91 (2001)
3. Yan, M.L., Wang, M.S., Yan, X.F., Zhang, D.G.: Study on the calculation method of multi-type-pile composite foundation. Chin. J. Geotech. Eng. J. **3**, 352–355 (2003)
4. Yan, M.L., Zhang, D.G.: CFG Pile Composite Foundation Technology and Engineering Practice, pp. 23, 144–158. China Water & Power Press, Beijing (2006)
5. Tong, J.X., Sun, X.H., Yang, X.H., Wang, M.S., Luo, P.F.: Experimental study on pile-soil bearing behavior and thickness-diameter ratio of composite foundation with long and short piles. Chin. J. Geotech. Eng. J. **35**(05), 955–960 (2013)
6. Ge, X.S., Gong, X.N., Zhang, X.M.: FEM analysis and design of long-short-pile composite foundation. J. Build. Struct. J. **4**, 91–96 (2003)
7. Deng, C.: The bearing capacity and settlement calculation of the long-short-pile composite foundation. MA thesis, Zhejiang University (2002)
8. Lu, W.Z.: FEM analysis for consolidation behavior of composite foundation with long-short-piles. MA thesis, Zhejiang University (2005)
9. Shi, S.W.: Study of engineering characteristics of composite foundation with long and short piles for deep soft ground. MA thesis, Zhejiang University (2006)
10. Chen, C.F., Xiao, S.J., Niu, S.S.: Optimization design method of long-short-pile composite foundation. J. Eng. Geol. **2**, 229–232 (2006)
11. Sun, X.H.: Effect of foundation rigidity and cushion thickness on rigid pile composite foundation bearing capacity. PhD dissertation, China Academy of Building Research (2010)
12. Deng, C., Gong, X.N.: Application of long-short-pile composite foundation to high-rise building. Build. Construct. J. **1**, 18–20 (2003)

13. Zhao, Z.P., Tong, J.X., Sun, X.Z., Feng, G.R.: Design and application of a composite foundation with two different pile types. Build. Sci. J. **32**(Suppl. 2), 277–281 (2016)
14. Technical Code for Building Pile Foundations (JGJ94-2008), p. 19. China Architecture & Building Press, Beijing (2012)
15. Code for Design of Building Foundation (GB50007-2011), p. 28. China Architecture & Building Press, Beijing (2012)

Research on Improvement Calculation Method of Design Thrust of Anti Slide Pile

Xiaoqiang Hou[1(✉)], Jierui Liu[1], Xinfei Wang[1], Zhongren Zhou[1], and Honglu Jia[2]

[1] School of Civil Engineering, Lanzhou Jiaotong University, Lanzhou 730070, China
34821998@qq.com
[2] China Construction Road and Bridge Group Co. Ltd., Shijiazhuang 050001, China

Abstract. At present, based on the transfer coefficient method, most of the anti-slide pile design thrusts are calculated by the overload method and the strength reserve method respectively. Many algorithms only consider the remaining sliding force behind the pile and the safety factor that meets the requirements of the design conditions. Generally, the safety factor is the safety factor of the sliding slope behind the pile after the anti slide pile is reinforced. For the entire landslide, there are two safety factors before and after the pile, which is not the design safety factor target value, and there is a big difference between the safety factor and the treatment goal required by the specification. Through the study of the pile-soil interaction of anti-slide piles, it is believed that in addition to the active residual sliding force transmitted by the blocks behind the pile, the anti-slide piles are simultaneously subjected to the passive residual anti-sliding force transmitted upwards by the blocks in front of the pile. The stress analysis shows that: Firstly, according to the different active and passive properties of anti-sliding force transmission and sliding force transmission, the mechanical model of anti-sliding force transmission is studied, and the calculation formula of anti-sliding force transmission coefficient is derived; Secondly, It is believed that the anti-slide pile provides horizontal thrust to the landslide, and two components of the sliding surface direction and the vertical sliding surface direction are generated. The balance equation is established and the overload method and the strength reserve method of anti-slide pile thrust calculation formula are derived; Thirdly, according to the principle of setting piles in the anti-slip section, the optimal location of anti-slide piles are proposed; Fourthly, after verification of cases, the safety factors before and after the piles calculated by the overload method are basically equal, and consistent with the design safety factors. Calculation result shows that the strength reserve method to calculate the safety factor before the pile is accurate and reliable, and the result of the safety factor behind the pile is relatively small.

Keywords: Anti-sliding pile · Design thrust · Anti-sliding force transmission coefficient · Location selection condition · Safety factor

1 Introduction

In recent years, China has carried out large-scale construction of infrastructure projects such as highways, water conservancy and railways. During the construction process,

© The Author(s) 2022
G. Feng (Ed.): ICCE 2021, LNCE 213, pp. 180–194, 2022.
https://doi.org/10.1007/978-981-19-1260-3_17

engineering excavation was carried out near the foot of the hillslope, resulting in a decrease in the shear resistance of the lower part of the slope. Sliding outside could cause landslide hazard and directly threaten project construction and the service capacity after construction. In terms of landslide treatment, if more anti-slide piles are placed in the position of the shear outlet, it will directly affect the space of other engineering construction. Therefore, the anti-slide piles arranged in the anti-slip section are usually selected as an effective method. This retaining structure is in the form and widely used, which has little impact on the construction site and can ensure that the slope is in a safe and stable state within the design period in the future. Because the anti-slide pile is subjected to the combined effect of the landslide thrust in front of the pile and the soil resistance behind the pile, the pile-soil relationship is complicated. The calculation methods of design thrust of anti-slide piles are diverse, and the results are quite different. At present, most of them are calculated according to the cantilever anti-slide pile thrust at the shear outlet, which affects the accuracy of the anti-slide pile design technical index, such as the bending moment, shear force, section size, length of the anti-slide pile. The anchorage depth of the anti-slide pile in the stable soil below the sliding surface layer is also influenced. Different design methods effect the difficulty of the anti-slide pile construction and the engineering cost. Therefore, it is worthy of further study the calculation method of the anti-slide pile thrust.

Many experts have proposed different calculation methods for the design thrust of anti-slide piles, and the calculation results of each method are quite different. T.Ito (1981) [1] and Reese (1992) etc. [2] respectively proposed to calculate the anti-sliding force of anti-sliding piles under design conditions which based on Fellenius's Swedish section method and simplified Bishop method. S. Zeng, R. Liang [3] (2002) and the design code for anti-slide pile reinforcement of landslides by the Ohio Department of Transportation (GB7: Drilled Shaft Landslide Stabilization Design 2014) [4, 5] puts forward a calculation formula to calculate the lateral load of piles, which is similar to the commonly used transfer coefficient method in China and proves the scientificity and applicability of the transfer coefficient method to calculate the anti-slide pile thrust. Recently, many domestic experts and scholars have applied the idea of transfer coefficient to carry out a lot of research work on the calculation of the anti-slide pile thrust based on the overload method and the strength reserve method. According to the sliding force curve between the engineering design state and the benchmark working condition, Wang Liangqing (2005) et al. [6] chooses the smaller sliding force as the anti-slide pile thrust. Although this method highlights the economics of anti-slide pile design, it just considered the horizontal component of the remaining sliding force of the anti-slide pile, which induced the result of horizontal thrust calculation is too large. Zheng Yingren (2006) et al. [7] proposed that the landslide thrust should be calculated with the strength reserve, but the calculation of the anti-slide pile thrust was not further clarified. Although He Haifang et al. (2008) [8] considered the anti-sliding effect of the anti-slip section at the front edge of the slope, but it did not consider the transmission of the landslide resistance to the rock and soil in front of the pile through the supporting pile. There is a certain discrepancy between thrust calculation and actual thrust. Hu Mingjun et al. (2011) [9] proposed that the anti-slide pile can be directly installed at the shear exit, and the residual sliding force of the shear exit can be used as the design thrust of the anti-slide pile by loading back pressure. However, in reality many landslide shear exits are not stabilized. It may be necessary to install other engineering facilities, so loading back pressure and laying anti-slide piles have certain limitations in space. Based on the design thrust is equal to the difference between the two calculated values in the horizontal projection, Wang Peiyong (2010) et al. [10] utilized the overload method to calculate the remaining sliding force and the remaining anti-sliding force of the slope, and the design thrust on the contact surface of each block of the slope body is obtained. It is unreasonable to use the sliding force calculation method for the anti-sliding force transmission. The design specification was promulgated in 2015 [11] and the design thrust was calculated by multiplying by $\cos\theta$ to the above two calculated landslide thrust, but it is not actually the design thrust of the anti-slide pile. According to transfer coefficient strength reserve method, Zhao Shangyi et al. [12] further studied that the horizontal reaction force of the anti-slide pile. The horizontal reaction force was added to the load system of the bar, and then the sliding surface was adjusted according to the design safety factor. The strength parameters were reduced so that the remaining sliding force of the last bar was exactly equal to 0. Li Huanhuan [13] further studied the algorithms of the overload method and the strength reserve method, and only the anti-sliding and anti-sliding effect of the anti-sliding pile were studied systematically. Zhao Shangyi et al. and Li Huanhuan did not consider the effect of the remaining anti-sliding force in front of the pile.

Based on the above, the anti-slide pile is simultaneously affected by the remaining sliding force behind the pile and the anti-sliding force in front of pile. At present, many algorithms only consider the remaining sliding force behind the pile and meet the safety factor required by the design conditions, usually referred to as the safety factor. There are two safety factors for the landslide before and after the pile for the entire landslide, which is not the target value of the design safety factor, and the safety factor required by the code and various technical standards is quite different. Through the principle of landslide section division, the authors use the transfer coefficient method to study the overload method and the strength reserve method based on the above-mentioned problem, and further improves the transfer coefficient overload method and the strength reserve method to calculate the horizontal thrust of the anti-slide pile.

2   Pile-Soil Interaction of Anti-slide Pile

For the design of landslide anti-slide piles, the design position of anti-slide piles is usually selected in the anti-slide section, and its position selection has a greater influence on the effect of pile-soil, and ultimately affects the issue of whether the safety factors of the two parts of the slope are the same before and after the pile. Due to the different nature of the front and rear forces of the pile, there is a residual sliding force along the sliding surface behind the pile, and there is a residual anti-sliding force in front of the pile along the sliding surface in front of the pile. The residual sliding force is usually active under the action of gravity. The direction is always along the sliding surface, and the remaining anti-sliding force only shows up when the sliding force acts. The anti-sliding force has a passive nature. Both of these two force transmission processes have distinct active and passive characteristics. It is concluded that there are three possible working conditions for the front and rear effects of the pile (see Fig. 1). In the first case, the anti-slide pile is setted at A, and the remaining sliding force P_1 behind the pile and the remaining anti-sliding force R_1 in the landslide part in front of the pile are safe. The safety factor is greater than 1 and less than the design safety factor, indicating that this is the sliding section and is not the best position, and the anti-slide pile needs to be moved down to B; In the second case, the anti-slide pile is set at B which locates at the anti-slide section. The safety factor of the landslide part is greater than or equal to the design safety factor. In addition to the residual sliding force P_2 after the pile, there is also a residual anti-sliding force R_2 in front of the pile along the sliding surface. When the anti-slide pile is designed, the remaining anti-sliding force in front of the pile and the remaining sliding force behind the pile must be considered, which can meet the same safety factor before and after the pile to ensure the uniqueness of the safety factor of the landslide. In the third case, the anti-slide pile is set at C. The anti-sliding force in front of the pile is zero, and the safety factor is zero. Only the remaining sliding force P_3 after the existing pile can be considered. Ultimately, it can also ensure that the overall landslide safety factor reaches the design safety factor and is unique. According to above, the second and third case are the best location for anti-slide pile treatment. From the study of anti-slide pile horizontal thrust, it is more reasonable to start with the calculation of residual sliding force and residual anti-sliding force. Based on overload method and strength reserve method, the following part will follow the idea of transmission coefficients method to study the calculation method of residual sliding force and residual anti-sliding force respectively.

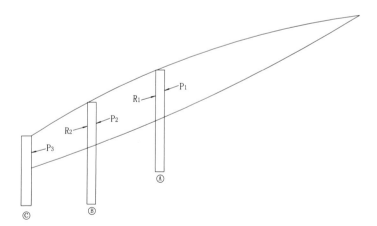

Fig. 1. Three working condition of fore-and-aft action of piles

3 Transmission Coefficients Method

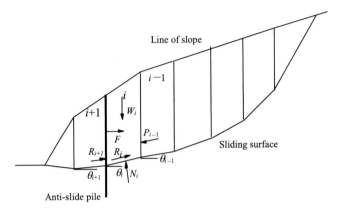

Fig. 2. Landslide block model

3.1 The Assumption of Transmission Coefficients Method

The transfer coefficient method is a common method for calculating the remaining sliding force and stability of a landslide for a broken line sliding surface or a complex sliding surface that combines a circular arc and a broken line. This method is widely used in roads, railways, water conservancy, land and other industries. The calculation of the transfer coefficient method has the following six assumptions [14] (see Fig. 2):

(1) The problem of landslide stability is regarded as a plane strain problem.
(2) The sliding force is dominated by the shear stress parallel to the sliding surface and the normal stress perpendicular to the sliding surface, and the stress is concentrated on the sliding surface.
(3) Regarding the landslide body as an ideal rigid plastic material, it is believed that the landslide body will not undergo any deformation during the entire loading process. Once the shear stress along the sliding surface reaches its shear strength, the landslide body begins to produce shear deformation along the sliding surface.
(4) The failure of the sliding surface obeys the Mohr Coulomb failure criterion.
(5) The direction of the remaining sliding force is consistent with the inclination angle of the sliding surface, and the difference between adjacent inclination angles is less than 10°. When it is a negative value, the remaining sliding force transmitted is 0.
(6) The static balance condition is satisfied along the entire sliding surface, but the moment balance condition is not satisfied.

3.2   Overload Method and Strength Reserve Method to Calculate the Remaining Sliding Force Calculation Formula

The overload method calculates the remaining sliding force, mainly by multiplying the remaining sliding force of each block by the design working condition safety factor Ks, increasing the sliding force of each block, and its anti-sliding force remains unchanged, and calculating the remaining sliding force.

$$P_i = K_s T_i - R_i + P_{i-1} \psi_{i-1} \tag{1}$$

$$T_i = W_i \sin \theta_i \tag{2}$$

$$R_i = W_i \cos \theta_i \tan \phi_i + c_i L_i \tag{3}$$

$$\psi_{i-1} = \cos(\theta_{i-1} - \theta_i) - \sin(\theta_{i-1} - \theta_i) \tan \varphi_i P_i = K_s T_i - R_i + P_{i-1} \psi_{i-1} \tag{4}$$

Similarly, the strength reserve method calculates the remaining sliding force, which is mainly to divide the anti-sliding force of each block by the preset safety factor Ks. The anti-sliding force of each block is reduced by Ks times, and its sliding force remains unchanged.

$$P_i = T_i - R_i/K_s + P_{i-1} \psi_{i-1} \tag{5}$$

$$\psi_{i-1} = \cos(\theta_{i-1} - \theta_i) - \sin(\theta_{i-1} - \theta_i) \tan \phi_i/K_s \tag{6}$$

P_i and P_{i-1} are the remaining sliding force (kN) of the i-th and $i-1$-th sliding bodies, respectively, and when $P_i < 0$ (i < n) P_i equals 0; F are the design values of the horizontal landslide thrust per linear meter, assuming the anti-slide pile is arranged on the left side of the first block; W_i is the self-weight per unit width of the i-th block (kN/m); c_i is the cohesive force of the block along the sliding surface (kPa); φ_i is the i-th block The internal friction angle along the sliding surface (°); θ_i is the angle between the bottom surface of the first sliding body and the horizontal plane (°); L_i is the length of the first sliding body along the sliding surface (m); $E_n = 0$ is the first block to the first calculation block The transfer coefficient; KS is the safety factor of the landslide anti-sliding design.

3.3   Derivation of the Residual Anti-sliding Force Calculation Formulae by the Overload Method and the Strength Reserve Method

When the force is transmitted in a landslide, most of them only consider the sliding force actively transmitted from the top of the slope to the shear outlet at the bottom of the slope, ignoring the anti-sliding force generated by the sliding surface that is passively transferred from the shear outlet at the bottom of the slope to the top of the slope. If the sliding force of each block is greater than the anti-sliding force in the sliding section, it indicates that the anti-sliding force provided by the block itself can not meet the sliding force of the block, and the active residual sliding force will be generated; In the anti-slip section, the sliding force of each block is less than or equal to the anti-slip force. At the same time, the remaining anti-slip force transmitted along the sliding surface and in the opposite direction prevents the block from sliding down. This anti-slip force is only sliding It only manifests passively when the force is applied.

Calculation idea: When the remaining sliding force of the first block is actively transmitted downwards, at the same time the first block produces a thrust to the first block that transmits upwards along the sliding surface. This thrust is the remaining anti-sliding force passively generated by the first block. The direction of the remaining sliding force on the contact surface of the bars is not on the same straight line as the remaining sliding force of the first bar, and the included angle is the difference between the inclination angles of two adjacent bars (see Fig. 3).

Remaining anti-sliding force calculation steps: the calculation sequence is opposite to the remaining sliding force direction, then the calculation is calculated from the toe to the top of the slope sequentially along the sliding surface, and the anti-sliding force direction is consistent with the inclination of the sliding surface until the remaining sliding force is 0, indicating Just in the critical state.

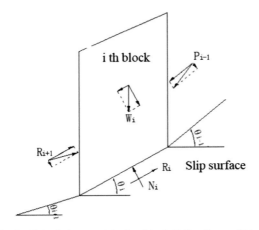

Fig. 3. Calculation model of residual sliding force of block

According to the force model in Fig. 3, the remaining sliding force transmission coefficient ψ_{i-1} is not the same as the remaining anti-sliding force coefficient ψ'_{i+1}. It can be seen from the landslide section that the difference between the two blocks is mainly due to the two different directions of transmitting the remaining force. In the process, the inclination angle θ_i is usually lager than θ_{i+1}. Therefore, when each bar transmits force, it is not in the same direction, and there are force components in other directions. When the residual anti-slip force of the i + 1th block is transmitted along the sliding surface to the i block, it is composed of two component forces, one of which is the same as the normal direction compressive stress on the sliding surface of the ith block to produce frictional resistance, and the other is divided into The force direction is upwards along the sliding surface, and the two components play an anti-sliding effect. Therefore, the residual anti-sliding force transfer coefficient is the sum of the two component directions. Similarly, it can be known that the remaining sliding force is the $i - 1$th to the ith. When the bars are transmitted, the direction of the force component of the $i - 1$th bar is perpendicular to the direction of the sliding surface to generate friction to reduce the sliding force, and the remaining anti-sliding force transmission coefficient is the difference between the two components.

$$R'_i = W_i \cos\theta_i \tan\phi_i + c_i L_i - W_i \sin\theta_i \\ + R_{i+1}(\cos(\theta_i - \theta_{i+1}) + \sin(\theta_i - \theta_{i+1})\tan\phi_i) \tag{7}$$

According to (1) and (2), we can get

$$R'_i = R_i - T_i + R_{i+1}\psi'_i \tag{8}$$

$$\psi'_{i+1} = \cos(\theta_i - \theta_{i+1}) + \sin(\theta_i - \theta_{i+1})\tan\phi_i \tag{9}$$

According to the overload method, the calculation of the remaining anti-sliding force also increases the sliding force of the bar by K_s times, and the calculation formula is:

$$R'_i = R_i - K_s T_i + R_{i+1}\psi'_{i+1} \tag{10}$$

$$\psi'_{i+1} = \cos(\theta_i - \theta_{i+1}) + \sin(\theta_i - \theta_{i+1})\tan\phi_i \tag{11}$$

According to the strength reserve method, the calculation of the remaining anti-sliding force also only reduces the anti-sliding force of the block by K_s times, and the calculation formula is:

$$R'_i = R_i/K_s - T_i + R_{i+1}\psi'_{i+1} \tag{12}$$

$$\psi'_{i+1} = \cos(\theta_i - \theta_{i+1}) + \sin(\theta_i - \theta_{i+1})\tan\phi_i/K_s \tag{13}$$

R'_i represents the residual anti-skid force; ψ'_{i+1} is the residual anti-sliding force transmission coefficient of the i-th block.

Based on the above formula, it can be seen that the residual anti-sliding force transmission and the remaining sliding force transmission coefficient of the landslide resistance section are completely different, indicating that the overload method and the strength reserve method are used to calculate the remaining anti-sliding force using the original residual sliding force transmission coefficient calculation formula.

4   Improved Calculation Method of Anti-slide Pile Thrust

The key to the thrust of the anti-slide pile is the choice of the position of the anti-slide pile. Usually choose the anti-slip section, and the best position must be the ratio of the remaining anti-slip force and the remaining sliding force before the pile of the anti-slip section is equal to the design safety factor, in order to ensure that the overall landslide meets the design safety factor requirements. Under normal circumstances, the remaining sliding force behind the pile is provided by the thrust provided by the pile to meet the design safety factor, and at the same time it is appropriate to ensure that the remaining sliding force in front of the pile is greater than or equal to the safety factor of the design working condition. The main action of the pile is used for the front block of the pile. At this time, the anti-sliding force of the front block of the pile is used for the anti-slide pile, and the force system of the anti-slide pile system (see Fig. 4).

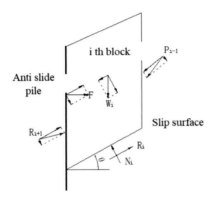

Fig. 4. Stress model between anti-slide pile and bar

The anti-slide pile is not only affected by the remaining sliding force transmitted downward from the back of the pile, but also by the remaining anti-sliding force transmitted upward from the front of the pile. When calculating the horizontal thrust of the anti-slide pile as the resultant force, it can be decomposed into two forces in the direction of the sliding surface and the normal direction of the sliding surface (see Fig. 4). The normal balance equation of the sliding surface of the block and the normal balance equation of the sliding surface are established to solve the problem. Two thrust calculation formulas are deduced separately: overload method and strength reserve method.

$$R_{i+1}\cos(\theta_i - \theta_{i+1}) + F\cos\theta_i + N_i\tan\phi_i \\ +c_iL_i = W_i\sin\theta_i + P_{i-1}\cos(\theta_{i-1} - \theta_i) \tag{14}$$

$$R_{i+1}\sin(\theta_i - \theta_{i+1}) + F\sin\theta_i + W_i\cos\theta_i \\ +P_{i-1}\sin(\theta_{i-1} - \theta_i) = N_i \tag{15}$$

It is obtained by combining the two Eqs. (14) and (15),

$$F = \frac{W_i\sin\theta_i - W_i\cos\theta_i\tan\phi_i - c_iL_i}{\cos\theta_i + \sin\theta_i\tan\phi} \\ + \frac{P_{i-1}(\cos(\theta_{i-1} - \theta_i) - \sin(\theta_{i-1} - \theta_i)\tan\phi_i)}{\cos\theta_i + \sin\theta_i\tan\phi} \\ - \frac{R_{i+1}(\cos(\theta_i - \theta_{i+1}) + \sin(\theta_i - \theta_{i+1}))}{\cos\theta_i + \sin\theta_i\tan\phi_i} \tag{16}$$

Substituting (2) from (3) and (9) into the calculation formula for the horizontal thrust of the anti-slide pile under the reference condition is

$$F = \frac{T_i - R_i + P_{i-1}\psi_i - R_{i+1}\psi'_{i+1}}{\cos\theta_i + \sin\theta_i\tan\phi_i} \tag{17}$$

In the same way, the calculation formula for the horizontal thrust of the overload method anti-slide pile under the design conditions is

$$F = \frac{K_s T_i - R_i + P_{i-1}\psi_i - R_{i+1}\psi'_{i+1}}{\cos\theta_i + \sin\theta_i \tan\phi_i} \tag{18}$$

Similarly, the formula for calculating the horizontal thrust of the anti-slide pile by the strength reserve method under the design conditions is

$$F = \frac{T_i - R_i/K_s + P_{i-1}\psi_i - R_{i+1}\psi'_{i+1}}{\cos\theta_i + \sin\theta_i \tan\phi_i/K_s} \tag{19}$$

When the residual anti-skid force is equal to zero by the three calculation formulas (17), (18) and (19), it is the second case. It can be seen that the horizontal thrust of the anti-slide pile is simply the horizontal component of the residual sliding force, and the horizontal projection of the residual sliding force difference between the design conditions and the reference conditions is not reasonable, but is related to the design safety factor. The transfer coefficient of the remaining sliding force behind the pile, the transfer coefficient of the remaining anti-sliding force in front of the pile, the horizontal angle of the sliding surface where the pile is installed and other factors are related.

After the slope is reinforced, it is verified whether the safety factor of the entire landslide is equal to the safety factor of the design working condition. According to the definition of "anti-sliding force/sliding force = safety factor", according to the force system (see Fig. 3), the anti-slide pile is used as the reference point. The anti-sliding force is composed of the two forces vectorized by the anti-sliding force of all the bars and the horizontal thrust of the anti-slide pile. The sliding force is vectorized by the sliding force of all the bars. The formula for calculating the safety factor after the pile (20) and Calculation formula of safety factor in front of pile (21)

$$K_h = \frac{R_i + F_i\cos\theta_i + R_{i+1}\psi'_{i+1} + F_i\sin\theta_i\tan\phi}{T_i} \tag{20}$$

$$K_q = \frac{\sum_{i+1}^{n}(W_i\sin\theta_i\tan\phi_i + c_iL_i) - R_{i+1}\psi'_{i+1}}{\sum_{i+1}^{n}W_i\cos\theta_i} \tag{21}$$

K_q is the calculated safety factor, T_i is the remaining sliding force of the i-th block, R_i is the remaining anti-sliding force of the i-th block, F_i is the horizontal thrust of the anti-slide pile at the i-th block, ψ'_{i+1} is the transfer coefficient of the residual anti-sliding force at the i + 1-th block, ψ_{i+1} is the first The remaining sliding force transmission coefficient at the i + 1-th block.

5   Case Study

Project overview: The landslide is an accumulation layer landslide, the roadbed is located at the front edge of the landslide, the anti-slide pile is located on the inner side of the roadbed, and above the sliding surface is humus soil and silty clay soil with a small amount of gravel, $\gamma_1 = 16.8$ kN/m^3, $\Phi_1 = 16.2°$, C $= 9.6$ kPa. Below the slip surface is Neogene sandy mudstone. According to the terrain conditions and slip surface conditions, 16 blocks are divided, and the angle difference between the blocks is less than 10°. The landslide section is shown in Fig. 5 and Table 1.

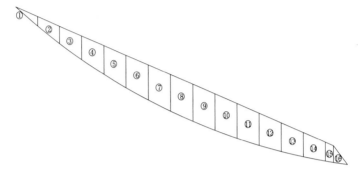

Fig. 5. Landslide calculation section

According to the actual example, the two remaining sliding forces of overload method and strength reserve method are calculated respectively. The safety factor of the selected design working condition for landslide treatment is 1.2, regardless of whether the overload method or the strength reserve method is adopted. As can be seen from Fig. 6, Article 12 to Article 16. If the block safety factor is greater than 1.2, anti-slide piles can be deployed, which can not only meet the anti-slide pile supporting effect, but also meet the safety factor of the design working condition for the blocks in front of the pile.

Table 2 is a comparison of calculation results obtained by various design thrust methods using the overload method. According to Table 2, it can be seen that Li Huanhuan's method, design code method, Wang Qingliang's method and the method in this paper are the same as those calculated by Li Huanhuan's method when anti-slide piles are installed at the exit of the landslide. The anti-skid force is 0, and the other calculated values are large, the maximum difference is 16%.

Table 3 is a comparison of calculation results obtained by various design thrust methods using the strength reserve method. According to Table 3, it can be seen that the design code method, Zhao Shangyi's method, Wang Qingliang's method, Wang Peiyong's method and the method in this paper are basically the same as those calculated by Wang Peiyong's method when anti-slide piles are installed at the landslide shear exit. The main reason is that the horizontal angle difference between the bars is small, which can be ignored. The numerical values of the design code method, Zhao Shangyi method, and Wang Liangqing method are quite different.

Analysis of the reason for the calculation difference: Zhao Shangyi's method did not consider the residual anti-sliding force, the lower block anti-sliding force of the anti-slip section was relatively large, and the design code method did not consider the residual anti-sliding force and the horizontal thrust of the anti-slide pile on the second block behind the pile. The friction caused by the positive pressure on the sliding surface, Wang

Table 1. Main geotechnical mechanical parameters

Block No	Volumetric weight γi (kN/m^3)	Force of cohesion Ci (kpa)	Friction angle φi (°)	Length of sliding surface Li (m)	Inclination angle of slip surface αi (°)	Weight of block Qi (KN)
1	16.8	9.6	15.9	9.9429	41.4	195.2
2	16.8	9.6	15.9	9.6624	34.1	553.2
3	16.8	9.6	15.9	9.4196	31.9	848.8
4	16.8	9.6	15.9	9.2074	29.7	1086.5
5	16.8	9.6	15.9	9.0214	27.5	1270.3
6	16.8	9.6	15.9	8.8578	25.4	1403.3
7	16.8	9.6	15.9	8.7141	23.4	1488.2
8	16.8	9.6	15.9	8.5877	21.3	1527.5
9	16.8	9.6	15.9	8.4770	19.3	1522.9
10	16.8	9.6	15.9	8.3976	17.7	1480.2
11	16.8	9.6	15.9	8.2815	15.0	1392.9
12	16.8	9.6	15.9	8.2246	13.4	1262.3
13	16.8	9.6	15.9	8.1637	11.5	1097.1
14	16.8	9.6	15.9	8.1131	9.6	894.4
15	16.8	9.6	15.9	3.0346	8.3	276.8
16	16.8	9.6	15.9	5.0465	7.3	216.2

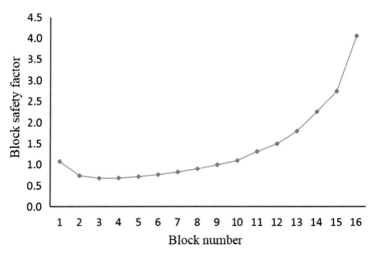

Fig. 6. Safety factors for each block

Liangqing's method did not consider the friction caused by the positive pressure on the sliding surface of the first block in front of the pile. The remaining sliding force transfer coefficient is calculated. The difference in the inclination angle of the sliding surface of the main slip resistance section of the slope is small, and the difference in the two transfer coefficients is small.

Table 2. Comparison of results

Block no	Proposed method (KN)	Li Huanhuan method (KN)	Difference %	Design code method (KN)	Difference %	Wang Liangqing method (KN)	Difference %
12	1075.6	1489.5	38.5%	984.9	−8.4%	1012.6	−5.9%
13	1065.6	1353.6	27.0%	1024.8	−3.8%	1045.8	−1.9%
14	1057.8	1193.8	12.9%	1050.0	−0.7%	1064.8	0.7%
15	1053.8	1129.8	7.2%	1054.3	0.0%	1065.4	1.1%
16	1051.2	1051.2	0.0%	1056.8	0.5%	1065.4	1.4%

Note: The block number refers to the location of the anti-slide pile, as shown in Fig. 5.

Table 3. Comparison of results

Block no	Proposed method (KN)	Design code method (KN)	Difference %	Zhao Shangyi method (KN)	Difference %	Wang Liangqing method (KN)	Difference %	Wang Peiyong method (KN)	Difference %
12	912.5	1263.2	38%			771.9	−15%	844.1	−7.5%
13	904.3	1156.0	28%			831.3	−8%	795.3	−12.1%
14	897.8	1024.6	14%			880.5	−2%	764.4	−14.9%
15	894.5	971.8	9%			893.7	0%	839.6	−6.1%
16	892.3	905.1	1%	892.3	0	908.5	2%	824.6	−7.6%

Notes: 1. The Li Huanhuan method is the same as the Zhao Shangyi method in the strength reserve method, which only satisfies the calculation of the anti-slide pile at the shear exit. 2. The comparison results are based on the results of proposed method in this paper

Substituting the calculation results into Eqs. (19) and (20) to verify the safety factor. From the comparison results in Table 4, it can be seen that the safety factor after overloading method piles is basically equal to the safety factor of 1.2, and the safety factor after strength reserve method is less than or equal to 1.2. And there is a big difference. The calculation results of the first two methods of the pile show that the safety factor is 1.2. It can be seen that the calculation of the overload method is more accurate and reliable, and the safety factor before and after the pile is consistent.

Table 4. Comparison of factor of safety.

Block number	Overload method				Strength reserve method			
	After pile	Difference %	Before pile	Difference %	After pile	Difference %	Before pile	Difference %
12	1.21	0.83	1.2	0	1.2	0	1.2	0
13	1.21	0.83	1.2	0	1.19	0.83	1.2	0
14	1.2	0	1.2	0	1.18	1.67	1.2	0
15	1.2	0	1.2	0	1.18	1.67	1.2	0
16	1.2	0	1.2	0	1.17	2.50	1.2	0

6 Conclusion

Using the idea of transfer coefficient method, the two methods of overload method and strength reserve method are studied, and the following conclusions are drawn:

(1) When studying the calculation of anti-slide pile thrust, in addition to the remaining sliding force behind the pile, the remaining anti-sliding force in front of the pile must also be considered.
(2) The calculation formula of the anti-skid force transfer coefficient and the calculation formula of the remaining anti-skid force are proposed and deduced.
(3) Considering the remaining sliding force before the pile and the remaining sliding force behind the pile at the same time, the calculation method of the anti-slide pile thrust is improved, and the optimal position selection condition of the anti-slide pile is proposed.
(4) It has been verified that this method uses the overload method to calculate the safety factor before and after the pile is basically the same, and is consistent with the design safety factor. The strength reserve method calculates the safety factor before the pile is accurate and reliable, but the result of the safety factor behind the pile is relatively small.

References

1. Ito, T., Matsui, T., Hong, W.P.: Design method for stabilizing piles against landslide–one row of piles. Soils Found. **21**(1), 21–37 (1981)
2. Reese, L.C., Wang, S.T., Fouse, J.L.: Use of drilled shafts in stabilizing a slope. Proc. Stabil. Perform. Slopes Embankments, 1318–1332 (1992)
3. Zeng, S., Liang, R.Y.: Stability analysis of drilled shafts reinforced slope. Soils Found. **42**(2), 93–102 (2002)
4. Liang, R.Y.: Field instrumentation, monitoring of drilled shafts for landslide stabilization and development of pertinent design method, Publication FHWA/OH–2010/15 (2010)
5. Liang, R.Y., Li, L.: Design method for slope stabilization using drilled shafts. Proc. Int. Conf. Ground Improve. Ground Control, 1475–1480 (2012)

6. Wang, L.Q., Tang, H.M., Hu, X.L., Liang, Y.: Application of residual pushing force curve to reservoir landslide anti-slide pile design. Rock Soil Mech. **16**(12), 2019–2022 (2005). (in Chinese)

7. Zheng, Y.R., et al.: Study on several problems of landslide stability analysis in the three Gorges Reservoir area. Chongqing Archit. **6**, 6–17 (2005). (in Chinese)

8. He, H.F., et al.: Calculation method of designed thrust for anti-slide pile. J. Eng. Geol. **16**(5), 694–698 (2008)

9. Hu, M.J., Liu, G.H., Huang, X.: The application of the remaining sliding force curve in landslide management design. J. Chongqing Jiaotong Univ. (Nat. Sci. Ed.) **12**(6), 587–589 (2011). (in Chinese)

10. Wang, P.Y., et al.: A new method on ascertaining designed thrust for timbering pile based on transfer coefficient method. J. Sichuan Univ. (Eng. Sci. Ed.) **42**(6), 73–78 (2010). (in Chinese)

11. The Professional Standards Compilation Group of People's Republic of China, JTG D30—2015 Code for design of highway subgrade, China Communications Press (2015) (in Chinese)

12. Zhao, S.Y., Zheng, Y.R., Ao, G.Y.: Calculation method of landslide thrust considering pile reaction force and design safety factor-implicit method of transfer coefficient. Chin. J. Rock Mech. Eng. **35**(8), 1668–1676 (2016). (in Chinese)

13. Li, H.H.: Calculation method and failure warning study of pile spanning and pile spanning design. Ph.D. Thesis, Chang'an University (2018)

14. Shi, W.M., Zheng, Y.R., Tang, B.M.: Discussion on evaluation method of landslide stability. Chin. J. Rock Soil Mech. **24**(4), 545–548 (2003). (in Chinese)

Treatment of Leakage Disease of a Deep Foundation Pit

Hui Liu[1(✉)] and Xinjie Zhang[2]

[1] Department of Architecture and Civil Engineering, Yunnan Land and Resources Vocational College, Kunming 65000, Yunnan, China
1587671510@qq.com
[2] YCIH Foundation Engineering Co. Ltd., Kunming 65000, Yunnan, China

Abstract. Due to the influence of groundwater runoff, the local water stop curtain of a deep foundation pit of a project failed, resulting in leakage of the foundation pit, which seriously threatens the safety of the foundation pit and surrounding buildings. After several rounds of site survey and demonstration, measures were adopted to "treat the leakage utilizing external sealing and internal blocking, a combination of dredging and blocking, and local strengthening", and an emergency rescue team was established to promptly eliminate potential safety hazards.

Keywords: Deep foundation pit · Water stop curtain · Leakage · Treatment

1 Introduction

Waterproof curtain is the key process for the success of deep foundation pit support scheme. Triaxial mixing pile is a common waterproof curtain method because of its ease to construct and good integration. However, due to the complex and changeable geological and hydrological conditions, the waterproof curtain often fails, resulting in leakage of the foundation pit. Leakage is one of the biggest construction risks of deep foundation pit. It seriously affects the safety of the foundation pit and construction quality [1].

Triaxial mixing piles were used for the waterproof curtain of a deep foundation pit. When the construction reached 1880 elevation, there were many water and sand gushing on the side wall of the foundation pit, causing local failure of the triaxial waterproof curtain of said foundation pit, resulting in large settlement, deformation and cracking of buildings and roads around it. It created imminent danger to personal and property safety. This research status summarizes the disease treatment scheme adopted by the foundation pit, which is intended to provide experience reference for similar projects.

The shortcomings of the research status is that the causes of foundation pit water leakage are not discussed. The reason is that the causes of foundation pit water leakage are complex, and no agreement has been reached in many rounds of expert argumentation.

G. Feng (Ed.): ICCE 2021, LNCE 213, pp. 195–208, 2022.
https://doi.org/10.1007/978-981-19-1260-3_18

2 Project Overview

The total land area of the project is about 7644.08 m² (about 11.5 mu). Two 35-storey high-rise buildings with a height of 99.7 m and a shear wall structure were proposed. The total building area is about 73827 m², and the underground area is about 17827 m². The buildings had rotary pile foundations with three-story basements. The foundation pit has an excavation depth of 16.8–17.4 m and is rectangular in shape, approximately 340 m in perimeter, 99 m in length and 72 m in depth. The east side of the foundation pit is adjacent to Yongle Road, a municipal road with many pedestrian and vehicular traffic. There are municipal water supply, drainage pipelines, communication cables and power cables under the road. The south side of the foundation pit is adjacent to a six-story concrete residential building, and the excavation boundary of the foundation pit is 10.51 m away from the outer wall of the building. The west side of the foundation pit is adjacent to a 16-storey concrete commercial and residential building with a basement. The excavation sideline of the foundation pit is 8.97 m away from the outer wall of the commercial and residential building. The north side of the foundation pit is adjacent to Yunnan Baiyao Pharmaceutical Factory, which is home to many old brick-concrete buildings. The excavation boundary of the foundation pit is only 5.23 m away from the nearest external wall of those old buildings. The current situation around the foundation pit is complex, so the settlement and deformation of the foundation pit need to be strictly monitored and controlled.

3 Geological and Hydrological Profile

3.1 Geological Conditions

According to the survey report, the stratum of the proposed site is mainly composed of artificial fill layer and Quaternary alluvial-proluvial and alluvial-lacustrine stratum, and the stratum involved in foundation pit support from top to bottom is described as follows:

① Miscellaneous fill: grayish white, gray, loose, slightly wet, mainly composed of gravel, broken bricks, concrete blocks, etc., with loose structure, low strength and uneven soil quality, distributed in the whole site;

② Clay: brown yellow, locally mixed with brown gray and gray, hard plastic state, slightly wet, medium compressibility, general soil uniformity, distributed in the site;

③ Silty clay: gray, brown gray, locally brown yellow, gray with brown yellow, plastic, slightly wet, medium compressibility, medium dry strength and toughness, distributed in some sections of the site;

④ Silt: gray, light blue gray, locally gradually changed into silty sand, wet, slightly dense, medium compressibility, slight shaking response, distributed in the site;

⑤ Gravelly sand: light gray, light blue gray, local gray, purple gray, saturated, slightly dense–medium dense. It contains a small amount of round gravel, the content is generally 20%–40%, the particle size is 0.5–2.0 cm, the main components are moderately weathered sandstone and basalt, and most of them are round and sub-round; Calcareous cementation layer and thin layer of cohesive soil and silt are occasionally intercalated between layers; It is distributed in the site;

⑥ Silty clay: locally clay, light gray, purple gray, light blue gray, slightly wet, plastic state, medium toughness and dry strength, medium compressibility, general soil uniformity, distributed in some sections of the site;

⑦ Silt; Light gray, purple gray, light blue gray, wet, medium dense, low dry strength, slight shaking response, medium compressibility, distributed in some sections of the site;

⑧ Round gravel; Gray, light gray, brown gray, locally blue gray, saturated, moderately dense, containing a small amount of pebbles, the content of gravel and pebbles is uneven, generally 50%–60%, locally up to 70%, the particle size is 0.2–2.0 cm, locally 2.0–5.0 cm, the parent rock is composed of moderately weathered sandstone and basalt, filled with silt, locally filled with silty clay, mostly round and sub-round. Calcareous cementation layer is occasionally intercalated between layers and distributed in some sections of the site;

⑨ Clay: locally silty clay, gray and dark gray, locally purple gray and light gray, slightly wet, plastic, locally hard plastic, medium toughness and dry strength, medium compressibility, distributed in some sections of the site.

3.2   Hydrological Overview

During the survey, the depth of the mixed stable groundwater level in the site measured by drilling is 0.6–1.8 m, the elevation is 1887.07–1888.32 m, and the water level variation difference is 1.25 m. The groundwater type of the proposed site is mainly pore phreatic water and the miscellaneous fill distributed on the surface contains a small amount of perched water. Pore water mainly exists in ③, ④ silt; ⑤ gravel sand and ⑧ round gravel. It is replenished by the infiltration of atmospheric precipitation and surface water.

There is pressured underground water on the site, and the top of the water table is about 6 m below ground surface. According to the results of the foundation pit pumping test, the inrush volume of the foundation pit is 1737.3 m^3/d, and the underground water has a great influence on the construction of this foundation pit, so it is necessary to protect the foundation pit from underground water during the design and construction phase.

4   Foundation Pit Support Design

According to the engineering conditions, excavation depth and surrounding conditions of the foundation pit, it was determined that the safety level of the retaining structure of the foundation ditch is Level I, and the structural importance coefficient is 1.1.

Foundation pit support scheme: "triaxial support curtain + prestressed combined steel support + local concrete support"; Drainage and drainage scheme: Φ850 × 850 × 850 triaxial mixing piles form a curtain, and local triaxial mixing piles can not be applied to the ground.See Fig. 1 Foundation Pit Support Plan, Fig. 2 Support Layout Plan and Fig. 3 Drain off water and Fig. 4 Foundation Pit Support Profile for details of the scheme of "drainage of blind ditch at the bottom of the pit + drainage of intercepting ditch at the top of the pit + drainage of collecting well + monitoring and recharge of recharge well outside the pit".

During the excavation of the foundation pit, a temporary collection well shall be created to allow water to be pumped and discharged. The bottom of the foundation pit shall be provided with a 300 × 300 mm blind drain along the excavation line of the foundation pit bottom; 800 × 800 × 800 mm dirt collection wells shall be set every 25 m, and 300 × 300 dirt intercepting ditches shall be set at the top of the foundation pit to guide the dirt on the ground to the periphery of the foundation pit, so as to ensure that no dirt accumulates inside the foundation pit during construction. The surrounding surface tritium can be discharged in time to minimize the amount of tritium in the seepage stratum. In the process of dewatering and drainage of the foundation pit, the level outside the pit shall be controlled at 2.0 m below the ground surface, and the level inside the pit shall be controlled at 0.55 m below the bottom of the foundation pit. 17 recharge wells with a depth of 26 m shall be arranged outside the pit, and one well shall be arranged at 20 m along the periphery of the excavation.

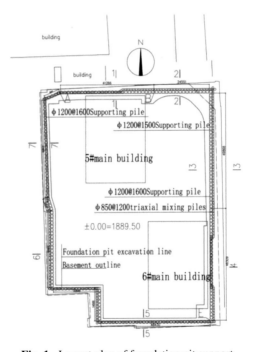

Fig. 1. Layout plan of foundation pit support

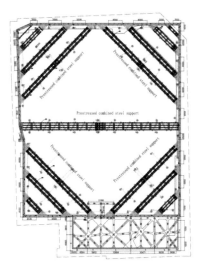

Fig. 2. Support layout plan

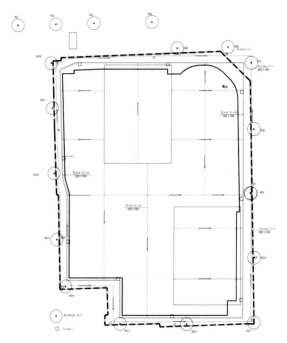

Fig. 3. Drain off water

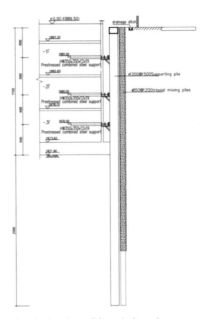

Fig. 4. Section of foundation pit support

5 Leakage of Foundation Pit

During the excavation at 1880.00 m, a great amount of water and sand mixture gushed in through the side wall of the foundation pit, causing the local failure of the original triaxial waterproof curtain of the foundation pit, and resulting in large amount of settlement, deformation and cracking of buildings and roads around the foundation pit. Data collected through the monitoring program data reflected the process of foundation pit deformation, as shown in Figs. 5, 6, 7, and 8. At this stage, the cumulative value and rate of settlement and deformation of the foundation pit exceeded the warning value. Especially on April 14–15 and June 22–23, two major dangerous situations occurred. The horizontal displacement of some crown beams increased suddenly, the groundwater level dropped rapidly, and the surrounding buildings subsided obviously.

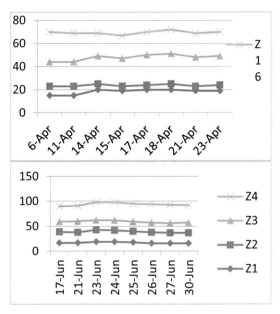

Fig. 5. Horizontal displacement of crown beam

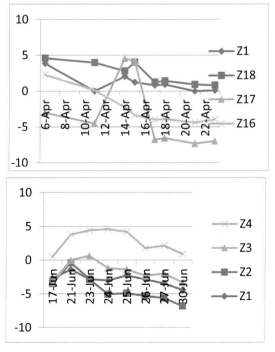

Fig. 6. Vertical displacement of crown beam

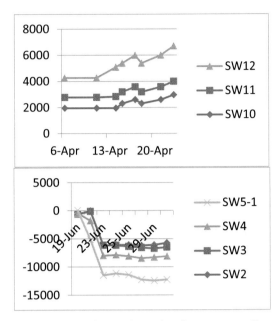

Fig. 7. Change of groundwater in adjacent surroundings

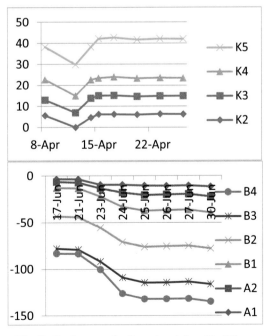

Fig. 8. Settlement of adjacent peripheral buildings

6 Cause Analysis of Leakage Disease

After the occurrence of water and sand gushing in through the side wall of the foundation pit, well-known experts in the province were invited to analyze, survey and demonstrate many times. At first, there were different opinions on the causes of foundation pit leakage, and no consensus could be reached. Through half a year's observation and several rounds of thematic meetings for analysis, investigation and demonstration, we have gradually reached a consensus that the leakage of foundation pit was caused by the influence of groundwater runoff and the loss of some cement slurry during construction, which resulted in the local failure of the original triaxial waterproof curtain of foundation pit and leads to leakage.

7 Construction Measures for Treatment of Leakage Disease

Due to the serious leakage of water from the side wall of the foundation pit, the roads and buildings around the foundation pit obviously settled and cracked. Geophysical prospecting showed that the local geological structure around the foundation pit had changed, and a large cavity had appeared, which endangered the safety of the surrounding people and property. Therefore, it was urgent to control the damages. See Fig. 9. Cracking of roads and buildings.

Fig. 9. Cracking of roads and buildings

After the demonstration of the special meeting, because the project support had been completed and the foundation pit backfilling conditions were not available, the treatment scheme of "external sealing and internal blocking, combination of dredging and blocking, local strengthening" has been adopted. After the implementation, the leakage was stopped in time, the deformation of the foundation pit was reduced, and the safety of the foundation pit was ensured. Specific measures are as follows:

7.1 Treatment of the Position Outside the Foundation Pit Corresponding to the Water Leakage Point

High-pressure rotary jet grouting is carried out on the outer side of the top beam of the foundation pit at the water leakage point between the corresponding supporting piles. A proper amount of water glass is added. The depth of high-pressure rotary jet grouting is 5 to 10 m deeper than the water leakage position. If the high-pressure jet grouting drilling encounters underground obstacles and cannot be lowered, the geological drilling rig shall be used to guide the hole before the high-pressure jet grouting construction, and the depth of the hole shall be determined according to the actual situation of the site.

7.2 Treatment of the Inner Side of the Foundation Pit Corresponding to the Leakage Point

A drainage tube is first made at a water leakage hole, two galvanized pipes are buried in advance, the water leakage hole is plugged by watertight, polyurethane is injected by a gear pump after the watertight strength of the plugging is reached, and finally the drainage tube is plugged to play a role of quickly stopping water and polyurethane foaming can also fill empty land, and then a double-liquid grout machine is adopted to spray cement slurry and wat glass aiming at that water leakage area, so that the cement slurry is quickly solidified, and then the wat leakage area is quickly subjected to mesh spraying construction [2].

If water gushing is accompanied by sand gushing, in order to avoid holes and ground subsidence, it would be necessary to block the water leakage port with cotton quilt to filter the sand and reduce the loss of sand, and use sandbags to backpressure the water leakage area in layers, with the backpressure height 2–5 m higher than the leakage point to the upper purlin.

If the water leakage of the foundation pit is accompanied by sand and holes occur, the outside of the foundation pit shall be filled with graded gravel while grouting, so as to reduce the settlement around the foundation pit caused by water leakage.

7.3 Treatment of Underground Cavity Location

Holes have appeared under the ground surface on the east and south sides of the foundation pit, and the buildings on the ground have been inclined. The first thing is to evacuate the people in the ground building immediately, close the area with a cordon, and prompt the passing people and vehicles to detour. Secondly, the soil shall be reinforced by densely distributed high-pressure jet grouting piles with a diameter of φ 500 @ 1000 mm and a depth of 26 m [3].

8   Construction Technology

8.1   Watertight and Polyurethane

Main construction parameters: the watertight adopts the quick-setting type; the polyurethane adopts a hydrophilic single-liquid PU polyurethane foaming agent.

Construction procedures: clean up the leakage point → bury the aqueduct → fill the water-tight plugging agent → inject the polyurethane foaming agent → plug the aqueduct → moisturize and cure → check and repair the leakage → hang the net and spray the surface.

(1) Clean up the leakage point

1) When cleaning the leakage point, the soft soil, silt and loose sand around the leakage point must be completely cleaned up. The treated base surface shall be fully wet to saturation, but without visible water, and shall be firm, clean and flat. The uneven part can be leveled with water without leakage.

2) The whole process of cleaning the leakage point should be timely and rapid;

3) During cleaning, pay attention to the change of soil around the leakage point at all times, and take emergency measures.

(2) Filling watertight

1) Watertight materials are easy to get damp and should be sealed and stored in a dry place before use;

2) When mixing watertight materials, special rubber gloves shall be worn to prevent skin burns;

3) The temperature of the water used for watertight mixing is generally controlled at 15–30 °C. When the ambient temperature is lower than 5 °C, nitrous acid accounting for 1–3% of the weight of the powder shall be added, and warm water (about 50 °C) shall be used for mixing below 10 °C;

4) The sequence of watertight filling should be from top to bottom, and it is not suitable to fill from the periphery to the center, otherwise the water plugging effect is not good, and the watertight material is wasted. The thickness of watertight filling should not be less than 10 cm;

5) During the use of watertight plugging, it is not allowed to add water twice;

6) When the construction environment is under the scorching sun, humidification and curing are required.

(3) Embedding of aqueduct

1) The length of the aqueduct should not be too long, so as not to affect the construction of the next process such as earthwork excavation, and the length should be 0.6 m;

(2) The number and diameter of buried aqueducts can be adjusted according to the amount of water leakage, generally 1–5 aqueducts are suitable, and the diameter is 0.15 m–0.3 m;

3) a ditch is dug around the water leakage point, and the water flowing out of the aqueduct is introduced into the ditch;

4) The accumulated water in the foundation pit shall be pumped away in time to reduce the impact of the accumulated water on the stability of the foundation pit.

(4) Filling polyurethane foaming agent under high pressure

1) Before drilling with an electric drill, ensure that the watertight has been completely solidified (10 min after laying the watertight), drill holes on the watertight after final setting, and lock the water stop needle;
2) before that mix liquid is injected into the wall body through the wat stop needle, the power supply is started to extrude the residual diluent in the material placing barrel and the pipe until the mix liquid is sprayed out;
3) During the matching and spraying of polyurethane and polyurethane paint, operators must wear special gloves to avoid contact with human body;
4) Polyurethane plugging agent is a water-swellable material. Wear protective equipment such as gloves and sunglasses when working. If it splashes into the eyes, wash it with clean water immediately and then send it to the hospital.

(5) Key points for operation of high pressure plugging and filling machine

1) Operate the high-pressure leak-stopping filling machine in strict accordance with the instructions;
2) Continuous injection of polyurethane should not exceed 0.5 h to prevent parts from being worn due to overheating of the fuselage;
3) When the injection of polyurethane is stopped for more than 30 min or the construction is completed, the machine shall be cleaned in time with diluent. The cleaning method is to pump all the slurry in the charging barrel and high-pressure pipe back to the container, and then pour 300 ml of diluent into the charging cup to push out the polyurethane in the high-pressure pipe. After the polyurethane is completely ejected, 300 ml of diluent is poured in, and the filling material is put into a material placing cylinder for circular cleaning; After cleaning for 2 to 3 min, the cleaning agent is pumped into the container, and then a proper amount of engine oil is poured into the container for circulation for maintenance and lubrication, and the cleaning is completed;
4) The filling machine shall be checked regularly before use (the gearbox shall be filled with grease regularly), and any abnormality shall be repaired immediately to prevent failure during construction.

8.2 Hanging Net and Spraying Wall

Construction procedures: earth excavation → manual cleaning → drilling and planting reinforcement on supporting pile → hanging of finished steel mesh → spraying fine aggregate concrete → concrete curing → fabrication and installation of reinforcement mesh → welding of reinforcing bar → spraying fine aggregate concrete → concrete curing.

(1) Preparation before net hanging
Remove the loose rocks, dangerous rocks and floating soil between the supporting piles, and fill the larger cracks and pits to make the slope smooth and tidy.

(2) Fixing of reinforcement mesh

Firstly, according to the construction drawings, drill holes (depth of 250 mm) at 1500 mm/1600 mm horizontally (corresponding to the spacing of the supporting piles in this section) and 1500 mm vertically on the supporting piles for rebar planting, and anchor C14 L-shaped mesh rebar with a total length of 800 mm;

A 600 mm long C14 reinforcement nail is driven into the center of the soil between the supporting piles (the same as the horizontal position of the L-shaped mesh reinforcement) to fix the reinforcement mesh between the piles;

The reinforcement mesh (two-way φ 6 @ 150 mm × 150 mm) is then fixed by welding (10 d on one side) the C14 transverse tie bars (i.e., stiffeners, @ 1500 mm/1600 mm × 1500 mm) to the L-shaped mesh reinforcement embedded in the pile, pressing the entire mesh.

(3) Raw materials and mix proportion

Raw materials: ① PO42.5 ordinary Portland cement; ② Silty sand shall be used, and the water content of the sand shall be 2%–3%; ③ Adopt melon seed stone as that stone; ⑥ Clean water is used for water; ⑤ That mix amount of the accelerator is 5% of the cement amount; ⑥ Watertight.

Selection of C20 fine aggregate concrete mix proportion (mass ratio) (dry material): cement: sand: gravel: water: accelerator: watertight = 1: 1.6: 2.8: 0.55: 0.05: 0.1.

8.3 Double-Liquid Grouting

1) Grouting pipe embedding

When the drainage pipe is buried at the leakage point, the galvanized pipe shall be buried. The buried depth of the galvanized pipe shall be as deep as possible according to the leakage situation of the leakage point. One end shall be sealed with adhesive tape to prevent silt from blocking the grouting pipe.

2) Grout configuration

Before slurry preparation, field test shall be carried out according to cement slurry concentration, water glass parameters and water temperature. The initial setting time of the double-liquid slurry is usually 20 s to 40 s, and the ratio of the slurry before each grouting is determined (1:1 to 1:2).

3) Double-liquid grouting

Before grouting, double pipes shall be injected with water to ensure that the embedded grouting pipe is unblocked. After confirming that the grouting pipe is unblocked, water glass and cement slurry shall be injected at the same time. At the beginning of grouting, the drainage pipe will flow out of the mixture of cement slurry and water glass. Continue grouting until the drainage pipe is blocked. At the same time, continue grouting. The amount of grouting shall be judged according to the time and size of water leakage.

9 Conclusion

(1) The geological structure is local and the geological conditions are unevenly distributed, so sometimes the geological survey report cannot fully reflect the geological conditions. Before construction, the geological and hydrological conditions around the plot should be fully understood and studied [4].

(2) The complexity of geological structure shall be fully considered in the design of waterproof curtain, the design scheme shall be put forward pertinently, the possible problems and remedial schemes shall be clarified in key links, and the feasibility of the design scheme shall be fully verified by relevant experiments before implementation.

(3) In case of abnormal conditions during construction, analyze the causes in time to avoid blind construction, and fully verify the reliability of the previous process before special process replacement. Emergency plans shall be prepared before implementation, and solutions shall be put forward and implemented in time when unexpected situations occur.

References

1. Ren, B.X., Guo, W.P.: Leakage treatment of a deep foundation pit after failure of waterproof curtain. Construct. Technol. **42**(16), 105–110 (2013)
2. Tian, Z.J.: Treatment methods and preventive measures for leakage of foundation pit in underground engineering. Build. Construct. **11**, 57 (2019)
3. Qu, S., Ye, H.S.: Remedial measures for water leakage of deep foundation pit support. Construct. Supervision 184–187 (2020)
4. Hou, X.Y., Liu, J., Xue, B.F., Huo, Y.X.: Cause analysis and treatment measures of underground continuous wall leakage in subway foundation pit. Build. Technol. **48**(9), 972–975 (2017)

Vibration-Based Damage Joint Identification Method for Superstructure and Substructure of Piles-Supported Frame Structure

Zhengang Zhou$^{(\boxtimes)}$, Dejun Liu, and Xiujie Lv

College of Civil Engineering and Architecture, Jiaxing University, Jiaxing, China
zhou_zhengang@126.com

Abstract. In order to jointly identify the damage locations of superstructure and substructure of the piles-supported frame structures, a damage identification method based on vibration is proposed. Firstly, the high-efficiency modes which are sensitive to the damage of the piles-supported frame structures are determined. Then, the element modal strain energy difference functions of the corresponding high-efficiency modes are calculated before and after the damage, and finally the damage locations are identified by the average values of the absolute values of the wavelet transform coefficients of the element modal strain energy difference functions of high-efficiency modes. The effectiveness of the method is studied by numerical simulation. Numerical results show that the method can identify the damage location of the single damage or multiple damage of the piles-supported frame structures. Although the adjacent effect exists, the damage areas can be effectively located. At the same time, the method can effectively identify the damage locations of the hidden pile foundation.

Keywords: Frame structure · Pile foundation · Damage identification · Modal strain energy · Wavelet transform

1 Introduction

Frame structures are widely used in multi-story buildings, high-rise buildings and stadiums. Due to the multiple influencing factors, such as long-term effect of loads, environmental corrosion, aging of structural materials, earthquake action, and typhoon action, etc., the accumulative damage will occur in frame structures within their service period, even worse, may lead to engineering accidents. Therefore, it is necessary to study the problem of damage identification of frame structures. Vibration-based structural damage identification method is one of the effective ways to solve this problem. The basic principle of this method is as follows: As the structural damage appears, the physical parameters of the structure change [1–5]. Therefore, combining certain identification techniques, it is possible to identify structural damage by using the measured responses or the indexes indirectly calculated from the measured responses, such as natural vibration frequencies [6], modal shapes [7], modal curvatures [8], residual forces [9], flexibility matrix [10, 11], modal strain energy [12], etc.

G. Feng (Ed.): ICCE 2021, LNCE 213, pp. 209–225, 2022.
https://doi.org/10.1007/978-981-19-1260-3_19

In the past three decades, several notable achievements had been made in the research of vibration-based damage identification methods of frame structures. The damage identification of the superstructural members (such as beams and columns or joints) of two-dimensional plane frame structures was firstly studied by adopting vibration-based damage identification methods. These research works were mainly based on the measured responses or the indexes indirectly calculated from the measured responses, such as strain modal shapes [13], modal frequencies and displacement modal shapes [14–21], modal strain energy [22], the measured displacement responses [23, 24], the measured acceleration responses [25–27]. In these works, some can only effectively identify the damage location of the superstructure members of two-dimensional plane frame structures [13, 14, 16, 23, 26], while others can effectively identify the damage location and damage degree both [15, 17–22, 24, 25, 27]. The above research works showed that the vibration-based damage identification methods can identify the damage of the superstructural members of two-dimensional plane frame structures. However, the actual frame structures are relatively complex three-dimensional spatial structures. The effectiveness of the vibration-based damage identification methods for the damage identification of three-dimensional frame structures needs further verification. Therefore, the damage identification of the superstructural members of three-dimensional frame structures was further studied by adopting vibration-based damage identification methods. These research works were also mainly based on the measured responses or the indexes indirectly calculated from the measured responses, such as modal frequencies and modal shapes [28–33], modal strain energy [34], the measured displacement responses [35], the measured acceleration responses [36–38]. In these works, some can only effectively identify the damage location of the superstructure members of three-dimensional plane frame structures [32, 34, 36, 38], while others can effectively identify the damage location and damage degree both [28–31, 33, 35, 37]. The above research works showed that the vibration-based damage identification methods can identify the damage of the superstructural members of three-dimensional plane frame structures. It seems that the above research can solve the problem of damage identification of actual frame structures. However, the existing research works on vibration-based damage identification methods of frame structures mainly focused on the damage identification of beams and columns or joints of frame structures, but the damage identification of floor slabs in frame structures has not been considered. At the same time, the existing methods are proposed under the rigid base assumptions, without considering the influences of the soil-foundation-structure interaction (SSI) effect. However, the SSI effect will make the dynamic characteristics and responses of the structural system great differences with the rigid foundation assumption situation. It will inevitably affect the accuracy of the vibration-based damage identification methods. In addition, the existing methods for damage identification of frame structures are merely applicable to the damage identification of its superstructural members, but not applicable to the damage identification of its hidden substructural members. Therefore, it is necessary to study the damage identification method of soil-foundations-frame structure as a whole, so as to realize the joint identification of superstructural and substructural damage. As the authors' knowledge, there has been no research on this issue.

In this paper, a vibration-based method to jointly identify the damage locations of the superstructure and substructure of the piles-supported frame structures was proposed. Firstly, the high-efficiency modes which are sensitive to the damage of the piles-supported frame structure were determined. Then, the element modal strain energy difference functions of the high-efficiency modes were calculated before and after the damage, and finally the damage locations were identified by the average values of the absolute values of the wavelet transform coefficients of the element modal strain energy difference functions of high-efficiency modes. Through numerical simulation, the feasibility of the proposed vibration-based method to jointly identify the damage locations of the superstructure and substructure of the piles-supported frame structures was preliminarily verified.

2 The Description of the Proposed Damage Identification Method

In this paper, the proposed damage identification method for piles-supported frame structure is introduced from the following aspects: damage identification process, high-efficiency modes determination, element modal strain energy and strain energy difference functions calculation, wavelet transform analysis, and damage identification index.

2.1 Damage Identification Process

The identification process of the proposed damage identification method is shown in Fig. 1. Specific identification steps are illustrated as follows:

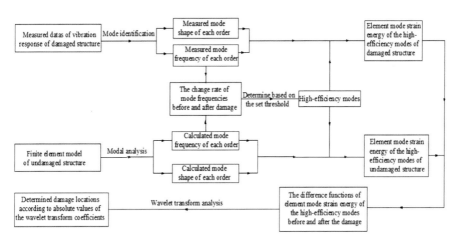

Fig. 1. Flow chart of damage identification of piles-supported frame structure.

Firstly, the finite element model of undamaged piles-supported frame structure will be established based on the available engineering data. Through the modal analysis, the modal frequencies and modal shapes of undamaged piles-supported frame structure will be obtained.

Secondly, by means of environmental excitation, the vibration response data of the damaged piles-supported frame structure will be obtained under the operating state. Through the modal identification, the measured modal frequencies and modal shapes of the damaged piles-supported frame structure will be obtained.

Thirdly, the high-efficiency modes will be determined by calculating the change rate of modal frequency of each order mode of the piles-supported frame structure before and after damage.

Fourthly, the element modal strain energy of the corresponding high-efficiency modes of the undamaged and damaged piles-supported frame structure will be calculated respectively. Then, the difference functions of element modal strain energy of the high-efficiency modes will be calculated before and after the damage.

Finally, the damage locations of corresponding members (such as beams, columns, plates and piles) can be determined by the average values of the absolute values of the wavelet transform coefficients obtained from wavelet transform analysis of the difference functions of element modal strain energy of the high-efficiency modes.

2.2 High-Efficiency Modes Determinations

According to the change rate δ of modal frequency of each mode before and after damage, the high-efficiency modes used for structural damage identification will be selected. The change rate δ_i of modal frequencies of the ith mode before and after damage can be calculated by Eq. (1).

$$\delta_i = \frac{f_i^{u} - f_i^{d}}{f_i^{u}} \times 100\% \tag{1}$$

Where f_i^{u} and f_i^{d} are the modal frequencies of the structure's ith mode before and after damage, respectively, in Hz.

The existence of structural damage can be judged according to δ_i. The larger the δ_i value is, the mode is more sensitive to the structural damage.

The high-efficiency modes for subsequent damage identification will be selected according to the following criteria: Firstly, the mode which has the largest δ value will be selected. Secondly, the modes that their δ values are not less than 80% of the largest δ value will be selected. If the number of selected modes is only one order, the selection criteria can be appropriately relaxed to ensure the number of the selected modes is not less than two orders.

2.3 Element Modal Strain Energy and Strain Energy Difference Functions Calculation

The element modal strain energy can be calculated by modal shapes and stiffness matrix. Since the element modal strain energy can reflect the change of local characteristics of the structure and it is sensitive to the local structural damage which is much higher than that of the modal shapes. Therefore, the element modal strain energy will be used as the basic quantity to determine the locations of structural damage.

The element modal strain energy of the ith high-efficiency mode of the jth element can be calculated as follows [39–44]:

$$MSE_{i,j}^{u} = (\Phi_{i}^{v})^{T}\mathbf{K}_{j}^{u}\Phi_{i}^{u} \tag{2}$$

$$MSE_{i,j}^{d} = (\Phi_{i}^{d})^{T}\mathbf{K}_{j}^{d}\Phi_{i}^{d} \tag{3}$$

Where $MSE_{i,j}^{u}$ and $MSE_{i,j}^{d}$ are the element modal strain energy of the ith high-efficiency mode of the jth element before and after damage respectively; $\mathbf{K}_{j}^{u}$ is the element stiffness matrix of the jth element before damage; $\mathbf{K}_{j}^{d}$ is the element stiffness matrix of the jth element after damage. Since the element stiffness after damage cannot be measured, $\mathbf{K}_{j}^{u}$ is generally adopted to replace $\mathbf{K}_{j}^{d}$.

The difference function of the element modal strain energy of the ith high-efficiency mode of the jth element before and after damage, $MSEC_{i,j}$, can be obtained by Eq. (4).

$$MSEC_{i,j} = MSE_{i,j}^{u} - MSE_{i,j}^{d} \tag{4}$$

2.4 Wavelet Transform Analysis

Once the structure is damaged, the $MSEC$ value of the element which located in the structure damage area will be small obvious change. Due to the wavelet transform can analyze the change of signal data well [23]. Therefore, in order to obtain better identification effect of damage localization, the one-dimensional continuous wavelet transform (CWT) will be used to analyze the $MSEC$ value of the element of each high-efficiency mode. The biorthogonal spline wavelet function bior6.8 will be used as the wavelet basis function. The sequence constituted by $MSEC_{i,j}$ values will be taken as the real-valued input signal, and the element number which is corresponding to location of the element will be taken as the time variable. The absolute values of the wavelet transform coefficient of the element modal strain energy difference function of the ith high-efficiency mode of the jth element, $MSECD_{i,j}$, will be obtained by wavelet transform. According to the peak of $MSECD_{i,j}$ values with the number of elements, the locations of the structure damage can be determined.

2.5 Damage Identification Index

In order to reduce the influence of random noise, the n order high-efficiency modes will be used to identify damage locations of the structure. The average value of the absolute values of the wavelet transform coefficients of the element modal strain energy difference functions of high-efficiency modes, $MSECM$, will be used as the damage identification index. The average value of the absolute values of the wavelet transform coefficients of the element modal strain energy difference functions of high-efficiency modes of the jth element, $MSECM_{j}$, can be calculated by Eq. (5). The damage locations of the structure will be identified according to the peak of the above damage identification index along with the locations of the elements.

$$MSECM_{j} = \frac{1}{n}\sum_{i=1}^{n}MSECD_{i,j} \tag{5}$$

3 Damage Identification Numerical Simulation

3.1 Numerical Modeling of Piles-Supported Frame Structure Test Model

The numerical simulation study had taken the test model (as show in Fig. 2 and Fig. 3) of the subsequent model test as the simulation object. The model test has not been carried out yet, and the feasibility of the proposed method was preliminarily discussed in this paper based on the numerical simulation.

The finite element models of undamaged and damaged cases of the piles-supported frame structure test model were established.

The integral finite element model of soil-piles-frame structure under undamaged case was established by using ANSYS software, as shown in Fig. 4. The finite element model of piles-frame structure is shown in Fig. 5 (soil, soil box and concrete base are not shown).

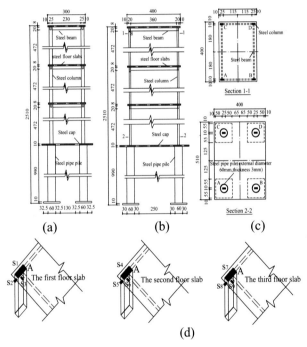

Fig. 2. The design of test model (unit: mm). (a) Front elevation. (b) Side elevation. (c) Section. (d) Layout of measuring points.

The corresponding dimensions of beams, columns, piles, floor slabs, cap, soil field, soil box and concrete base are same as the test model, as shown in Fig. 2 and Fig. 3.

In the numerical simulation study, the assumed values of material parameters are as follows: The elastic modulus, density and Poisson ratio of steel are 2.1×10^5 MPa, 7850 kg/m^3 and 0.33, respectively. The elastic modulus, density and Poisson ratio of loose sand are 25 MPa, 1900 kg/m^3 and 0.35, respectively. The elastic modulus, density

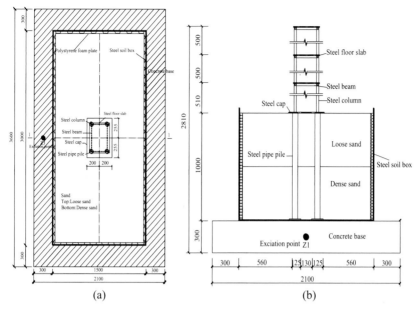

Fig. 3. The design of test soil box and the layout of the test model (unit: mm). (a) Layout of test model in the soil box. (b) Section 1–1.

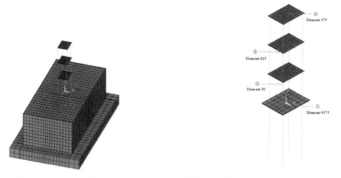

Fig. 4. The finite element model of the undamaged soil-piles-frame structure system.

Fig. 5. The finite element model of piles-frame structure system.

and Poisson ratio of dense sand are 120 MPa, 2000 kg/m^3 and 0.3, respectively. The elastic modulus, density and Poisson ratio of Polystyrene foam plate are 7 MPa, 30 kg/m^3 and 0.3, respectively. The elastic modulus, density and Poisson ratio of concrete are 3 × 10^4 MPa, 2500 kg/m^3 and 0.2, respectively.

The steel beams, steel columns and steel pipe piles were simulated by beam188 elements. The steel floor slabs and soil box were simulated by shell63 elements. The steel cap, soil, Polystyrene foam plate and concrete base were simulated by solid45 solid elements. Each steel beam, steel column and steel pipe pile was all divided into

10 elements. Each steel floor slabs was divided into 144 elements. Steel soil box was divided into 1520 elements. The steel cap, soil, Polystyrene foam plate and concrete base were divided into 36 elements, 4680 elements, 920 elements and 1536 elements respectively.

At the bottom of the concrete base, the solid boundary was adopted. As the structural responses under environmental excitation are generally small, the nonlinear effects of soil and the separation and slip between piles and soil have no obvious influence on the structural responses. Therefore, the nonlinear effects mentioned above were ignored in the numerical simulation.

3.2   Damage Identification Numerical Simulation Cases

The damage cases of the numerical simulation study are shown in Table 1. The damage location numbers are shown in Fig. 5, where ① represents the damage of pile top element, ② represents the damage of the element at the top of the first layer of the column, ③ represents the damage of the element at the end of the second floor side beam, and ④ represents the damage of the element at the mid-span of the third floor slab. In the numerical simulation, the stiffness of the corresponding damage element was reduced to simulate the damage of the structure. For each damage cases, only by modifying the material parameters at the corresponding damaged element (reducing the elastic modulus) on the basis of the undamaged piles-supported frame structure finite element model, the damaged structure finite element model under the corresponding damage cases were obtained.

Table 1. Damage identification numerical simulation cases

Case numbers	Damage locations	Damage degrees
1	①	25%
2	②	10%
3	③	10%
4	④	50%
5	①, ②	25%, 10%
6	①, ③	25%, 10%

3.3   Modal Analysis and High-Efficiency Modes Selection

The modal shapes calculated from the above six damaged cases models are similar to the undamaged case model, with the mode numbers corresponding to each other. Figure 6 shows the modal frequency change rates of the first 50 modes in each damaged case. According the Fig. 6 and the selection criteria of high-efficiency mode described in Sect. 2.2, the high-efficiency modes can be determined for each damaged case, as shown in the Table 2.

The high-efficiency modes determined in Table 2 are shown in Fig. 7 (soil, soil box and concrete base are not shown). The modal frequencies of the modes corresponding to Fig. 7 calculated from the models of the undamaged and damaged cases are shown in Table 3.

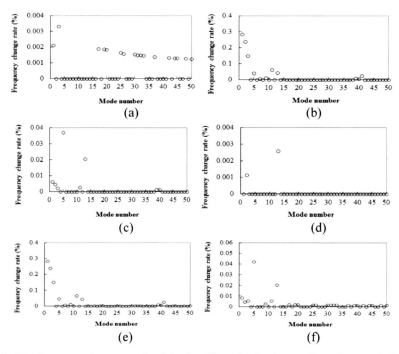

Fig. 6. The frequency change rate δ_i of the first 50 modes for damaged. (a) Case 1. (b) Case 2. (c) Case 3. (d) Case 4. (e) Case 5. (f) Case 6.

Table 2. Selection of high-efficiency modes for damaged cases

Case numbers	High-efficiency modes' numbers
1	1, 3
2	1, 2
3	5, 13
4	2, 13
5	1, 2
6	5, 13

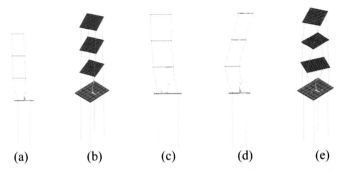

(a)   (b)   (c)   (d)   (e)

Fig. 7. The modal shapes of high-efficiency modes. (a) 1-order. (b) 2-order. (c) 3-order. (d) 5-order. (e) 13-order.

Table 3. The modal frequencies of the high-efficiency modes of undamaged and damaged models (unit: Hz)

Mode numbers	Undamaged model	Damaged models					
		Case 1	Case 2	Case 3	Case 4	Case 5	Case 6
1	4.7482	4.7481	4.7348	4.7479	4.7482	4.7347	4.7478
2	8.9094	8.9094	8.8883	8.9090	8.9093	8.8882	8.9090
3	9.0357	9.0354	9.0223	9.0355	9.0357	9.0220	9.0352
5	18.9900	18.9900	18.9820	18.9830	18.9900	18.9810	18.9820
13	38.8470	38.8470	38.8300	38.8390	38.8460	38.8300	38.8390

4 Results and Analysis of the Damage Identification Numerical Simulation

4.1 Results and Analysis of the Single Damage Identification

In order to verify the validity of the proposed method in this paper for damage location identification of a single damaged case, the damage identification algorithm mentioned above was used to identify the damage locations of damaged cases 1 to 4. Figure 8, Fig. 9, Fig. 10 and Fig. 11 show the damage location identification results of the above four damaged cases.

Figure 8 shows that the damage identification results for damaged case 1. The average values of the absolute values of the wavelet transform coefficients of the element modal strain energy difference functions of high-efficiency modes (namely *MSECM* values) of pile element No. 9375 and its adjacent pile element No. 9376 have an obvious peak. The *MSECM* values of column elements, beam elements and slab elements change small. Furthermore, the *MSECM* value of the pile element No. 9375 is the largest of all. Therefore, it is known that there may be damage at and near pile element No. 9375.

Figure 9 shows that the damage identification results for damaged case 2. The *MSECM* values of column element No. 30 and its adjacent column element No. 29

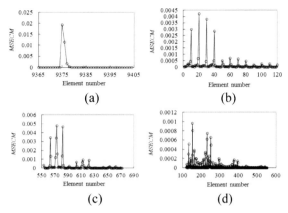

Fig. 8. The damage identification results of the damaged case 1. (a) Pile elements. (b) Column elements. (c) Beam elements. (d) Slab elements.

and beam elements No. 563, 572 to 573 and 582 have an obvious peak. The *MSECM* values of pile elements and slab elements change small. Furthermore, the *MSECM* value of the column element No. 30 is the largest of all. Therefore, it is known that there may be damage at and near column element No. 30.

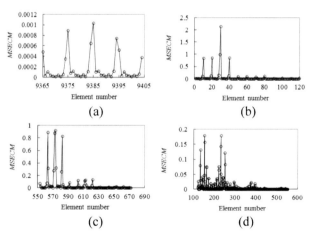

Fig. 9. The damage identification results of the damaged case 2. (a) Pile elements. (b) Column elements. (c) Beam elements. (d) Slab elements.

Figure 10 shows that the damage identification results for damaged case 3. The *MSECM* values of beam element No. 623 and its adjacent beam element No. 624, slab elements No. 385 to 387 and 406 to 408, column elements No. 61, 70, 101 and 110 have an obvious peak. The *MSECM* values of pile elements change small. Furthermore, the *MSECM* value of the beam element No. 623 is the largest of all. Therefore, it is known that there may be damage at and near beam element No. 623.

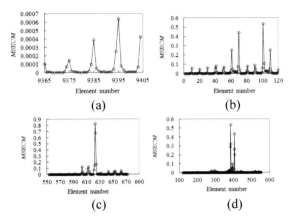

Fig. 10. The damage identification results of the damaged case 3. (a) Pile elements. (b) Column elements. (c) Beam elements. (d) Slab elements.

Figure 11 shows that the damage identification results for damaged case 4. The *MSECM* values of slab element No. 475 and its adjacent slab elements No. 474 to 476 have an obvious peak. The *MSECM* values of pile elements, column elements and beam elements change small. Therefore, it is known that there may be damage at and near slab element No. 475.

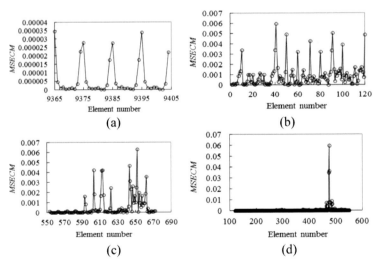

Fig. 11. The damage identification results of the damaged case 4. (a) Pile elements. (b) Column elements. (c) Beam elements. (d) Slab elements.

Through the numerical simulation results of the above four single damaged cases, it can be seen that the proposed damage location identification method in this paper can better identify the damage location of single damage that may exist in different parts of the piles-supported frame structure. Despite the influence of adjacent effect exists, but the damage area of the piles-supported frame structure still can be effectively located.

4.2 Results and Analysis of the Multiple Damage Identification

In order to verify the validity of the proposed method in this paper for damage location identification of a multiple damaged case, the damage identification algorithm mentioned above was used to identify the damage locations of damaged cases 5 to 6. Figure 12 and Fig. 13 show the damage locations identification results of the above two damaged cases.

Figure 12 shows that the damage identification results for damaged case 5. There are two main locations where the *MSECM* values have an obvious peak: (1) Pile element No. 9375 and its adjacent pile element No. 9376. (2) Column element No. 30 and its adjacent beam elements No. 563, 572 to 573 and 582. The *MSECM* values of slab elements change small. Furthermore, the *MSECM* value of the pile element No. 9375 is the largest in the pile elements. The *MSECM* value of the column element No. 30 is the largest in the column elements. Therefore, it is known that there may be damage at and near pile element No. 9375 and column element No. 30.

Figure 13 shows that the damage identification results for damaged case 6. There are two main locations where the *MSECM* values have an obvious peak: (1) Pile element No. 9375 and its adjacent pile element No. 9376. (2) Beam element No. 623 and its adjacent beam element No. 624, slab elements No. 385 to 387 and 406 to 408 and column elements No. 61, 70, 101 and 110. Furthermore, the *MSECM* value of the pile element No. 9375 is the largest in the pile elements. The *MSECM* value of the beam element No. 623 is the largest in the beam elements. Therefore, it is known that there may be damage at and near pile element No. 9375 and beam element No. 623.

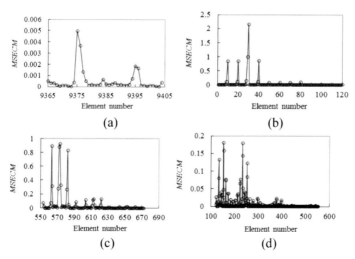

(a)   (b)

(c)   (d)

Fig. 12. The damage identification results of the damaged case 5. (a) Pile elements. (b) Column elements. (c) Beam elements. (d) Slab elements.

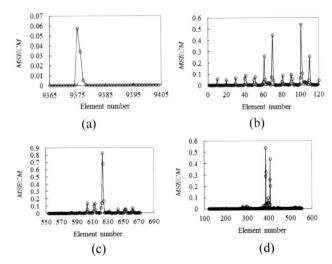

Fig. 13. The damage identification results of the damaged case 6. (a) Pile elements. (b) Column elements. (c) Beam elements. (d) Slab elements.

Through the numerical simulation results of the above two multiple damaged cases, it can be seen that the proposed damage location identification method in this paper can better identify the damage locations of multiple damage that may exist in different parts of the piles-supported frame structure. Despite the influence of adjacent effect exists, but the damage areas of the piles-supported frame structure still can be effectively located.

5 Conclusion and the Prospect of Further Research

A vibration-based damage identification method for the damage location jointly iden-tification of superstructure and substructure of the piles-supported frame structure was proposed, which takes the soil-piles-frame structure as a whole for damage identifica-tion. The influence of soil-piles- structure interaction on the damage identification of piles-supported frame structure was considered. The validity of the proposed method was preliminarily verified by numerical simulation. The results show that the method can identify the damage location of the piles-supported frame structure under the cases of single damage or multiple damage, and can effectively locate the damage area in spite of the adjacent effect exist. In addition, the damage locations of hidden pile foundation can be identified by this method.

It should be pointed out that the above research work is a preliminary numerical simulation study under the assumption that the modal parameters of the damaged struc-ture have been obtained and the influence of random noise is not considered. In order to apply this method to damage identification and health monitoring of piles-supported frame structures, the following problems need to be further studied:

Firstly, how to predict the acceleration response data of unknown measurement points of damaged structure based on the acceleration response data of limited known measurement points of damaged structure? This problem can be solved by predicting the acceleration response data of the unknown measurement points of the damaged structure according to acceleration response data of the known measurement points of the damaged structure and the acceleration response transitivity function of the known and unknown measurement points of the undamaged structure.

Secondly, how to identify the modal frequencies and mode shapes of the damaged structure according to the acceleration response data of measuring points (including the acceleration response data of the known measuring points and the predicted acceleration response data of unknown measuring points)? This problem can be solved by using stochastic subspace method to identify the modal frequencies and mode shapes of the damaged structure according to the acceleration response data of the damaged structure.

Thirdly, how to effectively identify the damage degree of damaged structure when the location of structural damage is known? This problem can be solved by using support vector machine (SVM) method to identify the damage degree of damaged structure.

Finally, the further numerical and experimental verification of actual feasibility of damage location and damage degree identification of soil-pile-frame structure should be carried out. Combined with the results of the above three further research work and considering the acceleration response of random noise effects, further numerical simulation and model tests research work will be carried out to verify the effectiveness and robustness of the joint identification method for the damage of superstructure and substructure of piles-supported frame structure.

The vibration-based joint identification method for the damage of superstructure and substructure considering soil-foundation-structure interaction is a challenging work, which can further improve the structural health monitoring research work of actual engineering. The purpose of this research report in this paper also hopes to attract more relevant researchers to adopt more effective methods and means to study this problem.

Acknowledgments. This work was supported by the Zhejiang Provincial Natural Science Foundation of China (LQ20E080025) and (LY19E080016).

References

1. Salawu, O.S.: Detection of structural damage through changes in frequency: a review. Eng. Struct. **19**(9), 718–723 (1997)
2. Doebling, S.W., Farrar, C.R., Prime M.B.: A summary review of vibration-based damage identification methods. Shock Vib. Dig. **30**(2), 91–105 (1998)
3. Farrar, C.R., Jauregui, D.A.: Comparative study of damage identification algorithms applied to a bridge: I. Experiment. Smart Mater. Struct. **7**(5), 704–719 (1998)
4. Yan, Y.J., Cheng, L., Wu, Z.Y., Yam, L.H.: Development in vibration-based structural damage detection technique. Mech. Syst. Signal Pr. **22**, 2198–2211 (2007)
5. Fan, W., Qiao, P.Z.: Vibration-based damage identification methods: a review and comparative study. Struct. Health Monit. **9**(3), 83–111 (2010)
6. Cawley, P., Adams, R.D.: The location of defects in structures from measurements of natural frequencies. J. Strain Anal. Eng. **14**(2), 49–57 (1979)
7. Yuen, M.M.F.: A numerical study of the eigenparameters of a damaged cantilever. J. Sound Vib. **103**(3), 301–310 (1985)

8. Pandey, A.K., Biswas, M., Samman, M.M.: Damage detection form changes in curvature mode shapes. J. Sound Vib. **145**(2), 321–332 (1991)
9. Ricles, J.M., Kosmatka, J.B.: Damage detection in elastic structures using vibratory residual forces and weighted sensitivity. AIAA J. **30**(9), 2310–2316 (1992)
10. Pandey, A.K., Biswas, M.: Damage detection in structures using changes in flexibility. J. Sound Vib. **169**(1), 3–17 (1994)
11. Pandey, A.K., Biswas, M.: Experimental verification of flexibility difference method for locating damage in structures. J. Sound Vib. **184**(2), 311–328 (1995)
12. Shi, Z.Y., Law, S.S., Zhang, L.M.: Structural damage detection from modal strain energy change. J. Eng. Mech. **126**(12), 1216–1223 (2000)
13. Yao, G.C., Chang, K.C., Lee, G.C.: Damage Diagnosis of steel frames using vibrational signature analysis. J. Eng. Mech. **118**(9), 1949–1961 (1992)
14. Morassi, A., Rovere, N.: Localizing a notch in a steel frame from frequency measurements. J. Eng. Mech. **123**(5), 422–432 (1997)
15. Li, G.Q., Hao, K.C., Lu, Y.: Two-step approach for damage identification of frame structures. J. Tongji Univ. **26**(5), 483–487 (1998)
16. Lam, H.F., Ko, J.M., Wong, C.W.: Localization of damaged structural connections based on experimental modal and sensitivity analysis. J. Sound Vib. **210**(1), 91–115 (1998)
17. Wang, B.S., Ni, Y.Q., Gao, Z.M.: Input parameters for artificial neural networks in frame connection damage identification. J. Vib. Eng. **13**(1), 137–142 (2000)
18. Yun, C.B., Yi, J.H., Bahng, E.Y.: Joint damage assessment of framed structures using a neural networks technique. Eng. Struct. **23**(5), 425–435 (2001)
19. Qu, W.L., Chen, W., Li, Q.S.: Two-step approach for joints damage diagnosis of frame structures by artificial neural networks. Chin. Civil Eng. J. **36**(5), 37–45 (2003)
20. Pathirage, C.S.N., Li, J., Li, L., Hao, H., Liu, W.Q., Ni, P.H.: Structural damage identification based on autoencoder neural networks and deep learning. Eng. Struct. **172**, 13–28 (2018)
21. Ding, Z.H., Li, J., Hao, H., Lu, Z.R.: Structural damage identification with uncertain modeling error and measurement noise by clustering based tree seeds algorithm. Eng. Struct. **185**, 301–314 (2019)
22. Shi, Z.Y., Law, S.S., Zhang, L.M.: Improved damage quantification from elemental modal strain energy change. J. Eng. Mech. **128**(5), 521–529 (2002)
23. Ovanesova, A.V., Suárez, L.E.: Application of wavelet transforms to damage detection in frame structures. Eng. Struct. **26**(1), 39–49 (2004)
24. Xu, B., Song, G., Masri, S.F.: Damage detection for a frame structure model using vibration displacement measurement. Struct. Health Monit. **11**(3), 281–292 (2012)
25. Shiradhonkar, S.R., Shrikhande, M.: Seismic damage detection in a building frame via finite element model updating. Comput. Struct. **89**(23), 2425–2438 (2011)
26. Döhler, M., Hille, F.: Subspace-based damage detection on steel frame structure under changing excitation. Struct. Health Monit. **5**, 167–174 (2014)
27. Qin, Y., Li, Y.M.: Damage detection considering uncertainties based on interval analysis. J. Chongqing Univ. **38**(6), 107–114 (2015)
28. Zapico, J.L., Worden, K., Molina, F.J.: Vibration-based damage assessment in steel frames using neural networks. Smart Mater. Struct. **10**(3), 553–559 (2001)
29. Chen, S.W., Li, G.Q.: A multi-hierarchical damage identification approach based on BP network for frame structures. Earthq. Eng. Eng. Vib. **22**(5), 18–23 (2002)
30. Li, L.: Numerical and experimental studies of damage detection for shearing buildings. Ph.D. Dissertation, Huazhong Univ. Sci. Tech., Wuhan, China (2005)
31. Park, S., Bolton, R.W., Stubbs, N.: Blind test results for nondestructive damage detection in a steel frame. J. Struct. Eng. **132**(5), 800–809 (2006)
32. Ji, X.D., Qian, J.R., Xu, L.H.: Damage diagnosis of a two-storey spatial steel braced-frame model. Struct. Control Hlth **14**(8), 1083–1100 (2007)

33. Chellini, G., Roeck, G.D., Nardini, L., Salvatore, W.: Damage analysis of a steel-concrete composite frame by finite element model updating. J. Constr. Steel Res. **66**(3), 398–411 (2010)
34. Li, H.J., Yang, H.Z., Hu, S.L.J.: Modal strain energy decomposition method for damage localization in 3D frame structures. J. Eng. Mech. **132**(9), 941–951 (2006)
35. Loh, C.H., Chan, C.K., Chen, S.F., Huang, S.K.: Vibration-based damage assessment of steel structure using global and local response measurements. Earthq. Eng. Struct. D. **45**(5), 699–718 (2016)
36. Sun, Z.S., Zhang, B., Fan, K.J.: Damage identification research of frame structure based on lifting wavelet. World Earthq. Eng. **26**(4), 25–30 (2010)
37. Zhou, Y.L., Figueiredo, M.E.N., Perera, R.: Damage detection and quantification using transmissibility coherence analysis. Shock Vib. **2015**(4), 1–16 (2015)
38. Yatim, N.H.M., Muhamad, P., Abu, A.: Conditioned reverse path method on frame structure for damage detection. J. Telecom. Elec. Comput. Eng. **9**(1–4), 49–53 (2017)
39. Cornwell, P., Doebling, S.W., Farrar, C.R.: Application of the strain energy damage detection method to plate-like structures. J. Sound Vib. **224**(2), 359–374 (1999)
40. Hu, H.W., Wu, C.B.: Development of scanning damage index for the damage detection of plate structures using modal strain energy method. Mech. Syst. Signal Pr. **23**(2), 274–287 (2009)
41. Guo, H.Y., Li, Z.L.: Structural damage detection based on strain energy and evidence theory. Appl. Mech. Mater. **48–49**, 1122–1125 (2011)
42. Shi, Z.Y., Law, S.S., Zhang, L.M.: Structural damage localization from modal strain energy change. J. Sound Vib. **218**(5), 825–844 (1998)
43. Fan, W., Qiao, P.Z.A.: strain energy-based damage severity correction factor method for damage identification in plate-type structures. Mech. Syst. Signal Pr. **28**, 660–678 (2012)
44. Wei, Z.T., Liu, J.K., Lu, Z.R.: Damage identification in plates based on the ratio of modal strain energy change and sensitivity analysis Inverse. Probl. Sci. Eng. **24**(2), 265–283 (2016)

Research and Development and Pilot Application of Innovative Technology of Prefabricated Concrete

Jindan Zhang, Jun Cai, Ying Su$^{(\boxtimes)}$, Qingyou He, and Xinyue Lin

College of Architecture and Electrical Engineering, Hezhou University, Hezhou 542800, China
544010279@qq.com

Abstract. Relying on the prefabricated construction project of a university in Guangxi, the standardization design of the building structure is proposed for the low standardization of prefabricated building design and serious conflicts in on-site construction, and the reinforced U-shaped ring is especially proposed for the problem of node collision in site construction. Advanced connection technologies such as buckle connection and steel slot connection and new components such as laminated truss floor slabs and corrugated pipe through-hole prefabricated columns. At the same time, BIM technology is used to verify the technical scheme. The result shows: the standardization degree of the integrated design of building structure proposed in the article High, can significantly improve the design and production efficiency, and the use of new connection technology and new components can effectively reduce the collision of steel bars and facilitate construction.

Keywords: Prefabricated buildings · BIM technology · Deepening design

1 Introduction

The 19th National Congress of the Communist Party of China clearly established the concept of green development and promoted the development of new industrialization. Prefabricated buildings and green buildings are important carriers. However, there are still some problems restricting development in the process of popularization and practice of prefabricated buildings.

The prefabricated building is different from the cast-in-place structure, but the design principle adopted is "equivalent to cast-in-place" [1]. Therefore, the design idea of "first overall analysis and then split design" is often adopted, which easily leads to a wide variety of prefabricated components that are split. Recombination after splitting is also prone to problems such as steel bar collision. The core feature of prefabricated buildings is standardized design and assembly construction. However, experts and scholars have done a series of studies on the design and construction of prefabricated buildings and pointed out that the development of domestic prefabricated buildings lacks a standard system [2–4], and standardized design is not yet available. Mature, lack of innovation ability [5], and there are many problems in on-site construction, which unreflect the advantages of prefabricated buildings [6]. Therefore, the study of standardized design

© The Author(s) 2022
G. Feng (Ed.): ICCE 2021, LNCE 213, pp. 226–237, 2022.
https://doi.org/10.1007/978-981-19-1260-3_20

methods and technologies that are conducive to site simplification and high-efficiency assembly is of great significance to the development of prefabricated buildings.

This paper aims at the insufficient design standardization and construction conflicts in the development of prefabricated buildings. Based on the case of a prefabricated building project in a university in Hezhou, the prefabricated building and structure are analyzed from the aspects of building type selection, component optimization, structural calculation, and deepening design. The whole design process is analyzed, the key issues in the design and implementation of the prefabricated building project are comprehensively introduced, and the BIM forward design and some cutting-edge prefabricated new technologies are adopted, hoping to be useful for similar projects in the future.

2 Project Introduction

This project is located in a university in Hezhou City. The design service life is 50 years. The fire resistance rating of the building is Class II, the roof waterproof rating is Class I, and the seismic fortification intensity is 6 degrees. The building base has a total land area of 1138 m^2, a total construction area of 3436 m^2, a total of 3 floors, the first and second floors are 4.8 m high, the three floors are 3.9 m high, and the total height of the building is 13.5 m. It adopts a prefabricated integrated frame structure, the prefabrication rate of single buildings reached 62.9%, and the assembly rate reached 70.1%.

2.1 Architectural Design

The design of prefabricated buildings is different from cast-in-place structures. The design of prefabricated concrete buildings must not only meet the functional requirements of buildings in various parts of our country, but also meet the requirements of industrialized construction. Based on the principle of "less specifications, more combinations", follow the modular coordination Standards, to carry out standardized, modular, and integrated architectural design.

2.1.1 Graphic Design

The nature of the building is a teaching building, and its main function is physical teaching and scientific research model display. Considering structural safety and economic rationality, it adopts a large bay and deep plan layout. The size of the column net is 7800 mm × 7800 mm, and the plan layout is regular and flat. It is more conducive to earthquake resistance and cost saving. In addition, the large-bay and large-depth plan layout reduces the types and quantity of prefabricated components, increases the repetition rate of components, and improves economic benefits. The graphic design is based on the different use functions of the room, and the depth is extended with the basic modulus. A variety of basic plane shapes are determined to form different personalized planes. However, the central location of the large bay and the deep depth is unfavorable for daylighting. Therefore, the internal atrium is opened, and the openings on each floor are gradually increased to increase the sense of hierarchy of the building. The atrium is equipped with a steel staircase and the top roof is a glass roof.

2.1.2   Facade Design

The building facade is a direct display of architectural artistic style. This project comprehensively coordinates the architectural effects and architectural functions for the facade design. The wall beams of this project are flat, highlight the column lines, and do not have too many external decorative components, which conforms to the industrial production characteristics of prefabricated building components with few specifications and high repetition rate. The façade wall adopts a new type of wall panel produced locally in Hezhou, and is designed as an integrated exterior wall. This wall panel integrates enclosure, waterproofing, heat preservation and heat insulation, and is assembled on site. The building façade enhances the façade effect through the combination of the spaced arrangement of doors and windows. The overall structure is simple and individual. The door and window openings and façade partitions all meet the requirements of modular design, presenting a standardized and diversified door and window enclosure system. And the component specifications are unified to meet the requirements of industrialization of prefabricated buildings.

This project is a regular symmetrical model, and the grid size is 7800 mm × 7800 mm. In order to increase the reuse rate and reduce the types of components, the project considers deepening design and component splitting in the architectural and structural design of the project, designing and merging components with similar sizes, reinforcements, and structures, and prefabricating components based on the standardization of component dimensions. Standardized and modular design. Finally, one column size is 600 mm × 600 mm, two main beam sizes (330 mm × 650 mm, 330 mm × 850 mm), one secondary beam size (250 mm × 600 mm), and one floor size (120 mm).

2.2   Structural Design

The project uses PKPM software for structural design, and the structural parameters are analyzed and calculated according to the actual situation of the project. Finally, the construction drawings of the structure were revised and perfected, considering the prominent problem of steel reinforcement collision in prefabricated buildings, and manual adjustments were made according to the results of the reinforcement calculated by the system. To reduce the collision problem of steel bars to a certain extent, and secondly, try to unify the reinforcement of the components as far as possible. The reinforcement should be controlled within ±5% according to the calculation area of the structural software, and the reinforcement area should be as large as possible to be larger than the calculation area to reduce the types of prefabricated components and improve the prefabrication of the factory. Production efficiency.

2.2.1   Checking Calculation of Bending Bearing Capacity

The basic components of prefabricated building beams and columns are prefabricated in the factory, transported to the construction site and then cast on site. According to the calculation results of reinforcement, many prefabricated beam components need to be equipped with multiple rows of steel bars at the bottom. When pouring on-site, it is easy to have too much steel bars in the node area, which is inconvenient for the connection, placement and pouring of the steel bars. The negative bending moment at the support is relatively large, and the lower bending moment is small, so consider a beam with multiple rows of steel bars. Under the effect of meeting the bearing capacity, part of the steel bars can be prevented from extending into the support. According to the concrete structure design specification GB50010-2010 [7], the flexural bearing capacity of the support can be checked. The check calculation results show that for beam members with multiple rows of steel bars, only the bottom row of steel bars can meet the requirements of flexural bearing capacity (Fig. 1). Therefore, the steel bars are cut according to the requirements of the plan atlas, and some steel bars do not extend into the support to reduce Anchor the steel bars at the cast-in-situ nodes to avoid serious collision problems caused by too dense steel bars in the node area, and the concrete is not densely poured.

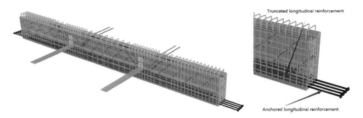

Fig. 1. Schematic diagram of truncated longitudinal bars

2.2.2 Checking Calculation of Shear Capacity

According to JGJ1-2014 [8] connection design requirements of the prefabricated concrete structure technical specification, the joint surface of the prefabricated component and the post-pouring concrete, grouting material, and setting material should be provided with a rough surface and a keyway. For the prefabricated slab components, it is sufficient to set the rough surface with a depth of not less than 4 mm and the post-cast concrete laminate layer to be combined. The rough surface should be less than 6 mm. The end face of the precast beam and the bottom of the precast column should also be designed with keyways according to the regulations, and their shear capacity should be checked. Therefore, the beam ends of this project are equipped with long keyways and checked according to the regulations. For parts that do not meet the requirements for shear resistance The required cross-section is added with shear-resistant steel bars in the post-cast part (Fig. 2).

3 Pilot Application of Prefabricated Technology

In order to improve the construction efficiency of prefabricated buildings and solve the quality problems of prefabricated buildings such as steel bar collision and improper concrete pouring, this project has applied many systematic comprehensive technologies of prefabricated concrete structures with high technical content and high economic benefits. The floor slab of the project adopts "reinforcement-free" laminated truss floor slabs. The columns in the dotted line range of the figure are prefabricated columns with

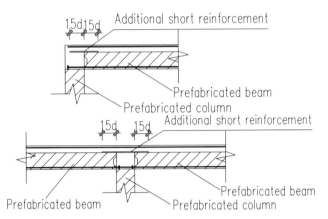

Fig. 2. Additional short ribs on parts that do not meet the shear requirements

corrugated pipes. The beam-column connection and the main and secondary beam connections adopt the new U-shaped steel bar ring buckle connection technology and steel card grooves.

3.1 Laminated Truss Floor Slab

Components of the laminated floor slab are made in the factory, prefabricated slabs are required to have lapped steel bars to be spliced on-site to form a force-bearing whole. The current conventional method is "beard rib" connection, but because of its "reinforced" structure that it has caused an increase in the process of opening holes, pouring and leak-proof grout sealing, etc. which is not convenient for industrial production. The "beard tendons" are also easy to bend and deform during transportation and hoisting, and even cause precast concrete damage and missing corners. Affecting the appearance and quality of the laminated board [9]. Formwork needs to be installed at the "beard ribs" of adjacent floor slabs on the construction site, and collisions also bring a lot of complicated procedures to the construction. Therefore, this project adopts the method of "no ribs"[10] in the joints between the laminated slabs (Fig. 3). At the same time, the composite wire mesh technology is used to improve the crack resistance of the slabs, so that the thickness of the precast laminated slabs is reduced to 30 mm, which is better than ordinary concrete. The laminated board is reduced by half, saving costs, and can be free of formwork and plastering, which greatly improves the efficiency of construction and installation.

3.2 U-Shaped Steel Bar Ring Buckle Connection Technology

On the third floor of the project, the ⑥ axis y direction AB and BC span reinforcements are 4 ⏀25, 2 ⏀25+2 ⏀22 respectively, and the left and right x direction beams are 2 ⏀25 reinforcements. At the intersection of the ⑥ axis and the B axis, the lower part of the joint is thick, and the amount is large, which is easy to produce Rebar collision problem. In order to reduce the collision problem of steel bars in the node area and ensure the

Fig. 3. Closely assembled laminated floor slab

effective transmission of node force, this project adopts U-shaped steel bar ring buckle connection technology. As shown in Fig. 4, the upper part of the laminated beam is the post-cast part, and the upper longitudinal steel bars of the beam can penetrate the nodes or supports, because the node reinforcement is too dense, the lower beams and the column reinforcements collide, and the lower longitudinal reinforcements of the beam undergo U-shaped bending. After folding, anchor into the beam concrete without extending into nodes or supports. Then arrange the closed circular steel bars and the U-shaped steel bars at the beam end between the longitudinal bars in the core area of the beam-column node to stagger each other, thereby forming a pair of U-shaped ring buckles at the beam end, and place four ring buckles at the four corners of the ring buckle. Root short-inserted steel bars, U-shaped steel bar ring buckle connection section stirrups are encrypted. The U-shaped steel bar ring buckle connection technology forms a "pin type" connection through the core area concrete surrounded by the ring buckle steel bar. The ring buckle steel bar can effectively transmit the tension and pressure required for the connection, ensuring the safety of the connection and effectively improving the on-site construction Efficiency [11].

The U-shaped steel ring buckle connection technology is simple in structure and convenient in construction. It can avoid the collision of the prefabricated beam steel bars with the joints or the steel bars in the supports, and solves the problem of complex and difficult anchoring of the steel bars in the core area of the beam-column of the traditional post-cast integral connection node. At the same time, the length of the post-cast section at the beam end can be shortened, and the standardization of prefabricated beam components can be realized.

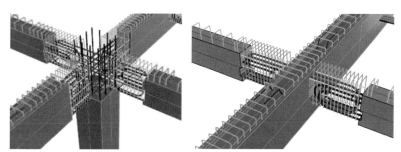

(a) U-shaped beam-column connection (b) U-shaped connection of primary and secondary beams

Fig. 4. U-shaped node connection technology

3.3 Overlap Connection of Primary and Secondary Beams and Steel Slot Connection Technology

The assembled monolithic concrete prefabricated secondary beam adopts the same super-imposed form as the main beam, and the upper longitudinal steel bars are tied together on-site and poured together with the upper steel bars of the floor. For the connection of the lower reinforcement, that is, the connection of the primary and secondary beam nodes, the integral cast or shelving connection is often used [12]. Partially vacant sections are reserved in the prefabricated main girder using the cast-in-place method, which causes the main girder to be discontinuous and increases the difficulty of prefabrication and hoisting. However, the shelving type is not connected to the lower secondary beam steel bar, which belongs to the "hinged connection" and is not suitable for structures that bear dynamic loads and large spans. It can be seen that no matter whether the cast-in-place or shelving connection method is used, there are construction drawbacks. The design of this project The primary and secondary beam connections not only have no gaps in the primary beam but also connect the lower longitudinal ribs of the secondary beam. In addition to the U-shaped connection technology described above, there are also primary and secondary beam lap connections and steel slot connections.

The lap connection of the primary and secondary beams is to set a keyway on the side of the main beam and the part where the secondary beam is connected, and at the same time reserve the steel bars that overlap with the lower part of the secondary beam, and connect with the secondary beam reinforcement through the post-cast belt, as shown in Fig. 5. This lap method can not only avoid gaps in the main beam, but also realize the complete force of the primary and secondary beams, but the reserved lap steel bars for the main beam are not convenient for transportation and consume more steel. At the same time, the reserved post-pouring belt is longer and there are more wet operations.

In order to reduce the wet work of the post-cast belt, the project also adopted a new type of steel slot connection technology, that is, the steel slot is pre-embedded on the side of the main beam at the junction of the primary and secondary beams, and the stressed steel bars at the bottom of the secondary beams can be directly extended into the steel. The slot is anchored, mortar is poured, and concrete is finally poured to form a whole, as shown in Fig. 6. The construction difficulty of the steel slot connection is greatly reduced, and because of its special structure, when the connecting steel bar is stressed, the mortar or concrete poured in the gap between the steel slot and the connecting steel anchor head is in a three-way compression state. Its bearing capacity and ductility have been greatly improved to ensure that damage does not occur in the steel slot [13].

3.4 Bellows Through Hole Prefabricated Column

In prefabricated concrete buildings, in order to ensure the reliability of vertical force transmission, longitudinal prefabricated column reinforcement is generally connected by grouting sleeves, but the grouting sleeve connection not only requires high manufacturing precision of the prefabricated columns, but also detects the compactness of the mortar inside the sleeves. The difficulty is also higher, and the construction cost is also higher [14]. Therefore, this project uses a bellows through-hole precast column, that is, a metal corrugated pipe with the same height as the column is inserted at the position of the

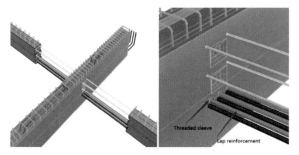

Fig. 5. Lap connection of main-secondary beams

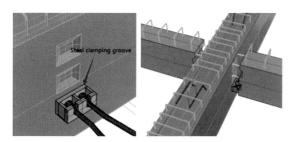

Fig. 6. Connection of main-secondary beam steel slots

steel bar inside the precast column, and the steel wire rope is tied and fixed with the column stirrup skeleton, and then concrete is poured, as shown in Fig. 7. When the upper and lower columns are connected, the longitudinal ribs of the upper and lower columns are connected by mechanical connection or welding at the nodes first, and then the prefabricated columns with the through-hole of the bellows are hoisted to the installation position, and finally mortar or special purpose is poured along the through-holes of the bellows. Grout until it is below the height of the bellows.

Fig. 7. Bellows through hole prefabricated column

The bellows through-hole prefabricated column has a large bellows cavity, which is much larger than the diameter of the column longitudinal ribs. When the beam-column steel bars collide, the longitudinal steel bars in the corrugated pipe have a larger space that can be moved, which greatly increases the fault tolerance rate, thereby reducing the precision requirements of the prefabricated column components and improving the construction efficiency. The longitudinal ribs of the upper and lower columns are welded or mechanically connected, and the integrity of the longitudinal ribs is good, which improves the load-bearing capacity of the precast column, and does not require the use of sleeves and high-strength mortar, which further reduces the cost.

4   Bim Collaborative Forward Design

The "language" of traditional building structure communication is mainly in the form of two-dimensional plan drawings, which has certain requirements for the space imagination ability of design and construction personnel. In addition, many problems cannot be seen on the plan drawings, but they cannot be carried out in actual construction, resulting in more on-site changes. In order to adapt to the increasing requirements of the construction industry for building structure design, BIM technology came into being. As a new comprehensive technology for building model design, BIM technology abstracts the specific data of architectural design through software Data processing is expressed in the form of 3D models [15]. Especially in prefabricated buildings, the components are prefabricated in the factory, and the joints are poured on site. The reinforcing bars of various components are overlapped or anchored at the joints, resulting in dense reinforcing bars at the joints and significant collision problems. If you continue to display with flat drawings, it will be difficult to deal with the collision problems Therefore, this project uses 3D modeling Rhino software to build 3D models of all components used in this project, especially for the node areas with dense steel bars and new connection technology. The modeling structure shows that all nodes using the new connection technology have not occurred. Joint collision problem, construction is convenient, and the joint steel bar with collision problem is adjusted, the steel bar is anchored by the combination of hook and anchor plate, and the colliding steel bar is bent, and the bending ratio is not more than 1:6. A structural reinforcement is added to the bent steel bar. The collision example is shown in Fig. 8. The two front and rear beams in figure a collide with the two outermost steel reinforcement bars as shown in Fig. 8 (a) (Beam End). The two ends of the steel bars are bent to the middle but collide with the column reinforcement as shown in Fig. 8 (b) (Middle bend). Therefore, the final solution is to bend upwards and add structural steel bars to the bending parts as shown in Fig. 8 (c)(Bends upwards).

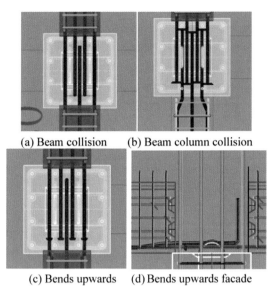

(a) Beam collision   (b) Beam column collision

(c) Bends upwards   (d) Bends upwards facade

Fig. 8. BIM technology application examples

5 Conclusion

With the advancement of industrialization, the development of prefabricated buildings is an inevitable trend. However, there are still some problems that restrict development, mainly including insufficient standardization in design, various conflicts and collision problems in construction affect the integrity of structural connections, in order to promote the industrialization of prefabricated buildings, a standardized design system must be established, and research and development Various new connection technologies and new components optimize the structural system.

The design of prefabricated concrete buildings is integrated with the structural splitting. Under the premise of meeting the building functions, the principle of "less specifications, more combinations" and the modular coordination standard are followed to carry out standardized, modular and integrated architectural design. And the structure can also reduce the types of components through reinforcement merging, etc., and carry out standardized design of node connections.

The use of new connection technologies, such as U-shaped buckle steel bar connection technology and steel card slot connection technology, can effectively improve the problems of steel bar collision and construction inconvenience in the node area, and improve construction efficiency; at the same time, the use of new components can not only solve some of the current construction Problems, superior performance, low cost, easy to popularize and use, and meet the requirements of building industrialization.

Acknowledgments. Foundation Items: Research Foundation for Young and Middle-aged Teachers in Guangxi Universities(2019KY0730); Science Research and Technology Development Project of Hezhou City, (Hezhou technology 20001); Innovation and Entrepreneurship Training Program for College Students (202011838065).

References

1. Zhang, J.D., et al.: Technical problems and solutions of composite beam connection in fabricated frame structure. Construct. **43**(10), 2056–2059 (2021)
2. Huang, X.K., Tian, C.Y., Wan, M.L., Li, R.: Research and practice of fabricated concrete structure in China. Build. Sci. **34**(09), 50–55 (2018)
3. Liu, Q., Li, X.M., Xu, Q.F.: Research and application status of prefabricated concrete structure. Construct. Technol. **43**(22), 9–14+36 (2014)
4. Yu, P.: Research on the Standardization of prefabricated building promoting the comprehensive development of housing Industrialization. China Housing Facilities (12), 116–117 (2017)
5. Ye, H.W., Fan, Z.S., Zhou, C., Liu, C.W.: Research on engineering application of standardized design method of prefabricated buildings. J. Shandong Jianzhu Univ. **33**(06), 69–74+84 (2018)
6. Xu, P.P., Wang, J., Liu, G.W.: Problems and Countermeasures of design standardization in China's architectural industrialization. Building economics **39**(3), 5–8 (2018)
7. National Standard of the People's Republic of China: Code for Design of Concrete Structures GB 50010–2010. China Architecture and Architecture Press, Beijing (2010)
8. Industrial Standard of the People's Republic of China: Technical Specification for Prefabricated Concrete StructureJGJ1-2014. China Architecture and Building Press, Beijing (2014)
9. Li, B.Y., Chen, K.P., Zhang, M.S.: Technical problems and solutions of reinforced concrete composite slab construction. Eng. Technol. Res. **5**(12), 1–2 (2020)
10. Chen, Y.H., et al.: Study on the new dense connection technology of bidirectionally stressed laminated floor slab. Industrial Construction **50**(05), 31–35 (2020)
11. Chen, Y.H., Lu, D., Zhang, M., Xie, G.X.: Study on the performance of U-shaped reinforcement ring buckle connection of fabricated concrete beam column joints. J. Guangxi Univ. **44**(06), 1552–1561 (2019)
12. Xu, J.M., Bai, R., Ma, H.B.: Analysis on the connection technology of prefabricated building nodes. Sichuan Architecture **37**(05), 177–178 (2017)
13. Tian, W., Lu, D., Zong, B.: Study on construction grouting technology of new reinforcement connection in prefabricated concrete. Construct. Technol. **47**(12), 84–86 (2018)
14. Yao, D.D., Lu, X.H.: Defect detection methods and research progress of sleeve grouting in prefabricated buildings. Concr. Cem. Products **6**(6), 85–90 (2021)
15. Wang, Q., Wang, W., Zhu, W.: Application research of prefabricated building based on BIM technology. J. Hunan Univ. Arts Sci. (Nat. Sci. Ed.) **30**(04), 55–58 (2018)

Symplectic Elastic Solution of Multi-layer Thick-Walled Cylinder Under Different Interlayer Constraints

Zhongyu Jiang$^{(\boxtimes)}$ (ID), Yajun Zhang, Huaqing Liu, and Xuanxuan Li

Key Laboratory of Mechanics, School of Architecture and Civil Engineering, Anhui Polytechnic University, Wuhu 241000, China

jiangzhy@ahpu.edu.cn

Abstract. Multi-layer thick-walled cylinder is a common supporting structure in engineering, which is widely used in various engineering fields. Considering the complex boundary conditions and the different interlayer constraints, it is difficult to solve the theoretical solution of multi-layer thick walled cylinder. In this paper, the general solution expressions of displacement and stress of multi-layer thick-walled cylinder are derived in Hamiltonian mechanics system. The complex boundary conditions are transformed into the form of algebraic sum by Fourier series expansion, and the complex boundary problems are solved by superimposing the special solutions of each order expansion term. At the same time, according to the characteristics of different interlayer constraints, the corresponding conditions of interlayer continuous smooth are proposed. Combined with the boundary conditions of thick-walled cylinder, the linear equations with undetermined coefficients are established. By solving the equations, the mechanical problems of multi-layer thick-walled cylinder are finally solved. By comparing the mechanical responses of multi-layer thick-walled cylinder under different constraint conditions, it is concluded that the overall mechanical performance of the tight interlayer connection is better, and the circumfluence stress component is more prominent than other stress components. Finally, the influence of lateral pressure coefficient and elastic modulus ratio on the circumferential stress of multi-layer thick-walled cylinder is discussed. These research results provide the necessary theoretical basis for solving the mechanical problems of multi-layer thick-walled cylinders.

Keywords: Multi-layer thick-walled cylinder · Hamiltonian mechanics · Smooth interlayer contact · Tight interlayer connection · Complex boundary conditions

1 Introduction

Thick-walled cylinder is a common engineering structure, which is widely used in mining, hydropower, chemical, military and other fields [1]. Thick-walled concrete cylinder is often used in mine engineering to ensure that the wellbore is still in elastic state under large load conditions. However, when the thickness of the wellbore is already large, simply relying on increasing the thickness of the wellbore will not only increase the

© The Author(s) 2022
G. Feng (Ed.): ICCE 2021, LNCE 213, pp. 238–253, 2022.
https://doi.org/10.1007/978-981-19-1260-3_21

time and labor cost, but also fail to significantly improve the elastic ultimate bearing capacity of the wellbore [2, 3]. It is a common economic and reasonable method to adopt multi-layer composite cylinder, which can effectively reduce the wall thickness and improve the ultimate bearing capacity of the structure [4].

On the basis of considering the different tensile and compression elastic modulus of the material, Wang Su et al. [5] established the stress expression of double-layer thick-walled cylinder under uniform internal pressure. The elastic limit solution corresponding to the internal pressure is obtained, and the influence of the cylinder parameters on the elastic limit is discussed. Lu, A. et al. [6] discussed the optimal design method of double-layer thick-walled cylinder by using the mixed penalty function method. Under the known uniform external load conditions, the minimum wall thickness of the cylinder, and the optimum thickness ratio and the elastic modulus ratio of the inner and outer layers are calculated. The disadvantage of this method is that it can only solve the problem of multi-layer thick-walled cylinder under uniform load with the help of classical elastic mechanics solution. Qiu J et al. [7] proposed an analytical method to solve the pressure and stress between multiple contact pairs by using the theory of multi-layer thick-walled cylinder, and applied it to the design of interference fit for engine crankshaft bearing. It mainly uses the finite element method to analyze the stress of the multi-layer thick-walled cylinder, which can not give the relationship expression between the variables accurately. Wu, Q. et al. [8] studied the plane strain problem of double-layer thick-walled cylinder by using the power series method of complex variable function, and obtained the analytical solution of stress of two-layer cylinder under the condition of complete contact. It is difficult to determine the complex potential function during the derivation of this method, and it needs to be recalculated for different external boundary conditions, which lacks generality. Abbas Loghman et al. [9] studied the magneto-thermo-elastic response of a double-layer thick-walled cylinder. The minimum effective stress distribution and the minimum radial displacement can be obtained by selecting the appropriate uniformity parameters under thermal-magnetic mechanical load.

With the development of modern mathematical and physical methods, Zhong [10] and Yao et al. [11] introduced the concept of symplectic geometry into Hamiltonian mechanical system, and established a new solution system of elasticity. It has shown great advantages in dealing with complex elastic mechanics problems. Zhou Jianfang et al. [12] extended the method of separating variables in Hamilton mechanics system, and applied it to the elasticity problem of non-homogeneous boundaries in polar coordinates. This method successfully solves the problem of thick-walled cylinder subjected to non-uniform hydrostatic pressure, which fully shows the superiority of Hamilton mechanical system. Based on the Hamiltonian state space method, Tseng and Tarn [13] discussed the theoretical solution of the stress field around a circular hole when the elastic plate with a hole is subjected to unidirectional tension by the method of variable separation and symplectic eigenfunction expansion. It can be seen that the symplectic elastic mechanics method has a wide application prospect in the field of basic research. It avoids the subjectivity of stress function selection, and uses rational logical derivation to solve the problem. Therefore, it is of great significance to study the theoretical solution of multi-layer thick-walled cylinder structure by symplectic elasticity in polar coordinate system.

2 Hamiltonian Mechanics in Polar Coordinates

2.1 The Mixed State Equation of Sector in Polar Coordinates

Define new variables: $S_\rho = \rho\sigma_\rho$, $S_\varphi = \rho\sigma_\varphi$, $S_{\rho\varphi} = \rho\tau_{\rho\varphi}$, in a typical sector region (Fig. 1) $R_1 \leq \rho \leq R_2$, $\alpha \leq \varphi \leq \beta$. Then perform variable substitution $\xi = \ln\rho$, that is $\rho = e^\xi$; and $\xi_1 = \ln R_1$, $\xi_2 = \ln R_2$. In the variational principle, φ is simulated as time coordinates, ξ is horizontal, and the S_ρ lateral force factors is eliminated. Then the mixed energy variational principle of the Hamiltonian system under polar coordinates is obtained [14].

$$\delta \int_\alpha^\beta \int_{\xi_1}^{\xi_2} \left\{ S_{\rho\varphi} \frac{\partial u_\rho}{\partial \varphi} + S_\varphi \frac{\partial u_\varphi}{\partial \varphi} + S_\varphi \left(u_\rho + v\frac{\partial u_\rho}{\partial \xi} \right) - S_{\rho\varphi} \left(u_\varphi - \frac{\partial u_\varphi}{\partial \xi} \right) + \right.$$
$$\left. \frac{1}{2}E\left(\frac{\partial u_\rho}{\partial \xi}\right)^2 - \frac{1}{2E}\left[(1-v^2)S_\varphi^2 + 2(1+v)S_{\rho\varphi}^2\right] \right\} d\xi d\varphi = 0 \tag{1}$$

The mixed energy variational principle is expanded, and the sign $(\cdot) = \partial/\partial\varphi$ is used to represent the derivative of φ, then the Hamiltonian regular equations can be written as

$$\begin{pmatrix} \dot{u}_\rho \\ \dot{u}_\varphi \\ \dot{S}_{\rho\varphi} \\ \dot{S}_\varphi \end{pmatrix} = \begin{bmatrix} 0 & 1 - \frac{\partial}{\partial\xi} & \frac{2(1+v)}{E} & 0 \\ -1 - v\frac{\partial}{\partial\xi} & 0 & 0 & \frac{1-v^2}{E} \\ -E\frac{\partial^2}{\partial\xi^2} & 0 & 0 & 1 - v\frac{\partial}{\partial\xi} \\ 0 & 0 & -1 - \frac{\partial}{\partial\xi} & 0 \end{bmatrix} \begin{pmatrix} u_\rho \\ u_\varphi \\ S_{\rho\varphi} \\ S_\varphi \end{pmatrix} \tag{2}$$

The total-state vector $v = \left(u_\rho\ u_\varphi\ S_{\rho\varphi}\ S_\varphi \right)^T$ and operator matrix H are introduced, and the Hamiltonian regular equations Eq. (2) become

$$\dot{v} = Hv \tag{3}$$

It can be proved that the operator matrix H is a Hamiltonian operator matrix in symplectic geometric space [15]. The Hamiltonian operator matrix is written in block form.

$$H = \begin{bmatrix} A & D \\ B & -A^* \end{bmatrix} \tag{4}$$

where

$$A = \begin{bmatrix} 0 & 1 - \frac{\partial}{\partial\xi} \\ -1 - v\frac{\partial}{\partial\xi} & 0 \end{bmatrix}, \quad B = \begin{bmatrix} -E\frac{\partial^2}{\partial\xi^2} & 0 \\ 0 & 0 \end{bmatrix}, \quad D = \begin{bmatrix} \frac{2(1+v)}{E} & 0 \\ 0 & \frac{1-v^2}{E} \end{bmatrix},$$

$$A^* = \begin{bmatrix} 0 & -1 + v\frac{\partial}{\partial\xi} \\ 1 + \frac{\partial}{\partial\xi} & 0 \end{bmatrix}.$$

The homogeneous boundary conditions on both sides $\xi = \xi_1$ and $\xi = \xi_2$ are:

$$E\frac{\partial u_\rho}{\partial \xi} + vS_\varphi = 0,\ S_{\rho\varphi} = 0, \tag{5}$$

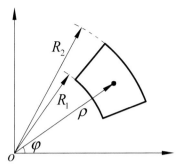

Fig. 1. Mechanical problems of sector in polar coordinates

2.2   Basic Eigensolutions of Homogeneous Equations

Solving the Hamilton regular equations Eq. (2) under the homogeneous boundary condition Eq. (5), the general solution of the equation is only related to the properties of the Hamilton operator H [16]. Therefore, the eigenvalues are $\mu = 0, \pm i$ respectively.

When the eigenvalue $\mu = 0$, the eigensolution and its Jordan-type eigensolution are expressed as

$$\psi_0^0 = \begin{pmatrix} 0 \\ e^\xi \\ 0 \\ 0 \end{pmatrix};$$

$$\psi_0^1 = \begin{pmatrix} c_1 e^\xi + c_2 e^{-\xi} + \frac{1-v}{2}\xi e^\xi \\ 0 \\ 0 \\ \frac{E}{1-v}c_1 e^\xi + \frac{E}{1+v}c_2 e^{-\xi} + \frac{E}{2}e^\xi\left(\xi + \frac{2-v}{1-v}\right) \end{pmatrix}$$

(6)

where $c_1 = -\frac{1}{2} - \frac{1-v}{2}\frac{R_2^2 \ln R_2 - R_1^2 \ln R_1}{R_2^2 - R_1^2}$; $c_2 = -\frac{1+v}{2}\frac{R_2^2 R_1^2}{R_2^2 - R_1^2}\ln\left(\frac{R_2}{R_1}\right)$.

When the eigenvalue $\mu = \pm i$, the eigensolution and its Jordan-type eigensolution are respectively expressed as

$$\psi_i^0 = (1\ i\ 0\ 0)^T$$
$$\psi_{-i}^0 = (1\ -i\ 0\ 0)^T;$$
$$\psi_i^1 = \left(u_\rho^1 i\ u_\varphi^1\ S_{\rho\varphi}^1\ S_\varphi^1 i\right)^T$$
$$\psi_{-i}^1 = \left(-u_\rho^1 i\ u_\varphi^1\ S_{\rho\varphi}^1\ -S_\varphi^1 i\right)^T$$

(7)

where

$$
\begin{cases}
u_\rho^1 = \dfrac{1}{2}(1 - v)\xi + \alpha(1 - 3v)e^{2\xi} + \beta(1 + v)e^{-2\xi} \\[2mm]
u_\varphi^1 = -\dfrac{1}{2}[1 + v + (1 - v)\xi] + \alpha(5 + v)e^{2\xi} + \beta(1 + v)e^{-2\xi} \\[2mm]
S_{\rho\varphi}^1 = E\left(\dfrac{1}{2} + 2\alpha e^{2\xi} - 2\beta e^{-2\xi}\right) \\[2mm]
S_\varphi^1 = E\left(\dfrac{1}{2} + 6\alpha e^{2\xi} + 2\beta e^{-2\xi}\right)
\end{cases}
;
$$

$$
\alpha = \frac{-1}{4\left(R_1^2 + R_2^2\right)}.
$$

$$
\beta = -\alpha R_1^2 R_2^2
$$

The original problem corresponding to the eigensolution and its Jordan eigensolution is solved as follow

$$
v_\mu^n = \sum_{m=0}^{n} \frac{1}{m!} \varphi^m \psi_\mu^{n-m} \, (n = 0, 1) \tag{8}
$$

2.3 Special Solutions of Non-homogeneous Boundary Conditions

It is usually difficult to solve the elastic mechanics problem under complex boundary conditions. Considering that the geotechnical material is a small deformation elastic body, the principle of linear elastic superposition is applicable. Any complex boundary function can be Fourier transformed and decomposed into simple regular triangular series. Therefore, the complex boundary loads are decomposed into a series of triangular series loads, and the stress and deformation laws under each order of triangular series loads are obtained, then the original problems are solved by superposition of the calculation results of each order [17].

In polar coordinates, the boundary load is expanded by Fourier series as:

$$
f(\varphi) = \frac{a_0}{2} + \sum_{k=1}^{\infty} (a_k \cos k\varphi + b_k \sin k\varphi) \tag{9}
$$

where $a_0 = \frac{1}{\pi}\int_{-\pi}^{\pi} f(\varphi)d\varphi$, $a_k = \frac{1}{\pi}\int_{-\pi}^{\pi} f(\varphi)\cos k\varphi d\varphi$, $b_k = \frac{1}{\pi}\int_{-\pi}^{\pi} f(\varphi)\sin k\varphi d\varphi$.

According to the property of the Hamiltonian operator matrix [18], the block operator A has an orthogonal eigenfunction system in the Hilbert space $X \times X$, and the eigenvalue and eigenfunction system of A can be expressed as:

$$
U_k = \begin{pmatrix} \cos k\varphi \\ \sin k\varphi \end{pmatrix}, \ (k = 0, \pm 1, \pm 2, \cdots) \tag{10}
$$

Block operator $-A^*$ has orthogonal eigenfunction systems in Hilbert space $X \times X$ and the eigenvalues and eigenfunction systems of $-A^*$ can be expressed as:

$$
\hat{U}_k = \begin{pmatrix} \sin k\varphi \\ \cos k\varphi \end{pmatrix}, \ (k0 \pm 1 \pm 2 \cdots) \tag{11}
$$

First, a simple case is considered, that is, the inner side is homogeneous boundary and the outer side is only subjected to k-order normal cosine boundary load. The specific form is:

$$\begin{aligned} \text{When } \xi = \xi_1,\ E\frac{\partial u_\rho}{\partial \xi} + vS_\varphi = 0, \qquad\qquad S_{\rho\varphi} = 0 \\ \text{When } \xi = \xi_2,\ E\frac{\partial u_\rho}{\partial \xi} + vS_\varphi = P_k \cos k\varphi,\ S_{\rho\varphi} = 0 \end{aligned} \tag{12}$$

where, P_0 is the k-order normal load coefficient, k is the coefficient of series term. Especially when $k = 0$, the boundary load is a constant.

According to the eigenfunctions Eqs. (10) and (11) of the block operators A and $-A^*$, the special solution of the equation satisfying the non-homogeneous boundary conditions Eq. (12) can be written as

$$\begin{cases} \tilde{u}_\rho = P_k\left(A_1 e^{\lambda_1 \xi} + A_2 e^{\lambda_2 \xi} + A_3 e^{\lambda_3 \xi} + A_4 e^{\lambda_4 \xi}\right) \cos k\varphi \\ \tilde{u}_\varphi = P_k\left(B_1 e^{\lambda_1 \xi} + B_2 e^{\lambda_2 \xi} + B_3 e^{\lambda_3 \xi} + B_4 e^{\lambda_4 \xi}\right) \sin k\varphi \\ \tilde{S}_{\rho\varphi} = P_k\left(C_1 e^{\lambda_1 \xi} + C_2 e^{\lambda_2 \xi} + C_3 e^{\lambda_3 \xi} + C_4 e^{\lambda_4 \xi}\right) \sin k\varphi \\ \tilde{S}_\varphi = P_k\left(D_1 e^{\lambda_1 \xi} + D_2 e^{\lambda_2 \xi} + D_3 e^{\lambda_3 \xi} + D_4 e^{\lambda_4 \xi}\right) \cos k\varphi \end{cases} \tag{13}$$

where A_i, B_i, C_i and D_i are undetermined coefficients.

These constants in the above formula are not completely independent, they should also satisfy equation Eq. (3). Based on the relationship between these constants, the quartic equation of the eigenvalue λ and the coefficient k of the series term can be derived.

$$\lambda^4 - 2\left(k^2 + 1\right)\lambda^2 + \left(k^2 - 1\right)^2 = 0 \tag{14}$$

Solve the equation and get

$$\lambda_{1,2} = \pm(k + 1);\quad \lambda_{3,4} = \pm(k - 1) \tag{15}$$

Thus, the specific expression of the special solution is determined under k-order normal outward load. Then the specific values of undetermined coefficients A_i, B_i, C_i and D_i are determined accosrding to the boundary condition Eq. (12).

Similarly, the inner side is considered as homogeneous boundary, while the outer side is only affected by k-order tangential sine boundary load. The specific form is as follows:

$$\begin{aligned} \text{When } \xi = \xi_1,\ E\frac{\partial u_\rho}{\partial \xi} + vS_\varphi = 0,\ S_{\rho\varphi} = 0 \\ \text{When } \xi = \xi_2,\ E\frac{\partial u_\rho}{\partial \xi} + vS_\varphi = 0,\ S_{\rho\varphi} = T_k \sin k\varphi \end{aligned} \tag{16}$$

where, T_0 is the k-order tangential load coefficient, k is the coefficient of series term.

By the same method, a special solution of the equation that satisfies the boundary condition Eq. (16) can be calculated. Finally, according to the principle of linear elastic superposition, the theoretical solution which satisfies the complex boundary condition of the original problem can finally be obtained.

3 Stress Analysis of Multi-layer Thick-Walled Cylinder

3.1 Problem Description

Considering the multi-layer thick-walled cylinder, the inner and outer radii of the i-th layer are R_i and R_{i+1}, respectively, and the material constants are E_i and v_i. The stresses in the x and y directions on the outside of the thick-walled cylinder are σ_x and σ_y (as shown in Fig. 2). The lateral pressure coefficient $K_0 = \sigma_y/\sigma_x$ is defined according to the stress values in the x and y directions. According to the theory of stress state, the force on the outside of the structure is transferred to the outermost thick-walled cylinder, and the stress boundary conditions remain unchanged along the axis of the multi-layer thick-walled cylinder. In this way, the problem can be simplified as the plane strain problem.

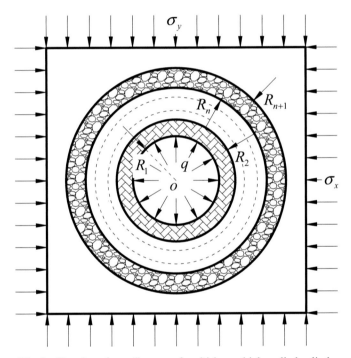

Fig. 2. The plane force diagram of multi-layer thick-walled cylinder

3.2 Boundary Conditions and Continuous Smooth Conditions

The inner and outer boundary conditions of the multi-layer thick-walled cylinder can be expressed as:when $\rho = R_1$,

$$\sigma_\rho^1(R_1) = 0, \ \tau_{\rho\varphi}^1(R_1) = 0; \tag{17}$$

when $\rho = R_{n+1}$,

$$
\begin{cases}
\sigma_\rho^n(R_{n+1}) = \dfrac{1 + K_0}{2}\sigma_x + \dfrac{1 - K_0}{2}\sigma_x \cos 2\varphi \\[3mm]
\tau_{\rho\varphi}^n(R_{n+1}) = \dfrac{1 - K_0}{2}\sigma_x \sin 2\varphi
\end{cases}
\tag{18}
$$

There are usually many constraint modes between layers of multi-layer thick-walled cylinder, among which two limit modes are smooth contact and tight connection, while the others are between them. The radial stress and displacement of the interlayer are continuous with different constraint modes, but the difference is the equilibrium condition between layers. The continuous smooth conditions of the two limit modes are listed below.

- Smooth interlayer contact

When $\rho = R_i, i = 2, \cdots, n$, the continuous condition is:

$$
\begin{cases}
u_\rho^{i-1}(R_i) = u_\rho^i(R_i) \\[2mm]
\sigma_\rho^{i-1}(R_i) = \sigma_\rho^i(R_i)
\end{cases}
\tag{19}
$$

At the same time, the stress equilibrium condition is:

$$
\tau_{\rho\varphi}^{i-1}(R_i) = \tau_{\rho\varphi}^i(R_i) = 0
\tag{20}
$$

- Tight interlayer connection

The continuity condition between layers is the same as formula (19). The equilibrium condition of stress and displacement between layers is as follows:

$$
\begin{cases}
u_\varphi^{i-1}(R_i) = u_\varphi^i(R_i) \\[2mm]
\tau_{\rho\varphi}^{i-1}(R_i) = \tau_{\rho\varphi}^i(R_i)
\end{cases}
\tag{21}
$$

According to formula (13), the general solution of displacement and stress of thick-walled cylinder of each layer can be written:

$$
\begin{cases}
u_\rho^i = \left(A_1^i e^{\lambda_1\xi} + A_2^i e^{\lambda_2\xi} + A_3^i e^{\lambda_3\xi} + A_4^i e^{\lambda_4\xi}\right)\cos k\varphi \\[2mm]
u_\varphi^i = \left(B_1^i e^{\lambda_1\xi} + B_2^i e^{\lambda_2\xi} + B_3^i e^{\lambda_3\xi} + B_4^i e^{\lambda_4\xi}\right)\sin k\varphi \\[2mm]
S_{\rho\varphi}^i = \left(C_1^i e^{\lambda_1\xi} + C_2^i e^{\lambda_2\xi} + C_3^i e^{\lambda_3\xi} + C_4^i e^{\lambda_4\xi}\right)\sin k\varphi \\[2mm]
S_{\varphi i} = \left(D_1^i e^{\lambda_1\xi} + D_2^i e^{\lambda_2\xi} + D_3^i e^{\lambda_3\xi} + D_4^i e^{\lambda_4\xi}\right)\cos k\varphi
\end{cases}
\tag{22}
$$

Where, the relationship between undetermined coefficients of each layer is

$$
\begin{cases}
A_j^i = \dfrac{\left[(1 - v\lambda_i)(1 + \lambda_i) - k^2\right]}{Ek\lambda_i^2}C_j^i \\[4mm]
B_j^i = \dfrac{\lambda_i^3 + \lambda_i^2 + (vk^2 - 1)\lambda_i + k^2 - 1}{Ek^2\lambda_i^2}C_j^i \quad j = 1, 2, 3, 4 \\[4mm]
D_j^i = \dfrac{1 + \lambda_i}{k}C_j^i
\end{cases}
$$

According to the statistical data, the undetermined coefficients of multi-layer thick-walled cylinder are $4n$ in total, and the continuous smooth conditions between layers are $4 \times (n - 1)$ equations, plus the 4 equations of the inner and outer boundary conditions, a total of $4n$ linear equations, from which all $4n$ undetermined coefficients can be calculated. By substituting the solved undetermined coefficients into formula (22), the stress and displacement fields of thick-walled cylinder of each layer can be calculated.

3.3 Example

The following parameters are selected for analysis and discussion of the calculation example: the radius of double-layer thick-walled cylinder from inside to outside is $R_1 = 3\,m$, $R_2 = 4\,m$ and $R_3 = 5\,m$ respectively. The outside of the thick-walled cylinder is subjected to $\sigma_x = 7\,MPa$ and the lateral pressure coefficient is taken as $K_0 = 0.6$. The elastic modulus of the inner and outer layers are taken as $2E_1 = E_2 = 40\,GPa$ respectively. Poisson's ratio is the same, take $v_1 = v_2 = 0.25$. According to the symmetry of the force on the multi-layer thick-walled cylinder, only a quarter model is taken for study, that is, $\theta \in [0°, 90°]$. There are two kinds of constraint modes between thick-walled cylinder: 1. Smooth interlayer contact, i.e. zero shear stress between layers; 2. Tight interlayer connection. The effects of two kinds of constraint modes on the stress and displacement fields of multi-layer thick-walled cylinder are compared.

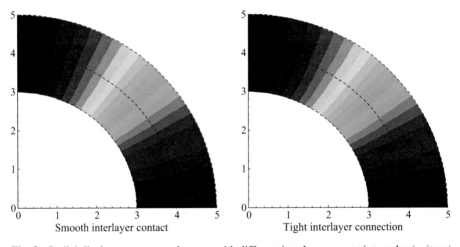

Smooth interlayer contact   Tight interlayer connection

Fig. 3. Radial displacement u_ρ nephogram with different interlayer constraint modes (unit: m)

Figures 3, Fig. 4, Fig. 5, Fig. 6 and Fig. 7 describe the nephogram of stress and displacement field under two different interlayer constraint modes. It can be seen from the comparison that the distribution patterns of radial displacement u_ρ and radial stress σ_ρ under two different interlayer constraints are basically the same, and smooth contact will cause the radial displacement value to be larger. However, the two constraint modes have great influence on the circumpolar displacement u_φ, the circumpolar stress σ_φ and the shear stress $\tau_{\rho\varphi}$. The distribution of nephogram under different constraints is completely

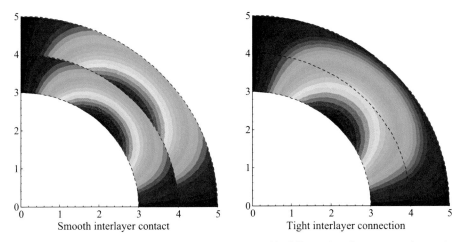

Fig. 4. Circumferential displacement u_θ nephogram with different interlayer constraint modes (unit: m)

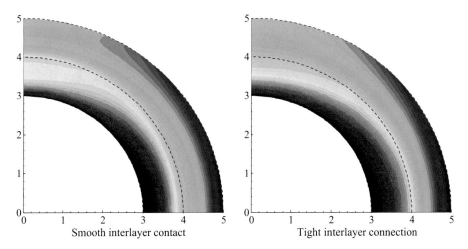

Fig. 5. Radial stress σ_ρ nephogram with different interlayer constraint modes (Unit: MPa)

different. The thick-walled cylinder with smooth interlayer contact is bounded by layers, and the inner and outer layers nephogram are distributed independently with similar distribution rules; while for the thick-walled cylinder with tight interlayer connection, the inner and outer nephogram are obviously continuous distribution. In terms of numerical value, the extreme value of stress and displacement under tight interlayer connection is smaller than that under smooth interlayer contact. Comprehensive analysis shows that the tight interlayer connection of multi-layer thick-walled cylinder increases the interlayer constrained force, the deformation is more coordinated, and the anti-deformation ability of thick-walled cylinder is improved. The overall mechanical properties are better than that of the thick-walled cylinder with smooth interlayer contact.

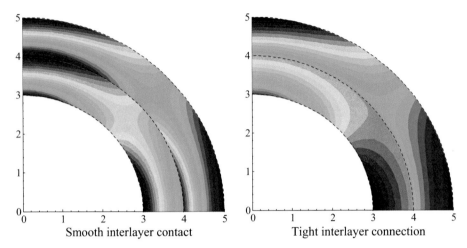

Fig. 6. Circumferential stress σ_φ nephogram with different interlayer constraint modes (Unit: MPa)

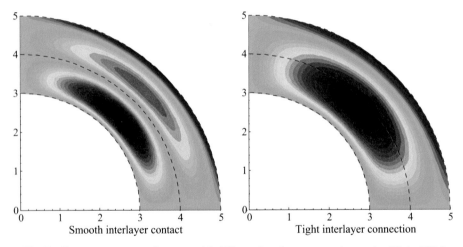

Fig. 7. Shear stress $\tau_{\rho\varphi}$ nephogram with different interlayer constraint modes (Unit: MPa)

3.4 Analysis of Influencing Factors

The influences of elastic modulus E, lateral pressure coefficient K_0 and other factors on the stress field distribution of thick-walled cylinder are discussed below. Because the mechanical properties of the tight interlayer connection of multi-layer cylinder are better than the smooth interlayer contact, only the tight interlayer connection of multi-layer thick-walled cylinder is considered. See the above example for other calculation parameters. In order to better study the relative relationship between the stress components of thick-walled cylinders and the influencing factors, and to obtain the general rule, each stress component is firstly dimensionless. The dimensionless values σ_ρ/σ_x, σ_φ/σ_x and

$\tau_{\rho\varphi}/\sigma_x$ are defined, and the relationship between the dimensionless stress values and the elastic modulus ratio of the inner and outer cylinders E_1/E_2 (0.1 ~ 10) and the lateral pressure coefficient K_0 (0 ~ 1) are analyzed. The variation rule of each stress field along the circumference direction of the cylinder and along the thickness direction of the cylinder is discussed.

- Influence of elastic modulus ratio

The calculation model takes the lateral pressure coefficient $K_0 = 0.6$, and the elastic modulus ratio E_1/E_2 of the inner and outer cylinders varies between (0.1 ~ 10). By comparing the nephogram of each stress component in the above example, it can be seen that the circumferential stress value is much larger than other stress components, and the extreme value of circumferential stress appears on the inner and outer circumference of the cylinder respectively. Therefore, the following focuses on the distribution of circumferential stress σ_φ along the inner and outer sides of the two-layer cylinder.

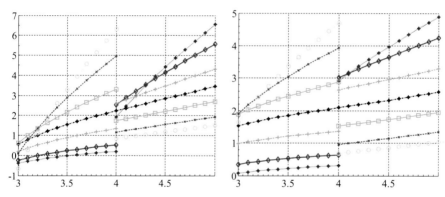

Fig. 8. Variation of σ_φ/σ_x along the cylinder thickness 0° direction

Fig. 9. Variation of σ_φ/σ_x along the cylinder thickness 30° direction

Figure 8, Fig. 9, Fig. 10 and Fig. 11 respectively describes the variation rule of circumferential stress σ_φ/σ_x along different directions of the cylinder thickness with different elastic modulus ratio. It can be seen from the figure that when $E_1 \neq E_2$, the circumferential stress is discontinuous at the interface of thick-walled cylinder, which is caused by boundary conditions; only when $E_1 = E_2$, the circumferential stress is continuous at the interface of thick-walled cylinder. The closer the elastic modulus ratio E_1/E_2 approaches 1, the smaller the difference of circumferential stress on both sides of the cylinder interface is; otherwise, the greater the difference is. The closer the ratio of elastic modulus E to 1, the smaller the difference of circumferential stress on both sides of the cylinder interface is, and vice versa. It can also be seen that when the direction angle is relatively small, the circumferential stress is an increasing function along the thickness of the cylinder, that is, for the same layer of cylinder, the stress of the outer cylinder is greater than that of the inner cylinder, and the stress extreme occurs in the outer cylinder. When the direction angle is relatively large, the circumferential stress is

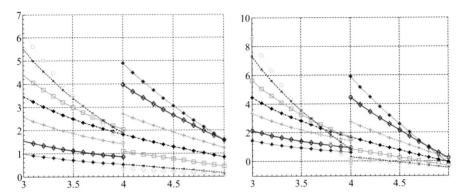

Fig. 10. Variation of σ_φ/σ_x along the cylinder thickness 60° direction

Fig. 11. Variation of σ_φ/σ_x along the cylinder thickness 90° direction

a decreasing function along the thickness of the cylinder, and the stress extreme occurs in the inner cylinder.

- Influence of lateral pressure coefficient K_0

The calculation model takes the elastic modulus ratio $E_1/E_2 = 0.5$ of the inner and outer cylinders, and the lateral pressure coefficient K_0 varies between (0 ~ 1). The distribution of circumferential stress ratio σ_φ/σ_x along the thickness direction of thick wall cylinder is discussed.

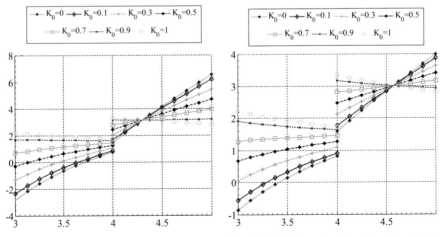

Fig. 12. Variation of σ_φ/σ_x along the cylinder thickness 0° direction

Fig. 13. Variation of σ_φ/σ_x along the cylinder thickness 30° direction

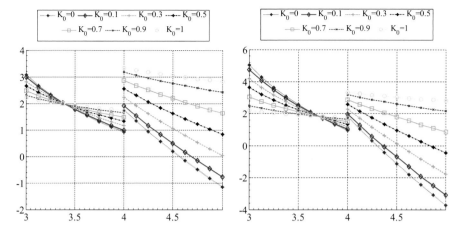

Fig. 14. Variation of σ_φ/σ_x along the cylinder thickness 60° direction

Fig. 15. Variation of σ_φ/σ_x along the cylinder thickness 90° direction

Figure 12, Fig. 13, Fig. 14 and Fig. 15 respectively describes the variation rule of circumferential stress σ_φ/σ_x along different directions of the cylinder thickness with different lateral pressure coefficients. It can be seen from the figures that (1) The circumferential stress is discontinuous at the interface of the multi-layer thick-walled cylinder. (2) When the direction angle is relatively small, there is an intersection point of circumferential stress in the outer cylinder corresponding to different lateral pressure coefficient K_0; while when the direction angle is relatively large, the intersection point of the circumferential stress appears in the inner cylinder. (3) When the direction angle and the lateral pressure coefficient K_0 are relatively small, the inside circumferential stress of the inner cylinder will be negative, while the outer cylinder will not be negative; when the direction angle is relatively large and the lateral pressure coefficient K_0 is relatively small, the negative circumferential stress value will appear on the outside of the outer cylinder. (4) In the 0° direction, there is a minimum value of circumferential stress inside the inner cylinder; there is a maximum value of circumferential stress outside the outer cylinder. In the 90° direction, there is a maximum value of circumferential stress inside the inner cylinder; there is a minimum value of circumferential stress outside the outer cylinder.

4   Conclusion

- The analytical expressions of displacement field and stress field of multi-layer thick-walled cylinder are derived by symplectic method of Hamiltonian mechanics.
- For the multi-layer thick-walled cylinder, the mechanical properties of tight interlayer connection are better than that of smooth interlayer contact. The circumferential stress component is much larger than other stress components, which is the main cause of structural failure.

- In the multi-layer thick-walled cylinder, as the ratio of inner and outer elastic modulus increases, the circumferential stress value of the inner layer cylinder gradually increases, while the circumferential stress value of the outer layer cylinder gradually decreases. It reflects that the softer the cylinder material is, the smaller the stress value is, and the harder the cylinder material is, the greater the stress value is. The extreme value of circumferential stress tends to appear in the cylinder region with larger elastic modulus.
- In the multi-layer thick-walled cylinder, when the lateral pressure coefficient approaches 1, the circumferential stress in the inner and outer layers changes gently and the numerical range becomes narrower. When the lateral pressure coefficient approaches 0, the change of the inner and outer layers circumferential stress is steeper and the numerical range is wider.

These conclusions provide specific theoretical guidance and technical support for solving the mechanical problems of multi-layer thick-walled cylinder.

Funding. This research was funded by the National Natural Science Foundation of China (Grant No. 42102082); and Key Research Program of Anhui Polytechnic University (Grant No. KZ42020043); National College Student Innovation and Entrepreneurship Training Program (Grant No. 202010363122).

Conflicts of Interest. On behalf of all authors, the corresponding author states that there is no conflict of interest.

References

1. Timoshenko, S.P., Goodier, J.N.: Theory of Elasticity. McGraw-Hill, New York (1970)
2. Gao, Y.T., Wu, Q.L., Lü, A.Z.: Stress analytic solution of a double-layered thick-walled cylinder with smooth contact interface subjected to a type of non-uniform distributed pressures. Eng. Mech. **30**(10), 93–99 (2013). https://doi.org/10.6052/j.issn.1000-4750.2012.06.0439
3. Jiang, B.S.: Elastic analysis of the composite shaft linings. J. China Coal Soc. **22**(4), 397–401 (1997). CNKI:SUN:MTXB.0.1997-04-012
4. Jiang, Z.L., Zhao, J.H., Lü, M.T., Zhang, L.: Unified solution of limit internal pressure for double-layered thick-walled cylinder based on bilinear hardening model. Eng. Mech. **35**(S1), 6–12 (2018). https://doi.org/10.6052/j.issn.1000-4750.2017.05.S025
5. Wang, S., Zhao, J.H., Jiang, Z.L., Zhu, Q.: Solution of ultimate bearing capacity for a double-layered thick-walled cylinder with different tension and compression characteristics. Chinese Q. Mech. **40**(03), 603–612 (2019). https://doi.org/10.15959/j.cnki.0254-0053.2019.03.19
6. Lu, A., Xu, G., Zhang, L.: Optimum design method for double-layer thick-walled concrete cylinder with different modulus. Mater. Struct. **44**(5), 923–928 (2011). https://doi.org/10.1617/s11527-010-9676-7
7. Qiu, J., Zhou, M.: Analytical solution for interference fit for multi-layer thick-walled cylinders and the application in crankshaft bearing design. Appl. Sci. **6**(167), 1–20 (2016). https://doi.org/10.3390/app6060167
8. Wu, Q.-L., Lü, A.-Z., Gao, Y.-T., Wu, S.-C., Zhang, N.: Stress analytical solution for plane problem of a double-layered thick-walled cylinder subjected to a type of non-uniform distributed pressure. J. Cent. South Univ. **21**(5), 2074–2082 (2014). https://doi.org/10.1007/s11771-014-2156-4

9. Loghman, A., Parsa, H.: Exact solution for magneto-thermo-elastic behaviour of double-walled cylinder made of an inner FGM and an outer homogeneous layer. Int. J. Mech. Sci. **88**, 93–99 (2014). https://doi.org/10.1016/j.ijmecsci.2014.07.007
10. Zhong, W.X.: New Solution System for Theory of Elasticity. Dalian University of Technology Press, Dalian, China (1995)
11. Yao, W.A., Zhong, W.X., Lim, C.W.: Symplectic Elasticity. World Scientific, Singapore (2009)
12. Zhou, J.F., Zhuo, J.S.: A new solution of elasticity in polar coordinate. Acta. Mech. Sin. **33**(6), 839–846 (2001). https://doi.org/10.3321/j.issn:0459-1879.2001.06.015
13. Tseng, W.D., Tarn, J.Q.: Three-dimensional solution for the stress field around a circular hole in a plate. J. Mech. **30**(6), 611–624 (2014). https://doi.org/10.1017/jmech.2014.48
14. Wu, Q.L., Lü, A.Z.: Stress analytical solution for plane problem of a thick-walled cylinder subjected to a type of non-uniform distributed pressures. Eng. Mech. **28**(6), 6–10 (2011)
15. Guo, X., Hou, G.: On the block basis property of the off-diagonal Hamiltonian operators and its application to symplectic elasticity. The Eur. Phys. J. Plus **131**(10), 1–8 (2016). https://doi.org/10.1140/epjp/i2016-16368-y
16. Jiang, Z.Y., Zhou, G.Q.: Comparative study of wellhole surrounding rock under nonuniform ground stress. Adv. Civ. Eng. **2019**, 1–9 (2019). https://doi.org/10.1155/2019/7424123
17. Jiang, Z., Zhou, G., Jiang, L.: Symplectic elasticity analysis of stress in surrounding rock of elliptical tunnel. KSCE J. Civ. Eng. **24**(10), 3119–3130 (2020). https://doi.org/10.1007/s12205-020-1810-7
18. Eburilitu Completeness of the eigenfunction systems of infinite dimensional Hamiltonian operators and its applications in elasticity, Inner Mongolia University (2011)

Research on Construction Process of Steel Beam Incremental Launching Based on Finite Element Method

Yancai Xiao[1], Kun Fu[2], Zhuang Li[1,3(✉)], Zhiping Zeng[1,4], Jian Bai[5],
Zhibin Huang[1,6], Xudong Huang[1], and Yu Yuan[7]

[1] School of Civil Engineering, Central South University, Changsha, Hunan, China
1877868843@qq.com
[2] Kunming Metro Construction Management Co., Ltd., Kunming, Yunnan, China
[3] China Railway Electrification Engineering Group Co., Ltd., Beijing, China
[4] MOE Key Laboratory of Engineering Structures of Heavy Haul Railway (Central South University), Changsha, Hunan, China
[5] Luoyang Kebosi New Material Technology Co., Ltd., Luoyang, Henan, China
[6] Southeast Coastal Railway (Fujian) Co., Ltd., Fuzhou, Fujian, China
[7] Information Engineering School, Nanchang University, Nanchang, Jiangxi, China

Abstract. In order to ensure the normal operation of the traffic under the bridge, reasonable calculation methods and construction techniques should be adopted for the construction of the newly added railway station. This paper establishes a structural calculation finite element model to calculate and analyze the various construction stages of the steel beam incremental launching construction of the newly-added Gaoping station on the Yichuang-Wanzhou Railway, and systematically study the mechanical properties of the steel beam in the process. The results show that: (1) The deflection of each rod can meet the requirements of the railway bridge steel structure construction specification. However, when the length of the front cantilever of the steel beam reaches 11.4 m, the maximum deflection of the upper and lower chord bars is close to the limit. (2) The load-bearing capacity of each member of the steel beam meets the requirements, which indicates that the structural design of the steel beam and the incremental launching construction plan are reasonable. (3) In view of the complexity and uncertainty of the incremental launching construction process, real-time monitoring of the construction process is required, and the beam should be dropped in time when abnormal conditions occur to ensure the safe operation of the existing line.

Keywords: Newly added station · Incremental launching construction · Steel beam structure · Finite element · Mechanical properties

1 Introduction

Along with the demand of socio-economic and transportation development, the bridge construction technology has also been developed tremendously [1–5]. And a common construction methods called incremental launching has received wide attention from

G. Feng (Ed.): ICCE 2021, LNCE 213, pp. 254–262, 2022.
https://doi.org/10.1007/978-981-19-1260-3_22

scholars. Lou Song et al. analyzed the mechanical behavior of large tonnage steel truss beam during step incremental launching and sliding construction based on MIDAS Civil software modeling [6]; Zhou Jianting et al. proposed a construction control method for incremental launching of large span rail steel box laminated beam bridge in the context of the north bank approach bridge of Nanjimen rail special bridge [7]; Liu Junhua used finite element method to study the facility construction method and incremental launching safety analysis method of a large span arch beam combination system bridge incremental launching construction [8]; Based on the law of lateral deflection of steel channel beam tangential incremental launching, Yang Zengquan proposed the construction process of steel channel beam step incremental launching without temporary pier and pier side support in the middle [9]; Shi Xiaoye et al. proposed a guide beam strengthening scheme with double rows of longitudinal stiffening ribs by analyzing the local instability and buckling deformation characteristics of the guide beam [10].

In this paper, a structural computational finite element model is established to calculate and analyze each construction stage of steel beam incremental launching construction by taking the newly added flyover at Gaoping station on Yichuang-Wanzhou Railway as a relying project, and calculating the steel beam displacement and stress respectively to verify the safety of steel beam structure during incremental launching construction.

2 Project Overview

Gaoping station is a new station on the Yichuang-Wanzhou Railway, which is one of the main skeleton of China's "eight vertical and eight horizontal" high-speed railroad network, with up to 96 high-speed trains running every day. In order to reduce the impact on the operation of existing lines and ensure the safety of the entire construction process, the best construction method of the station flyover is incremental launching method [11–13].

The bridge is equipped with 4 sets of step incremental launching equipment with longitudinal center spacing of $12.6 + 14.4 + 12.8 + 7.9 + 5.5$ m and transverse center spacing of 9 m. The steel beam push can be divided into three stages. The first stage is to assemble a 27 m steel beam, and then push 6 m integrally to the end of the steel beam to reach the top D7 and D8 of the walker; the second stage is to continue to assemble a 6 m steel beam. Then push it integrally for 10 m; the third stage is to continue to assemble the remaining 10.075 m steel beam, push it 9.6 m integrally until the steel beam is pushed to the design position.

3 Finite Element Modeling

3.1 Calculation Parameters

The steel used for the steel beam of the flyover at Gaoping station includes Q345qD and Q235. The allowable stress for compression and bending of Q345qD steel is 250 MPa, and the allowable stress for shear is 140 MPa; the allowable stress for compression and bending of Q235 steel is 170 MPa, and the allowable shear stress is 100 MPa [14].

3.2 Calculation Load

- Self-weight of steel beam

    Taking into account the self-weight of steel beams, detailed structures such as bridge decks, diaphragms and stiffeners are not included in the finite element model. Therefore, the self-weight coefficient is increased to balance the gravity of the detailed structure, and the self-weight coefficient is taken as 1.37.
- Self-weight of incremental launching temporary piers

    The self-weight of the temporary pier support and the assembled support is automatically loaded by the program. Taking into account that the welds and stiff plates are not included in the model, the self-weight coefficient is taken as 1.2.
- Lncremental launching horizontal force

    Horizontal loads in the transverse and longitudinal directions are applied to the top of the temporary piers, and the load is taken as 5% of the reaction force of the vertical fulcrum.
- Construction load

    Consider the construction load of 2.0 kN/m^2 on the working platform on the top of the temporary pier.

3.3 Calculation Conditions

The temporary piers supporting the steel beam structure at different positions during the entire incremental launching construction process will change accordingly, and the whole process will be divided into 17 working conditions (calculated once every 2 m of incremental launching). The reaction force of temporary pier support, steel beam deflection and stress are calculated under each working condition. The calculation models at different stages are shown in Figs. 1, 2 and 3, and the description of the controlled working conditions and the length of the front and rear cantilever during the pushing process of the steel beam are shown in Table 1.

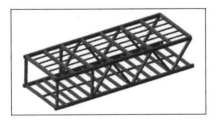   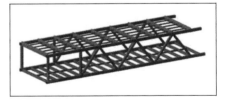

Fig. 1. The first stage.   **Fig. 2.** The second stage.

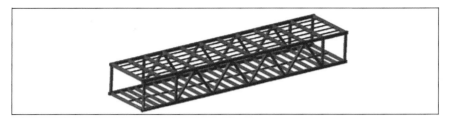

Fig. 3. The third stage.

Table 1. Controlled construction steps.

Working condition	Front cantilever (m)	Rear cantilever (m)	Description
1	1.4	5	Assemble 27 m-long steel beam sections
2	3.4	3	Push the assembled steel beam forward integrally by 2 m
3	5.4	1	Continue to push the assembled steel beams forward integrally for 2 m
4	7.4	6.7	Continue to push the assembled steel beams forward integrally for 2 m
5	7.4	5	Continue to assemble steel beam sections for 6 m
6	9.4	3	Push the assembled steel beam forward integrally by 2 m
7	11.4	1	Push the assembled steel beam forward integrally by 2 m
8	13.4	6.7	Push the assembled steel beam forward integrally by 2 m
9	14.4	5.7	Push the assembled steel beam forward integrally by 1 m
10	0	5.7	The front end of the steel beam is put on the top D1 and D2
11	3	2.7	Push the assembled steel beam forward integrally by 3 m
12	3	5	Continue to assemble the remaining 10.075 m steel beam section
13	5	3	Push the assembled steel beam forward integrally by 2 m
14	7	1	Push the assembled steel beam forward integrally by 2 m

(*continued*)

Table 1. (*continued*)

Working condition	Front cantilever (m)	Rear cantilever (m)	Description
15	9	6.7	Push the assembled steel beam forward integrally by 2 m
16	11	4.7	Push the assembled steel beam forward integrally by 2 m
17	13	2.7	Push the assembled steel beam forward integrally by 2 m

4   Calculation and Analysis of Steel Beam Pushing Process

4.1   Steel Beam Displacement Calculation

The maximum displacement of each member of the steel beam under various controlled construction conditions is shown in Fig. 4. The deflection of the lower chord (LC) and the upper chord (UC) refers to the vertical displacement, and the web member deflection (WM) refers to the horizontal displacement.

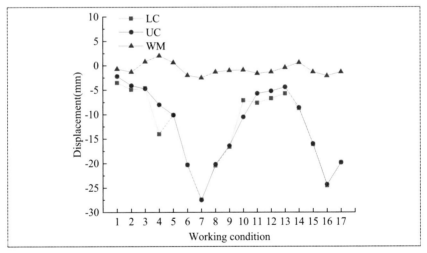

Fig. 4. Maximum deflection of incremental launching steel beam member under various controlled construction conditions.

It can be seen from Fig. 4 that under different working conditions, the maximum deflection of the upper chord and the lower chord of the steel beam are basically the same, and both are greater than the deflection of the web. The maximum deflection of each member appears in working condition 7 (the front cantilever length is 11.4 m). At this time, the maximum vertical deflection of the lower chord is −27.4 mm, the maximum vertical deflection of the upper chord is −27.5 mm, and the maximum horizontal deflection of the web member is −2.5 mm. Figure 5 shows the displacement cloud picture for working condition 7.

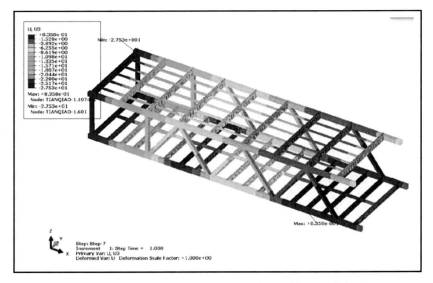

Fig. 5. Steel beam deflection cloud map for working condition 7.

According to "Steel Structure Design Standard", the allowable value of deflection for working condition 7 is 28.5 mm [15]. From the above analysis, it can be seen that although the deflection of each member can meet the requirements of the railway bridge steel structure construction code, the maximum deflection of the upper chord and the lower chord is close to the limit value, indicating that the steel beam structure has made full use of some chords' flexural stiffness.

4.2 Steel Beam Stress Calculation

Figure 6 shows the maximum normal stress of the upper chord (UC), lower chord (LC), vertical web member (VWM), and inclind web member (IWM) of steel beams under various controlled construction conditions.

As can be seen from the above figure, the maximum value of the normal stress appears in the upper chord during the whole process of the steel beam incremental launching construction. The phenomenon is particularly significant in working condition 7 and working condition 16. The maximum normal stress of the lower chords of the steel beam appears in the working condition 7, which is 107.9 MPa. The maximum normal stress of the upper chords and the vertical web members both appear in the working condition 16, which are 52.4 MPa and 65.7 MPa, respectively. And the maximum normal stress of the inclind web members appears in the working condition 17, which is 54.2 MPa. The steel used in the steel beam is marked with Q235, whose allowable stress is 250 MPa.

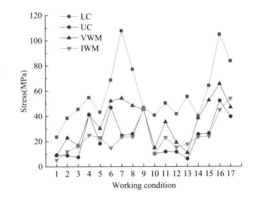

Fig. 6. Maximum normal stress of incremental launching steel beam member under various controlled construction conditions.

Therefore, the load-bearing capacity of each member of the steel beam can meet the requirements during the incremental launching process, and the structure is safe and controllable.

5   Conclusion

Basing on the above researches, conclusions and suggestionscan be drawn as follows: (1) In the whole process of steel beam incremental launching construction, the deflection of each member can meet the requirements of the railway bridge steel structure construction code. But when the length of the front cantilever of the steel beam reaches 11.4 m (working condition 7), the maximum deflection of the upper chord and the lower chord is close to the limit. (2) During the construction of the steel beam incremental launching, the load-bearing capacity of each member of the steel beam meets the requirements. It shows that the design of steel flyover structure and the design of the incremental launching construction plan are reasonable, and the structure of the incremental launching process is safe and controllable. (3) In view of the complexity and uncertainty of the incremental launching construction process, real-time monitoring of the construction process is required, and the beam should be dropped in time when abnormal conditions occur to ensure the safe operation of the existing line under the bridge.

Acknowledgments. The research is financially supported by the Science and Technology Research Development Project of China Railway Nanchang Group Co., Ltd. (grant CRNG202007), Natural Science Foundation of Hunan Province, China (Grant 2019JJ40384), High-Speed Railway Foundation Joint Fund Project (Grant U1734208), Scientific Research Test Project of China Railway Corporation (Grant SY2016G001), which is gratefully acknowledged by the authors.

References

1. Zeng, Z., Yu, Z., Cheng, X., Zhao, G.: Analysis on spatial vibration of train-turnout-continuous frame bridge with train running through turnout branch. J. Vib. Shock **028**(002), 40–44 (2009)
2. Zhao, H., Reng, W., Gao, J., Jin, L., Wang, H.: Simulation research on the whole process of cantilever erection of long-connected large-span continuous steel truss girders. Rail. Standard Design **65**(11), 6–11 (2021)
3. Zeng, Z., Song, S., Luo, J., Rao, H., Xie, H., Yin, H.: Research on the limited value of the first pier height for simply supported beam bridges of high-speed railway. J. Railw. Eng. Soc. **35**(05), 24–29 (2018)
4. Wu, K., Gao, Y.: Study of pylon construction schemes for mingyuexia changjiang river bridge. World Bridges **49**(05), 20–26 (2021)
5. Zeng, Z., Zhang, X., Sun, Y., Wang, X., Chen, X.: Experimental study of beam and rail longitudinal displacement of cwr track on bridge in budongquan region along qinghai-tibet railway. J. Cent. South Univ. (Sci. Technol.) **045**(002), 638–642 (2014)
6. Lou, S., Wu, F., Jiang, Y., Wang, J.: Mechanical analysis of walking-type incremental launching and sliding construction of large-tonnage steel truss girder. Brdg. Construct. **51**(01), 66–73 (2021)
7. Zhou, J., Li, X., Wu, Y., Li, X., Ding, P.: Incremental launching construction control method of long-span track steel box composite girder bridge. World Bridges **49**(03), 64–71 (2021)
8. Liu, J.: Research on mechanical performance of arch rib pushing of long-span beam-arch combined bridge. Highway **66**(04), 176–181 (2021)
9. Yang Z.: Key Technology of long-span curved steel trough beam incremental launching construction. J. Highway Transp. Res. Dev. **38**(3), 56–62+72 (2021)
10. Shi, X., Liang, Y., Wang, D., Chen, H., Fang, L.: Research on buckling analysis and strengthening measures of guide beam for bridge incremental launching construction. J. Zhengzhou Univ. (Eng. Sci.) **42**(05), 74–78 (2021)
11. Fu, M., Liang, Y., Long, Y.: Optimization analysis of platform for incremental launching method construction of continuous steel box girder bridge. Building Structure **51**(S1), 2360–2364 (2021)
12. Zhao, R., Zhang, S.: Research status and development trend on incremental launching construction of bridges. China J. Highw. Transp. **29**(2), 32–43 (2016)
13. Ye, J.: Key techniques for incremental launching construction of steel tub girders in north approach bridge of Oujiang River north estuary bridge. Br Construct. **50**(S2), 115–120 (2020)
14. Code for Design on Steel Structure of Railway Bridge. TB 10091-2017
15. Standard for Design of Steel Structures. GB 50017-2017

Numerical Simulation and Experimental Study on Concrete Block Fabricated Foundation

Huiyuan Liu[1], Xibin Yang[1], Jie Li[1], Mintao Ding[2(✉)], Congyue Song[3], and Yongping Li[4]

[1] Inner Mongolia Power (Group) Co., Ltd., Inner Mongolia, China
[2] China Electric Power Research Institute, Beijing, China
`dingmintao@yeah.net`
[3] Ordos Electric Power Bureau, Ordos, China
[4] Inner Mongolia Electric Power Survey and Design Institute Co., Ltd., Inner Mongolia, China

Abstract. The fabricated foundation can be prefabricated in the factory, so it is easy to control the size, weight and quality of components, and it's easy to transport since the size is controllable. The operation links such as on-site concrete pouring and maintenance, on-site formwork and steel bar binding are omitted, the opening time of foundation pit is reduced, and the construction time is greatly shortened. In this paper, two different types of concrete block fabricated foundation (Foundation-A and Foundation-B) are designed. The foundation is divided into concrete blocks one by one, which are connected by bolts and steel plates on site. In order to study the connection performance and bearing capacity of the two types of foundation, the in-situ test and the numerical simulation analysis of fabricated foundations is carried out. The research shows that both foundation types can meet the requirements of bearing capacity, but Foundation-B is better than Foundation-A in bearing performance, integrity, processing difficulty and construction difficulty.

Keywords: Fabricated foundation · Concrete block · Bearing capacity

1 Instruction

The fabricated foundation has strong social and economic benefits in transmission line projects with severe natural conditions such as high altitude, lack of water, difficult sand and gravel collection, long transportation distance or urgent time limit [1–3]. Therefore, further research on fabricated foundation has important practical and engineering significance. Since the 1960s, scholars have studied the bearing mechanism of aeolian sand foundation by indoor model test, numerical analysis and field test, and deduced the theoretical calculation formula of ultimate bearing capacity [4–9]. The ultimate bearing capacity of foundation is influenced by the geometric dimension of foundation, buried depth and ground overload [10, 11]. Liu [12] studied the bearing mechanism of straight column extended bottom foundation on eolian sand foundation under uplift load based on field tests. Qian [13] carried out field test research on inclined column extended foundation of aeolian sand foundation under the combined load of uplift horizontal force. Lu

G. Feng (Ed.): ICCE 2021, LNCE 213, pp. 263–272, 2022.
https://doi.org/10.1007/978-981-19-1260-3_23

[14] conducted a compressive test on the prefabricated concrete assembled foundation of Qinghai-Tibet AC750 kV/DC ±400 kV frozen soil power grid interconnection project. Under compression load, the displacement of the ring flange connecting the column, the bottom slab, and the joint seam of the two consolidated bottom slabs conformed with that of the top plane of the foundation. The displacement and the stress of the different connected parts redistributed until the new balance developed. As a result, the different connected parts of the precast concrete assembly foundation co-operated to carry the compression load. The joint working performance of foundation components under external load is the key and difficult point of design, as well as an important guarantee for the safety and stability of foundation. At present, there are few relevant specifications and experience at home and abroad [15–17]. Compression Behavior on Precast Concrete Assembly Spread Foundation for Qinghai-Tibet AC/DC Grid Interconnection Project.

2   Model Design

In this paper, two types of fabricated foundation are designed, including Foundation-A and Foundation-B, both are formed by concrete blocks of different sizes. The two types of foundation are shown in Fig. 1.

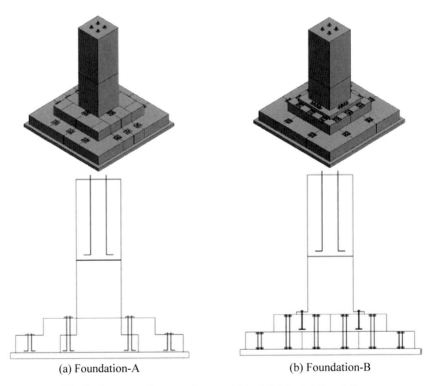

(a) Foundation-A   (b) Foundation-B

Fig. 1. Structure diagram of concrete block fabricated foundation

Foundation-A is composed of special-shaped concrete blocks, and the concrete blocks are overlapped into a Z-shape. Foundation-B adopts laminated type, and the two layers of concrete blocks are staggered and tower overlapped with each other to form the whole foundation. The overall structure of the two foundation types is similar to the spread foundation with two steps, with a buried depth of 3.5 m, a bottom plate width of 3 m, and a square section with side length of 1 m in the middle column. In order to increase concrete area at the junction of the upper and lower layers of special-shaped blocks of Foundation-A, the step of Foundation-A is narrower than that of Foundation-B(The step of Foundation-A is 400 mm wide and 300 mm high, and the step of Foundation-B is 600 mm wide and 300 mm high).

The in-situ test is adopted in this research. The test foundation is prefabricated according to the above sizes. Uplift and compression static load test shall be conducted for Foundation-A and Foundation-B respectively.

3 In Situ Test

In this test, two kinds of in-situ static load tests of foundations were carried out, including two test conditions of uplift and compression. In the uplift test, it can be observed that a circular soil failure surface is formed around the foundation, as shown in Fig. 2. After the test, dig out the foundation was digged out to observe the foundation connection, as shown in Fig. 3.

(a) Foundation-A   (b) Foundation-B

Fig. 2. Failure surface of uplift test

From the uplift failure mode of the test, it can be seen that the failure of Foundation-A and Foundation-B is mainly due to the shear failure of the soil. While observing the connection of the test foundation, it can be seen that the concrete blocks and connections are not damaged, which means that the two types of foundation structures can meet the requirements of bearing capacity. The load-displacement curve of two foundations is shown in Fig. 4.

(a) Foundation-A   (b) Foundation-B

Fig. 3. Connection of foundation

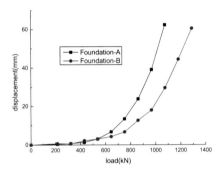

Fig. 4. Load-displacement curve

The test results show that the bearing performance of Foundation-B is better than that of Foundation-A. In the in-situ test, displacement rods (Fig. 5, composed of steel bars welded on the upper surface of the foundation, the steel bars extend vertically out of the soil surface, and the steel bars and the soil are separated by PVC tubes. During the test, the displacement of the foundation at this position can be known by putting the displacement monitoring on top of the steel bar extended out of the surface). The displacement rods are respectively set at the four corners of two concrete steps, It is used to measure the displacement of each step in the test. The displacement of the displacement rods is shown in Fig. 6. In the figure, the displacement of the top of the foundation is measured by the sensor at the top of the foundation, the displacement of the upper step and the lower step are measured by the displacement rods, and the displacement rods is the average value of the displacement of the upper and lower steps.

It can be seen from the displacement rods:

(1) From the data of the compression load test, the displacement difference between two steps of Foundation-A is less than that of Foundation-B. In terms of mechanical structure, Foundation-A is a special-shaped structure, and the upper and lower steps belong to the same concrete block, so the displacement difference is less than that of

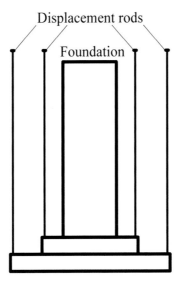

Fig. 5. Load-displacement curve

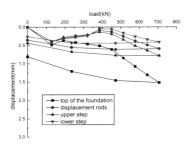

(a) Compression of Foundation-A

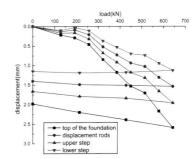

(b) Compression of Foundation-B

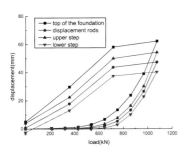

(c) Uplift of Foundation-A

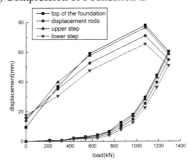

(d) Uplift of Foundation-B

Fig. 6. Load-displacement curve of displacement rods

Foundation-B. However, the displacement difference between the top and the lower part of Foundation-A is larger than Foundation-B, indicating that the connection between the surrounding concrete blocks and the middle column of Foundation-A is weak under the compression load, the punch-resistant capability is weak under the compression load. Therefore, the connection performance of Foundation-B is better than that of Foundation-A.

(2) From the perspective of uplift condition, the displacement dispersion of each position of Foundation-A is significantly greater than that of Foundation-B. The structural analysis of two types of foundation shows that the bolts is mainly sheared during the uplift of Foundation-A, the upper step of concrete block is tensioned under the action of the bolts. Meanwhile, the bolts is mainly tensioned during the uplift test due to the stacking structure of Foundation-B and the concrete block has obvious compression effect under the action of bolts. The tensile performance of bolts is better than the shear performance, while the compressive performance of concrete is better than the tensile performance. Therefore, the overall mechanical performance of Foundation-B is better than that of Foundation-A.

(3) According to the above analysis, the mechanical performance of Foundation-B is better than that of Foundation-A under both compression and uplift load tests. It is also found that the difficulty of formwork and construction of Foundation-B are less than that of Foundation-A, so the structural form of Foundation-B is better than that of Foundation-A.

4 Numerical Simulation

The calculation model of the foundation is divided into two steps, the first step is the global finite element model, which includes the foundation and nearby foundation soil. The joint reaction force of the contact surface between foundation and soil can be obtained according to the whole finite element model. The second step is the substructure element model, the foundation substructure model includes the concrete block of the foundation, the connecting bolt between the blocks and the equivalent connecting key. During the calculation, the reaction force of the foundation boundary node calculated by the global finite element model in the first step is applied as the input, and the force of the connecting bolt and the equivalent connecting key is calculated. The finite element model is shown in the Fig. 7.

The global finite element model adopts Solid185 element, with a total of 576875 nodes and 565104 elements. Materials are divided into soil and concrete. Concrete has a calculated density of 2400 kg/m³ and an elastic modulus of 28 GPa, while soil has a calculated density of 1600 kg/m³ and an elastic modulus of 200 MPa. Friction contact (TARGE170 element and CONTA174 element) was adopted for the foundation and foundation, and the friction coefficient was set as 0.5.

As shown in the Fig. 8, the Foundation-A substructure model adopts solid185 element (concrete) and Link180 element (simulating bolt and steel plate). The model consists of 65,674 nodes and 65,148 units. The Foundation-B consists of 67,263 nodes and 62,920 units. Friction contact was adopted between the blocks (TARGE170 element and CONTA174 element), and the friction coefficient was set as 0.5.

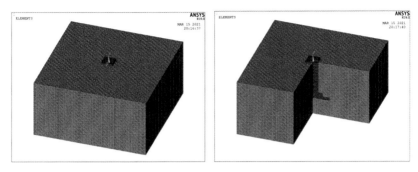

Fig. 7. Finite element model (overall foundation model)

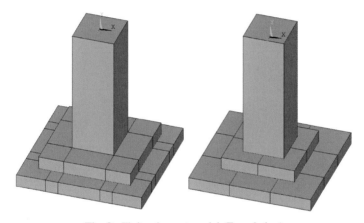

Fig. 8. Finite element model (Foundation)

According to the calculation results, the stress of Foundation-A is smaller under the upward drawing condition and larger under the downward pressing condition. In order to maintain the stability of the foundation under the uplift condition, the stress should be greater than 0. Therefore, the uplift load that Foundation-A can bear is smaller. With the same foundation bearing capacity, Foundation-A can bear a smaller down load. Therefore, from the perspective of foundation bearing capacity, Foundation-B is a better choice.

Foundation-A is connected by 32 vertical bolts, Foundation-B is connected by 12 vertical bolts for layer 1 flange, and 128 bolts for layer 2 flange. Table 1 shows the maximum tension of vertical bolts and the corresponding bolt specifications. It can be seen from the table that the vertical connecting bolt is most dangerous for Foundation-A under the action of downward pressure, and the vertical connecting bolt is most dangerous for Foundation-B under the condition of upward pulling. According to the calculation of the corresponding bolt area based on the maximum bolt axial force, it can be seen that the vertical bolt of the 1st floor flange of Foundation-B has the highest dimension requirement, and M27 bolt is required at least. This is because the total number of bolts is small and the bending moment caused by transverse load is greatly affected under the pulling up condition. However, there are more bolts in the second layer, so the stress of the single bolt is smaller.

Table 1. Bearing capacity of Bolts

Foundation	Foundation-A	Foundation-B	
	Vertical bolt	Flange bolt	Vertical bolt
Uplift	3.36 kN	72.6 kN	13.6 kN
Compression	28.2 kN	4.38 kN	2.90 kN
Bolt area	177 mm^2	454 mm^2	85 mm^2
Bolt specification	M18 and above	M2 and above	M14 and above

Link element is used for equivalent simulation of the steel plate. The sum of the axial forces of two adjacent LINK elements is the force of the steel plate. For the upper connecting steel plate, the steel plate is pulled under the pulling condition. For the lower side connecting steel plate, the steel plate is pulled under the lower pressing condition. Table 2 shows the maximum tensile stress of the connecting key, the corresponding specification of the connecting bolt and the thickness of the steel plate. The thickness of the steel plate is calculated according to the minimum bolt size. When the bolt size is larger, the thickness of the steel plate can be reduced appropriately, so the calculation results in the table are more safe and conservative. As can be seen from the table, although the steel plate in Foundation-B is subjected to Foundation-A larger tension, the bolt area and the corresponding steel plate thickness obtained in Foundation-B are smaller because the number of stressed bolts in basic scheme B is twice that in Foundation-A on the same steel plate.

Table 2. Bearing capacity of Connector

Foundation	Foundation-A		Foundation-B	
	Upper steel plate (uplift)	Lower steel plate (compression)	Upper steel plate (uplift)	Lower steel plate (compression)
Maximum tension	38.9 kN	28.9 kN	66.6 kN	36.2 kN
Bolt area	122 mm^2	90 mm^2	104 mm^2	57 mm^2
Bolt specification	M16 and above	M14 and above	M16 and above	M12 and above
Steel plate thickness	8.6 mm	7.4 mm	7.4 mm	5.5 mm

5 Conclusion

In this paper, two types of fabricated foundation are designed. The connection performance and bearing capacity of the two types of foundation are verified by in-situ test, and the test results are analyzed by numerical simulation.

(1) No structural failure occurred in the two foundation in the test. The failure of two foundations is mainly due to the shear failure of the soil. Both types of foundation can meet the requirements of bearing capacity.
(2) The structural form of Foundation-B is superior to Foundation-A, considering bearing performance, integrity, processing difficulty and construction difficulty.
(3) The conclusion of in-situ test is also verified by numerical simulation analysis.

References

1. Qian, Z.Z., Lu, X.L., Ding, S.J.: Experimental study of assembly foundation for transmission line tower in Taklimakan desert. Rock Soil Mech. **32**(8), 2359–2365 (2011)
2. Cheng, Y.F., Ding, S.J.: Prototype tests of assembly foundation of transmission line in aeolian sand area. Rock Soil Mech. **33**(11), 3230–3236 (2012)
3. Sieira, A., Gerscovich, D., Sayao, A.S.F.J.: Displacement and load transfer mechanisms of geogrids under pullout condition. Geotext. Geomembr. **27**(4), 241–253 (2009)
4. Murray, E.J., Geddes, J.D.: Uplift behaviour of plates in sand. J. Geotech. Eng. Div., ASCE **113**(3), 202–215 (1987)
5. Dichin, E.A., Leung, C.F.: The influence of foundation geometry on the uplift behavior of piles with enlarged bases in sand. Can. Geotech. J. **29**(3), 498–505 (1992)
6. Birch, A.J., Dickin, E.A.: The response to uplift loading of pyramid foundations in cohesionless backfill. Comput. Struct. **68**(1–3), 261–270 (1998)
7. Dichin, E.A., Laman, M.: Uplift response of strip anchors in cohesions soil. Adv. Eng. Softw. **38**(8), 618–625 (2007)
8. Ilamparuthi, K., Dickin, E.A., Muthukrishnaiah, K.: Experimental investing of the uplift capacity of circular plate anchors in sand. Can. Geotech. J. **39**, 648–664 (2002)
9. Kouzer, K.M., Kumar, J.: Vertical uplift capacity of two interfering horizontal anchors in sand using an upper bound limit analysis. Comput. Geotech. **36**(6), 1084–1089 (2009)
10. Hjiaj, M., Lyamin, A.V., Sloan, S.W.: Bearing capacity of a cohesive-frictional soil under non-eccentric inclined loading. Comput. Geotech. **31**(6), 491–516 (2004)
11. Paolucci, R., Pecker, A.: Soil inertia effects on the bearing capacity of rectangular foundations on cohesive soils. Eng. Struct. **19**(8), 637–643 (1997)
12. Liu, W.B., Zhou, J., Su, Y.H., Liu, L.: Load displacement behaviors for uplift of spread foundations in aeolian sand with geogrid reinforcement. Chinese J. Geot. Eng. **25**(5), 562–566 (2003)
13. Qian, Z.Z., Lu, X.L., Ding, S.J.: Full scale tests on pad and chimney foundation subject to uplift combined with horizontal loads in aeolian sand. Rock Soil Mech. **30**(1), 257–260 (2009)
14. Lu X. L., Qian Z. Z., Tong R. M.: Compression behavior on precast concrete assembly spread foundation for Qinghai-Tibet AC/DC grid interconnection project. intelligent system design and engineering applications (ISDEA). In: 2013 Third International Conference on. IEEE, 2013

15. Kulhawy, F.H., Beech, J.F.: Transmission line structure foundation for uplift-compression loading. Electric Power Research Institute (1983)
16. Trautmann, C.H., Kulhawy, F.H.: Uplift load-displacement behavior of spread foundations. J. Geot. Eng. ASCE **114**(2), 168–184 (1988)
17. Stewart, H.E., Kulhawy, F.H.: Field Evaluation of Grillage Foundation Uplift Capacity. Electric Power Research Institute, Palo Alto (1990)

Pavement Crack Detection and Quantification Based on Scanning Grid and Projection Method

Zhaoyun Sun, Lili Pei[✉], Bo Yuan, Yaohui Du, Wei Li, and Yuxi Han

School of Information Engineering, Chang'an University, Xi'an, Shaanxi, China
peilili@chd.edu.cn

Abstract. Pavement cracks are difficult to monitor and quantify due to their complex texture and easy to be disturbed by noise and illumination. To solve this problem, a road crack monitoring and quantification method based on vehicle video is proposed. First, a method for extracting morphological features of dynamic road cracks is proposed. Combine automated vehicle-mounted equipment with GPS signals to obtain crack images with location information. Then, a calculation algorithm of crack parameters based on the combination of UK scanning grid and projection method is proposed, which uses the reverse engineering principle of perspective transformation to correct the image and divides the entire image into grid blocks. Finally, based on the analysis of different crack grades, the crack distress evaluation method is improved. The experimental results show that the proposed method has strong reliability and adaptability and achieves high-frequency and wide-range road detection.

Keywords: Pavement cracks · Video image detection · Feature extraction · Crack quantification

1 Introduction

Crack detection technology is becoming more and more mature, providing a lot of technical support for pavement maintenance. Scholars from all over the world have gradually developed a variety of pavement detection system based on intelligent detection technology [1], designed the double connectivity detection of pavement crack detection algorithm [2], image measurement method of asphalt pavement damage [3]. Asphalt pavement crack detection based on mathematical morphology [4] based on Prim pavement crack connection algorithm of minimum spanning tree [5]. The processing algorithm in HARRIS pavement damage detection system will be affected by environmental factors. Therefore, the elements of a sidewalk automatic measurement and 3D methods, based on 3D data of pavement crack detection algorithm [6, 7] is proposed.

The crack images are easily influenced by texture features and there's a lot of noise [8], so the next step is to filter the image, this method is often combined with neural network classifier, for distress image segmentation [9]. For early pavement management and repair, the circular Radon transform RGB camera image is used to propose a set of better road infrastructure evaluation indicators [10]. When using cameras and deep

G. Feng (Ed.): ICCE 2021, LNCE 213, pp. 273–281, 2022.
https://doi.org/10.1007/978-981-19-1260-3_24

learning networks for crack detection, there is a problem of insufficient filtering of the road background. In order to improve the detection efficiency as a whole, a lightweight Tiny-Darknet is combined with YOLOv3, and open VINO is used for model optimization and reasoning to accelerate the detection of road cracks. But this method slows down the number of frames detected on the video image [11].

Traditional detection methods are time-consuming, while pavement detection vehicles are expensive [12]. In this paper, the improved UK scanner method divides video frame image into grid block, which can not only calculate the number of cracks, but also calculate the critical rate of cracks, and increase the accuracy of crack detection and statistics.

2   Crack Image Acquisition and Feature Extraction

Perform image extraction and a series of image processing on the road video, and then correct the image and remove duplicate parts. The process of crack image acquisition and feature extraction is shown in the Fig. 1.

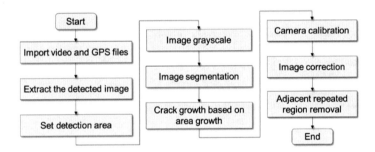

Fig. 1. The process of crack image acquisition and feature extraction.

2.1 Image Acquisition

A video data acquisition system is designed to collect images of pavement cracks, including a lightweight waterproof high-resolution GARMIN motion camera and a vehicle. Set a specific angle and height to the camera fixed on the vehicle to ensure complete collection of road images. Use GPS locator to record the longitude, latitude, altitude and time of the vehicle during the video capture process. The camera acquires 20 to 40 frames per second. According to the principle of dynamic photography, a four-point calibration method is used to calibrate the detection area of the road video frame image.

2.2 Feature Extraction

This paper uses the region growth algorithm to further process the image. First, enhancing non-scale filtering and adaptive threshold segmentation, the contour of the crack is obtained. However, this contour is not a complete crack but a part of the real crack.

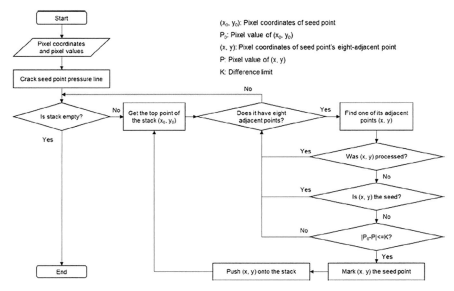

Fig. 2. Crack growth algorithm implementation flow based on region growth.

Second, calculate the complete crack size according to the extracting cracks image features. The crack growth algorithm based on regional growth is shown in Fig. 2.

The traditional region growth algorithm can only grow adjacent pixels. If a crack is not a seed point and is not adjacent to other crack seeds, these cracks cannot be detected. The proposed crack growth algorithm optimizes the traditional region growth algorithm, which the search scope is extended to the n*n region. The condition is that the absolute difference between pixel values is less than or equal to the threshold K. Experiments are carried out based on the optimization region of n*n region convolution. Let n be 3, 5, 7, 9, 11, 13, 15, 17, 19, 21, 23, 25, 27, 29, 31. The threshold K is set from 1 to 50. After comparative analysis, when n = 15, K = 15, the effect is best.

2.3   Frame Correction and the Overlapping Part Removal

When acquiring video frame image of road surface, the detection area in the image is irregular quadrilateral due to the difference in Angle of view and acquisition distance. It leads to the errors of different pixel points corresponding to the actual coordinate system values, which directly affects the calculation accuracy of geometric parameters of crack distress. Therefore, the image detection area shown in Fig. 3(a) needs to be corrected by reverse-engineering method of perspective transformation to restore to the real rectangular road surface image shown in Fig. 3(b). Then, by using the image crack feature extraction algorithm given in Sect. 2.2, the obtained binary image is shown in Fig. 3(c).

There is inevitable overlap between the two frames in the detection area. By analyzing the longitude, latitude, altitude and time information recorded every second, the method of removing overlap in the detection area is shown in Fig. 4.

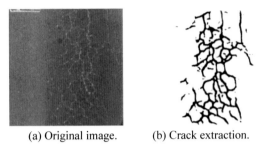

(a) Original image.   (b) Crack extraction.

Fig. 3. Crack feature extraction.

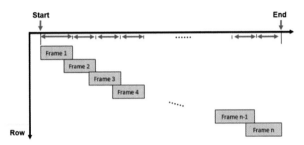

Fig. 4. Remove overlap schematic diagram.

Figure 4 shows only the first frame image can be fully used, and the actual length of pavement cracks in other frame images can be calculated according to Eq. (1). L_1 is the total length of road surface, l is the corresponding length of road surface in the first video frame image, and N is the number of frames.

$$len = \frac{L_1 - l}{N - 1} \tag{1}$$

3   Crack Image Distress Parameters Calculation and Assessment

The geometrical parameters of crack include crack area, crack length and crack width. As long as the grids are small enough, the grids can be approximately regarded as external rectangles of cracks. For each sub-block crack image, the image is scanned. If there is crack pixel, the sub-block is marked as a crack region sub-block. Otherwise, the sub-block will be not marked.

Set the initial pixel area of the crack $A_x = 0$ and scan each pixel in the sub-block from left to right from top to bottom. If the pixel is a crack, increase the pixel area, $A_x = A_x + 1$. If the pixel is the background, continue to scan the next pixel point until all the pixels in the sub-block are scanned, and the pixel area of the sub-block can be obtained. According to the actual length and width of each pixel, the actual area of each pixel can be calculated. The crack area in this sub-block A can be obtained by multiplying the number of pixels A_x.

Initially set the length and width of crack projection $T_c = 0, T_k = 0$. Scan each column pixel in the sub-block from left to right. If a column pixel contains crack pixels, the projection width will increase, that is $T_k = T_k + 1$, otherwise no operation will be carried out and continue to scan the next column pixel until all the columns are scanned, then the projection width of crack T_k can be obtained. Similarly, scan each row pixel in the sub-block from top to bottom to calculate the projection length T_c.

The camera correction information for each pixel is calculated based on the actual length and width. Projection length T_c and projection width T_k become an actual length L_c and an actual width L_k. Since the sub-blocks of the whole pavement are small, the crack length can be approximated according to the Pythagorean theorem, as shown in Eq. (2).

$$L = \sqrt{L_c^2 + L_k^2} \tag{2}$$

The crack width can be approximately divided by the actual crack area by the actual length.

The UK Scanner pavement crack damage assessment method is shown in Fig. 5. The left picture is the original pavement image, and the right picture is the crack map measured after grid division.

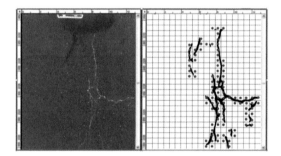

Fig. 5. UK Scanner pavement crack damage assessment schematic.

The UK scanner method divides the pavement crack grade into three categories: light, medium and heavy. Different grades of cracks are different in the evaluation of pavement condition. The measurement method of DMI (distress measurement index) for relevant research and evaluation is shown in Eq. (3).

$$DMI = \frac{1 * N_1 + 2 * N_2 + 3 * N_3}{(1 + 2 + 3) * N} \tag{3}$$

In Eq. (3), N_1 is the number of light crack distresses, N_2 is the number of medium crack distresses, N_3 is the number of heavy crack distresses, N is the total number of cracks, and 1, 2, 3 are the specified coefficients of cracks of different grades.

The crack width standard is k_1 mm, k_2 mm, where $k_1 \leq k_2$, the crack width $D \leq k_1$ is the light crack, the crack width $k_1 < D \leq k_2$ is the medium crack, the crack width $D > k_2$ is heavy cracks. Calculation of the pavement condition index (DMI): The different grades of cracks have different degrees of distress to the pavement. Set the light crack coefficient as, the medium crack coefficient as k_M, and the heavy crack coefficient as k_L. Calculate the test pavement index DMI as shown in Eq. (4), and calculate the pavement distress index. The crack data is processed, classified, calculated and stored in the database.

$$DMI = \frac{k_S * N_S + k_M * N_M + k_L * N_L}{N_S + N_M + N_L} \tag{4}$$

4 Test Results Analysis

In order to test the accuracy of crack feature extraction algorithm, 20 image samples were extracted from different video samples collected, and two methods were used for processing. 1) The grid was divided with 15*6 (pixel) as the side length, and introduce 'N$_1$' as the pseudo-ground truth. 2) The image is detected by crack feature extraction algorithm. The grid is divided into 15*6 (pixel) edges, and the number N_2 of the grid containing cracks is counted in the background of the system. The relative error calculation formula obtained by comparing the detection results of the two methods is shown in Eq. (5). The statistical results are shown in Table 1.

$$\delta = \frac{|N_1 - N_2|}{N_1} \times 100\% \tag{5}$$

Table 1. Statistical table of cracks relative errors.

Sample number	N_1	N_2	$\delta(\%)$
1	1344	1032	23.21
2	2683	1879	29.97
3	461	613	32.97
...	...	...	...
19	639	872	36.46
20	1893	1259	33.49
average	/	/	31.57

The average relative error of the 20 samples is 31.57%, and the error between the manual detection value and the system extraction value is analyzed as follows: $N_1 > N_2$ means that the grid containing cracks has not been detected, and $N_1 < N_2$ means that many interferences have been mistakenly identified as cracks, such as water and shadows. As the upper half of the perspective image has less image information than the lower half, the incomplete and missing part of the corrected image leads to blurring, which affects the detection results.

The length of the segment taken from the collection video is 4′25″52 s, and the detection distance is 6213 m. All cracks were divided into three degrees according to the principle that the crack width less than 3 mm was considered as mild crack, the crack width between 3 and 6 mm was considered as moderate crack, and the crack width greater than 6 mm was considered as severe crack.

The number of crack grids, the total number of crack grids, the total number of grids and the percentage of crack grids in each interval are respectively counted. The test data are counted every 1000 m, and the number of crack grids with different damage degrees in each interval is analyzed and compared, as shown in Fig. 6. The blue part is the number of mild crack grids, the orange part is the number of moderate crack grids, and the gray part is the number of severe crack grids.

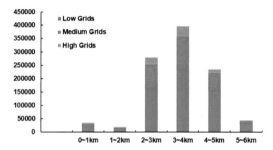

Fig. 6.   Crack level grid number statistics.

The actual damage degree of crack distress in test sections is shown in Fig. 7. It can be seen that the distress degree of the measured section is the most serious in the range of 3000–4000 m.

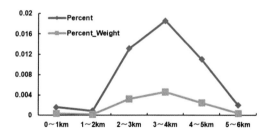

Fig. 7.   Pavement distress assessment distribution.

5   Conclusions

This paper proposes methods to solve the problem of crack feature extraction and parameter quantification in video. The main conclusions can be summarized in the following two aspects:

1)   The image is segmented using automatic thresholds, and non-crack interferences are filtered according to the area and crack morphological characteristics. The traditional

area growth algorithm is improved to solve the problem of the algorithm's inability to grow across areas and identify the characteristics of cracks.

2) A crack quantification method based on the combination of UK Scanner grid method and projection method is proposed. Count the crack information of all grid sub-blocks and use the improved UK Scanner pavement crack damage evaluation method to evaluate the detected road.

Acknowledgments. This work was supported by the National Natural Science Foundation of China (51978071); Project the Chang'an University Ph.D. Candidates' Innovative Capacity Development Grant Program (Grant No. 300203211241).

References

1. Sun, Z.Y., et al.: Research on pavement potting crack detection method based on improved faster R-CNN. J. South China Univ. Technol. (Nat. Sci. Edition) **48**(2), 84–93 (2020)
2. Peng, B., et al.: A denoising algorithm for pavement cracking images based on bi-layer connectivity checking. J. Highw. Transp. Res. Dev. **10**(3), 18–25 (2016)
3. Zhou, X.L., et al.: Measurement method for mean texture depth of asphalt pavement based on laser vision. China J. Highw. Transp. **27**(3), 11–16 (2014)
4. Wei, L.I.: Image detection algorithm research for asphalt pavement crack. Comput. Eng. Appl. **48**(19), 162–163 (2012)
5. Liang, R., et al.: Pavement crack connection algorithm based on prim minimum spanning tree. Comput. Eng. **41**(1), 31–36 (2015)
6. Li, W., et al.: An innovative primary surface profile-based three-dimensional pavement distress data filtering approach for optical instruments and tilted pavement model-related noise reduction. Road Mater. Pavement Des. **2017**(11), 1–19 (2017)
7. Li, W., et al.: Pavement cracking detection based on three-dimensional data using improved active counter model. J. Transp. Eng. **144**(2), 1–18 (2018)
8. Frighetto-Pereira, L., et al.: Shape, texture, and statistical features for classification of benign and malignant vertebral compression cracks in magnetic resonance images. Comput. Biol. Med. **73**(C), 147–156 (2016)
9. Viswanath, K., Gunasundari, R.: Analysis and implementation of kidney stone detection by reaction diffusion level set segmentation using xilinx system generator on FPGA. VLSI Design **2015**, 1–10 (2015)
10. Ouma, Y.O., Hahn, M.: Wavelet-morphology based detection of incipient linear cracks in asphalt pavements from RGB camera imagery and classification using circular Radon transform. Adv. Eng. Inf **30**(3), 481–499 (2016)
11. Zhang, X., Chen, X.J., Liu, R.K., et al.: Intelligent road crack detection system based on OpenVINO model optimization. Inf. Technol. **44**(07), 62–68 (2020)
12. Liu, H., et al.: Camera calibration method exploiting reference images and roadway information for traffic applications. J. Highw. Transp. Res. Dev. **9**(4), 58–63 (2015)

Test and Evaluation for Performance of Composite Pavement Structure

Zhihao Yang[1], Linbing Wang[2(✉)], Dongwei Cao[3], Rongxu Li[3], and Hailu Yang[1]

[1] University of Science and Technology Beijing, National Center for Materials Service Safety, No. 30 Road Xueyuan, Haidian District, Beijing, China
[2] Joint USTB Virginia Tech Lab on Multifunctional Materials, Department Civil and Environmental Engineering, USTB, VA Tech, Blacksburg, Virginia 24061, USA
wangl@vt.edu
[3] China-Road Transportation Verification and Inspection Hi-Tech Co. Ltd., No. 8 Road Xitucheng, Haidian District, Beijing, China

Abstract. Perpetual pavement has become an important research field of highway development in China. Reasonable selection of pavement structure and ensuring the durability of the structure are one of the necessary measures to build perpetual pavements. The inverted asphalt pavement structure can not only provide high strength and good bearing capacity of semi-rigid base, but also make use of the graded crushed stones for restraining the reflection cracks of semi-rigid base. This paper presented a study on three pavement structures are, namely, a semi-rigid asphalt pavement and two inverted asphalt pavements. The performances of the three pavement structures after one million loading repetition are obtained. Taking rutting depth, deflection and dynamic response as evaluation indexes, the feasibility of inverted asphalt pavement structure as perpetual pavement structure is evaluated. It is found that the composite asphalt pavement structure with permeable asphalt mixture of large particle size as base and cement stabilized macadam as subbase has the best performance as perpetual pavement.

Keywords: Composite asphalt pavement structure · Accelerated pavement test · Rut · Subgrade settlement · Dynamic response

1 Introduction

Perpetual asphalt pavement is one of the hotspots in the field of highway development in China. Semi-rigid base or flexible base structure is usually used in perpetual pavement in China. The pavement structure with flexible material between asphalt layer and semi-rigid subbase is called inverted asphalt pavement structure [1]. Semi-rigid base has high stiffness and stress diffusion capacity, but its bottom is prone to produce temperature shrinkage cracks and dry shrinkage cracks when the temperature and humidity of the surrounding environment changes. Such cracks may gradually develop into bottom-up reflection cracks on the surface, which leads to the cracking of the pavement [2–4]. The inverted structure not only has the advantages of high strength and good bearing

© The Author(s) 2022
G. Feng (Ed.): ICCE 2021, LNCE 213, pp. 282–297, 2022.
https://doi.org/10.1007/978-981-19-1260-3_25

capacity of semi-rigid base, but also has the advantages that the flexible base can reduce the reflective crack [5, 6]. The lower layer of graded broken stone can bear the vertical load from the top to the bottom of the road surface and dissipate the strain energy from the bottom to the upper cement stabilized macadam base, and it can also make full use of the characteristics of the high compressive strength and low tensile strength of graded macadam base to reduce the damage of pavement structure caused by water erosion and overloading of semi-rigid materials [7].

Researchers in the United States, Australia, South Africa and other countries found that flexible interlayer can significantly reduce the asphalt pavement structure in the use of structural reflection cracks [8]. However, the mix of composite structures varies from country to country. The flexible interlayer of composite pavement structures in South Africa is usually 15 cm of graded or asphalt stabilized macadam, while graded macadam is commonly used in Australia and France [9]. The composite pavement structure consisting of asphalt stabilized macadam base course and granular subbase or cement stabilized subbase course is usually adopted in Japanese highways [10]. Since the modulus of the flexible interlayer of graded macadam is smaller than that of water-stabilized macadam, the stress transmitted by the base of water-stabilized macadam can be reduced, the inverted base asphalt pavement structure can give full play to the graded macadam base of the mechanical characteristics and drainage characteristics, so that the pavement structure has good durability [11].

According to the summary, the inverted structure with graded crushed stone flexible interlayer is rarely used at home and abroad. Moreover, most foreign studies are aimed at the composite pavement formed by paving asphalt mixture layer on cement concrete pavement, and there is a lack of research on cement stabilized macadam subbase. Through material design and accelerated loading test, this paper systematically compares and studies the performance differences between two kinds of Inverted Asphalt Pavement and semi-rigid asphalt pavement. Lay the foundation for the inverted asphalt pavement structure to be used as a perpetual pavement.

2 Pavement Structure and Material Design

2.1 Pavement Structure

According to the perpetual asphalt pavement structure in Hebei province and the perpetual pavement structure in Shandong Binda Expressway, three pavement structures are selected. In order to study the durability of composite base course, the control variable method was used. That is, the surface layer and the roadbed structure are consistent, but the roadbed is different. The base of Structure A is Large Stone Porous asphalt Mixes (LSPM) and cement stabilized macadam. The base of structure B is graded crushed stone and cement stabilized crushed stone. The base of structure C is two layers of cement stabilized macadam. Structure A and structure B were the experimental group, and the structure C was the control group. The structure is shown in Table 1.

Table 1. The structural forms in the accelerated pavement test section.

Structural layer	A Structural form		B Structural form		C Structural form	
	Thickness	Materials	Thickness	Materials	Thickness	Materials
Surface	4 cm	SMA-13	4 cm	SMA-13	4 cm	SMA-13
Middle course	8 cm	AC-16C	8 cm	AC-16C	8 cm	AC-16C
Lower layer	10 cm	AC-20C	10 cm	AC-20C	10 cm	AC-20C
Base	18 cm	Unbound aggregate	18 cm	LSPM-25	18 cm	Cement stabilized macadam
Subbase	20 cm	Cement stabilized macadam	20 cm	Cement stabilized macadam	20 cm	Cement stabilized macadam
Subgrade	Cement stabilized soil					
	Situ subgrade					

2.2 Materials Design

2.2.1 Mixture

SBS modifier/crumb rubber composite modified asphalt (SRA) was used as binder for asphalt mixture. To ensure the rationality of gradation design, the maximum nominal size of gradation increases from the surface layer down. The grading is shown in Table 2. In addition, the performance of the four kinds of asphalt mixture was studied. As shown in Table 3. The process of splitting strength test are shown in Fig. 1.

Table 2. Mix proportion of asphalt mixtures.

Passing rate of sieve (%)		Size (mm)												
		31.5	26.5	19.0	16.0	13.2	9.5	4.75	2.36	1.18	0.60	0.30	0.15	0.075
Gradation	SMA-13	100	100	100	100	97.7	59.6	24.5	20.8	18.2	15.2	13.4	12.2	8.9
	AC-16C	100	100	100	95.8	88.1	76.8	55.6	31.1	22.8	15.1	11.1	8.7	7.1
	AC-20C	100	100	97.3	86.9	77.3	58.3	35.0	26.0	20.1	14.6	8.5	6.2	3.9
	LSPM-25	100	94.7	68.7	58.6	40.7	29.7	18.0	10.4	7.5	4.6	3.1	2.1	1.2

As can be seen from Table 3, SMA-13 has the best high temperature stability and low temperature stability among all asphalt mixtures. And its dynamic modulus is 9777 MPa, which shows that its ability of resisting traffic load is the strongest. This also meets the functional requirements of SMA-13 as an anti-wear layer. However, according to the

Fig. 1. The process of splitting strength test.

Table 3. The performance of asphalt mixture.

Test projects	Unit	SMA-13	AC-16C	AC-20C	LSPM-25
Asphalt saturation	%	73.2	72.4	67.9	70.7
Marshall stability	kN	12.02	14.39	13.37	13.54
Flow value	mm	2.4	3.0	3.6	3.3
Loss of kentucky flying test	%	4.4	3.1	3.4	4.78
Loss of immersion kentucky flying test	%	5.92	5.1	5.0	7.6
Tensile strength ratio	%	94.7	90.2	86.4	82.8
Marshall remnant stability	%	85.7	89.0	85.6	77.3
60 °C Rutting dynamic stability	Times/mm	8777	8521	8646	7893
−10 °C Bending strain	$\mu\varepsilon$	2318	2241	2316	1932
Fatigue life (10 °C, 10 Hz, 400 $\mu\varepsilon$)	Times	397571	504970	481560	253759

fatigue test data, the fatigue life of SMA-13 is lower than AC-16C and AC-20C, but higher than LSPM-25. This is due to the discontinuous gradation of SMA-13 and its larger void ratio will reduce the bond strength between coarse aggregates, which reduces its fatigue life.

In summary, through the experimental verification, we can conclude that the asphalt mixture used in the test road has good road performance.

2.2.2 Unbound Aggregate

Unbound aggregate is a mixture of aggregates of different sizes obtained by gradation design. Graded macadam obtains its load-bearing capacity through the force between its internal aggregates. Mixture gradation was shown in Table 4. Although the graded macadam base can prevent the reflection crack very well, but its bearing capacity is usually less than that of asphalt or cement stabilized base. In order to enhance the shear strength and rutting resistance of graded broken stone, 1‰ polypropylene fiber was added to the mixture. The properties of graded broken stone are shown in Table 5.

Table 4. Mix proportion of unbound aggregate.

Passing rate of sieve (%)	Size (mm)												
	31.5	26.5	19.0	16.0	13.2	9.5	4.75	2.36	1.18	0.6	0.3	0.15	0.075
Gradation	100.0	98.1	77.2	69.0	60.7	47.7	35.9	23.5	14.7	9.1	5.9	3.6	2.7

Table 5. The performance of aggregate.

Test projects	Unit	Results	
		No fiber	Adding fiber
Optimum moisture content	%	4.9	5.6
Maximum dry density	g/cm^3	2.327	2.333
Standard CBR test	%	280.6	330.4
Immersion CBR test	%	230.4	275.7
Unconfined compressive strength	MPa	1.7	2.1
Sheer strength	kPa	991	1354
Splitting strength	kPa	15.1	18.7

By comparing the performance of graded broken stone with and without fiber, we can see that the performance of graded broken stone with fiber is greatly improved. The standard CBR value is increased by 17.75%, and the flooding CBR value is increased by 19.66%. The most remarkable thing is that the shear strength and splitting strength of graded broken stone increased by 36.63% and 23.84% respectively after adding fiber. This shows that its ability to resist wheel load is greatly improved.

2.2.3 Cement Stabilized Macadam

In order to obtain the best bearing capacity of cement-stabilized macadam, the influence of different cement-based binders on the performance of cement-stabilized macadam was studied during the design of cement-stabilized macadam. Three binders, 5% Portland cement, 5% super sulphur cement and 2% steel slag powder + 3% Portland cement, were compared in this test. The gradation and properties of cement stabilized macadam are shown in Tables 6 and 7. The process of splitting strength test are shown in Fig. 2.

From Table 7, it can be seen that the unconfined compression strength of super-sulphur cement stabilized macadam is the highest at 7 days, reaching 6.11 MPa under the same curing condition. Its 14 days splitting strength is 1.21 MPa. The results show that it has reached the design requirements of the intensity of the base of the expressway and the

Table 6. Mix proportion of cement stabilized macadam.

Passing rate of sieve (%)	Size (mm)												
	31.5	26.5	19.0	16.0	13.2	9.5	4.75	2.36	1.18	0.6	0.3	0.15	0.075
Gradation	100.0	97.8	83.5	72.9	62.7	54.6	30.7	24.5	17.6	12.0	9.3	7.4	6.0

Fig. 2. The process of splitting strength test.

Table 7. The performance of cement stabilized macadam.

Test projects	Unit	Results		
		5%portland cement	5% super sulfated cement	2% steel slag powder + 3% portland cement
Optimum moisture content	%	5.7	5.2	6.1
7d unconfined compressive strength	MPa	3.73	6.11	1.98
14d splitting strength	MPa	0.45	1.25	0.27
28d plitting strength	MPa	0.49	1.33	0.31
14d flexural-tensile strength	MPa	0.73	1.97	0.41

first-class highway under the extremely heavy traffic condition. The compressive strength of steel slag powder + Portland cement stabilized macadam is the lowest in 7 days, which is only 1.98 MPa. Its 14 days cleavage strength is also the lowest, only 0.27 MPa. When the curing period reached 28 days, the unconfined compressive strength of three kinds of cement stabilized macadam increased obviously. Among them, the unconfined compressive strength of super-sulfur cement stabilized macadam increased most significantly, reaching 10.73 MPa. However, compared with the unconfined compressive strength, the indirect tensile strength of three kinds of cement stabilized macadam did not increase obviously.

Super sulfated cement was selected as binder for cement stabilized macadam.

3   Experimental Design and Data Collection Scheme

3.1   Experimental Design

In this research, based on the above research on the structure and materials, a test road of perpetual asphalt pavement was built with an effective loading length of 12 m. Three kinds of pavement structures were tested by linear accelerated loading test equipment, as shown in Fig. 2. The test parameters are: axle load 100 kN and tire pressure 0.7 MPa. In addition, a constant temperature control device is used to control the test temperature at 25 °C. At the same time, through the sensors embedded in the test road in advance, the road surface ruts, deflections, the settlement of the subgrade, and the longitudinal and transverse tensile strain at the bottom of surface course and base course were monitored and studied for different loading amounts. The test road and accelerated loading equipment as shown in Figs. 3 and 4 shows the embedding position and mark number of each sensor.

Fig. 3. The test road and accelerated loading equipment.

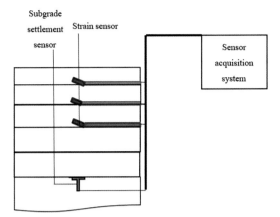

Fig. 4. Layout of sensor embedding position.

3.2 Data Collection Scheme

3.2.1 Rut Cross Section

The rutting was caused by the road surface deformation under the action of different amounts of axle loading. The durability of the different structures could be obtained by studying the change law of the rutting section for different load times amounts. In this experiment, a three-dimensional laser section instrument was used to scan the rutting sections for different amounts of axle loading. Each structure needed to collect five groups of sections, and after processing, the average value was taken as the rutting depth of each structure.

3.2.2 Deflection

In this experiment, a Falling Weight Deflectometer (FWD) was used to measure the deflection of the road surface under different amounts of axle loading. The deflection data of each structure were measured three points, and the deflection data of each measuring point were measured three times. The average value of three measured points was taken as the deflection data of the structure.

3.2.3 Strain

Tensile strain is an important index to control pavement cracking. In order to measure the tensile strain of the base and the surface layer, the transverse strain sensor, the longitudinal strain sensor and the vertical strain sensor are embedded. Firstly, the dynamic response of three kinds of asphalt pavement under different axle load is studied. The dynamic response of the road surface under the action of 80 kN, 100 kN and 120 kN is measured by the acceleration loading equipment. Secondly, the dynamic response of asphalt layer under different loading times is studied. It is an important data to reflect the internal fatigue failure of pavement structure and to judge the service condition of pavement structure under the action of axle load.

4   Results and Analysis

4.1   Rut Cross Section

Figure 5 presents the results of the rutting depth data of the three structures after different loading amounts.

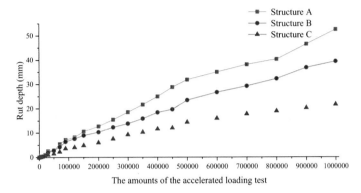

Fig. 5. Rut depth curve.

As can be seen from Fig. 5, the rutting depth curves of structure a and structure B are quite similar. Among them, the rutting depth of structure a increases evenly from 0 to 500,000 times and slows down from 500,000 to 800,000 times. But it began to grow rapidly after 800000 times. The rutting depth of C structure increases relatively slowly. The B structure is between the A and C structures. This is because the base of structure A and B structure is flexible, and the aggregate is further compressed and compacted by the wheel load at the early stage, which leads to the increase of rut depth. At this stage, the cement-stabilized macadam subbase is still in the state of cementation. This stage corresponds to the load number of 0 to 500000 times. At the end of the first stage, the cement stabilized macadam subbase with C structure will be broken gradually under the action of axle load. This is the transition phase, which corresponds to 500000 to 800000 times. After 800000 times, the cement stabilized macadam subbase of the C structure is treated as an equivalent granular state, leading to a further increase in rutting depth. However, the strength of B structure base is much higher than that of graded crushed stone. Therefore, the duration of the first phase is greater than that of the A structure. After one million times, the rutting depth of three pavement structures: structure A > structure B > structure C. In the composite structure, the rutting resistance of structure B is better than that of structure A.

4.2   Deflection

Figure 6 presents the results of the deflection test data after different loading amounts.

According to the curve chart of the deflection value change in Fig. 6, the deflection value of structure A was the largest, followed by the values of structure B, and the value of

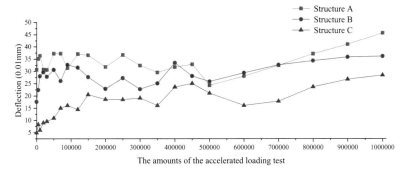

Fig. 6. Road surface deflection value change rule graph.

structure C was the smallest. In other words, the deflection value of the composite asphalt pavement with base layer of LSPM was generally lower than that of the composite asphalt pavement with base layer of unbound aggregate, but both were larger than the deflection value of the semi-rigid asphalt pavement. This showed that the bearing capacity of semi-rigid asphalt pavement is better than that of the composite asphalt pavement. In addition, in the composite asphalt pavement, the bearing capacity of LSPM base was better than that of the unbound aggregate base.

4.3   Dynamic Response of Pavement Structure Under Different Axle Load

In order to study the dynamic response of the three pavement structures under different axle loads, the axle loads of 80 kN, 100 kN and 120 kN are adopted. The applied loading speed is 22 km/h. The strain-time curves of the three structures under different axial loads are shown in Figs. 7, 8, 9, 10, 11, 12, 13, 14 and 15.

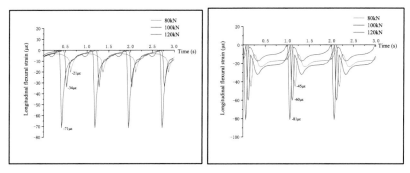

Fig. 7. Transverse (left) and longitudinal (right) flexural strains of surface layer of A structure at different loading times.

It can be seen that the strain in the pavement structure increases with the increase of axle load and there is obvious correlation between them. And the shape of the strain curve is related to the material properties of the structure and the depth of the monitoring point.

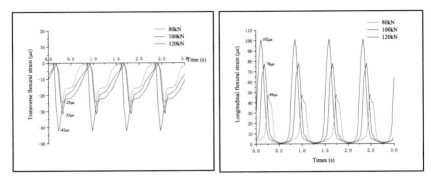

Fig. 8. Transverse (left) and longitudinal (right) flexural strains of middle course of A structure at different loading times.

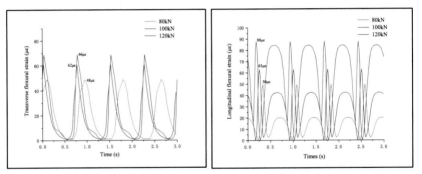

Fig. 9. Transverse (left) and longitudinal (right) flexural strains of lower layer of A structure at different loading times.

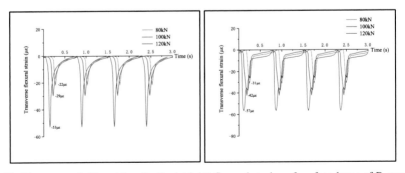

Fig. 10. Transverse (left) and longitudinal (right) flexural strains of surface layer of B structure at different loading times.

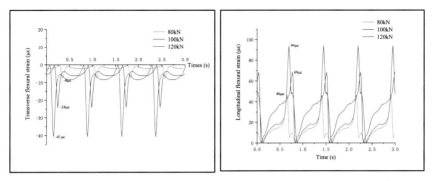

Fig. 11. Transverse (left) and longitudinal (right) flexural strains of middle course of B structure at different loading times.

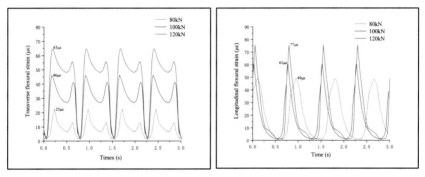

Fig. 12. Transverse (left) and longitudinal (right) flexural strains of lower layer of B structure at different loading times.

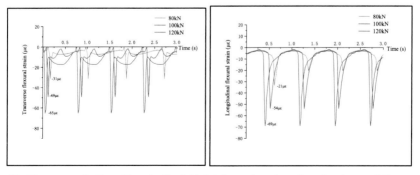

Fig. 13. Transverse (left) and longitudinal (right) flexural strains of surface layer of C structure at different loading times.

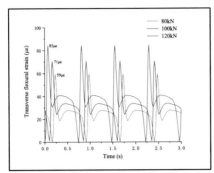
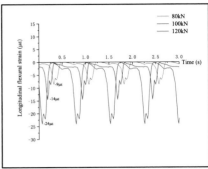

Fig. 14. Transverse (left) and longitudinal (right) flexural strains of middle course of C structure at different loading times.

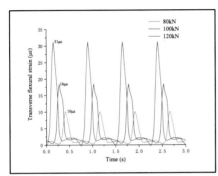
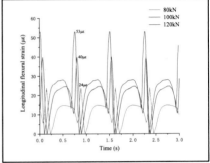

Fig. 15. Transverse (left) and longitudinal (right) flexural strains of lower layer of C structure at different loading times.

In the case of A structure, when the axial load is increased from 80 kN to 100 kN, the longitudinal and transverse tensile strains of the upper layer increase by 33.3% and 62% respectively. In addition, the transverse strain of the three structures is more sensitive to load magnitude than the longitudinal strain. This may imply that this asphalt pavement structure under heavy load may be sensitive to longitudinal cracks.

In structure C, the bearing capacity is strong because the base is made of semi-rigid materials. It is shown in the strain diagram that the strain of the lower layer of structure C is the smallest, indicating that its bending deformation is the smallest. It is found that the transverse and longitudinal strains of structure B are smaller than those of structure A. This is because the base of structure B using LSPM, has a stronger load-bearing capacity than graded crushed stone materials. The maximum transverse tensile strain of structure A and structure B appears in the lower layer, and the maximum longitudinal tensile strain appears in the middle layer.

4.4   Dynamic Response of Pavement Structure Under Different Loading Times

Figures 16 and 17 show how the longitudinal and transverse flexural tensile strains in the three asphalt layers vary with loading times.

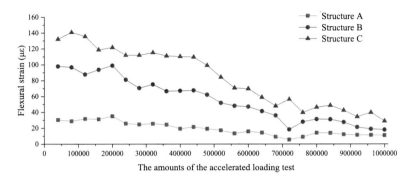

Fig. 16. Transverse bending strain at the bottom of asphalt layer at different loading times.

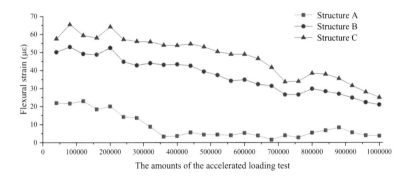

Fig. 17. Longitudinal bending strain at the bottom of asphalt layer at different loading times.

The longitudinal flexural-tensile strain of the three pavement structures is smaller than that of the transverse flexural-tensile strain at different loading times. In the course of the test, the strain value of the three pavement structures fluctuates up and down, which is mainly caused by the temperature change of the pavement layer during the loading. However, the bending and tensile strains of the three structures show a decreasing trend. Among them, the transverse and longitudinal flexural tensile strains appear at the bottom of the asphalt layer of structure C is the largest. This shows that the bottom of structure C asphalt layer is easy to incur Buttom-Up crack extending from bottom to top. In the inverted asphalt pavement structure, the bending strain of structure B is less than that of semi-rigid structure, but its changing trend is similar to that of structure C. However, the bending strain at the bottom of A structure asphalt does not increase during the loading. This indicates that the fatigue cracking resistance of structure A is the strongest. That is to say, the bending and tensile strain at the bottom of asphalt layer can be reduced obviously by setting graded broken stone base.

5 Results and Analysis

Through material design and accelerated loading test, this paper studies the performance of composite asphalt pavement structure and its feasibility as a perpetual pavement structure. The following conclusions are obtained:

(1) Through the study on the properties of graded broken stone, it is considered that the bearing capacity, wheel load resistance and rutting resistance of graded broken stone with fiber are improved. The rutting resistance increases the most. This material can greatly improve the performance and service life of composite asphalt pavement. In addition, the bearing capacity of cement stabilized macadam can be improved by using super sulfate cement.

(2) In three kinds of pavement structure, semi-rigid asphalt pavement has strong mechanical advantages. The inverted asphalt pavement with graded crushed stone base is the worst. The rutting resistance of semi-rigid asphalt pavement is better than that of composite asphalt pavement. In the composite structure, the rutting resistance of composite pavement with LSPM base is better than that with graded macadam base. Of the three pavement structures, structure A is the most prone to fatigue cracks. The pavement performance of structure C is still very good after loading one million times, and there is no structure fatigue phenomenon. This shows that improving the modulus of flexible base can improve the bearing capacity and rutting resistance of composite asphalt pavement.

(3) The transverse flexural-tensile strain of the three pavement structures is more sensitive to axle load than longitudinal flexural-tensile strain. And semi-rigid asphalt pavement is more sensitive to the load. The inverted asphalt pavement structure with graded macadam base is more adaptable to load, and its sensitivity to load is much less than semi-rigid asphalt pavement. This shows that graded macadam base can play the role of diffusion stress, can effectively extend the service life of pavement.

(4) By comparing the loading test results, it is concluded that the bearing capacity and rutting resistance of structure B are stronger than that of structure A. Its stress change is more similar to that of semi-rigid asphalt pavement. It can also restrain the reflective crack to a certain extent. Therefore, it is recommended to use LSPM base compound asphalt pavement structure for perpetual pavement.

Acknowledgements. The research is supported by grants from the National Key Research and Development Program of China (No. 2017YFF0205600, 2017YFF0205602) and the National Key R&D Program of China (2019YFE0117600). The sponsorship is gratefully acknowledged. The contents of this paper only reflect the views of the authors and do not reflect the official views or policies of the sponsors.

References

1. Liu, J.: Structural Behavior Analysis of Inverted Base Asphalt Pavement. School of Civil Engineering, Southwest Jiaotong University, Chengdu (2019)
2. Luan, L.: Study on fatigue crack propagation and life prediction of asphalt pavement with semi-rigid base. J. Civ. Eng. **50**(9), 118–128 (2017)
3. Zhou, X., Shi, J., Wang, X.: Experimental study on interlayer strain transfer of semi-rigid base asphalt pavement and analysis of influencing factors. Highway Traffic Technol. **34**(6), 1–6 (2017)
4. Liu, K., Xu, X., Zhang, L.: Study on dynamic response of semi-rigid asphalt pavement under dynamic load. For. Eng. **35**(2), 82–86 (2019)
5. Wang, X.: Study on Material and Structural Characteristics of Asphalt Pavement with Graded Crushed Stone Base. Chang'an University (2010)
6. Pandya, H., Weideli, T., Elshaer, M.: Performance Evaluation of Composite Pavements Using Long-Term Pavement Performance (LTPP) Database. Airfield and Highway Pavements (2019)
7. Hasan, I., Hussain, U.: Performance of cement treated base course in composite pavement. In: 1st Conference on Sustainability in Civil Engineering (2019)
8. Chen, C., Williams, R.C., Marasinghe, M.G.: Survival analysis for composite pavement performance in Iowa. In: Transportation Research Board Meeting (2014)
9. Liu, J.: Study on Structural Performance of Asphalt Pavement with Graded Crushed Stone Base for Expressway. School of Civil Engineering, Hebei University of Technology, Hebei (2014)
10. Ma, Z.: Three dimensional finite element analysis of inverted asphalt pavement structure under two-way load. Highway Transp. World Construct. Maint. Mach. **12**(6), 126–127 (2015)
11. Jiang, Z.: Study on mechanical properties of Inverted Asphalt Pavement Structure. School of Civil Engineering, Hunan University, Changsha (2012)

Parameter Sensitivity Analysis of Long Span PC Continuous Beam Bridge with Corrugated Steel Webs

Yu Jiang[✉] and Peiheng Long

Department of Civil and Traffic Engineering, Beijing University of Civil Engineering and Architecture, No. 15 Yongyuan Road, Daxing District, Beijing, China
320679216@qq.com

Abstract. In order to explore the regularity of alignment and stress variation of long-span corrugated steel web (CSW) continuous beam bridge during cantilever casting construction was taken as the engineering background. Based on numerical simulation, the sensitivity of parameters of alignment and stress of the main beam is carried out on cast-in-place section weight, modulus of elasticity and temperature gradient. The results show that the obvious influence of cast-in-situ section weight and temperature gradient on the alignment and stress is the key control parameters, while modulus of elasticity is the secondary control parameters. Therefore, it is necessary to monitor concrete dense and environmental temperature change in real time during construction, closure at a suitable temperature. Correct construction errors in time, ensure structural safety and smooth alignment.

Keywords: Corrugated steel web · Parameter sensitivity · Cantilever casting construction · Deflection · Stress

1 Introduction

With the continuous growth of China's economy, new materials and new technologies are also emerging. CSW replace concrete webs to form a new type of bridge structure, which has been widely used in China [1]. Since France built the world's first highway bridge with CSW-Cognac bridge, it have been deeply studied and developed because of their unique "accordion effect" [2], small axial stiffness, reduced constraints on the roof and floor, improved prestress efficiency, less cracks in the webs, great shear performance [3], good energy dissipation capacity and seismic performance [4]. However, there are many errors in the cantilever casting construction of long-span CSW bridge, and there are errors between the actual construction state and the design theoretical state. In order to reduce the influence of error, the parameter sensitivity of CSW has been studied. It is analyzed that the main beam quality, concrete elastic modulus, prestress loss, shrinkage and creep are linear sensitivity parameters [5]. It is recognized that concrete density, elastic modulus and prestress loss are the main parameters of linear control, and temporary load is the secondary parameter [6]. The least square method is used to identify the parameters, combined with the grey system theory to predict the construction deflection

© The Author(s) 2022
G. Feng (Ed.): ICCE 2021, LNCE 213, pp. 298–305, 2022.
https://doi.org/10.1007/978-981-19-1260-3_26

error, compare it with the measured value, and the feedback is timely corrected, so as to form an adaptive control [7]. The influence of temperature difference and temperature gradient on bridge deformation is studied by numerical simulation and actual measurement. The error between the measured value and the calculated value of the model is used to identify and correct the design parameters and guide the subsequent construction [8].

At present, the parameter sensitivity research of long-span CSW bridge mostly focuses on the main beam alignment at the completion stage, and there is a lack of sensitivity research on the construction process and main beam stress. This paper mainly analyzes the parameter sensitivity of the alignment and stress in the maximum cantilever stage and the completion stage, monitors the changes of stress and alignment, and corrects the deviation in time to ensure that the main beam reaches the ideal design state.

2   Project Overview

A bridge has a total length of 319.0 m, crosses the river valley, with a span of 83 m + 153 m + 83 m. It is a three span PC composite continuous box girder bridge with CSW. The design standard is a two-way four lane expressway, the design reference period is 100 years, Fig. 1 show general layout of bridge.

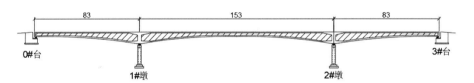

Fig. 1.   General layout of bridge (unit: m).

The upper structure of the bridge is single box single chamber variable section CSW box girder, the roof and floor are made of C50 concrete, the width of the roof is 13.1 m, the length of the cantilever on both sides is 3.3 m, the thickness of the roof is 0.3 m, the thickness of the cantilever plate end is 0.2 m, the width of the floor is 6.5 m, and the thickness of the floor is 0.3 m~1.1 m. The height of the fulcrum is 8.8 m, the height of the side span end and the middle beam are 3.5 m. The thickness of the box girder floor varies from the fulcrum to the middle of the span according to a parabola of 1.8 times, Fig. 2 show standard transverse section.

Figure 3 show dimension drawing of CSW. The type of CSWs used in the experiment is the BCSW-1600. The length of the flat subpanel and the horizontal projected length of the inclined are 430 mm and 370 mm, a height of 220 mm and the thickness is 10~22 mm.

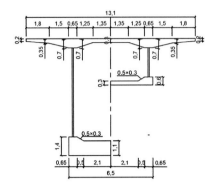

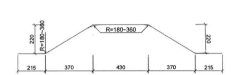

Fig. 2. Standard transverse section (unit: m).

Fig. 3. Dimension drawing of CSW (unit: mm).

3 Establishment of Finite Element Model

The bridge adopts Midas civil 2020 finite element software for numerical simulation calculation. Using the CSW section provided by the software, a total of 84 elements are established for the main beam. The single-side cantilever had 16 suspension casting sections with a length of 4.8 m, and the length of middle span and side span closure section was 3.2 m, adopting the method of side span first and then middle span closure, Fig. 4 show finite element model of the main bridge.

Fig. 4. Finite element model of the main bridge.

4 Parameter Sensitivity Analysis

In order to reasonably control the alignment and stress and guide the design and construction, parameter sensitivity is vital to the alignment and stress control of the completed bridge. Therefore, the ±10% variation of the reference value of each parameter is taken to analyze its influence on the alignment and stress of the main beam, Table 1 show variation range of each parameter.

4.1 Cast-In-Situ Section Weight

Due to the influence of concrete dense temperature variation, size error of formwork and other factors during cantilever pouring, the actual weight value of cast-in-place

Table 1. Reference value and variation interval of design variables.

Design parameters	Reference value	Variation range
Concrete dense (KN/m^3)	2600	[2340, 2860]
Modulus of elasticity (MPa)	34500	[31050, 37950]
Temperature gradient difference (°C)	8.5	[7.65, 9.35]

section always deviates from the theoretical value, resulting in the variation of section dead weight. Change the concrete dense to simulate the change of section weight, and monitor the change of main beam alignment and stress.

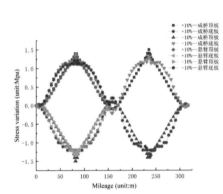

Fig. 5. Stress variation of main beam in bridge completion and cantilever stage under different concrete dense.

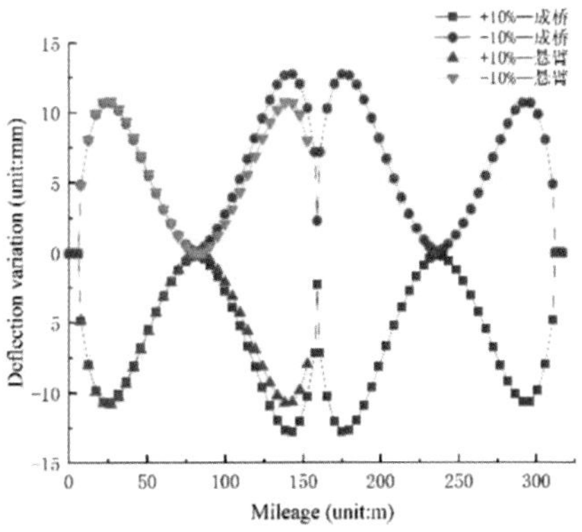

Fig. 6. Deflection variation of main beam in bridge completion and cantilever stage under different concrete dense.

It can be seen from Figs. 5 and 6 that the variation range of roof and floor stress is almost the same in the maximum cantilever stage and bridge completion, and the maximum variation of 0# block is 1.5 Mpa. With the increase of concrete dense the tensile stress of roof and compressive stress of floor increase. The maximum deflection variation is 12.5 mm at the quarter point of side span and middle span.

4.2 Modulus of Elasticity

The modulus of elasticity will increase with the pouring age, and the measured value of the modulus of elasticity is generally greater than the theoretical value after the completion of the bridge. it directly affects the stiffness of concrete, thus affecting the alignment and stress of the main beam, so the sensitivity analysis of the change of elastic modulus should be carried out.

It can be seen from Figs. 7 and 8 that the variation of modulus of elasticity in the maximum cantilever and bridge completion has no obvious effect on the stress, the maximum value is 0.33 Mpa, and the change of roof stress is greater than that of the floor. The maximum deflection in the maximum cantilever and bridge completion is only

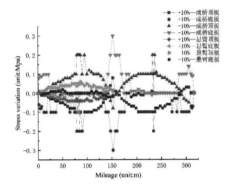
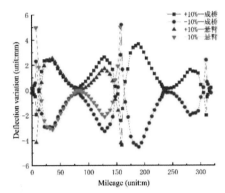

Fig. 7. Stress variation of main beam in bridge completion and cantilever stages under different modulus of elasticity.

Fig. 8. Deflection variation of main beam in bridge completion and cantilever stages under different modulus of elasticity.

5.0 mm, and the modulus of elasticity has little effect on the alignment and stress of the main beam.

4.3 Temperature Gradient

For the temperature load, seasonal temperature difference load and sunshine temperature difference load are mainly considered in the construction, but the seasonal temperature difference load has no obvious influence on the structural stress and deformation. This paper only studies the influence of temperature gradient change on the alignment and stress of main beam in the construction and bridge completion.

Table 2. Stress variation of roof and floor of 0# block under different temperature gradient in the maximum cantilever stage.

Temperature gradient variation	Roof stress variation (unit:Mpa)	Floor stress variation (unit:Mpa)
+10%	−0.248	−0.013
−10%	0.248	0.013

It can be seen from Table 2 that during the construction process, the cantilever structure is a static structure, the temperature secondary internal force does not change significantly, and only slight stress changes are generated in the cantilever 0# block [9].

It can be seen from Figs. 9 and 10 that the change of roof stress is greater than that of floor stress at bridge completion stage. The stress change of floor stress is within 1.2 Mpa, and the maximum change of roof stress can be up to 5.7 Mpa, the maximum deflection in the bridge completion stage is 9 mm. The change of temperature gradient has a significant impact on the stress and alignment of the main beam in the bridge completion stage.

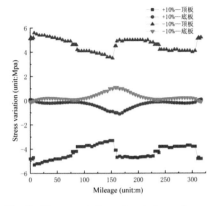

Fig. 9. Stress variation of main beam in bridge completion stage under different temperature gradient.

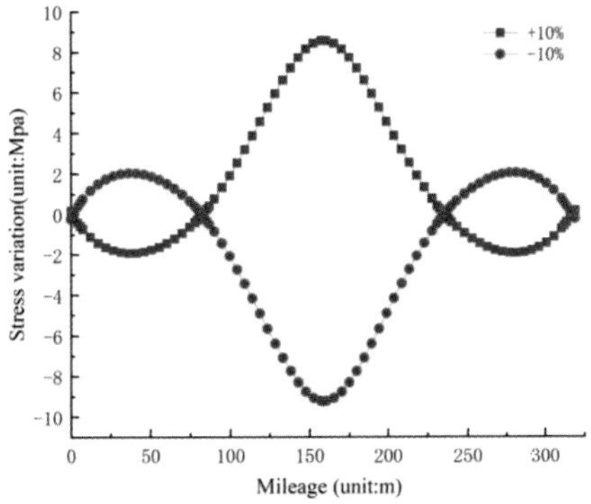

Fig. 10. Deflection variation of main beam in bridge completion under different modulus of elasticity.

5 Sensitivity Calculation

There is no clear functional relationship between design parameters and objective parameters, so it is difficult to obtain the analytical solution of objective function accurately. In order to facilitate parameter analysis and reasonably optimize the alignment and stress during construction. Parameter sensitivity is introduced to quantify the impact of each parameter on target parameters [10].

$$\eta = \lim_{\Delta x \to yields0} \frac{\Delta y/y}{\Delta x/x} \tag{1}$$

Where, y is the target control parameter, i.e. stress and deformation in cantilever state and bridge completion state. x is the design parameter.

According to Formula (1), take ±10% of the change of each parameter to calculate the sensitivity of it to alignment and stress. The calculation results are shown in the Table 3.

Table 3. Sensitivity of design parameters in bridge completion and maximum cantilever stages.

Design parameters	Bridge completion		Maximum cantilever	
	Stress sensitivity η_σ	Deflection sensitivity η_ω	Stress sensitivity η_σ	Deflection sensitivity η_ω
Concrete dense	0.895	0.918	1.150	1.092
Modulus of elasticity	0.226	0.527	0.026	0.331
Temperature gradient	1.624	1.479	0.017	0.429

Note: Higher sensitivity value means higher sensitivity of parameters

It can be seen from Table 3 that the sensitivity of concrete dense and temperature gradient is strong, and the sensitivity of modulus of elasticity is weak.

6 Conclusion

(1) The concrete dense and temperature gradient have great influence on the alignment and stress of the main beam, which is the key control parameter, and the elastic modulus has little influence, which is the secondary control parameter. The parameters of midspan closure section, midspan and side span quarter points and cantilever 0# blocks vary greatly, so the above positions should be monitored during construction to ensure that the stress and alignment of the bridge meet the design value.

(2) During construction, real-time monitoring of concrete dense changes, according to the specification of concrete vibration and formwork. Select the appropriate temperature to close the bridge, monitor the temperature field and temperature gradient effect, avoid the structure being affected by the temperature gradient, and ensure the smoothness of the bridge alignment.

(3) There are abrupt changes in stress and deflection corresponding to parameter changes at the closure of mid span and side span, which is related to the closure method, closure sequence, age and weight of closure section. There are abrupt changes in stress and deflection corresponding to parameter changes at the closure of mid span and side span, which is related to the closure method, closure sequence, age and weight of closure section.

Acknowledgments. I sincerely thank my tutor for my careful guidance and all the members of the research group for their opinions and findings.

References

1. He, J., et al.: The development of composite bridges with corrugated steel webs in China. J. Civ. Eng.: Bridge Eng. **174**, 28–44 (2021)
2. Shi, J., Liu, S.: Study on the improvement of prestressing efficiency by accordion effect of corrugated steel web box girder. J. Sci. Technol. Eng. **16**, 253–256 (2016)
3. Li, L., Hou, L., Sun, J.: Study on shear properties of corrugated steel webs. J. Hunan Univ. Nat. Sci. **42**, 56–63 (2015)
4. Wang, S., Liu, Y., Zhuang, W., Yao, H.: Experimental study on seismic behavior of composite beams with corrugated steel webs. J. Harbin Inst. Technol. **50**, 61–67 (2018)
5. Du, Y., Song, T., Wang, B.: Parameter sensitivity analysis of prestressed concrete continuous beam bridge with corrugated steel webs. J. Shandong Jiao Tong Univ. **29**, 58–63 (2021)
6. Jiang, K., Ding, Y., Liu, W., Wang, P., Zheng, F.: Sensitivity analysis for design parameters of cantilever casting pc box-girder with corrugated steel webs. J. Adv. Mater. Res. **430**, 1546–1550 (2012)
7. Liu, Z.: Research on construction monitoring technology of long-span corrugated steel web PC Continuous Box Girder Bridge Based on Grey System Theory. D. Southeast University. MS thesis (2015)

8. Zhao, X.: Research on key problems of construction control of PC composite box Girder bridge with corrugated Steel Webs. D. South China University of Technology. MS thesis (2020)
9. Liu, D., Qi, T.: Analysis of parameter sensitivity of construction control of multi span PC continuous beam bridge with corrugated steel webs. In: C. IOP Conference Series: Earth and Environmental Science, vol. 510, pp. 052094 (2020)
10. Kuang, Z., Li, X., Wu, X.: Analysis of linear control in construction of continuous girder bridge with corrugated steel Webs. J. Const. Technol. **45**, 250–255 (2016)

Preparation and Impact Resistance of Carbon Fiber Reinforced Metal Laminates Modified by Carbon Nanotubes

Zehui Jia[1], Lingwei Xu[2], Shuangkai Huang[1], Haoran Xu[1], Zhimo Zhang[1], and Xu Cui[1(✉)]

[1] College of Civil Aviation, Shenyang Aerospace University, Shenyang 110136, China
18842416952@163.com
[2] College of Aeroengine Academy, Shenyang Aerospace University, Shenyang 110136, China

Abstract. Fiber reinforced metal laminates (FMLs) are a kind of interlaminar hybrid composites made of metal sheets and fibers alternately stacked and cured at a certain pressure and temperature. In this paper, through the simulation of ABAQUS finite element software and recording the change of projectile velocity, the energy loss of projectile is calculated and the impact resistance is judged. Through the comparison of three groups of simulation experimental results, the energy absorbed by carbon fiber reinforced metal laminate is about 300 times that of aluminum alloy plate, which fully shows that carbon fiber reinforced metal composite has excellent impact resistance compared with aluminum alloy. After adding 1 wt% carbon nanotubes to carbon fiber reinforced metal laminates, the absorbed energy is about 10 times that of the original, which shows that carbon nanotubes improve the ultimate yield stress of resin and materials in epoxy resin and enhance the weakness that the composites are easy to delamination under impact load.

Keywords: FMLs · Carbon fiber · Finite element · Carbon nanotubes · Impact load

1 Introduction

Fiber reinforced metal laminates (FMLs) are formed by alternately stacking metal material layers and fiber reinforced composite layers, similar to sandwich structure. Each material layer is bonded by adhesive under certain temperature and vacuum conditions. It can show the excellent mechanical properties of fiber composites and metal materials, the high strength, high stiffness and fatigue resistance of carbon fiber reinforced composites, as well as the high toughness and excellent damage tolerance of metal materials [1–3]. As an excellent modifier and reinforcing material, carbon nanotubes have attracted extensive attention of scientists all over the world [4, 5]. After adding CNTs, the performance of FMLs is greatly improved, and the crack resistance, interlaminar fracture toughness and impact resistance of the matrix are improved [6].

© The Author(s) 2022
G. Feng (Ed.): ICCE 2021, LNCE 213, pp. 306–313, 2022.
https://doi.org/10.1007/978-981-19-1260-3_27

Avila et al. [7] studied the effect of adding nano clay to glass reinforced composites under low velocity impact. The results show that adding 5 wt% nano clay can increase the energy absorption by 48%. Khoramishad et al. [8] studied the effect of multi walled carbon nanotubes on the impact resistance of fiber metal laminates and found that adding 0.5 wt% CNT can improve the impact resistance of fiber metal laminates. Xian Xingjuan [9] studied the static and fatigue tensile failure characteristics of CFRP (epoxy resin matrix) composite laminates with 0° and 45° ply without notch and with straight edge notch through the analysis of a series of test results of 120 groups of specimens. Lin Xiaohong and others used the finite element analysis software ABAQUS/explicit for low-speed impact analysis of carbon fiber epoxy resin composites in the 0°/90° paving direction. The Johnson cook model parameters were used for aluminum alloy, and the Hashin damage criterion was used for fiber layer to establish the bonding layer element, analyze the loss of impact energy and observe and analyze the stress distribution diagram. Therefore, ABAQUS is used to simulate the results of FMLs under high-speed impact, and to simulate the impact resistance of FMLs with CNTs.

2   Preparation and Testing of Carbon Fiber Reinforced Metal Laminates Modified by Carbon Nanotubes

2.1   Preparation of Carbon Fiber Reinforced Metal Laminate

The carbon fiber reinforced metal laminate used in the experiment is composed of two metal plates and a layer of orthogonal carbon fiber woven cloth. The metal aluminum plates are aluminum alloy 2024-T3 aluminum commonly used in aircraft, with a thickness of about 0.3 mm, and the thickness of orthogonal carbon fiber woven step is about 0.25 mm. E51 epoxy resin is used between layers. CNTs were added to epoxy resin by ultrasonic method. Finally, the method of autoclave curing is used for molding (Table 1).

Table 1.   Shows the material parameters of 2024-T3 aluminum alloy.

Property	Value					
Elastic parameters	$E = 70$ GPa	$\mu = 0.3$				
Yieldsurface parameters	$A = 318$ MPa	$B = 673$ MPa	$C = 0.011$	$n = 0.4822$	$m = 0$	$\varepsilon 0 = 1$
Failure parameters	$d1 = 0.112$	$d2 = 0.123$	$d3 = 1.5$	$d4 = 0.007$	$d5 = 0$	
Fracture energy	$Gf = 8$ kJ/m^2					
Density	$\rho = 2700$ kg/m^3					

2.2 High Speed Impact Test

In order to explore the impact resistance of carbon fiber reinforced metal laminates modified by carbon nanotubes under high-speed conditions, this experiment mainly tested the energy absorption of aluminum plate, carbon fiber reinforced metal laminate and carbon fiber reinforced metal laminate modified by carbon nanotubes with the same thickness compared with that after complete breakdown of projectile, So as to judge the impact resistance. The experimental device is mainly composed of launcher, target box and high-speed photography system. First fix the test piece in the target box, and then spray the projectile with an air gun at high speed, and then puncture the test piece. Two high-speed cameras record the velocity of the projectile before and after passing through the test piece, as shown in Fig. 1.

The basic working principle of the experimental device is to use a high-speed camera to record the impact velocity VI of the projectile and the residual velocity VR after passing through the experimental plate. The evaluation criterion of impact resistance is the kinetic energy loss of projectile in the process of penetration Δ Ek:

$$\Delta E_k = \frac{1}{2}M_p\left(V\mathrm{i}^2 - V_r^2\right) \tag{1}$$

In this paper, three groups of test pieces were tested: pure aluminum plate, carbon fiber reinforced metal laminate without carbon nanotubes and carbon fiber reinforced metal laminate with 1 wt% carbon nanotubes.

Fig. 1. High speed impact test device.

3 Finite Element Simulation

3.1 Model Establishment and Grid Division

In the simulation, the plate size of the test piece is 60 mm * 60 mm, as shown in Fig. 2. Orthogonal carbon fiber interlayer bonding property parameters and carbon fiber aluminum alloy interlayer bonding property parameters are shown in Tables 2 and 3 respectively. The impact projectile is made of 45# steel, the density is 7.8 g/cm³, the elastic modulus is 210 gpa, the radius R is 5 mm and the mass is 4.16 g. Considering that the number of elements affects the computer computing speed, the approximate global size of each component is 1. In the load, the contact between the projectile and the target plate adopts general contact, the contact attribute is tangential contact, normal hard contact, and the friction coefficient is 0.2. The four sides of the carbon fiber reinforced composite are completely fixed, the degrees of freedom and rotation of the projectile in three directions are limited, and the initial velocity in the negative direction of the x-axis is defined as 200 m/s. The bonding layer is very thin, so it is defined by interlayer constraints. The modeling process of carbon fiber reinforced metal laminate and carbon nanotube fiber reinforced metal laminate is exactly the same. The difference is that the mechanical properties of the bonding layer of carbon fiber reinforced metal laminate with carbon nanotubes are about 15% higher than those without carbon nanotubes.

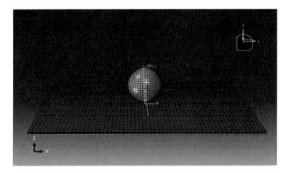

Fig. 2. High speed impact finite element model

Table 2. Interlaminar bonding property parameters of orthogonal carbon fiber plates

Parameter	Knn (GPa)	Kss (GPa)	Ktt (GPa)	t_n (MPa)	t_t (MPa)	t_s (MPa)
Parameter value	5.2	3.91	3.91	71	78	78

Table 3. Bonding property parameters between carbon fiber plate and aluminum alloy

Parameter	Knn (GPa)	Kss (GPa)	Ktt (GPa)	t_n (MPa)	t_t (MPa)	t_s (MPa)
Parameter value	3.5	3.5	35	35	39	39

3.2   Simulation Results and Analysis

Figure 3 shows the finite element simulation results of the three kinds of experimental parts after high-speed impact. It can be seen from the Fig. 3 that the aluminum plate has higher shape, smaller bullet holes and more regular edges after impact. Because of the high brittleness of carbon fiber plate, the damage of high speed impact is great, and the cracking phenomenon is great. Because of the dual properties of metal and composite material, the carbon fiber reinforced aluminum alloy composite has better impact resistance and smaller damage area after high speed impact.

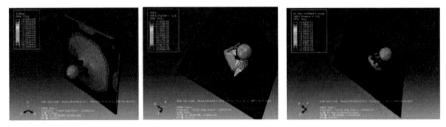

Fig. 3. Stress nephogram of breakdown of different materials under high speed impact (a) Aluminum plate (b) Carbon fiber composites (c) Carbon fiber reinforced aluminum alloy composites

Carbon fiber composite material stress and strain contours of each layer are shown in Fig. 4 below, the carbon fiber orientations can be found on the damage form of fiber reinforced metal laminates composite materials have great influence, therefore, in practice, according to the characteristics of different structure optimization design in the form of different bearing to meet different needs.

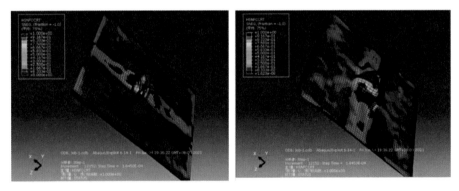

Fig. 4. Hashin Criteria for different Angle carbon fiber layer damage cloud (a) 0° (b) 90°

The output speed of the projectile in the simulation process is shown in Fig. 5 below, and the kinetic energy and energy loss of the three groups of projectile are calculated by this speed, and the results are shown in Table 4 below:

By comparing the results of three groups of simulation experiments, the energy absorption of carbon fiber reinforced metal laminates is about 300 times that of aluminum alloy plates, which fully demonstrates that the impact resistance of carbon fiber

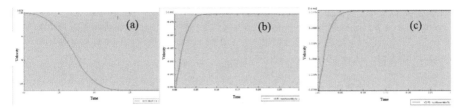

Fig. 5. The velocity curve of the projectile hitting different materials (a) Aluminum plate (b) Carbon fiber metal laminates (c) Carbon fiber metal laminates modified by carbon nanotubes

Table 4. Projectile velocity change and kinetic energy loss

Group	Al plate	CFMLs	CNTs-CFMLs
Initial velocity (m/s)	10	−200	−200
The residual speed (m/s)	9.83	−197.4	−172
The kinetic energy loss (J)	0.00701	2.14913	21.66528

reinforced metal composites has a great advantage compared with aluminum alloy. However, when 1 wt% carbon nanotubes were added to the carbon fiber reinforced metal laminates, the absorption energy was about 10 times of the original, indicating that carbon nanotubes increased the ultimate yield stress of the resin and the material in the epoxy resin, and alleviated the weakness that the composite was easy to delaminate under impact load.

4   Conclusion

The impact resistance of carbon fiber reinforced metal composites and mechanical properties of carbon nanotube modified carbon fiber epoxy resin matrix were studied, and the conclusions were as follows:

(1) Compared with traditional aluminum alloy materials, the impact resistance of carbon fiber reinforced metal composites has been greatly improved, but the delamination fracture of bond layer is still the weak part of carbon fiber reinforced metal composites.

(2) Adding a certain amount of nano-materials into epoxy resin matrix can improve the impact resistance of composite materials.

(3) The laying Angle of carbon fiber metal laminates has a great influence on the damage form of high-speed impact. In practical application, composite materials can be designed according to this characteristic to meet the actual demand.

Acknowledgements. This work was financially supported by the College Students' Innovative Entrepreneurial Training Plan Program (D202104022110228193), Aeronautical Science Foundation of China (2020Z055054002) and the Scientific Research Funds from Liaoning Education Department (JYT2020006 & JYT2020007).

References

1. Megahed, M., Abd El-baky, M.A., Alsaeedy, A.M., Alshorbagy, A.E.: An experimental investigation on the effect of incorporation of different nanofillers on the mechanical characterization of fiber metal laminate. Compos. B. Eng. **176**, 107277 (2019)
2. Yao, L., Sun, G., He, W., Meng, X., Xie, D.: Investigation on impact behavior of FMLs under multiple impacts with the same total energy: Experimental characterization and numerical simulation. Compos. Struct. **226**, 111218 (2019). https://doi.org/10.1016/j.compstruct.2019.111218
3. Bahari-Sambran, F., Eslami-Farsani, R., Arbab, C.S.: The flexural and impact behavior of the laminated aluminum-epoxy/basalt fibers composites containing nanoclay: an experimental investigation. J. Sandwich Struct. Mater. **22**, 1931–1951 (2018)
4. Zhang, X., et al.: Effect of multi-walled carbon nanotubes addition on the interfacial property of titanium-based fiber metal laminates. Polym. Compos. **39**, E1159–E1168 (2018)
5. Dhaliwal, G.S., Newaz, G.M.: Experimental and numerical investigation of flexural behavior of carbon fiber reinforced aluminum laminates. J. Reinf. Plast. Compos. **35**, 945–956 (2016)
6. Zhang, H., Gn, S.W., An, J., Xiang, Y., Yang, J.L.: Impact Behaviour of GLAREs with MWCNT Modified epoxy resins. Exp. Mech. **54**(1), 83–93 (2013). https://doi.org/10.1007/s11340-013-9724-7
7. Bahari-Sambran, F., Meuchelboeck, J., Kazemi-Khasragh, E., Eslami-Farsani, R., Arbab Chirani, S.: The effect of surface modified nanoclay on the interfacial and mechanical properties of basalt fiber metal laminates. Thin-Walled Struct. **144**, 106343 (2019)
8. Khoramishad, H., Alikhani, H., Dariushi, S.: An experimental study on the effect of adding multi-walled carbon nanotubes on high-velocity impact behavior of fiber metal laminates. Compos. Struct. **201**, 561–569 (2018)
9. Xingjuan, X.: Static and fatigue failure characteristics of carbon fiber reinforced epoxy composites with edge notch. Sci. China: Mathematics Physics Astronomy Tech. Sci. **02**, 183–192 (1984)

Comparative Analysis of the Displacement Dynamic Load Allowance and Bending Moment Dynamic Load Allowance of Highway Continuous Girder Bridge

Kaixiang Fan[✉]

School of Highway, Chang'an University, Xi'an 710064, Shaanxi, China
939224679@qq.com

Abstract. The dynamic load allowance (DLA) of the bridge structure is an important parameter in the bridge design. In order to study the variation law of displacement DLA and bending moment DLA of continuous girder bridge, taking 2×30 m continuous girder bridge and five-axis vehicles as the research object, the road roughness was simulated by the trigonometric series approach. With the help of ANSYS and APDL language, the influence of vehicle speeds, vehicle weights and road roughness on the displacement DLA and bending moment DLA are studied. The results show that the displacement DLA showed increasing trend with the increase of vehicle speed, and bending moment DLA showed increasing first and then decreasing; With the increase of body weight, the displacement DLA and bending moment DLA show a gradually increasing trend; Displacement DLA and bending moment DLA do have numerical differences. And the value of the displacement DLA is slightly larger than the value of the bending moment DLA. It is suggested that the displacement DLA and bending moment DLA should be distinguished in engineering design and dynamic load test. The research conclusion can provide reference for bridge structure engineering design and dynamic load test.

Keywords: Bridge engineering · Continuous girder bridge · Vehicle-bridge interaction · Dynamic load allowance · Time-history curve

1 Introduction

DLA is one of the important aspects of the application research of vehicle bridge coupling vibration analysis. There are many factors affecting the DLA of bridges, and the stress mechanism is very complex, so it has attracted more and more attention of researchers.

The DLA of the bridge structure was first a model test conducted by British and French engineers in 1844 to study the dynamic performance and bridge carrying capacity of the vehicle-bridge system. Many experts and scholars have done a lot of studies from many aspects: G G.Stokes [1], K.P.Chatterjee [2], V Kolousek [3] analyzed the coupling response of the bridge under the moving load, and derive the vehicle-bridge coupling

© The Author(s) 2022
G. Feng (Ed.): ICCE 2021, LNCE 213, pp. 314–320, 2022.
https://doi.org/10.1007/978-981-19-1260-3_28

vibration equation of the vehicle load when the vehicle load passes through the bridge structure; Song Yifan [4] found that the road roughness of bridge deck is the most significant factor affecting the dynamic action of vehicles based on analysis method of vehicle vibration response caused by pavement roughness; Shi Shangwei [5] analyzed the trend of difference and causes between the measured value and standard value of the DLA of the girder bridge. Jiang Peiwen [6] calculated the time history response of long-span continuous girder bridge under the action of multiple vehicles, and summarized the variation laws of displacement DLA and bending moment DLA based on the vehicle bridge coupling calculation method in ANSYS single environment; Zhou Yongjun [7] comprehensively considered the impact effect of vehicle load on the whole bridge, weighted the DLA at all peaks on the time history curve, and fully included the impact effect at multiple positions of the bridge structure; Denglu [8] found that the strain dynamic amplification factor is basically less than the deflection dynamic amplification factor by using the methods of theoretical derivation and numerical simulation.

In conclusion, many experts and scholars have carried out many researches on the DLA of bridges, and have made fruitful research results. But most of them are only for displacement DLA, and few researches on bending moment DLA are involved. In this paper, the coupling vibration model of the vehicle and bridge is established by ANSYS. The displacement time history curve and bending moment time curve of the bridge under different working conditions are analyzed. The change law of displacement DLA and bending moment DLA of continuous girder bridge under different vehicle speed, vehicle weight and road surface road roughness are studied.

2 Solution Method of Vehicle-Bridge Interaction

Through the coordination relationship of the action force and displacement at the contact point, the equation of vehicle-bridge interaction is established as follows [9]:

$$\begin{bmatrix} M_b & \\ & M_v \end{bmatrix} \begin{Bmatrix} \ddot{y}_b \\ \ddot{y}_v \end{Bmatrix} + \begin{bmatrix} C_b & C_{bv} \\ C_{vb} & C_v \end{bmatrix} \begin{Bmatrix} \dot{y}_b \\ \dot{y}_v \end{Bmatrix} + \begin{bmatrix} K_b & K_{bv} \\ K_{vb} & K_v \end{bmatrix} \begin{Bmatrix} y_b \\ y_v \end{Bmatrix} = \begin{Bmatrix} F_{br} + F_{vg} \\ F_{vr} \end{Bmatrix} \quad (1)$$

Where M, C, K are the quality, damping and stiffness matrix of the vehicle and bridge respectively; d is the vertical displacement, b and v are the bridge and vehicle respectively; F_{br} and F_{vr} are the interaction force of the vehicle-bridge system respectively, the subscript r represents the road roughness, and F_{vg} represents the vehicle gravity.

Based on the above principle, the time-variation equation is solved by the Newmark-β method. Using the transient analysis function of large-scale finite element program ANSYS, the vehicle bridge coupling vibration model is established by APDL language and displacement contact method, and the time-varying equation is solved by direct integration method.

3 Bridge, Vehicle and Road Roughness Model

3.1 Bridge Model

In order to study the displacement DLA and bending moment DLA of continuous girder bridge, 2×30 m continuous girder is selected the research object, the material is C50 concrete, and its section form is shown in Fig. 1.

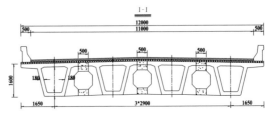

Fig. 1. Cross section of bridge (unit: cm)

3.2 Vehicle Model

1/2 vehicle model is selected. The vehicle layout diagram is shown in the Fig. 2, detailed parameters reference literature [10].

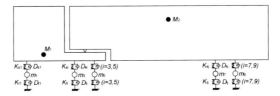

Fig. 2. 1/2 Vehicle model

3.3 Road Roughness Model

Road roughness is a very important excitation source of vehicle bridge coupling vibration. In practice, the statistical characteristics of road roughness are usually described by power spectrum, and the power spectral density of pavement roughness can be fitted as

$$G_x(n) = G_x(n_0)\left(\frac{n}{n_0}\right)^{-\omega} \tag{2}$$

Where n is the spatial frequency; n_0 is the reference spatial frequency; $G_x(n)$ is the displacement power spectral density value, $G_x(n_0)$ is the coefficient of road roughness; ω is the frequency index, $\omega = 2$.

4 The Method of DLA Calculation and the Working Condition Layout

4.1 Working Condition Layout

2×30 m continuous girder bridge is used for analysis. The effects of vehicle speed, vehicle weight and road roughness on the DLA of continuous girder bridge are considered. Among them, five speeds, five vehicle weights and three grades of road roughness are considered respectively. The specific working conditions are shown in the Table 1:

Table 1. Working conditions

Vehicle speed/$m \cdot s^{-1}$	Vehicle body weight/%	Road roughness
10,20,30,40,50	50,75,100,125,150	A, B, C

4.2 Calculation Method of the DLA

The DLA (μ) is defined under the current specification:

$$\mu = \frac{f_{d.\max}}{f_{s.\max}} - 1 \tag{3}$$

Where $f_{d.max}$ is the maximum dynamic response of the vehicle when passing the bridge; $f_{s.max}$ is the maximum static response corresponding to the bridge structure of the same vehicle. μ_d is the displacement DLA, and μ_M is the bending moment DLA.

5   Results Analysis

Based on the established calculation model and considering various influencing factors in the Table 1, the vehicle-bridge interaction calculation is carried out, and the effects of different grades of road roughness, vehicle speed and vehicle body weight on the displacement DLA and bending moment DLA in the mid-span of the first span of continuous girder bridge are analyzed. The relevant results are analyzed as follows.

5.1 Influence of Vehicle Speed on the Displacement DLA and Bending Moment DLA of the Continuous Girder Bridge

In order to study the influence of vehicle speed on the displacement DLA and bending moment, the road roughness is not considered in the analysis. The analysis results are shown in the following Figs. 3, 4 and 5:

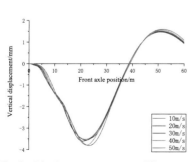

Fig. 3. Displacement curve at different speeds of smooth pavement

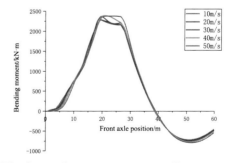

Fig. 4. Bending moment curve at different speeds of smooth pavement

The vehicle speed has a significant impact on the displacement DLA and the bending moment DLA of the continuous girder bridge. With the increase of the vehicle speed,

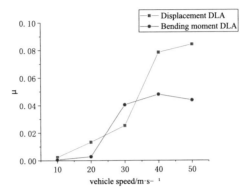

Fig. 5. Comparison of displacement DLA and bending moment DLA at different speeds

the change rules of the displacement DLA and the bending moment DLA are different. And the maximum value of the bending moment DLA is less than the maximum value of the displacement DLA. At the same time, the difference between the value of the displacement DLA and the value of the bending moment DLA changes with the increase of the vehicle speed. And the value of the former is about double that of the latter.

5.2 Influence of the Vehicle Body Weight on the Displacement DLA and Bending Moment DLA of the Continuous Girder Bridge

In order to study the influence of vehicle body weight on the displacement DLA and bending moment DLA of continuous girder bridge, the influence of road roughness is not considered in the analysis process, and the vehicle speed is set as $v = 20$ m/s. The analysis results are shown in the Fig. 6:

The weight of vehicle body has a significant impact on the displacement DLA and bending moment DLA of continuous girder bridge. With the increase of vehicle body weight, the variation law of displacement DLA and bending moment DLA is the same, showing a gradual increasing trend, and the displacement DLA is greater than the bending moment DLA. And the value of the former is about triple that of the latter.

5.3 Influence of Road Roughness on the Displacement DLA and Bending Moment DLA of the Continuous Girder Bridge

In order to study the impact of road roughness on displacement DLA and bending moment DLA of continuous girder bridge, the vehicle speed $v = 20$ m/s during the analysis. In order to avoid the influence of the randomness of the road roughness samples on the results, the DAL under 15 randomly generated road roughness samples is calculated for each working condition, and then the average value is calculated. The analysis results are shown in the Fig. 7.

The road roughness has a significant influence on the displacement DLA and bending moment DLA of the continuous girder bridge. Compared with grade A road roughness, the value of DLA under grade C road roughness is about 9 times that under grade A road

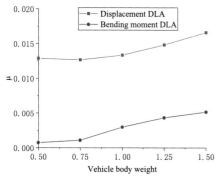

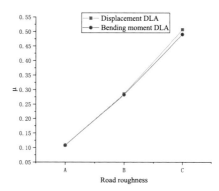

Fig. 6. Comparison between displacement DLA and bending moment DLA under different vehicle body weight

Fig. 7. Comparison between displacement DLA and bending moment DLA under different road roughness grades

roughness. With the increase of the grade of road roughness, the displacement DLA and bending moment DLA both increase gradually. At the same time, displacement DLA is slightly larger than bending moment DLA for the same grade of road roughness. And the difference between them increases with the increase of the grade of road roughness. The maximum difference between them is $\Delta = 0.016$.

6   Conclusion

Through the calculation and analysis of the finite element model, the influence of vehicle speed, vehicle body weight and road roughness on the displacement DLA and bending moment DLA of the continuous girder bridge are studied. The conclusion of comparative analysis as follows:

(1) The vehicle speed has a certain impact on the displacement DLA and bending moment DLA of the continuous girder bridge. With the increase of the vehicle speed, the displacement DLA gradually increases, and bending moment DLA first increases and then decreases.

(2) The vehicle body weight has a certain impact on the displacement DLA and bending moment DLA of the continuous girder bridge. With the increase of the weight of the vehicle body, the displacement DLA and bending moment DLA gradually increase.

(3) The road roughness has a significant impact on the displacement DLA and bending moment DLA of continuous girder bridge. With the increase of the grade of road roughness, the displacement DLA and bending moment DLA gradually increases.

(4) Compared with the displacement DLA and bending moment DLA, there is a numerical difference between them. And the value of the displacement DLA is slightly larger than the value of the bending moment DLA. It is suggested to distinguish the displacement DLA and bending moment DLA in engineering design and dynamic load test.

References

1. Stokes, G.G.: Discussions of a differential equation related to the breaking of railway bridges. Trans. Cambridge Phil. Soc. **8**(PartS), 12–16 (1896)
2. Chatterjee, P.K., et al.: Vibration of continues bridges under moving vehicles. J. Sound Vib. **169**(5), 619–632 (1994)
3. Kolousek, V., et al.: Civil Engineering Structures Subjected to Dynamic Load. SVTL Bratislava, pp. 54–58 (1967)
4. Song, Y.F., Chen, R.F.: Analysis method of vehicle vibration response caused by pavement roughness. J. Traffic Trans. Eng. **4**, 39–43 (2007)
5. Shi, S.W., Zhao, J., Shu, S.Y.: Analysis of the difference between measured dynamic load allowance and standard taking value of girder bridge. World Bridges **2**, 79–82 (2010)
6. Jiang, P.W., He, S.H., Song, Y.F., Wang, L.B., Zhou, Y.J.: Vibration response analysis of long span continuous girder bridge under multi-vehicle. J. Zhengzhou Univ. (Engineering Science) **32**(05), 91–95 (2011)
7. Zhou, Y.J., Cai, J.Z., Shi, X.W., Zhao, Y.: Computing method of bridge impact factor based on weighted method. J. Traffic Transp. Eng. **13**(4), 29–36 (2013)
8. Deng, L., Duan, L.L., Zou, Q.L.: Comparison of dynamic amplification factors calculated from bridge strain and deflection. J. Vib. Shock **5**(1), 126–135 (2018)
9. Song, Y.F.: Dynamic of Bridge Structures, pp. 65–83. China Communication Press, Beijing (2020)
10. Deng, L., et al.: Study on vehicle model for vehicle-bridge coupling vibration of highway bridges in china. Chin. J. Highw. Transp. **31**(7), 92–100 (2018)

Temperature Distribution and Transmission in Multilayer Porous Asphalt Courses

Hongqing Chen[1], Chunfa Jiang[1], Yunfei Su[1], and Xiaowei Wang[2(✉)]

[1] CCCC-SHEC Third Highway Engineering CO., LTD., Xi'an, China
[2] School of Civil Engineering, Xi'an University of Architecture and Technology, Xi'an, China
wswxw2011@163.com

Abstract. The objective of this paper was to investigate the temperature distribution and the law of temperature transmission in multilayer porous asphalt courses. Three types of multilayer porous asphalt courses are designed as the same thickness of actual pavements. In order to achieve this objective, experiments were conducted in an oven with different temperature, and samples were enwrapped with an efficient insulation material except the surface. The final temperature in samples is lower by 5~10 °C than ambient temperature and reduced gradually from top to bottom. The ability of heat transmission is related to the mixture's air void and ambient temperature. The temperature transmission rate in porous asphalt mixture is lower than traditional dense graded asphalt mixture. At last, temperature transmission formulas of porous asphalt mixture are given.

Keywords: Porous asphalt mixture · Multilayer porous asphalt courses · Temperature distribution · Temperature transmission · Temperature reduction

1 Introduction

Pavement distresses, such as rutting and moisture damage, were occurred in porous asphalt (PA) pavements during their service life because asphalt material is a typical temperature sensitive material. There is a closed relationship between pavement distresses and temperature [1–3]. When the solar radiation falls on the road surface, the surface start to warm up and heat begin to pass down. Gradually, the pavement structure was heated up to a certain temperature and forming a temperature gradient in pavement structure. The road materials and structure may appear damage in such environmental conditions for a long term. High temperature causes rutting damage, and great temperature gradient would cause cracking in pavement structures [4, 5]. Furthermore, frequent changes in temperature will exacerbate the possibility of such damage. In order to better prevent the emergence of these diseases, it is necessary to study the law of heat transfer in PA pavement.

　　PA pavement is a kind of open graded asphalt pavement which has a large air void (AV) content of 15%~25%, and skeleton-pore structure [6, 7]. PA pavement has the advantages of runoff mitigation, driving safety improvement, noise reduction, and et al. [8]. Asphalt types, aggregate type, porosity will affect heat-conducting property of road

© The Author(s) 2022
G. Feng (Ed.): ICCE 2021, LNCE 213, pp. 321–329, 2022.
https://doi.org/10.1007/978-981-19-1260-3_29

structure. Due to the high AV content, temperature in PA pavements is lower than traditional dense graded asphalt pavements, and PA pavements are used to mitigate the effects of urban heat island (UHI) [9]. In addition, the temperature has a significant effect on the bearing capacity and performance of PA pavement [10–12]. A variety of common damage is also directly or indirectly related to the distribution of road surface temperature. At present, many research concentrate on the temperature field under the comprehensive impact of solar radiation and temperature for dense graded asphalt pavements. In this paper, the temperature distribution and the law of temperature transmission in multilayer PA courses was investigated.

2   Materials and Experiment Design

2.1   Materials

PA mixture and dense graded asphalt mixtures (SUP) with different nominal maximum aggregate size (NMAS) was used in the paper, including PAC-13, PAC-16, PAC-25, SUP-20 and SUP-25. The details of these mixtures are illustrated in Table 1. Aggregates of basalt and limestone were used in this paper. Limestone was used in PAC-25, SUP-20 and SUP-25, and other mixtures used basalt. High viscosity binder (HVB) and SBS modified asphalt was used in the paper and the optimal asphalt contents is presented in Table 1. The technical properties of HVB and SBS modified asphalt are shown in Table 2.

Table 1. Aggregate gradations and mix design results.

Sieve size(mm)	Passing Percent (%)				
	PAC-13	PAC-16	PAC-25	SUP-20	SUP-25
31.5	100.0	100.0	100.0	100.0	100.0
26.5	100.0	100.0	98.1	100.0	97.0
19.0	100.0	100.0	70.5	97.6	85.5
16.0	100.0	91.6	52.3	84.1	80.0
13.2	95.3	83.4	48.1	71.4	68.9
9.5	61.6	56.8	29.1	64.6	53.1
4.75	23.3	20.1	17.7	44.9	37.4
2.36	16.3	12.3	11.9	28.7	24.0
1.18	13.2	10.0	9.1	17.8	16.8
0.60	10.7	8.1	7.0	11.3	12.5
0.30	8.8	7.1	5.7	7.8	8.7
0.15	7.4	6.0	4.5	6.7	7.2
0.075	5.4	4.7	3.3	4.6	5.6
Binder	HVB	HVB	HVB	SBS	SBS
Optimal binder content (%)	4.8	4.6	4.0	4.5	4.2

Table 2. Properties of HVB and SBS modified asphalt.

Properties		HVB	SBS
Penetration (25 °C, 100g, 5s) (0.1 mm)		47	56
Ductility (5 °C, 5 cm/min) (cm)		32	42
Softening point (Ring and ball method) (°C)		86	85
Viscosity of bitumen (60 °C) (Pa·s)		133105	21510
Solubility (Trichloroethylene) (%)		99.95	99.98
Elastic recovery (%)		96	88
RTFOT (163 °C, 5 h)	Mass loss (%)	0.14	0.08
	Penetration ratio (%)	83	78
	Ductility (5 °C, 5 cm/min) (cm)	21	28

2.2 Specimen Preparation

Specimen with a diameter of 150 mm was compacted by Superpave gyratory compactor (SGC) with the height control model to obtain the target air voids. As shown in Fig. 1, three types multilayer PA courses were designed. Multilayer PA courses consist of single-layer PA (type I), double layer PA courses (type II), and triple layer PA courses (type III). All of them consist of top layer, middle layer and bottom layer with a total thickness of 18 cm, and emulsified asphalt was used to glue each layer together. The thickness of top layer, middle layer, and bottom layer is 4 cm, 6 cm, and 8 cm, which is equal to the actual pavement.

In order to record the temperature inside the sample and study the law of heat transmission accurately, temperature sensor was put in the drill hole which is made in advance. The location is shown in Fig. 2.

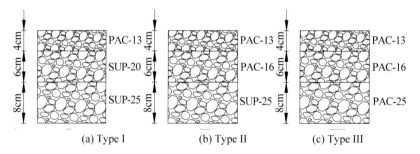

(a) Type I   (b) Type II   (c) Type III

Fig. 1. Types of samples.

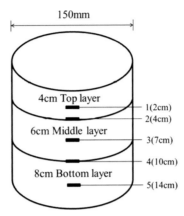

Fig. 2. The location of sensors.

2.3   Test Scheme

All the tests are performed in an oven, which can keep a constant temperature during the test. A temperature recorder is used to record the temperature through temperature sensor. In actual pavements, heat only transit from surface to bottom. In experiment, in order to simulate this situation, an efficient insulation material is applied to the sample except the top. Three temperatures of 40 °C, 50 °C and 60 °C are selected. When the oven reach the given temperature, put sample in and turn on the temperature recorder. The record interval of recorder is 10 min. The Termination condition is that the temperature change of 14 cm less than 1 °C in an hour.

3   Results and Discussion

3.1   Temperature at Depth of 2 cm

As we can see from the Fig. 3, firstly, the final temperature of 2 cm is lower about 5 °C than the ambient temperature no matter what temperature the sample I in. The explanation is that there will be heat losses in distribution of temperature because of the heat-transfer capability of mixture. Secondly, no matter what the temperature conditions, the early heating rate is very high and then become gentle slowly. But, the time of the first phases is different. The period in which sample warm up quickly is 150 min when the ambient temperature is 40 °C, while 50 °C is 200 min, 60 °C is 250 min. It means the ambient temperature which the mixture in is higher, the time which mixture will go through temperature changes is longer. Thirdly, the higher the temperature at which the sample is placed, the greater the rate of temperature rise. In that case, the ambient temperature of mixture is higher, it will undergo more sharp temperature changes.

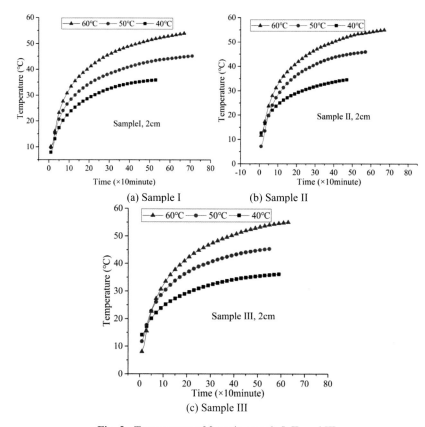

Fig. 3. Temperature of 2 cm in sample I, II, and III

3.2 Temperature at 10 cm

As shown in the Fig. 4, the following results can be achieved. When samples are placed in low temperature environment, the ability of temperature transmission of different aggregate gradations is equivalent. In the beginning, the PAC-16 mixture is better than SUP-20 mixture, gradually, the SUP-20 is better than PAC-16. When in medium or high temperature, the PAC-16 mixture shows a pretty excellent performance of temperature transmission, especially in medium temperature.

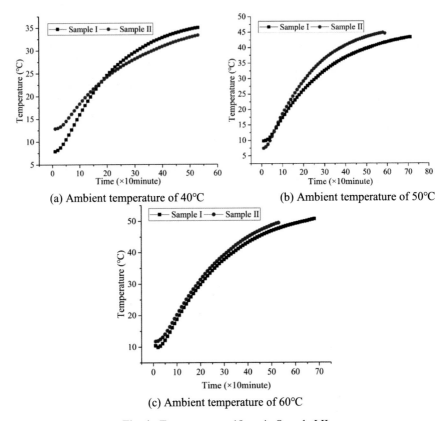

(a) Ambient temperature of 40°C   (b) Ambient temperature of 50°C

(c) Ambient temperature of 60°C

Fig. 4. Temperature at 10 cm in Sample I II

3.3 Temperature at 14 cm

The temperature distribution at 14 cm is shown in Fig. 5. Obviously, the temperature transmission ability of PAC-25 and SUP-25 is pretty equivalent. It has been concluded in the previous results. The temperature of bituminous concrete subsurface is a little low. In lower ambient temperature, the ability of SUP and PAC is pretty close.

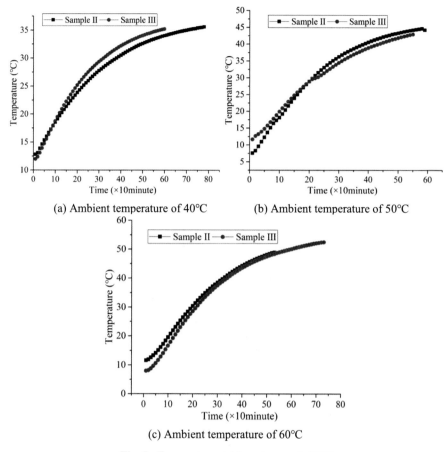

(a) Ambient temperature of 40°C   (b) Ambient temperature of 50°C

(c) Ambient temperature of 60°C

Fig. 5. Temperature at 14 cm in sample II III

3.4 The Temperature Transmission Law

In order to study the temperature transmission law, the following method is adopted. Firstly, the temperature of 2 cm in sample I, 10 cm in sample II and 14 cm in sample III is selected, and the corresponding ambient temperature is 40 °C 50 °C and 60 °C. Secondly, calculate the different value between ambient temperature and temperature of the above location and draw them in coordinate system. Then, connect them with a smooth curve. Lastly, fitting curve with cubic polynomials and obtain the temperature transmission formulas. Figure 6 (a) shows the temperature difference between 2 cm in sample I and ambient at the 40 °C and the fitting line of the temperature differences. Figure 6 (b) shows the temperature difference between 10 cm in sample II and 50 °C, and Fig. 6(c) shows the temperature difference between 14 cm in sample III and 60 °C.

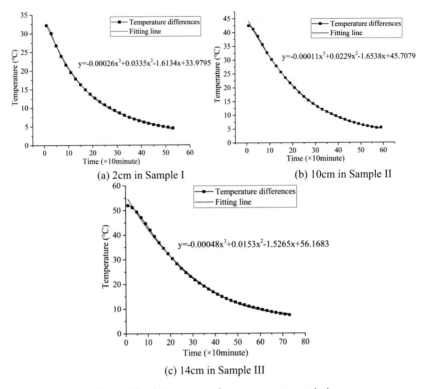

Fig. 6. The fitting curve of temperature transmission

All the formulas above is concluded from few limited date and there isn't every verification. At the same time, there are lots of effects which can influence the temperature transmission. As a result, the applicability and practicability of those formulas are limited. In this paper, they are adopted to reveal the sketchy low of temperature transmission, which can provide a reference for the further research.

4 Conclusions

The main conclusion of this paper can be summarized as followed:

(1) The higher ambient temperature the samples are placed in, the longer time samples experienced sharp temperature change. The final temperature in samples is lower 5~10 °C than ambient temperature and reduce gradually from top to bottom.
(2) The ability of temperature transmission is related to ambient temperature. Generally speaking, in low temperature, the ability of SUP mixture is closed to the PA mixture. But in medium and high temperature, the PA mixture possesses more outstanding performance than SUP.
(3) Through the analysis of the different value between ambient temperature and the temperature of particular location, temperature transmission formulas of PA mixture are given.

References

1. Ji X., Sun E., Zou H., Hou Y., Chen B.: Study on the multiscale adhesive properties between asphalt and aggregate. Const. Build. Mater. **249**, 118693 (2020)
2. Zhang Q., Zhang J., Li Z., Wen Z., Yang Y.: Fatigue-damage model of a pothole-repairing composite structure for asphalt pavement. J. Mater. Civ. Eng. **29**(11) (2017)
3. Wang, X., Gu, X., Hu, X., Zhang, Q., Dong, Q.: Three-stage evolution of air voids and deformation of porous-asphalt mixtures in high-temperature permanent deformation. J. Mater. Civ. Eng. **32**(9), 04020233 (2020)
4. Jiang, Y., Yuan, K., Deng, C., Tian, T.: Fatigue performance of cement-stabilized crushed gravel produced using vertical vibration compaction method. J. Mater. Civ. Eng. **32**(11), 04020318 (2020)
5. Ren, J.L., Xu, Y.S., Huang, J.D., Wang, Y., Jia, Z.R.: Gradation optimization and strength mechanism of aggregate structure considering macroscopic and mesoscopic aggregate mechanical behaviour in porous asphalt mixture. Constr. Build. Mater. **300**, 124262 (2021)
6. Wang, X., Ren, J., Hu, X., Gu, X., Li, N.: Determining optimum number of gyrations for porous asphalt mixtures using superpave gyratory compactor. KSCE J. Civ. Eng. **25**(6), 2010–2019 (2021). https://doi.org/10.1007/s12205-021-1005-x
7. Wang, X., Gu, X., Jiang, J., Deng, H.: Experimental analysis of skeleton strength of porous asphalt mixtures. Constr. Build. Mater. **171**, 13–21 (2018)
8. Wang, X.W., Ren, J.X., Gu, X.Y., Li, N., Tian, Z.Y., Chen, H.Q.: Investigation of the adhesive and cohesive properties of asphalt, mastic, and mortar in porous asphalt mixtures. Constr. Build. Mater. **276**, 122255 (2021)
9. Wang, X., Hu, X., Ji, X., Chen, B., Chen, H.: Development of water retentive and thermal resistant cement concrete and cooling effects evaluation. Materials **14**(20), 6141 (2021)
10. Wang, X., Gu, X., Dong, Q., Wu, J., Jiang, J.: Evaluation of permanent deformation of multilayer porous asphalt courses using an advanced multiply-repeated load test. Constr. Build. Mater. **160**, 19–29 (2018)
11. Ma X., Zhou P., Jiang J., Hu X.: High-temperature failure of porous asphalt mixture under wheel loading based on 2D air void structure analysis. Constr. Build. Mater. **252**, 119051 (2020)
12. Du, Y., Chen, J., Zheng, H., Liu, W.: A review on solutions for improving rutting resistance of asphalt pavement and test methods. Constr. Build. Mater. **168**, 893–905 (2018)

Research on Support Damage of Highway Bridge Based on Midas

Miaomiao Fang[1]($\boxtimes$), Yuqi Wang[2], Jiaxin Liu[1], and Fan Sun[1]

[1] Chang'an University, Xi'an, Shaanxi, China
1349633273@qq.com
[2] China Railway Shanghai Design Institute Group Co., Ltd., Shanghai, China

Abstract. During the operation of highway bridges, the bearing stiffness will decrease with the service life, and the mechanical properties will also change. In order to study the influence of stiffness damage on bearing. In this paper, a continuous beam bridge is selected for finite element model analysis, and the effects of stiffness damage on bearing force and bearing offset under the conditions of concrete shrinkage and creep and 30 °C temperature difference are comprehensively considered. The results show that the bearing stiffness damage has little influence on the vertical displacement, horizontal displacement and bearing capacity of the bearing, but has a great influence on the vertical compression deformation and durability of the bearing.

Keywords: Rubber bearing · Stiffness reduction · Mechanical performance · Support offset · Durability

1 Introduction

In recent decades, with the continuous development of highway bridges in China, the traffic mileage of expressway is also increasing. Plate rubber bearing is widely used because of its simple structure, simple processing, manufacturing and installation, simple maintenance and low cost. According to the survey, the terrain of Shanxi Province is complex and the temperature difference between day and night is large. Under the influence of coal transportation, the traffic volume increases suddenly. This makes the bridges in this area overloaded seriously, resulting in the stiffness problems of 95% of the bridge bearings, and the durability of the bearings will be affected.

Chen Yanjiang et al. [1] studied the influence of recycled rubber on the durability of plate rubber bearing from the perspective of material, and came to the conclusion that the amount of recycled rubber added into the branch seat rubber can be determined by detecting the mechanical index of compression shear modulus before and after bearing aging, so as to judge the early failure of bearing; Xu Yue et al. [2] studied the influence of plate rubber bearing stiffness on the stress and deformation of bridge structure through T-beam finite element analysis. It is concluded that the loss of bearing stiffness has little influence on internal force, but has great influence on structural deformation; Huang Yueping [3], studied the influence of uneven rubber layer thickness on the durability of

© The Author(s) 2022
G. Feng (Ed.): ICCE 2021, LNCE 213, pp. 330–337, 2022.
https://doi.org/10.1007/978-981-19-1260-3_30

plate rubber bearing, and concluded that uneven rubber layer thickness will lead to the improvement of its mechanical properties, and the processing and testing process should be strictly controlled during the production and inspection of rubber bearing.

The above studies are from the perspective of materials and bridge structure stress. However, in the actual bridge operation, under the influence of environmental factors such as load and climate, the bearing stiffness will be reduced, which will have a great impact on the bearing durability and bridge durability. This paper mainly studies the stiffness damage of bearing, taking a 2×40 m T-beam bridge as an example, the damage of the bearing is simulated, and then by changing the bearing stiffness, the stress, stress and offset are analyzed to obtain the impact of the reduction of bearing stiffness on the bearing.

2 Bearing Damage Simulation

There are two very important mechanical indexes of the bearing: compressive elastic modulus and shear elastic modulus, so the design of the bearing is inseparable from these two data. According to the above analysis, the damage of plate rubber bearing can be simplified as the change of shape coefficient and shear modulus of rubber bearing [4].

Support damage simulation can generally be carried out from the following two aspects: a. change the plane size of the support; b. Change the rubber shear modulus in the bearing. In the finite element software, the elastic connection is generally used to simulate the restraint effect of the bearing. Since the degrees of freedom in six directions SD_x, SD_y, SD_z, SR_x, SR_y, SR_z, in the elastic connection are determined according to the plane size and shear stiffness g of the rubber bearing, the damage degree of the bearing can be effectively simulated by adjusting the corresponding restraint stiffness [5–13].

The stiffness calculation formula of plate rubber bearing is:

Stiffness in x-axis direction of element local coordinate system: $SD_x = EA/L$;

Stiffness of element local coordinate system in Y, Z axis direction: $SD_y = SD_z = GA/L$; Rotational stiffness in x-axis direction of element local coordinate system: $SR_x = GI_p/L$; Rotational stiffness of element local coordinate system in Y-axis direction: $SR_y = EI_y/L$; Rotational stiffness of element local coordinate system in Y-axis direction: $SR_z = EI_z/L$.

3 Model Building

The supporting bridge is a river crossing bridge. The main beam adopts C50 concrete, and its superstructure is fabricated prestressed concrete T-beam. The bearing is teflon plate rubber bearing, its differential settlement is $\Delta = 5$ mm and the specification of the side fulcrum bearing is GJZF4400 $\times$ 400 $\times$ 85 mm, and the specification of the middle fulcrum support is GJZ F4550 $\times$ 550 $\times$ 130 mm. The bridge is located in the environment with large temperature difference between day and night. Rigid connection between main beam and pier; In order to comprehensively analyze the bearing stress, under the basic load combination, simulate the displacement and stress of the bearing under the combined overall temperature rise of 30 °C and overall temperature drop of 30 °C and concrete shrinkage and creep. The bridge model and model bearing reaction are shown in Figs. 1 and 2.

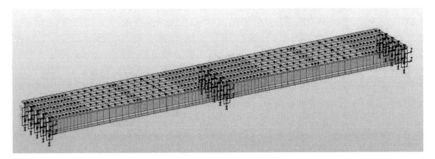

Fig. 1. Depends on the 2 × 40 m finite element model of the bridge

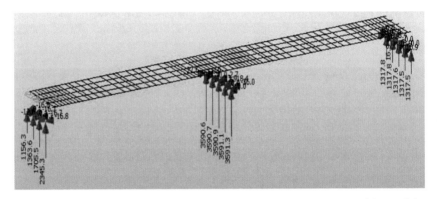

Fig. 2. Reaction value of rear support of bidirectional movable support of the model

4   Bearing Durability Analysis

4.1   Stress and Deformation Analysis

In order to better analyze the impact of stiffness reduction of two-way movable bearing, this paper takes longitudinal bearing, transverse bearing and two-way movable bearing for comparison, and takes a point for each type of bearing for analysis. In the environment with large temperature difference, the thermal expansion and cold contraction of concrete beam and the shrinkage and creep of concrete have a great impact on the bearing. Therefore, considering the shrinkage and creep of 10 years and the temperature difference of 30°, the stress and offset of the bearing are analyzed. The stress and offset of various bearings under different working conditions are shown in the Tables 1, 2and3 below.

It can be seen from the above table that the longitudinal displacement of the longitudinal movable bearing is large, and the temperature has a great impact on the longitudinal

Table 1. Forces and offsets of longitudinal movable supports under different working conditions

Working condition	DX/mm	DY/mm	DZ/mm	FX/kN	FY/kN	FZ/kN
Shrinkage and creep	−3.1769	0.0073	−1.2801	−18.63	−80.64	854.98
Rising 30 °C	−16.5867	−0.0073	−1.2809	6.09	80.64	855.49
Cooling 30 °C	13.4040	0.0073	−1.2793	−38.63	−80.64	854.48

Table 2. Forces and offsets of lateral movable supports under different working conditions

Working condition	DX/mm	DY/mm	DZ/mm	FX/kN	FY/kN	FZ/kN
Shrinkage and Creep	0.0491	−0.0276	−2.5840	−491.10	0.06	2038.26
Rising 30 °C	0.0491	−1.6395	−2.5827	−491.12	3.07	2037.22
Cooling 30 °C	0.0491	1.6164	−2.5853	−491.08	−3.02	2039.29

Table 3. Forces and offsets of bidirectional movable supports under different working conditions

Working condition	DX/mm	DY/mm	DZ/mm	FX/kN	FY/kN	FZ/kN
Shrinkage and creep	−3.1474	−0.0379	−1.2065	−18.63	0.06	800.66
Rise 30 °C	−16.5572	−1.6496	−1.2075	6.04	2.47	801.27
Cooling 30 °C	13.4038	1.6385	−1.2055	−38.63	−2.45	800.04

stress and displacement of the bearing. When the temperature rises by 30 °C, the longitudinal displacement of the bearing changes by 13.4 mm; When the temperature drops by 30 °C, the longitudinal displacement of the support changes by 16.58 mm.

The vertical displacement of the lateral movable bearing is large, but the temperature has a great influence on the lateral displacement. When the temperature rises to 30 °C, the lateral displacement of the bearing changes by 1.61 mm; When the temperature drops by 30 °C, the lateral displacement of the support changes by 1.64 mm.

The support reaction force of two-way movable bearing is small, the vertical displacement is small, and the longitudinal and transverse displacement is large. Temperature change has great influence on longitudinal and transverse displacement. The variation amplitude of longitudinal displacement is similar to that of longitudinal movable support, and the variation amplitude of transverse displacement is similar to that of transverse movable support.

According to the specification, for the support GJZF4400 × 400 × 85 mm, the limit value of two-way movable bearing along the bridge direction is 90 mm, the limit value of transverse bridge direction is 40 mm, the limit value of one-way movable bearing along the bridge direction is 90 mm, and the limit value of transverse bridge direction is 3 mm; For the support GJZF4550 × 550 × 130 mm, the limit value of unidirectional movable bearing along the bridge is 130 mm, and the limit value of transverse bridge is 3 mm. According to the results of the above analysis, the maximum longitudinal displacement of the support is 18% of the longitudinal limit and the maximum transverse displacement of the support is 55% of the transverse limit. Although the displacement meets the requirements of the specification, the lateral offset of the support is obviously larger than the longitudinal offset.

According to the above analysis results, the bearing reaction value is less than the maximum bearing capacity of the bearing, and there is nearly 30% surplus, which meets the requirements of the specification. However, with the passage of time, the bearing disease will reduce the rubber stiffness of the bearing, and the surplus in the design may be offset in the operation process, making the bearing reaction greater than the design value, resulting in the failure of the bearing.

4.2 Bearing Stiffness Damage Analysis

The above analysis of the force and deflection of the support under three different conditions shows that the force of the support has a certain margin in the design. However, during the operation of the bridge, the stiffness of the support will continue to weaken, causing the surplus to be gradually offset, which will have a great impact on the durability of the support. Next, for the support of the bridge, under three different conditions, the force and displacement of the support are analyzed under the conditions of constant stiffness of the support and reduction of 20%, 40%, and 60%. Because the force and deformation changes of the two-way movable support under different working conditions are relatively large, this article takes the two-way movable support as an example to analyze the influence of the change of the stiffness of the support on the force and deformation of the support. The 10-year shrinkage and creep of the concrete, the combined temperature rise of 30 °C, and the combined temperature drop of 30 °C support offset and force are shown in Figs. 3, 4 and 5.

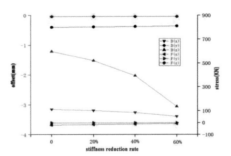

Fig. 3. 10-year shrinkage and creep of concrete

It can be seen in Fig. 3 that under the condition of shrinkage and creep of the concrete 10, the value of the support in the F(x) and F(y) directions is very small, which can be ignored and unchanged, and the support is in these three directions. The force changes are very small. However, the offset changes in the three directions of the support are different. The offset in the D(z) direction is very obvious, from -1.2065 mm to -3.0557 mm; the

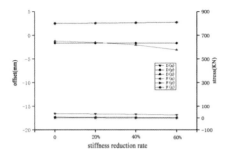

Fig. 4. Temperature rising 30 °C

support is in the D(x) direction. Before the stiffness of the seat is reduced to 40%, the offset changes more gently, but when the stiffness is reduced to 60%, the offset of the support is larger.

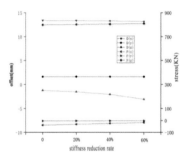

Fig. 5. Temperature cooling 30 °C

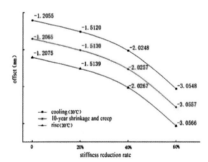

Fig. 6. Comparison of three working conditions

It can be seen from Fig. 5 and Fig. 6 that under the temperature difference of 30 °C, the stress change of the bearing is roughly the same as that under the shrinkage and creep of concrete, with little change; The variation range of bearing offset in D(x) and D(y) directions is very small, but the increase range of vertical displacement is very obvious, from −1.2055 mm to −3.0548 mm.

In conclusion, the reduction of bearing stiffness weakens the bearing durability. Under the three different conditions given above, the increase of stiffness damage of bidirectional movable bearing reduces the stress in x and y directions and increases the stress in z direction, but the increase range is small; The influence on the longitudinal and transverse deformation of the bearing is very small and can be ignored, but it has a great influence on the vertical displacement of the bearing. The vertical deformation change of support under three working conditions is shown in the figure below.

It can be clearly seen from Fig. 6 that under the condition of 10-year shrinkage and creep, the stiffness of the two-way movable bearing is reduced to 60%, which is 1.8492 mm higher than the vertical extrusion deformation without reduction, with an increase of about 60.52%; Under the condition of combined temperature rise of 30 °C, the vertical extrusion deformation of the support increases by 1.8491 mm, with an increase of about 60.49%; When the combined cooling temperature is 30 °C, the vertical extrusion deformation of the support increases by 1.8493 mm, with an increase of about 60.54%. Therefore, with the passage of time, the stiffness of the support is continuously reduced and the vertical displacement of the support is continuously increased, which is not conducive to the durability of the support, and the increase of the extrusion deformation of the support is more likely to cause bulging cracks.

5 Conclusion

(1) Concrete shrinkage and creep have little effect on the displacement of the support, but the change in temperature will greatly increase the longitudinal and lateral displacement of the support, and the longitudinal displacement will reach about 4–5 times of the original. The offset range is 60 times as much as before. Although the displacement of the support varies greatly, it still meets the standard support offset limit. The maximum longitudinal displacement of the support is 18% of the longitudinal limit, and the maximum lateral displacement of the support is 55% of the transverse limit.

(2) The stiffness damage of the support has little influence on the longitudinal and lateral displacement and force, but it has a greater influence on the vertical compression deformation of the support.

(3) When the rigidity of the plate rubber bearing is reduced to 40% of the previous one, the vertical compression deformation of the bearing is increased by 61%, which is likely to cause the bearing to produce bulging cracks, which will increase the rigidity of the bearing over time Reduction is not conducive to the durability of the support.

References

1. Chen, Y.J., Li, B.B., Zhao, Q.Y.: Experimental study on the effect of recycled rubber on the durability of plate rubber bearing. Highway Traf. Sci. Technol. (Applied Technology Edition) **14**(08), 120–122 (2018)

2. Xu, Y., Min, J.G., Wang, X.B.: Study on the influence of plate rubber bearing stiffness on bridge model calculation results. Highway Traf. Sci. Technol. (Applied Technology Edition) **15**(08), 147–148 (2019)
3. Huang, Y.P., Xu, M., Zhou, M.H.: Effect of uneven rubber layer thickness on the durability of plate rubber bearing. Modern Transportation Technology **2**, 29–32+54
4. Yin, J.N.: Numerical simulation analysis of bearing and hinge joint damage of fabricated reinforced concrete hollow slab bridge. Tianjin University (2009)
5. Tao, J.: Influence of bearing failure on bridge structure and failure criteria. Chang'an University (2015)
6. Liang, D., Zhang, C.Z., Liu, J., Chen, L., Chen, H.X.: Research on damage identification of beam bridge support based on Gaussian curvature modal correlation coefficient. Earthq. Eng. Eng. Vib. **40**(02), 23–32 (2020)
7. Wang, G.X., Ge, Z.Q., Qin, J.M.: Spring element stiffness coefficient calibration and shaking table test verification of simulated bridge bearing based on ANSYS. Highway **65**(2), 74–78 (2020)
8. Luo, K., Ou, K.K., Lei, X.Y.: Experimental study on the influence of bearing stiffness on the noise of track box girder structure. Proceedings of the National Acoustic conference in 2019.Chinese acoustic Society: Chinese acoustic society, 443–444 (2019)
9. Yan, W.W., Guo, F.J., Shi, Y.X., Deng, Q.Y., Wu, X.G.: Study on bearing force of fabricated wide continuous beam bridge. J. Shenyang University (Natural Science Edition) **31**(02), 141–145 (2019)
10. Wu, X.G., He, S.L., Zheng, P., Yin, Y., Guo, Z.Q.: Study on failure criteria of plate rubber bearings for highway bridges. J. Zhengzhou University (Engineering Edition) **40**(01), 67–71 (2019)
11. Yang C.S.: Application analysis and research of plate rubber bearings for highway bridges. Tianjin University (2007)
12. Xu, M., Huang, Y.P., Zhou, M.H.: Experimental study on the relationship between shape coefficient and damage of plate rubber bearing. Test Technol. Testing Machine **4**, 8–11 (2005)
13. Transportation industry standard of the people's Republic of China. Specification series of plate rubber bearings for highway bridges. JT / T 663–2006

Application Analysis of Several Basement Exterior Wall Construction Schemes in Compacted Pile Foundation Treatment Engineering

Yatuan Yan, Chen Chen[✉], Yufei Xu, Shoufu Li, Teng Wang, and Di Ding

China Construction Third Engineering Bureau Co. LTD. Northwest Branch, Xian 710065, Shaanxi, China
cctj_9126@126.com

Abstract. Collapsible loess exists in a large area in Northwestern China. When collapsible loess exists in the ground, it is usually necessary to treat it to eliminate the collapsibility before placing the building foundation. When compacting piles are used to treat the collapsible loess foundation, there are three different construction schemes for the construction of the basement exterior wall, which are the conventional excavation scheme, the outer frame edge excavation scheme, and the unilateral support form excavation scheme. Through the detailed description of the three construction schemes, and the comparative analysis from the aspects of construction difficulty, economy, construction period, etc. It can be concluded that the conventional excavation scheme has the highest cost, the highest construction efficiency, the longest construction period, and other aspects. The frame edge excavation scheme has the lowest construction efficiency, and the unilateral formwork excavation scheme has the lowest cost and the shortest construction period.

Keywords: Basement exterior wall · Collapsible loess · Compacted piles · One-sided support module · Construction plan

1 Introduction

Collapsible loess is a kind of soil with special properties. Microscopically, it is composed of structural units, cements and pores. The color is mostly light gray-yellow or yellow-brown, the soil quality is relatively uniform, good permeability, large pores, and easy disintegration, vertical joint development, and the mechanical properties are anisotropic. The landform and structure of loess are shown in Fig. 1. The characteristics of Collapsible loess are generally higher strength and lower compressibility when it is not wetted. When it is wetted by water under a certain pressure, the soil structure will be destroyed quickly, resulting in a large additional subsidence and a rapid decrease in strength.

Loess collapsibility is an unstable deformation with a large amount of subsidence and a fast subsidence rate. It can lead to uneven settlement of the foundation, which is

G. Feng (Ed.): ICCE 2021, LNCE 213, pp. 338–346, 2022.
https://doi.org/10.1007/978-981-19-1260-3_31

more harmful to the building. It will cause the tilt of the structure, the damage of the wall in the house and the cracking of the load-bearing structure such as beams and columns. In addition, the hazards to road engineering are mainly manifested in uneven settlement after encountering water, causing large-scale cracking of municipal roads, and sinking affects road construction quality and driving safety.

It is necessary to consider the importance of the building, the degree of the possibility of the foundation being wetted by water and the strictness of the uneven settlement limit during the use period when constructing in collapsible loess areas. In order to prevent the loess foundation from collapsing, the replacement cushion method, dynamic compaction method, impact compaction method, Compacted pile method and other measures can be used to strengthen the soil to improve the bearing capacity of the soil layer. Some foundation treatment methods are shown in Fig. 2.

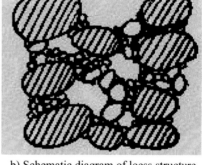

a) Loess landform   b) Schematic diagram of loess structure

Fig. 1. Landform and structure of loess

a) Replacement cushion method   b) Compacted pile method

Fig. 2. Some foundation treatment methods

Under normal circumstances, the construction of the basement exterior wall is to erect the frame after the excavation of the foundation pit reaches the specified elevation, and then support the formwork for concrete pouring. However, when there is collapsible loess in the construction zone, it is necessary to treat the collapsible loess foundation locally or wholly. When using compacted piles to wholly treat the collapsibility, the plane treatment range should exceed the outer edge of the building exterior wall by 2 m, and should not be less than 1/2 of the thickness of the treated soil layer [1]. At this time, if the outer edge of the foundation pit excavation is designed outside the compacted pile, the construction site needs a larger plane working surface. However, with the continuous advancement of the urbanization process in China, the old and new buildings are often close to each other, and measures should be taken to reduce the mutual influence between the buildings when the new buildings are under construction in the basement. However, with the continuous advancement of the urbanization process, the old and new buildings are often relatively close, and measures should be taken to reduce the mutual influence between the buildings during the construction of the new building basement [2]. For this kind of project with narrow construction site and limited space, reduce the excavation range of the foundation pit fertilizer trough, and directly carry out the compaction pile construction from the positive and negative zero elevation. It is a good way to use unilateral support for the construction of the basement exterior wall. A construction plan that effectively saves working surface and reduces the mutual influence between buildings. Based on a typical residential project in Xi'an, this article introduces in detail three construction schemes for the basement exterior wall when using plain soil compaction piles to treat collapsible loess foundations [3, 4]. Comprehensive analysis of several programs.

2   Overview of Construction Scheme

For a residential project in Xi'an City, the seismic grade of shear wall of the project is grade 2, the design grade of underground garage foundation is grade C, and the basement has two layers, each with a height of 3.7 m. Some foundation soils are of grade II heavy collapsibility. The collapsible loess foundation is treated as a whole with plain soil compaction piles. The diameter of compaction piles is 400 mm, the distance between piles is 500 mm. The collapsibility treatment range is 4 m outside the raft cushion, the foundation is 400 mm thick raft, and 170 mm thick foundation cushion is set under the raft.

2.1   Conventional Excavation Scheme

In order to ensure the effectiveness of foundation treatment, the conventional excavation scheme carries out pile row support construction at the edge of collapsibility treatment range, excavates all soil within collapsibility treatment range, and constructs compacted piles at elevation of pit bottom at −9.670 m after excavation. After construction of compacted piles, external frames are set up in turn, reinforcement bars of external walls are tied up, and external wall formwork is set up. Then the concrete construction of the basement outer wall is carried out [5, 6]. The schematic diagram of conventional excavation scheme is shown in Fig. 3.

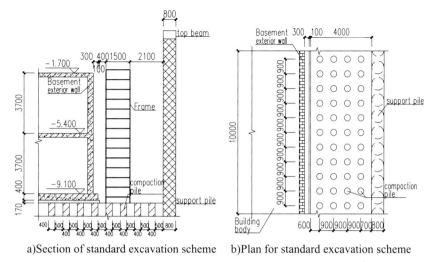

a)Section of standard excavation scheme   b)Plan for standard excavation scheme

Fig. 3. Schematic diagram of conventional excavation plan

2.2 Excavation Scheme for Outer Frame Edge

Considering the installation of scaffolding for basement outer wall construction, the row pile support in this scheme is set at 2.5 m away from the outer edge of the outer wall, reserving the erection space for the outer frame and excavating all the earth inside the support. The compacted piles outside the excavation scope are drilled from the zero elevation and compacted to the elevation of −9.670 m. The above part is backfilled with plain soil. The compacted piles on the side within the excavation scope are constructed from the bottom elevation of −9.670 m pit. Schematic diagram of excavation scheme for outer frame edge is shown in Fig. 4.

2.3 Excavation Scheme of One-Sided Formwork

Regardless of scaffolding during construction of basement exterior wall structure, unilateral formwork supporting technology is proposed for construction of basement exterior wall structure. In this scheme, row pile support is set 0.9 m away from outer edge of basement exterior wall, and excavation side line is close to inner side of support. Compacted piles with collapsibility treatment and expansion range are constructed from positive and negative zero elevation, compacted to −9.670 m elevation, and the above parts are backfilled with plain soil. Brick tyre film is used on the outside of the basement outer wall, and fine sand is used to fill the gap between the support and row of piles. The basement outer wall is constructed with one-sided formwork from the inside without setting up external scaffolding.

The schematic diagram of the one-sided support mode scheme is shown in Fig. 5.

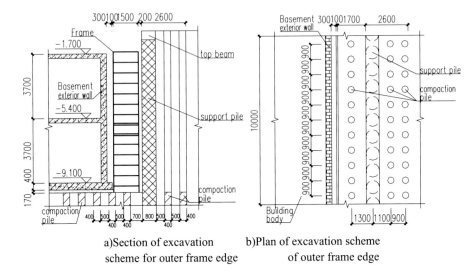

a)Section of excavation   b)Plan of excavation scheme
scheme for outer frame edge   of outer frame edge

Fig. 4. Schematic diagram of outer frame edge excavation plan

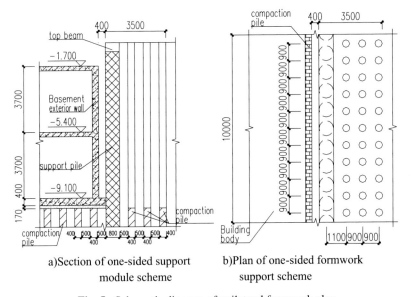

a)Section of one-sided support   b)Plan of one-sided formwork
module scheme   support scheme

Fig. 5. Schematic diagram of unilateral formwork plan

3   Overview of Construction Scheme

3.1   Economical Efficiency

The list of quantities of the three schemes is shown in Table 1 and the economic comparison diagram is shown in Fig. 6. From the chart, it can be seen that the foundation pit of conventional excavation scheme has a large scope of outlay. Even without considering the support form, large area earthwork excavation and backfilling will cause significant increase in cost and relatively high construction cost. Due to the small excavation scope,

Table 1. Quantities list of exterior wall construction plan

Programme	Name	Unit	Quantity of works	Reference unit price (Yuan)	Total Tax free price (Yuan)
Standard excavation scheme	Earthwork excavation and backfilling of fertilizer tank	m^3	46.5	100	4650
	Plain soil compaction pile (empty pile)	m	0	12	0
	Formwork works	m^2	14.3	110	1573
	Erection and disassembly of outer frame	m^2	7.8	60	468
	Subtotal				6691
Excavation scheme for outer frame edge	Earthwork excavation and backfilling of fertilizer tank	m^3	19.47	100	1947
	Plain soil compaction pile (empty pile)	m	55.62	12	667.5
	Formwork works	m^2	14.3	110	1573
	Erection and disassembly of outer frame	m^2	7.8	60	468
	Subtotal				4655.5
Excavation scheme of one-sided formwork	Earth excavation of fertilizer tank	m^3	3.44	100	344
	Plain soil compaction pile (empty pile)	m	74.16	12	889.9
	Brick mould	m^3	1.85	300	555
	Fine sand backfilling	m^3	1.11	210	233.1
	Single side formwork works	m^2	6.9	135	931.5
	Subtotal				2253.5

the one-sided formwork construction scheme basically does not need soil backfilling of the fertilizer tank, and the minimum amount of earth excavation and backfilling is required. At the same time, no external erection and double-sided formwork support are required. Only the tyre form of outer wall bricks is required, thus the construction cost is relatively low. The construction cost of the outer frame edge construction scheme is between the conventional construction scheme and the one-sided formwork construction scheme [7, 8].

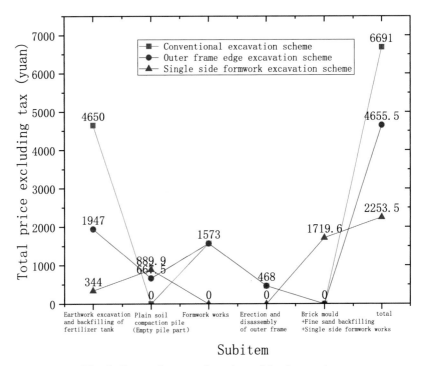

Fig. 6. Economic comparison chart of the three options

3.2 Difficulty of Construction

Conventional excavation schemes increase the amount of work due to earth excavation and backfilling within 4.4 m of the outer side of the foundation, and often due to site constraints, foundation pit support is affected after foundation treatment is expanded, sometimes it is necessary to strengthen or even change the form of support, resulting in passivity of the constructor, reinforcement, formwork and support for the outer basement wall. Concrete construction is the same as inner wall construction, which cannot effectively improve work efficiency. The excavation scheme of one-sided formwork avoids the problem of ineffective earthwork excavation and backfilling outside the basement [9]. Because the exterior expanded compacted piles are struck from the ground, the length of bore holes needs to be increased. This part of work is relatively small and controllable.

At the same time, the construction efficiency of the exterior wall structure is improved by using the single-sided formwork technology and the single-sided formwork frame can be rented, although it is necessary to increase the tyre form of exterior wall Costs have increased, but construction convenience has increased considerably. The foundation treatment of outer frame edge excavation scheme is partially excavated and then compacted piles are constructed. Some compacted piles are driven from the ground by the same measures as the one-sided formwork scheme, which improves the efficiency to a certain extent. However, the general formwork is still used to construct the basement outer wall structure with lower overall efficiency.

4   Conclusion

Conventional excavation schemes will be affected more by pollution control, haze reduction, rainy season and other factors due to more earthwork excavation and backfilling in the expanded part of foundation treatment, thus affecting the construction period. In addition, the basement exterior wall structure is constructed by traditional technology, which is relatively time-consuming and has the longest construction period. The main difference of the excavation scheme of single-side formwork is that the exterior wall uses brick tyre mould, while the basement exterior wall structure uses single-side formwork technology, which effectively improves labour efficiency. In addition, when the expanded part of compacted piles of foundation treatment are laid from the ground, a certain construction period can be saved and the construction period is relatively shortest. The construction period of the outer frame edge excavation scheme is between the above schemes.

References

1. Shaanxi Planning Committee. Building Code for Collapsible Loess Area. China Construction Industry Press (2004)
2. Ministry of Housing and Urban-Rural Construction of the People's Republic of China. Code for Design of Building Foundation Foundation. China Construction Industry Press (2012)
3. Dong, Y.Q.: Application of unilateral formwork scheme in construction of basement exterior wall. Shanxi Arch. **38**(018), 110–111 (2012)
4. Kong, W.J.: Comparison and analysis of construction schemes for basement roof of Hauli Hotel in Cambodia. Constr. Technol. (Phase S1), 791–794 (2019)
5. Feng, Z.L., Zhang, Y.S., Zhang, D.: Research on one-sided high-support formwork system of deep foundation pit. Constr. Technol. (Stage S1): 67–70 (2016)
6. Liang, H.B., Jiang, C., Huang, G., Fan, Y.H., Wei, Z.Y.: Construction technology of single-sided formwork of assembled aluminium alloy. Constr. Technol. (Phase 2), 101–104 (2016)
7. Jing, L.I., Chen, S.: Application of rotary jet grouting pile replacing single-side formwork technology of crown beam in urban renewal project. Constr. Technol. (Stage S1), 3–4 (2019)
8. Liang, Q., Qiu, Z., Zhu, Y., et al.: Research and application of single side formwork support system of basement external wall. Build. Constr. (Stage S10), 3–4 (2019)
9. Deng, T.J., Zhu, M.: Research on economical design of cast-in-place single-side concrete formwork templates and the support system. J. Hunan Univ. (Nat. Sci.) (2015)

Multilateral Boundary Blasting Theory of High and Steep Slope in Open Pit Mine and Its Application

Junkai Chen[1], Wenxue Gao[1(✉)], Xiangjun Hao[2], Zheng Wei[2], Xiaojun Zhang[1], and Zhaochen Liu[1]

[1] College of Architecture and Civil Engineering,
Beijing University of Technology, Beijing 100124, China
wxgao@bjut.edu.cn
[2] Inner Mongolia Kinergy Blasting Co., Ltd., Erdos 017000, China

Abstract. At present, the blasting theory of high and steep rock slope mainly focuses on flat terrain, ignoring the influence of micro-terrain boundary factors on blasting effect, which leads to excessive blasting energy and affects the stability of slope. Therefore, based on the theory of multilateral boundary rock blasting, this paper deduces the calculation formula of blasting charge for high and steep rock slope under multilateral boundary conditions, and verifies it with field test. The results show that: (1) The multilateral boundary charge calculation formula directly includes micro-topography boundary conditions and blasting effect, and the rock blasting theory is based on the interaction of blasting energy provided by explosives and potential energy in medium, which effectively improves the energy utilization rate of explosives. (2) The influence of surplus blasting energy on the surrounding environment under different boundary conditions is controlled, and the explosive explosion effect is effectively controlled, so that a stable high and steep slope of open pit mine is formed after blasting.

Keywords: Micro-topography · Multilateral boundary · Blasting charge · High and steep slope

1 Introduction

The application and research of engineering blasting technology in national economy and national defense construction has a long history. Since modern open pit mine stripping works are mostly carried out under the condition of high mountains and complex changes of wavy micro-topography, it is very important to establish blasting characteristics and blasting charge calculation theory under multilateral boundary conditions. The multilateral boundary condition refers to the boundary condition of micro-topography, which belongs to the shape geometry condition compared with the horizontal boundary condition [1].

Wang [2] systematically proposed the comprehensive theory of multilateral boundary stone blasting for the first time, and directly introduced the micro-topography into the

© The Author(s) 2022
G. Feng (Ed.): ICCE 2021, LNCE 213, pp. 347–357, 2022.
https://doi.org/10.1007/978-981-19-1260-3_32

blasting theory as a main condition. Chen et al. [3] studied quantitatively the influence of geological conditions and physical and mechanical properties of rock on the blasting effect and the effect of blasting on the stability of mountain or slope. Gu [4] proposed that the possibility of amplitude superposition should be considered when the interval time between charge bags is small, and the permissible distance of blasting vibration should be calculated according to equivalent charge. Chen et al. [5] improved the three-multiple blasting method and adopted multi-layer, multi-initiation and multi-face open terrain blasting to make the deep cutting take shape in one blast, which improved the blasting method of deep cutting facing open terrain in multi-boundary blasting theory. Xia et al. [6] made equivalent transformation of blasting vibration speed monitored on site according to Sadoski formula, and used linear regression control method to calculate blasting charge amount for subsequent construction. Huang et al. [7] uses high-speed photography technology and numerical calculation method to carry out research, and obtains that the maximum initial throwing velocity of slope rock increases, stabilizes and decreases in the direction of blasting propagation in the extended blast hole. Hu et al. [8] put forward the concept of equivalent path, which provides a new method for predicting the peak value of particle vibration velocity of bench blasting seismic wave. Gan et al. [9] built a theoretical model of iron ore crushing energy, which provides a theoretical basis for the calculation of ore crushing energy in iron mines. Yang et al. [10] proposed a liquid explosive fracturing technology based on deep-hole blasting of coal and rock, which enhanced the energy utilization rate of explosives in the direction of hard rock fracturing.

The above-mentioned research mainly focuses on the blasting theory and charge calculation of flat terrain. However, the change of boundary conditions of the micro-topography is not considered enough, or only the micro-topography is considered as a factor that has an influence on the blasting effect. Since the charge calculation formula and design, method did not consider the potential energy of the rock mass itself, it caused a waste of explosive charge, easily caused geological diseases, and caused slope instability. Based on the comprehensive blasting theory of multi-boundary stonework, this paper directly introduces the micro-topography boundary conditions into the charge calculation and blasting effect, and conducts deep-hole controlled blasting to study the action law of charge, terrain conditions and blasting effect under multi-boundary conditions.

2   Main Role of Explosive Under Multilateral Boundary Conditions

2.1   Upswing

Upswing is also called sublation action under horizontal boundary condition. The explosive wave generated by the explosive charge explosion and the thrust of the expansion of the explosive product make the medium rise upward and then blow out to form the blasting funnel. It can be seen that the size of the blasting funnel and the quality of the throwing effect depend on the kinetic energy generated by the explosive explosion. The larger the dosage and the higher the throwing, the larger the throwing amount obtained, and the stable blasting effect. Under the condition of horizontal boundary, the drop rate is proportional to the amount of charge.

$$E \propto Q \qquad (1)$$

Upswing is the basis for determining the volume of the visible blasting funnel and the throwing rate of the two types of boundary conditions, the horizontal and the concave pass.

2.2 Collapse

Under non-horizontal terrain conditions, the medium in the collapse funnel itself contains a certain amount of potential energy, and the blasting surface of the charge receiving bag collapses. It has a great effect on a number of technical indicators of engineering blasting, such as blasting hopper volume, throwing rate and unit consumption, etc. It is also the physical basis of multi-boundary blasting theory. According to the statistics of a large number of engineering practices, in different media, the ratio of the upper and lower damage radius is calculated by Eq. (2):

$$R_u = \left(1 + \frac{\alpha^2}{5000 \sim 7000}\right) R_l \tag{2}$$

In Eq. (2), R_u is the upper destruction radius of the blasting funnel (m). R_l is the lower destruction radius of the blasting funnel (m). α is the ground slope (°).

Under the inclined boundary condition, the reason why the upper destruction radius of the blasting funnel increases with the increase of the ground slope is essentially caused by the collapse. The upper damage radius increases with the slope of the ground, which is 1 to 2.6 times larger than the lower damage radius, and it is larger in soil and rock masses controlled by structural planes. From this, it can be seen that the collapse effect expands the damage range of the cartridge explosion.

2.3 Slump

The high throwing rate in steep terrain is due to the existence of slump. When the medicine packet is only broken, the slump rate of the rock block is Eq. (3):

$$E_o = 1 - \frac{1}{\xi}\left(\frac{\tan\theta}{1 - \cot\psi\,\tan\theta} \Big/ \frac{\tan\alpha}{1 - \cot\psi\,\tan\alpha}\right) \tag{3}$$

In Eq. (3), E_o is the slump rate of the rock block when the medicine packet is only broken (%). ξ is the looseness coefficient of the rock, generally taken as 1.3. θ is the average angle of repose of the rock, generally around 35°, not more than 40°. ψ is the angle formed by the designed step slope line and the horizontal plane.

According to Eq. (3), it is obtained that when the cartridge only serves to break the rock mass, the slump rate of the exploded rock mass increases sharply with the steepening of the natural ground slope. Therefore, in the terrain above the steep slope ($\alpha > 50°$), the rock mass only needs to be fully broken, and its slump rate will reach the best throwing rate of the slope terrain using throwing blasting, and it has a good effect on the stability of the steep slope. In cliff topography, the slump rate can exceed 70%. Under steep terrain, no need to use throwing blasting. Because the potential energy contained in the rock mass itself has replaced the explosive energy required to throw the rock block out of the blasting funnel, the potential energy contained in the rock mass itself increases the effective utilization of explosive energy.

2.4 Side Throwing

The effect of side throwing can be expressed by the increment of throwing rate, and the relationship with the natural ground slope is shown in Eq. (4):

$$\Delta E = -77.52 \lg f(\alpha) \tag{4}$$

In Eq. (4), ΔE is the increment of throwing rate (%). $f(\alpha)$ is the slump coefficient, and the calculation method is shown in Eq. (5):

$$f(\alpha) = \begin{cases} 1 - \alpha^2/7000 & \alpha < 30° \\ 26/\alpha & \alpha < 30° \end{cases} \tag{5}$$

Under multiple boundary conditions, the charge volume remains unchanged, and the increase in throwing rate increases with the steepness of the natural ground slope. The essence is that under multiple boundary conditions, the explosive kinetic energy and the potential energy of the rock are the result of the combined effect. The rock mass has a certain potential energy and a favorable throwing angle, so that the rock mass thrown from the side is not easy to fall back into the blasting funnel. Therefore, only a smaller throwing height and throwing distance are required to obtain a higher throwing rate.

3 Calculation Formula of Blasting Charge Under Multilateral Boundary Conditions

3.1 Basic Principles and Assumptions

(1) Under multiple boundary conditions, the mechanical energy required to slump a certain amount of the same medium is constant. In addition to the kinetic energy provided by the explosion of the cartridge, this mechanical energy also has the potential energy contained in the medium itself. In horizontal and steep terrain, kinetic energy and potential energy respectively play a major role in the blasting effect. The mutual transformation of kinetic energy and potential energy depends on the change of boundary conditions. Under general boundary conditions, the blasting effect is usually the result of the combined action of kinetic energy and potential energy.
(2) The potential energy of the medium itself is equivalent to an increase in the effective explosive energy of explosives.
(3) The throwing rate is used as the standard for evaluating the blasting effect. With horizontal boundary conditions, the throwing rate $E = 27\%$ is the standard state.
(4) In order to achieve a new balance, the rock blocks formed by blasting slid into piles outside the blasting funnel under the action of their potential energy. The slope of the piles is the angle of blasting repose.

3.2 Multilateral Boundary Dose Calculation Formula

(1) Theoretical calculation formula of multilateral boundary dose

According to the law of conservation of mechanical energy, the principle of functional balance and the above assumptions, the calculation formula for the multilateral boundary drug amount is obtained as follows:

$$Q = KW^3 F_\psi (E, \alpha) = KW^3 \frac{10^{0.0129E}}{\left(\sqrt{0.05\alpha} + 1\right)\left(1.11 - \frac{86.133}{E} \lg f(\alpha)\right)} \tag{6}$$

In Eq. (6), Q is the multi-boundary blasting charge (kg). K is the dosage per cubic meter of standard throwing blasting (kg/m^3). W is the minimum resistance line (m). $F_\psi (E, \alpha)$ is the theoretical drug packet property index, which can be calculated by Eq. (7):

$$F_\psi (E, \alpha) = \varphi(E) \cdot f_\psi (\alpha, E) \tag{7}$$

In Eq. (7), $\varphi(E)$ is the function of the throwing rate, generally $\varphi(E) = 0.45 \times 10^{0.0129E}$, for the collapse blasting, $\varphi(E) = 1$. $f_\psi (\alpha, E)$ is the topographic coefficient or the dose attenuation coefficient, which can be calculated by Eq. (8):

$$f_\psi (\alpha, E) = V(\alpha) \cdot E(\alpha) \tag{8}$$

In Eq. (8), $V(\alpha)$ is the collapse factor, calculated by Eq. (9). $E(\alpha)$ is the side throwing factor, calculated by Eq. (10):

$$V(\alpha) = 2 / (\sqrt{A\alpha} + 1) \tag{9}$$

In Eq. (9), A is the collapse coefficient. In the calculation of multi-boundary drug dose, $A = 0.05$, and $A\alpha \geq 1$.

$$E(\alpha) = 1 / \left[1 - \frac{77.52}{E} \lg f(\alpha)\right] \tag{10}$$

In Eq. (10), $f(\alpha)$ is the slump coefficient, and the calculation method is shown in Eq. (5).

According to Eq. (8)–(10), the topographic coefficient is shown in Eq. (11):

$$f_\psi (\alpha, E) = 1 / \left(\sqrt{0.05\alpha} + 1\right)\left[0.5 - \frac{38.76}{E} \lg f(\alpha)\right] \tag{11}$$

The Eq. (11) to calculate the theoretical property index of the drug pack is shown in Eq. (12):

$$F_\psi (E, \alpha) = \frac{10^{0.0129E}}{\left(\sqrt{0.05\alpha} + 1\right)\left(1.11 - \frac{86.133}{E} \lg f(\alpha)\right)} \tag{12}$$

(2) Empirical calculation formula of multilateral boundary dose

$$Q = KW^3 F(E, \alpha) \tag{13}$$

In Eq. (13), $F(E, \alpha)$ is the property index of the drug package.

$$F(E, \alpha) = \varphi(E) \cdot f(\alpha) \tag{14}$$

3.3 Comparison of Theoretical Formula and Empirical Formula

By comparing theoretical Eq. (6) and empirical Eq. (13), we can see that there are certain differences between the two, which are mainly manifested in the following two aspects:

(1) The terrain coefficient is different from the slump coefficient
    The theoretical topographic coefficient refers to the slump coefficient when considering the law of lateral tossing, and takes a step forward, so that the topography or slump coefficient has a clear physical and mechanical meaning, that is, the $f_\psi(\alpha, E)$ value reflects the effective utilization rate of blasting energy along with the topographic boundary The law of changing conditions. The terrain is favorable, the effective utilization rate of explosive energy is increased, and the amount of charge needs to be reduced. On the contrary, the terrain is unfavorable, and the utilization rate of explosive energy decreases, and the amount of charge needs to be increased like the terrain of a pass.
(2) The relationship between slump coefficient and terrain coefficient
    The current empirical formula may produce the following situations in production: when the design-throwing rate $E < 60\%$, the actual effect is higher and safer. When $E > 60\% \sim 70\%$, it is likely to be slightly lower, mainly in steep terrain. At this time, due to the sharp increase in slump, the actual blasting effect will not be affected. Therefore, it can be considered that the calculation Eq. (13) of the multi-boundary charge is consistent with the theoretically derived formula. In the steep terrain, the principle of slump should be obeyed, and slump blasting should be used.

3.4 Calculating Formula of Charge for Multilateral Boundary Deep Hole Blasting

Based on the analysis of the characteristics of rock mass blasting under multi-boundary conditions, especially the rock mass blasting of high and steep slopes, according to the calculation principle of formula (13), the formula for calculating the packing charge of deep-hole blasting columnar charge under multi-boundary conditions is obtained:

$$Q = KW^3 F(E, \alpha) = KF(E, \alpha)aWh \tag{15}$$

In Eq. (15), a is the hole distance (m). h is the drilling depth (m).
    The multilateral boundary charge calculation formula directly includes the terrain boundary conditions and the blasting effect. The blasting theory is based on the combination of the explosive energy provided by the explosive and the potential energy in

the medium to control the excess blasting energy and excessive explosive consumption under different terrain conditions. Established the functional relationship among the explosive quantity, terrain boundary conditions and blasting effect, which is convenient for practical inspection and application of engineering blasting, and has been widely used in engineering blasting in our country.

4 Project Example

4.1 Project Overview

Xingguang Coal Mine is located in the west wing of Zhuozishan anticline, and its geomorphic features belong to the low mountain and hilly area of the Inner Mongolia Plateau. The terrain is relatively flat, the thickness of the loose cap layer is small, and geological disasters such as landslides and mudslides are rarely occurred. Gray-white medium and coarse sandstone with a small amount of gray-green sandy mudstone are distributed in a large area in the area. The stratum has been extensively weathered and eroded and is in integrated contact with the underlying Shanxi Formation. Sandy mudstone has weak water richness, hardens after losing water, easy to crack, and has little water content. The softening coefficient of various types of rocks varies greatly, ranging from 0.21 to 0.83, with an average of 0.40. Except for coal, all kinds of rocks are basically easy to soften rocks. The mine is a gas mine, and there is no outburst of coal and gas.

4.2 Deep Hole Blasting Design Under Multilateral Boundary Conditions

According to the multilateral boundary blasting theory, the open-pit mine blasting design is carried out. The deep hole loosening millisecond delayed blasting construction is used to control the direction of the minimum resistance line of blasting, avoid facing the direction of residential houses and various buildings, and take safety protection measures. The environmental conditions of the blasting area are good, and the rock is drilled once and blasted to reach the design depth to ensure the construction progress. In order to control the blasting flying rocks, ensure the length of the blockage during construction, press sandbags at the orifices, and cover the wicker if necessary. At the same time, the millisecond delayed detonation technology is used to limit the maximum amount of detonating charge in a period to control the impact of blasting vibration and flying rocks on the surrounding environment. The blasting design parameters are as follows:

(1) The unit consumption of explosive is $q = 0.22$ kg/m^3, the density of linear charge is $q_1 = 5.7$ kg/m, and the single hole charge is $Q = 44$ kg.
(2) The height of step is $H = 16$ m, and the slope angle of step is 80°.
(3) The depth of step hole is $L = 16.5$ m, the ultra-depth is $H = 1$ m, and the resistance line of chassis is $W = 4$ m.
(4) The distance between rows of holes is $a \times b = 5$ m $\times$ 4 m, and the delay time is 50–100 ms between rows and 20–40 ms between holes.
(5) The drilling type is DTH drilling rig with a drilling diameter of 90 mm.

(6) The blasting workers measure the actual hole depth when charging, and then charge and plug.

(7) Charge. ANFO explosives are used for anhydrous holes, and Φ70 mm emulsion explosives are used for water holes.

(8) Plugging. The plugging is dense, plugging with blasting mud and tamping with blasting stick and the plugging length $L_2 = 2.8$ m.

4.3 Detonation Network

The network of digital electronic detonator is used in blasting, in which the outermost row near the side slope is firstly detonated, and then detonated hole by hole from outside to inside. Delay time setting: delay time between holes is 25 ms, and delay time between rows is 60 ms. The schematic diagram of initiation network is shown in Fig. 1.

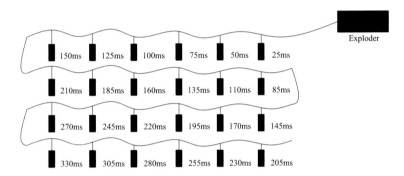

Fig. 1. Initiation network diagram of bench blasting.

4.4 Blasting Effect and Analysis

In this mining area, the multilateral boundary blasting theory is studied for deep-hole controlled blasting. The blasting of 76848 m^3 and 73438 m^3 is studied by multi-boundary design method and empirical design method respectively. Combined with the actual situation in the field, groups of measuring points are arranged at the horizontal distance of 70 m, 90 m, 110 m and 130 m respectively. Through the analysis of waveform diagram of vibration time history of four groups of measuring points (Fig. 2), the amplitude of blasting vibration velocity produced by multilateral boundary design method is less than that of empirical design. Under the same conditions, the multi-boundary design method uses less explosives, improves the energy utilization rate of explosives, makes the construction safer, and has less influence on the surrounding disturbance.

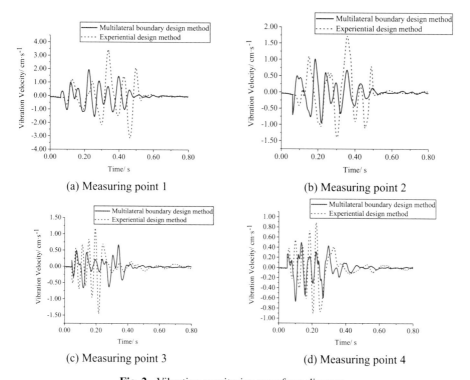

(a) Measuring point 1   (b) Measuring point 2

(c) Measuring point 3   (d) Measuring point 4

Fig. 2. Vibration monitoring waveform diagram.

Based on the theory of multi-boundary rock blasting and the formula of explosive charge calculation, the design and construction of rock mass blasting on high and steep slope of mine have been fully considered, and the blasting effect has been significantly improved. Field monitoring and camera show that blasting vibration and flying rocks have been effectively controlled; the fragmentation of blasting pile meets the design requirements, and all broken rock mass collapses within the design range (Fig. 3 and Fig. 4).

(a) Multilateral boundary design method   (b) Experiential design method

Fig. 3. Comparison of blasting fragmentation.

(a) Multilateral boundary design method   (b) Experiential design method

Fig. 4. High and steep slope formed after blasting.

5   Conclusions

Multi-boundary blasting theory considers the influence of topography and geological conditions on charge blasting. The blasting engineering practice of Xingguang Coal Mine in Inner Mongolia Kinergy Blasting Co., Ltd. shows that based on the analysis of blasting characteristics of charge under multi-boundary conditions, the following conclusions are drawn:

(1) The multi-boundary charge calculation formula directly includes micro-terrain boundary conditions and blasting effect, the theory of rock blasting is based on the interaction of explosive energy and potential energy in the medium, and controls the influence of surplus explosive energy on the surrounding environment under different boundary conditions, to effectively improve the energy utilization rate of explosives.

(2) The multi-boundary blasting theory considers the influence of topography and geological conditions on blasting at the same time, the millisecond delay controlled blasting inside and outside the hole is adopted to effectively control the explosive explosion, control the rock breaking effect of blasting and the harm of blasting vibration to the slope. All the broken rock mass collapses within the design range, effectively prevent blasting flying rocks, and stabilize the high and steep bench slope of open-pit mine after blasting.

References

1. Wang, H.Q.: Observation and comparison of blast effect characteristics under multi-boundary conditions. J. Beijing Univ. Technol. **10**(01), 41–51 (1984)
2. Wang, H.Q.: On the calculating formula of top radius of blast funnel in a multi-boundary blasting system. China J. Highway Transp. **5**(02), 26–31 (1992)
3. Chen, J.P., Gao, W.X., Liu, Y.T.: Study on the structure planes effect of rock mass blasting blasting. **22**(02) 30–33 (2005)
4. Gu, Y.C.: Discussion on the utilization of blasting vibration formula blasting. **26**(4) 78-80 (2009)
5. Chen, X.D., Xue, E.P., Zhang, L.J., et al.: Application research on rock blasting technology for subgrade. Constr. Technol. **43**(23), 105–108 (2014)

6. Xia, C.C., Liu, Z.F., Shan, G.Y., et al.: Explosive dosage calculation based on blasting vibration velocity and control value. Mod. Tunnel. Technol. **55**(04), 163–170 (2018)
7. Huang, Y.H., Liu, D.S., Li, S.L., et al.: Numerical simulation on pin-point blasting of sloping surface. Explos. Shock Waves **34**(04), 495–500 (2014)
8. Hu, X.L., Qu, S.J., Jiang, W.L., et al.: Attenuation law of blasting induced ground vibrations based on equivalent path. Explos. Shock Waves **37**(06), 966–975 (2017)
9. Gan, D.Q., Gao, F., Sun, J.Z., et al.: Energy-size relationship of ore comminution in iron mine. J. Harbin Inst. Technol. **51**(04), 163–170 (2019)
10. Yang, J.X., Yu, B., Kuang, T.J., et al.: Development and technical practice of liquid explosive based on deep-hole blasting problem. J. China Coal Soc. **46**(06), 1874–1887 (2021)

Laboratory Model Test of Eco-Concrete Slab Slope Protection

Hao Chen[✉], Fengchi Wang, Gang Xu, and Lilong Guo

School of Civil Engineering, Shenyang Jianzhu University, Shenyang, China
823114063@qq.com

Abstract. In order to study the protective effects of eco-concrete slope and the influencing factors of eco-concrete slope deformation. The displacement characteristics and ultimate bearing capacity of the slope model under different geometric parameters are obtained through the laboratory model test of eco-concrete slope protection, and the influence laws of slope deformation under different protection slope conditions are summarized, as well as the influence laws of soil compaction, soil moisture content and slope ratio on the horizontal displacement restraint capacity and stability of the slope. Compared with the unprotected slope, the ultimate load of the reinforced concrete slab slope and the ordinary concrete slab slope are increased by 2.2 times and 2.4 times respectively, and the horizontal displacement restraint capacity of the slope is increased by 29.3% and 51.6% respectively. The moisture content, compactness and slope gradient of slope soil have a certain influence on the deformation restraint capacity of slope protected by vegetation concrete slab.

Keywords: Eco-concrete · Moisture content · Compactness · Slope gradient

1 Introduction

In the traditional concrete slope protection project, ordinary concrete can effectively improve the stability of the slope, while also causing the balance of the slope ecological environment to be broken. In order to avoid sacrificing the ecological environment in exchange for urban development, in the early 1990s, many experts and scholars developed a new type of concrete-vegetable concrete that can maintain the high-strength properties of concrete while also taking into account the biocompatibility.

After continuous research by scholars at home and abroad, certain results have been achieved in the preparation of vegetation concrete [1–3], water permeability [4–6] and alkali reduction technology [7, 8]. However, the eco-concrete can protect the slope The extent of the impact still needs to be studied.

This test is designed as a scaled test to simulate the damage of the slope model, to study the protection effect of the eco-concrete slope and the factors affecting the deformation of the eco-concrete slope. Through the experiment, the displacement characteristics and ultimate bearing capacity of the slope model under different geometric parameters are obtained. According to the force, the slope failure load, the maximum settlement and the horizontal displacement of the slope are obtained, which provides a reference for the application of the eco-concrete slab on the slope.

© The Author(s) 2022
G. Feng (Ed.): ICCE 2021, LNCE 213, pp. 358–367, 2022.
https://doi.org/10.1007/978-981-19-1260-3_33

2 Test overview

2.1 Experiment Material

The cement used in the test is P.O 42.5 grade ordinary Portland cement, and the ultra-fine ore powder is obtained by crushing the ore of AnShan Iron and Steel Group. The water reducing agent is a polycarboxylic acid water reducing agent liquid, and the designed usage amount is 0.5%. The coarse aggregate of eco-concrete is recycled concrete aggregate.

2.2 Test Plan

The test includes the following four working conditions, and the specific scheme is shown in Table 1. The planting soil suitable for plant growth in the local suburbs of ShenYang was selected as the test soil sample to be filled into the slope model groove as the main filling material for the slope. The specific model slot is shown in Fig. 1.

Table 1. The model test scheme

Test conditions	Slope protection method	Compactness (%)	Soil water content (%)	Slope ratio
Condition1	No reinforce	90	12	1:1
	Eco-concrete	90	12	1:1
	Ordinary concrete	90	12	1:1
Condition2	Eco-concrete	85	12	1:1
	Eco-concrete	90	12	1:1
	Eco-concrete	95	12	1:1
Condition3	Eco-concrete	90	8	1:1
	Eco-concrete	90	12	1:1
	Eco-concrete	90	16	1:1
Condition4	Eco-concrete	90	12	1:1
	Eco-concrete	90	12	1:1.5

2.3 Specimen Production

Before the soil sample is prepared, the test soil sample is air-dried and crushed first, and the soil sample is sieved with a 10mm square hole sieve. Calculate the mass of the soil samples required for each group of test models according to the standard of 90% compaction, multiply it by the loss coefficient of 1.1, weigh it, and mix with water.

Randomly test the moisture content of the soil sample, and fill the model slope in layers when the moisture content is close to ±2% of the optimal moisture content

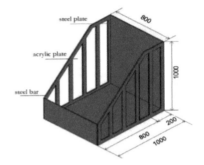

Fig. 1. The test model tank

value. The weight of the compaction hammer is M = 13.5 kg, the drop distance is H = 685.5 mm, the total length of the guide rod is L = 800 mm, and the diameter of the compaction backing plate is φ = 153 mm. The filling height of each layer is 20 cm, and 40 compaction points are set in the filling section of each layer, and each layer is compacted 3 times. Each compaction ensures staggered compaction.

2.4 Sensing Element Layout and Loading

During the test, the column tension and compression load cell was used to collect the vertical load, and the load cell range was 0–200 kn. The YHD displacement sensor is used to measure the slope displacement, and the comprehensive layout of the displacement sensor is shown in Fig. 2.

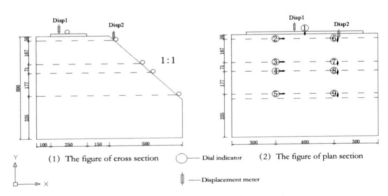

Fig. 2. The displacement sensor layout scheme

The loading phase is divided into the initial loading phase and the destruction loading phase. In the primary loading stage, the vertical load is applied at a loading speed of 5 KN/min. The vertical deformation of the specimen is detected by Disp1 and Disp2. When the values of Disp1 and Disp2 are stable and the difference between the two does not increase, the next First class load. In the loading stage of the destruction period, the loading speed is 1 KN/min. When the load reading appears "falling back", record the peak load stress at this time as the load design value. Stop loading until the first-level load value is reached, and collect the displacement count value after the compression deformation and stress redistribution inside the model soil (about 10 Min) are completed. Begin to apply the lower load. When there are obvious cracks in the slope, reduce the load loading level and increase the number of data observations until the slope is damaged on a large scale and the peak load no longer increases.

3 Effect Analysis of Eco-Concrete Slope Protection

For the unprotected slope test,eco-concrete slope test and ordinary concrete slope test, the relationship between displacement and load at different positions of the slope is shown in Fig. 3, 4 and 5.

From the point of view of the vertical settlement of slope, slope ecological concrete and ordinary concrete slope compared with no protective slope, slope vertical settlement aspects appear very big contrast, the vertical settlement of the slope changed little, considering this is the laboratory scale experiments, the actual situation of the ecological concrete slope and no protective slope vertical settlement will be bigger; Compared with the unprotected slope, the concrete slab will have greater vertical settlement due to its larger weight during the experiment. The maximum settlement measured in the experiment is 10 mm, which is acceptable compared to the model size.

From the point of view of the horizontal displacement of the slope, with the increase of the vertical load, the trend of the horizontal displacement of the slope with different disposal methods is roughly the same. As can be seen from Table 2, both ecological concrete slab and ordinary concrete slab can increase their ultimate bearing capacity and effectively control their horizontal displacement. The maximum horizontal displacement of ecological concrete slope and ordinary concrete slope is reduced by 29.3% and 51.6% respectively compared with unprotected variable slope.In terms of the control ability of slope horizontal displacement, eco-concrete slab is slightly lower than ordinary concrete slab, but with the growth of plant roots, the control ability of eco-concrete slab on slope horizontal displacement will be greatly improved [9, 10].

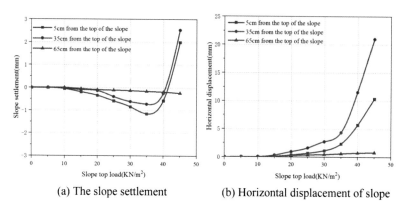

(a) The slope settlement   (b) Horizontal displacement of slope

Fig. 3. The load displacement curve of bare slope test

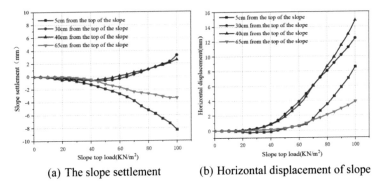

(a) The slope settlement   (b) Horizontal displacement of slope

Fig. 4. The load displacement curve of vegetation regenerated concrete test

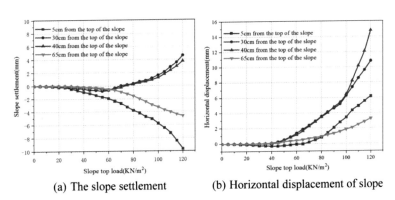

(a) The slope settlement   (b) Horizontal displacement of slope

Fig. 5. The load displacement curve of normal concrete slope test

Table 2. The ultimate load and maximum horizontal displacement of slopes on different slopes

Slope type	Ultimate load (KN/m²)	Maximum horizontal displacement of slope (mm)
Unprotected slope	45	21.01
Eco-concrete slope	100	14.86
Concrete slope	110	10.16

4 Analysis of Influencing Factors on Deformation and Stability of Ecoconcrete

4.1 The Influence of Soil Compaction Degree Change on Slope

Figure 6 shows the distribution of horizontal displacements at different positions of the slope surface under the action of different soil compaction degrees of the slope. The horizontal displacement of the slope surface gradually decreases with the increase of the slope soil compaction. When the slope soil compaction degree increases from 85% to 95%, the maximum horizontal displacement of the slope surface decreases by 11.3% and 19.8%, which is a large reduction; therefore, the degree of compaction is an important factor affecting the horizontal displacement of the slope surface.

Figure 7 shows the maximum settlement of the top of the slope when the slope of the soil with different compaction degrees is damaged. From the analysis of the relationship between the degree of compaction and the maximum settlement of the top of the slope, the maximum settlement of the top of the slope increases with the degree of soil compaction of the slope. And quickly decrease. The compaction of the slope soil increased from 85% to 95%. When the slope is damaged, the maximum settlement of the slope top also decreases by 1.7% and 4.1%. From the analysis of the relationship between the degree of compaction and the ultimate load of the slope, the ultimate load gradually increases with the increase of the degree of compaction; the compaction of the slope soil increases from 85% to 95%, the ultimate load of the slope It also increases by 11.1% and 21.5%. The compaction of the soil is the key influencing factor of the ultimate load of the slope. In the design of vegetation concrete slope protection, the degree of soil compaction should be increased as much as possible. The increase of the degree of compaction is conducive to reducing the deformation of the slope and improving the stability of the slope.

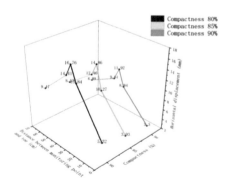

Fig. 6. The influence of compactness on slope horizontal displacement

4.2 Influence of Soil Water Content Change on Slope

Figure 8 shows the horizontal displacement distribution diagram of the slope soil under different water content conditions. The horizontal displacement of the slope gradually

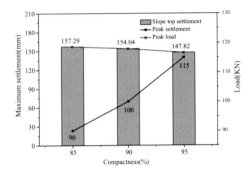

Fig. 7. The influence of compaction degree on the maximum settlement of slope top

increases with the increase of the water content of the slope soil. When the water content of the slope soil increased from 12% to 18%, the maximum horizontal displacement of the slope increased by 4.6% and 8.5%, and the increase in the horizontal displacement of the slope gradually increased. Therefore, the water content of the slope soil is an important factor affecting the horizontal displacement of the slope.

Figure 9 shows the maximum settlement of the slope top under different moisture content conditions. Based on the analysis of the relationship between the water content and the maximum subsidence of the top of the slope, the maximum subsidence of the top of the slope gradually decreases with the increase of the water content. The moisture content of the soil has increased from 12% to 18%, and the maximum settlement of the top of the slope has also been reduced by 8.1% and 1.9%. From the analysis of the relationship between the moisture content and the ultimate load of the slope, the ultimate load of the slope gradually decreases with the increase of the moisture content. The moisture content of the soil increased from 12% to 18%, and the ultimate load of the slope was also reduced by 15% and 11.8%. The moisture content of the soil has a greater impact on the ultimate load of the slope. When vegetation concrete is used for slope protection, special attention needs to be paid to the possibility of slope damage when the soil moisture content changes due to precipitation, changes in underground runoff and other factors.

4.3 Influence of Different Slope Ratios on Slope

Figure 10 shows the horizontal displacement distribution diagram of the slope under different slope ratios. As the slope ratio decreases, the horizontal displacement of the slope gradually decreases. When the slope ratio of the slope is reduced from 1:1 to 1:1.5, the maximum horizontal displacement of the slope is reduced by 7.8%. Therefore, the slope ratio of the slope has a certain influence on the horizontal displacement of the slope.

Figure 11 shows the relationship between the slope ratio and the maximum settlement of the top of the slope. From the analysis of the relationship between the slope ratio and the maximum settlement at the top of the slope, the maximum settlement at the top of the slope gradually increases as the slope ratio decreases; the slope ratio decreases from 1:1 to 1:1.5, the slope top when the slope is damaged The maximum settlement increased

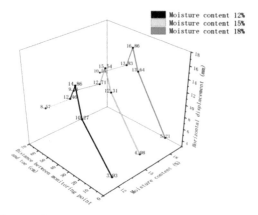

Fig. 8. The influence of water content on slope horizontal displacement of slope top

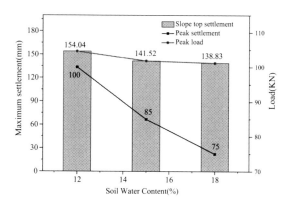

Fig. 9. The influence of water content on the maximum settlement of slope top

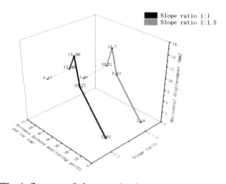

Fig. 10. The influence of slope on horizontal displacement of slope

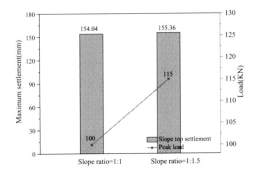

Fig. 11. The influence of slope ratio on the maximum settlement of slope top

by 0.9%, and the slope ratio had little effect on the maximum settlement of the top of the slope when the slope was damaged. From the analysis of the relationship between the slope ratio and the ultimate load of the slope, the ultimate load of the slope gradually increases with the decrease of the slope ratio; the slope ratio is reduced from 1:1 to 1:1.5, the ultimate load of the slope failure This has increased by 15%. In the design of eco-concrete slopes, the slope ratio should be reduced as much as possible, which greatly improves the ultimate load of eco-concrete slopes.

5 Conclusion

Under the action of slope top load, the deformation law of different slopes is basically the same, partial failure of slope occurs, slope protection structure is different, corresponding to the amplitude of each stage is also different.

Compared with the unguarded slope, the ecological concrete slope is significantly improved in ultimate bearing capacity and slope constraint capacity. In terms of ultimate load of slope, the ecological concrete slope and ordinary concrete slope are increased by 2.2 times and 2.4 times respectively. In terms of slope constraint capacity, the horizontal displacement of eco-concrete slope and ordinary concrete slope is reduced by 29.3% and 51.6%, respectively. Compared with the conventional concrete slope, there is still a gap in the protection effect of ecological concrete slope, but it has met the requirements of slope protection.

Slope in the case of ecological concrete protection, the deformation and stability of slope compaction, slope moisture content, slope ratio and other factors, in the design of ecological concrete slope, should consider the above factors on the deformation and stability of the slope.

Eco-concrete can restrain slope deformation and improve slope stability. By changing the above factors, the constraint ability of eco-concrete on slope deformation can be improved and the stability of slope can be improved, thus providing a certain reference for the application of eco-concrete in slope ecological restoration and reinforcement.

References

1. Xu, Y.L., Li, R.W., Tan, X.J., Xiao, P., Kuang, G.M.: Preparation and planting test of vegetation-type porous concrete. New Build. Mater. **36**(2), 5 (2009)
2. Li, O.Y., Wang, C.F., Ji, D.B., Yu, W.: Plant-compatible ecological concrete production and performance test. Environ. Sci. Manag. **023**(009), 149–151 (2007)
3. Xin, J.B.: Research on the preparation and performance of lightweight aggregate planted concrete. Jinan University, Guangzhou
4. Xian, G., Deng, W.W., Long, L.J., Shao, Y.Y.: Experimental study on the permeability of vegetation concrete under freeze-thaw cycles. Agric. Technol. **39**(337(20)), 56–58 (2019)
5. Qi, Y.G.: Research on the water permeability of plant-based recycled brick concrete. Doctoral dissertation, Shandong Agricultural University (2015)
6. Guo, Q.W., Hu, Y.Y., Zheng, B.H., Hu, C.M., Zhang, T.P., Wang, X.: Research on the permeability of planted concrete. Water Res. Hydropower Technol. **37**(9), 4 (2006)
7. Li, X.D., Li, K., Weng, S.F.: Plant growth test, alkali reduction and salt reduction research of porous concrete vegetation brick. New Build. Mater. **44**(1), 4 (2017)
8. Tan, S.Q., Jiao, C.J.: Research on the strength and alkalinity of planted concrete based on orthogonal experiment. Concrete (10), 5 (2020)
9. Wang, F., Sun, C., Ding, X., Kang, T., Nie, X.: Experimental study on the vegetation growing recycled concrete and synergistic effect with plant roots. Materials **12**(11), 1855 (2019)
10. Ji, X.L.: Research on ecological slope stability based on the distribution of vegetation root system. Doctoral dissertation, Nanjing Forestry University (2013)

Compressive Strength Performance of Additives for Cement-Based Grouting Material with Low Water-Binder Ratio by Response Surface Methodology

Jiaxu Guo[1]([⊠]), Shaowei Hu[1], Xuan Zhao[2], Xiu Tao[2], and Ying Nie[2]

[1] School of Civil Engineering, Chongqing University, Chongqing 400030, China
guo_jx95@163.com
[2] CISDI Engineering Co., Ltd., Chongqing 401122, China

Abstract. In order to research the influence and the function mechanism of calcium formate and defoaming agent on the compressive strength of cement-based grouting material with low water-binder ratio at different ages, quadratic polynomial regression models were established by RSM, and the mix proportion was optimized. The function mechanism of additives was analysed by macroscopic mechanical properties and microstructure. The results indicated that the response surface method is scientific in optimizing the mix proportion of cement-based grouting material. The optimal mix proportion was obtained as fallow: the calcium formate was 0.64%, the water-binder ratio was 0.21 and the defoaming agent was 0.26%, with taking 1d, 3d, 28d compressive strength as the optimization objective. Calcium formate is highly significant for the early compressive strength of cement-based grouting materials with low water-binder ratio, while the water-binder ratio and defoaming agent are highly significant for that of the middle and late period. Calcium formate promotes the formation of CSH gel and $Ca(OH)_2$ crystallization in the early period, and the defoaming agent can effectively reduce macropores. The results can provide an optimization method for the mix proportion design of cement-based grouting material and a theoretical reference for its mechanical properties.

Keywords: Cement-based grouting material · Response surface method · Compressive strength · Calcium formate · Defoaming agent · Mix proportion

1 Introduction

Cement-based grouting material is a kind of building material prepared in a professional factory, which has the characteristics of early strength, high strength, self-levelling and micro-expansion after mixing with water in a prescribed proportion at using sites. With the rapid advance of infrastructure construction, cement-based grouting materials have been widely used in the fields of bolt anchorage, reinforcement and reconstruction of concrete, and prefabricated construction projects. At the same time, higher requirements are placed on the performance of cement-based grouting materials.

© The Author(s) 2022
G. Feng (Ed.): ICCE 2021, LNCE 213, pp. 368–379, 2022.
https://doi.org/10.1007/978-981-19-1260-3_34

Cement-based grouting material should adopt low water-binder ratio to obtain excellent later strength, but it will lead to slow early hydration of cement and insufficient early strength [1]. Early strength agent is often added to dry powder system to satisfy the early strength requirements of cement-based grouting materials in practical engineering applications. Calcium formate has been widely used in dry powder system as early strength agent due to its excellent coagulation promotion ability and solubility. Calcium formate solution is weakly acidic, and it improves the hydration rates of C_2S and C_3S. And the ionized Ca^{2+} accelerate the crystallization of AFT owing to ion effect, it promotes the early strength development [2–4]. However, the incorporation of calcium formate will lead to the loss of fluidity, and it is harmful to working performance [5, 6]. As a surfactant, defoaming agent can promote the rupture of bubbles in slurry, reduce the porosity, and effectively improve the strength of cement-based grouting materials [7, 8].

Response surface methodology (RSM) is an analytical method to establish an accurate prediction model with limited experimental data which is obtained from standard tests. RSM uses second-order standard polynomial to fit the response value and different factors, and can draw response surfaces that intuitively reflects the influence of factors on the response value. RSM has been widely used in the process formulation design of grain, oil and food, chemical engineering, and biological engineering [9], but it is rarely used in the related fields of construction and civil engineering [10–12].

There are few reports on the influence of calcium formate and defoaming agent on the mechanical properties of cement-based grouting materials with low water binder ratio. In this paper, calcium formate content, water-binder ratio and defoaming agent content were selected as experimental factors, and the compressive strength of 1d, 3d and 28d were response values. Box-Behnken design (BBD) was used to design the compressive strength test of cement-based grouting materials with low water-binder ratio, and the functional relationships between the response values and the factors were established to analyse the influence of various factors and their interactions on the compressive strength, so as to optimize the mix proportion. And the mechanism of agents was analysed of action combined with macroscopic mechanical microscopic morphology.

2 Experiment

2.1 Materials and Instruments

P · II 52.5R portland cement (PC), II fly ash (FA) and S95 slag powder (SL) were used as cementitious materials, and the chemical composition is shown in Table 1.

Table 1. The chemical composition of cementitious materials

	CaO	SiO_2	Al_2O_3	Fe_2O_3	MgO	SO_3^-	Loss
PC	62.83	21.08	4.49	3.41	2.11	2.04	
FA	38.09	31.15	15.64	1.02	8.67	0.78	1.22
SL	3.71	52.73	28.68	7.29	1.38	0.62	5.24

Concrete admixtures used in the experiment included industrial grade calcium formate (purity > 95%), P803 powder defoaming agent, low alkali u-type expansive agent, QH-100 plastic expansive agent, QS-8020 polycarboxylate superplasticizer. Other materials included water and ISO standard sand.

The compressive strength test adopted NELD-CH2000 electro-hydraulic servo pressure testing machine (Beijing Nierde Intelligent Technology Co., Ltd), structural equation model (SEM) analysis adopted Gemini300 scanning electron microscope (Carl Zeiss AG, German).

2.2 Compressive Strength Test Design

According to preliminary test results, the basic mix proportion is shown in Table 2.

Table 2. The basic mix proportion

S/B	FA	SL	U-type expansive agent	Superplasticizer	Plastic expansive agent
1.0	5%	5%	8%	0.5%	0.05%

BBD method was used in the experimental design and data analysis would be completed. Calcium formate ratio, water-binder ratio and defoaming agent ratio were expressed as x_1, x_2 and x_3, respectively. Each independent variables was set at low level, central point and high level, which were encoded as -1, 0 and $+1$, respectively. The level of each factor and the corresponding coding value are shown in the Table 3.

Table 3. Coding and level of independent variables

Coding and level	Independent variables		
	$x_1(\%)$	x_2	$x_3(\%)$
-1	0	0.17	0.10
0	0.5	0.22	0.20
$+1$	1.0	0.27	0.30

2.3 Specimen Forming and Test Method

The dry powder was weighed according to the proportion during the molding process of the specimen. After mixing evenly, water was added in proportion and stirred at a low speed for 2 min, stopped for 15 s, and stirred at high speed for 2 min. The mortar was poured into the triple mold, demoulded after curing for 24 h, and maintained to the target age. The loading speed of compressive strength test was controlled at 2.4 ± 0.2 k N/s. The blocks with particle size less than 5 mm were taken on the new section in SEM analysis, and the microstructure of hydration products was observed after drying, spraying gold, fixing and vacuuming.

3 Results and Analysis of Compressive Strength Test

3.1 Results and Fitting Models

The test results of 1d, 3d, 28d compressive strength are shown in Table 4. A total of 17 groups were set up, of which 5 groups were at the central point of each factor values, repeated to evaluate the test deviation.

Table 4. Design and results of Box-behnken test

Sample	Independent variable level			Compressive strength/MPa		
	x_1	x_2	x_3	1d	3d	28d
1	−1	−1	0	24.65	52.66	78.33
2	1	−1	0	28.68	55.82	78.56
3	−1	1	0	26.84	49.70	73.13
4	1	1	0	31.62	53.51	73.82
5	−1	0	−1	25.70	52.47	75.80
6	1	0	−1	31.77	53.94	76.47
7	−1	0	1	27.05	55.41	77.23
8	1	0	1	33.07	56.19	78.02
9	0	−1	−1	29.41	54.91	77.34
10	0	1	−1	31.09	51.69	70.90
11	0	−1	1	32.11	55.88	78.78
12	0	1	1	32.20	54.39	77.03
13	0	0	0	31.57	57.86	80.54
14	0	0	0	32.64	57.64	78.93
15	0	0	0	32.34	59.06	80.75
16	0	0	0	32.56	59.17	79.40
17	0	0	0	31.19	58.60	80.98

The second-order standard polynomials fitting of Table 4 test data were carried out by RSM. The simulation models of 1d compressive strength (S_1), 3d compressive strength (S_2) and 28 d compressive strength (S_3) are Eq. (1)–(3).

$$S_1 = -2.77 + 15.46x_1 + 232.46x_2 + 13.84x_3 + 7.5x_1x_2 - 0.25x_1x_3 + 79.50x_2x_3$$
$$\bowtie -11.84x_1^2 - 461.50x_2^2 + 29.62x_3^2$$

$$(1)$$

$$S_2 = +0.38 + 12.08x_1 + 467.41x_2 + 47.14x_3 + 6.50x_1x_2 - 3.45x_1x_3 + 86.50x_2x_3$$
$$\bowtie -10.52x_1^2 - 1165.70x_2^2 - 133.42x_3^2$$

$$(2)$$

$$S_3 = +41.35 + 6.05x_1 + 347.89x_2 + 25.05x_3 + 4.60x_1x_2 + 0.60x_1x_3 + 234.50x_2x_3$$
$$\bowtie -6.59x_1^2 - 1005.50x_2^2 - 159.38x_3^2$$

(3)

Table 5 shows variance analysis of regression models. F-value represents the test index of obviousness and P-value represents the probability, in which the smaller the P-value is, the stronger the significance of the model is and the higher the simulation accuracy is. The P-value of lack of fit reflects the significant degree that the experimental data is not related to the model. If the value is less than 0.05, the item is significant. and when the value is less than 0.05, the item is highly significant. The P-values of S_1, S_2 and S_3 are 0.0004, 0.0005 and 0.0010, respectively, which are not greater than 0.01, indicating that the simulation models are highly significant. As for lack of fit, the P-values of each model are 0.2141, 0.2347 and 0.4019, which are greater than 0.05, indicating that the mismatch term is not significant and the errors of simulation models are small. The fitting equation is highly consistent with the actual.

Table 5. Variance analysis and of regression models

Sourse	df	Mean square			F-value			P-value		
		S_1	S_2	S_3	S_1	S_2	S_3	S_1	S_2	S_3
Model	9	12.30	12.64	12.81	18.77	17.59	14.24	0.0004	0.0005	0.0010
x_1	1	54.60	10.63	0.71	83.34	14.79	0.79	< 0.0001	0.0063	0.4045
x_2	1	5.95	12.45	41.09	9.08	17.33	45.65	0.0196	0.0042	0.0003
x_3	1	5.22	9.81	13.91	7.96	13.66	15.46	0.0257	0.0077	0.0057
x_1x_2	1	0.14	0.11	0.053	0.21	0.15	0.059	0.6572	0.7128	0.8154
x_1x_3	1	6.250E−004	0.12	3.600E−003	9.540E−004	0.17	4.000E−003	0.9762	0.6961	0.9513
x_2x_3	1	0.63	0.75	5.50	0.96	1.04	6.11	0.3587	0.3414	0.0427
x_1^2	1	36.86	29.11	11.41	56.26	40.52	12.68	0.0001	0.0004	0.0092
x_2^2	1	5.60	35.76	26.61	8.56	49.78	29.56	0.0222	0.0002	0.0010
x_3^2	1	0.37	7.50	10.69	0.56	10.43	11.88	0.4771	0.0144	0.0107
Residual	7	0.66	0.72	0.90						
Lack of Fit	3	0.97	1.04	1.02	2.35	2.17	1.25	0.2141	0.2347	0.4019
Pure Error	4	0.42	0.48	0.81						

It can be seen from Table 6 that $P(x_1) > P(x_2) > P(x_3)$ for the 1d compressive strength model S_1. It indicates that the water-binder ratio, the amount of calcium formate and the amount of defoaming agent all have a significant effect on the 1d compressive strength of the cement-based grouting material, and the main factor is calcium formate.

For the 3d compressive strength model S_2, $P(x_2) < P(x_1) < P(x_3) < 0.01$. The influence of water-binder ratio, calcium formate and defoaming agent on the 3d compressive strength of the cement-based grouting material is highly significant. The influence order is $x_2 > x_1 > x_3$, and water-binder ratio is main factor.

The order of influence on the 28d compressive strength model S_3 is $x_2 > x_2 > x_1$, where the P-value of x_1 is greater than 0.05. The influence of water-binder ratio and defoaming agent on 28d compressive strength of grouting material is highly significant, among which water-binder ratio is still the most important factor, but calcium formate will not have a significant impact.

In summary, the incorporation of calcium formate plays a significant role in the early strength development of low water-binder ratio cement-based grouting materials, but contributes less to the later strength. The effects of water-binder ratio and defoaming agent on the compressive strength of cement-based grouting materials with low water-binder ratio at all ages cannot be ignored, especially for the strength contribution in the middle and late stages.

The closeness of R^2 and Adj R^2 can be used to verify the fitting degree of the simulation models, and the smaller the C.V is, the Adeq Precisior is greater than 4, indicating that the reliability and accuracy of the test are higher. It can be seen from Table 6 that the C.V of each model are 1.97%, 3.57% and 1.45%, respectively, and the Adeq Precisior of each model are 7.527, 9.524 and 6.199, respectively, indicating that the simulation models has high reliability and accuracy.

Table 6. Model reliability test analysis

Group	Std. Dev/MPa		R^2	Adj R^2	Pred R^2	Press	C.V/%	Adeq. precisior
S_1	0.81	30.26	0.9602	0.9090	0.5715	49.38	2.67	13.120
S_2	0.85	55.23	0.9577	0.3032	0.5554	52.80	1.53	12.470
S_3	0.95	77.41	0.9482	0.8816	0.5567	53.92	1.23	12.184

3.2 Response Surface Interaction Analysis

The response surface diagrams of 1d, 3d and 28d compressive strength were established as shown in Fig. 1, 2 and 3. The response surface reflects the effect of the interaction of two factors on the compressive strength of each age when the other factor is at the central point.

Figure 1 shows the 1d compressive strength had a process of first increasing and then decreasing with the increase of calcium formate ratio and water-binder ratio from −1 level and the increase of defoaming agent ratio promoted the continuous growth of 1d compressive strength, but the growth was limited under high ratio. Maximum 1d compressive strength was attained at around 0.72% calcium formate, 0.23 water-binder ratio and 0.30% defoaming agent. Table 6 shows that the P-values of the interaction of two factors in the 1d compressive strength model are 0.6572, 0.9762 and 0.3587, respectively, which are greater than 0.05. It indicates that the interaction of any two factors does not contribute significantly to the 1d compressive strength.

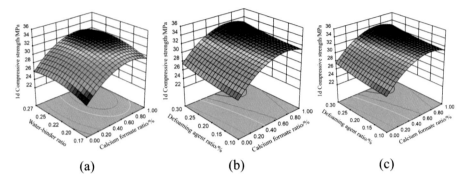

Fig. 1. Effect of two-factor interaction on 1d compressive strength

The compressive strength of 3d and 28d increased first and then decreased significantly with the increase of calcium formate ratio, water-binder ratio and defoaming agent ratio, as is shown in Fig. 2 and Fig. 3. Maximum 3d compressive strength was attained at around 0.60% calcium formate, 0.21 water-binder ratio and 0.24% defoaming agent, while Maximum 28d compressive strength was attained at around 0.54% calcium formate, 0.20 water-binder ratio and 0.23% defoaming agent. The compressive strength decreased slightly with the increase of the dosage, after the defoaming agent ratio reached the optimal solution, but the decrease was limited. Plot of Table 6 reveals the P-values of the interaction of the two factors in the 3d compressive strength model are 0.7128, 0.6961 and 0.3414, respectively, which are greater than 0.05, indicating that the interaction of any two factors has a low contribution to the 3d compressive strength. While, the P-value of the interaction of defoaming agent ratio and water-binder ratio in the 28d compressive strength model is 0.0427, which is less than 0.05, indicating that there is a strong dependence of 28d compressive strength on the interactive pattern between defoaming agent and water-binder ratio.

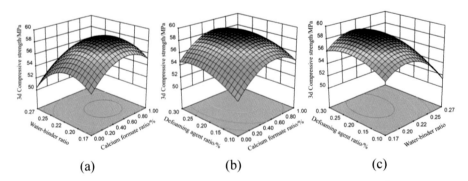

Fig. 2. Effect of two-factor interaction on 3d compressive strength

With the increase of calcium formate ratio, water-binder ratio and defoaming agent ratio, the compressive strength of each age generally had a process of first increasing and then decreasing. The proper calcium formate ratio promotes the hydration of C_3S,

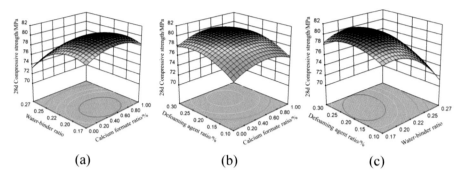

Fig. 3. Effect of two-factor interaction on 28d compressive strength

C_2S and the formation of CSH gel, which is helpful for the early hydration of slurry. However, the high content leads to excessive loss of fluidity, which is not conducive to the formation of dense structure, and it causes the reduction of compressive strength [13]. Higher water consumption contributes to the early hydration reaction of cement particles, but the residual free water inside the slurry evaporates to form pores with the increase of age. The pores will increase with the increase of water-binder ratio, resulting in strength loss. With the increase of defoaming agent ratio, the compressive strength of 3d and 28d first increased and then decreased, and the compressive strength of 1d continued to increase, which was consistent with the early research [14]. The defoaming agent weakens the effect of water reducing agent, especially in the case of high dosage, resulting in the loss of fluidity. It is not conducive to the formation of dense structure, and it induce the decrease of strength.

3.3 Parameter Optimization and Verification

Taking the compressive strength of each age as the optimization object, the mix proportion of cement-based grouting material was optimized, and the optimized mix proportion scheme was obtained as follows: 0.64% calcium formate, 0.21 water-binder ratio and 0.26% defoaming agent. The predicted values and test values of the regression model were compared, and the relative error was used for characterization. The results are shown in Table 7.

Table 7. Comparison of predicted value and test value after parameter optimization

Predicted compressive strength/MPa			Test compressive strength/MPa			Relative error /%		
1d	3d	28d	1d	3d	28d	1d	3d	28d
33.05	58.81	80.45	33.84	59.93	83.68	2.33	1.67	3.86

The experimental results show that the compressive strength of cement-based grouting material prepared by optimized mix ratio is Relatively high, and the relative error between predicted value and test value is small. It is reliable and accurate to optimize the mix proportion of cement-based grouting materials by response surface method.

4 Analysis of the Action Mechanism of Additives

Calcium formate and defoaming agent are additives widely used in cement-based grouting materials. In order to analyse the mechanism of action of the two, five groups of samples were taken to observe the type, morphology and quantity of hydration products of cement-based grouting materials by scanning electron microscope. The typical morphology diagrams are shown in Fig. 4.

Figure 4a shows that a large number of needle-like AFT crystals were formed on the surface of cement, mineral powder and fly ash particles after hydration for 1 day, and the network structure was constructed by alternating AFT, with 1% calcium formate and 0.3% defoaming agent. Increased magnification factor from 10k to 30k, Fig. 4b shows that there was a small amount of CSH gel in the network structure, forming a relatively stable hydration structure, but the overall hydration degree of the slurry was low and the structure was loose. It can be seen from Fig. 4c that when the slurry was hydrated to 3 days, the AFT crystallization was more robust, and the constructed network structure was obviously filled with a large number of flocculent CSH gels. The lamellar $Ca(OH)_2$ crystals were obviously observed on the surface of cement, mineral powder and fly ash particles, and the hydration degree was significantly improved. The overall structure was more stable than that of 1 day, forming a relatively dense hydration structure. When the slurry is hydrated to 28 days, the microstructure is shown in Fig. 4f. The network structure constructed by AFT crystallization was basically filled and compacted by hydration products, and a stable structure formed between cement, mineral powder and fly ash particles. At this time, the hydration reaction has been basically completed and the microstructure was dense.

Comparing Fig. 4c (3d, 0.1% calcium formate) with Fig. 4d (3d, 0% calcium formate), the sample without calcium formate still had a network structure formed by alternating overlap of AFT, but the CSH gel was significantly reduced compared with the calcium formate group. There are no lamellar $Ca(OH)_2$ crystals were observed, and the hydration structure was loose. The above phenomenon shows that the incorporation of calcium formate into cement-based grouting materials helps the slurry to form.

$Ca(OH)_2$ crystallization and CSH gel in the early stage, thereby improving the early compressive strength.

It can be seen from Fig. 4c (3d, 0.3% defoaming agent) and Fig. 4e (3d, 0.1% defoaming agent), that when the ratio of defoaming agent decreased from 0.3% to 0.1%, a large number of flocculent CSH gels are still filled into the network structure constructed by AFT crystallization, and a large number of flake $Ca(OH)_2$ crystals are precipitated as well, and there was no obvious increase in the gap between particles. Therefore, it indicated that the significant influence of defoaming agent is not based on changing the microstructure of hydration products to improve compressive strength. Compared with Fig. 5a (3d, 0.3% defoaming agent) and Fig. 5b (3d, 0.1% defoaming agent), when the dosage of defoaming agent ratio was 0.1%, there were a large number of visible pores on the cross section of the sample. With the increase of the dosage to 0.3%, the pores on the

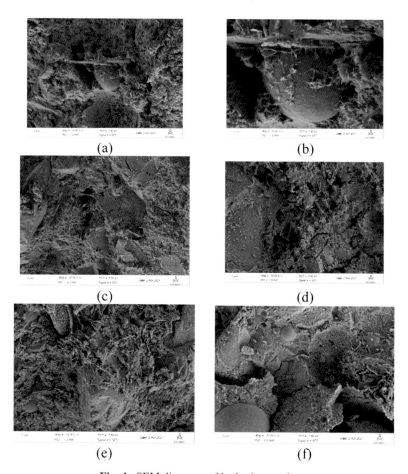

(a)   (b)

(c)   (d)

(e)   (f)

Fig. 4. SEM diagrams of hydration products

(a)   (b)

Fig. 5. Section of test block

cross section decreased significantly and the structure was denser. The above phenomena show that the defoaming agent can improve the compressive strength of cement-based grouting materials, because it can promote the bubble rupture in the slurry and reduce the macro pores, rather than having no significant influence on the microstructure of hydration products.

5 Conclusion

The effects of calcium formate and defoamer on the mechanical properties of cement-based grouting materials were studied through compressive strength test and microstructure analysis. The results show that:

(1) The models established by RSM can accurately predict the results, providing a reliable method for the optimization design of the mix proportion of cement-based grouting materials.
(2) The water-binder ratio and the defoaming ratio have significant effects on the compressive strength of cement-based grouting materials at different ages, but it is more significant in the middle and later stages. While the dosage of calcium formate only has significant effects on the compressive strength of 1d and 3d. The 28d compressive strength was significantly affected by the interaction of water binder ratio and defoaming agent.
(3) Comprehensively considering the compressive strength of each age, the optimal mix proportion of cement-based grouting materials with low water-binder ratio was obtained by response surface method, namely, the calcium formate content was 0.64%, the water-binder ratio was 0.21, and the defoamer content was 0.26%.
(4) Calcium formate is beneficial to the formation of CSH gel and $Ca(OH)_2$ crystallization in the early slurry, which is beneficial to the rapid formation of dense hydration structure. The defoamer mainly acts on reducing macro pores and has no significant effect on the microstructure.

Acknowledgments. This work was financially supported by Technology Innovation and Application Development Special Key Program of Chongqing, China (Grant No. cstc2019jscx-gksbX0013) and Innovation Group Science Foundation of the Natural Science Foundation of Chongqing, China (Grant No. cstc2020jcyj-cxttX0003).

References

1. Wu, F.F., Dai, K.B., Dong, S.K., et al.: Effects of admixtures and hydration products on hydration products and mechanical properties of cement-based materials. Trans. Chin. Soc. Agric. Eng. **32**(4), 119–126 (2016)
2. Heikal, M.: Effect of calcium formate as an accelerator on the physicochemical and mechanical properties of pozzolanic cement pastes. Cement Concr. Res. **34**(6), 1051–1056 (2004)

3. Yu, J.J., Zhang, X.P., Sun, C.Z.: The impact on performance about the compound early strength agent and sand ratio to super high early strength grouting material. J. Shenyang Jianzhu Univ. (Nat. Sci.) **30**(02), 298–304 (2014)
4. Xu, F.T., Chen, R.B., Gu, K.: Application of calcium formate early strength agent in dr powder mortar construction. Wall Innov. Build. Energy-Saving **2**, 56–58 (2008)
5. Yang, X.: Effect of hardening accelerating admixture on the early-age shrinkage and cracking of mortar **3**, 24 (2014)
6. Zhou, C.G.: Experimental research on a super-early strength cement-based grouting material. **5**, 51 (2019)
7. Huang, H., Yuan, Q., Deng, D.H., et al.: Relationship between air content stability and rheological properties of cement paste with chemical admixtures. J. Chin. Ceram. Soc. **47**(11), 1593–1604 (2019)
8. Leng, D., Zhang, X., Shen, Z.L., et al.: Effects of main components on the performance of cementitious grout China. Concr. Cement Prod. **5**, 12–16 (2008)
9. Xu, X.H., Hua, M.Z.: Experimental design, application of design. Expert SPSS **6**, 146 (2010)
10. Lv, G.J., Ji, D.: Mechanical properties of ternary polymer mortar based on response surface methodology. J. Build. Mater. (2021) https://kns.cnki.net/kcms/detail/31.1764.TU.20210309.0910.004.html
11. Liu, B.M., Chu, M., Bai, C.X., et al.: Zeolite synthesized with oil shale ash and optimized by response surface method. J. Chin. Ceram. Soc. **48**(08), 1317–1324 (2020)
12. Liu, Y.H., Chen, T.H., Wang, C., et al.: 4A zeolite derived from weathered washing sand tail mud as a high-efficiency NH^+_4 -N adsorbent. J. Chin. Ceram. Soc. **49**(07), 1429–1438 (2021)
13. Yang, Y.B., Zhang, J.S., Li, L.H., et al.: Preparation and characterization of early strength accelerator. J. Liaoning Shihua Univ. **32**(3), 29–32 (2012)
14. Gao, P.: Research on properties of sleeve grouting material for assembly building. **4**, 36 (2020)

Study on Seismic Performance of Integrated Support of Steel Arch and Anchor in Loose Surrounding Rock of Highway Tunnel

Zhigang Zhang[1], Xiyuan Liu[2(✉)], Chong Zhang[2], Jing Zhang[2], and Dongqiang Xu[2]

[1] Hebei Provincial Expressway Yan Chong Construction Office, Zhangjiakou, China
[2] School of Civil Engineering and Transport, Hebei University of Technology, Tianjin, China
`liuxiyuan@hebut.edu.cn`

Abstract. In order to study the seismic performance of the new support structure with the integration of steel arch and anchor rod under the bulk surrounding rock of highway tunnel, this paper takes Xinglinpu tunnel of Yanchong expressway as the engineering background and conducts the dynamic time-history analysis of the deformation, stress and plastic zone distribution of the tunnel second lining structure under the original support and the new support according to the specification and with the help of FLAC3D finite difference software. It is concluded that the seismic performance of the original support and the new support structure is basically the same. The new support structure reduces the number of anchors, the disturbance to the surrounding rock and the cost, so the new support structure is worth promoting.

Keywords: Highway tunnel · Scattered surrounding rock · Seismic performance · Dynamic response · Numerical simulation

1 Introduction

After the 21st century, the average annual growth rate of road tunnels in China was 20% [1]. With the increase in tunnel mileage, the number of tunnels built in mountain ranges in high-intensity seismic zones had reached a significant percentage. In 2008, 33 of the 56 tunnels near the earthquake area of the Wenchuan earthquake (magnitude 8.0) were damaged. The tunnel structure in the fault fracture zone and lining defect section was the most damaged, and it was difficult to repair after tunnel failure [2]. Therefore, it is necessary to carry out a systematic study of the seismic performance of road tunnels.

Qiangqiang Sun [3] carried out a dynamic non-linear time course numerical simulation of tunnel construction to investigate the effect of the construction-induced initial stress state on the seismic response of the tunnel. Longqi Yan [4] applied a two-dimensional transient dynamic finite element simulation technique to investigate the seismic response of a soil-structure interaction system under the oblique incidence of P- and SV-waves. Wang Mingnian [5] and Cui Guangyao [6] pointed out that the seismic damping of underground structures in high-intensity seismic zones can be done by

© The Author(s) 2022
G. Feng (Ed.): ICCE 2021, LNCE 213, pp. 380–393, 2022.
https://doi.org/10.1007/978-981-19-1260-3_35

changing their structural properties or by installing a damping layer. By conducting a large shaking table model test study on the cavern section of the Galongla tunnel, Jiang Shuping [7] concluded that attention should be paid to the prevention and control of surrounding rock instability during the earthquake process to reduce the damage produced by the surrounding rock to the tunnel. Wang Qjuyi [8] proposed a comprehensive seismic measure of adding a damping layer between the initial support and the second lining and appropriately increasing the reinforcement of the second lining. Liang Bo [9] analyzed the dynamic response of road tunnel lining in mega-section and pointed out that the second lining could play a major role as a safety reserve during earthquakes. By analyzing different grouting schemes for highway tunnels, Based on the concrete plastic damage model, He Zegan [10] established a three-dimensional nonlinear finite element model of the tunnel-surrounding rock system and pointed out that the arch shoulder and arch waist of the lining structure are the weak parts of its seismic resistance. By performing dynamic time-history analysis of road tunnels, Liu Liyu [11] concluded that the acceleration response of the tunnel structure decreases when located on soft ground and poor surrounding rock, but larger stresses are generated.

The seismic performance of tunnels based on the new practical patent technology "Integrated support structure of steel arch and prestressed anchor rod for highway tunnel" invented by Xu Dongqiang [12] has not been studied in the above-mentioned existing tunnel seismic studies. This new support structure optimizes the arrangement of anchor rods (pipes), so that the steel arch, anchor rods (pipes), shotcrete and surrounding rock form an integrated structure, thus greatly improving the stability of the surrounding rock and support structure.

Therefore, this paper presents a dynamic time-history analysis of the deformation, stress and plastic zone distribution of the tunnel second lining structure under the original support and the new support.

2 Project Overview

The Xinglinpu tunnel is a highway separated long tunnel, located in the north of Xinglinbao, Huailai County, Zhangjiakou, which is one of the more difficult tunnels to construct with poor surrounding rock conditions in Yanchong expressway. The left and right sides of the Hebei section of the tunnel are both 1520 m. The average proportion of grade V surrounding rock is 44.5%, and the average proportion of grade IV surrounding rock is 43.8%. The broken bulk surrounding rock occupies 90% of the total length of the tunnel. The maximum burial depth of the tunnel is 480 m. The area to which the Xinglinpu tunnel belongs is a temperate subarid zone in the continental monsoon climate. The groundwater is localized by bedrock fracture water and dominated by pore diving. It is located in the middle mountainous area of igneous rocks. The stratigraphic rocks are mainly andesite and coarse andesite with developed joints and fissures, and quartz sandstone is locally present.

According to the *Seismic ground motion parameters zonation map of China* (GB18306–2015) [13] issued by the National Seismological Bureau, the area is a Class II site, with a 50-year exceedance probability of 10%, a seismic intensity of 8 degrees, a design basic ground vibration peak acceleration of 0.20 g and a characteristic period of 0.40 s for the ground vibration response spectrum.

3 Simulation Support Scheme

The original support structure scheme (hereinafter referred to as the original scheme) under grade V surrounding rocks is 28 hollow grouting system anchors per bay of steel arch, 2 hollow grouting locking foot anchors on each left and right arch waist, and grouting small conduits within 120° of the arch roof, with a grouting reinforcement radius of 1 m, as shown in Fig. 1(a).

The new support structure scheme for the tunnel under grade V surrounding rock is 2 rows of hollow grouting locking foot anchors per steel arch at the left and right arch shoulders of the tunnel, 1 row of hollow grouting foot locking anchors at the left and right arch waist and arch foot, two of them in each row. The grouting small conduit is laid at 120° at the top of the arch, and the radius of grouting reinforcement is 1m. Due to the poor condition of the surrounding rock, reinforced support is required at the top of the arch crown, so two different reinforced support schemes are used. As shown in Fig. 1(b), the new support structure scheme 1 (hereinafter referred to as new scheme 1) has 8 additional anchors of the hollow grouting system within 60° of the arch crown. As shown in Fig. 1(c), the new support structure scheme 2 (hereinafter referred to as new scheme 2) has 14 additional anchors of the hollow grouting system within 120 of the arch crown.

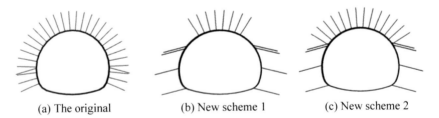

(a) The original   (b) New scheme 1   (c) New scheme 2

Fig. 1. Support structure scheme

4 Establishment of Tunnel Model and Selection of Parameters

4.1 Establishment of Calculation Model

The simulations were carried out using FLAC3D for the dynamic time-history analysis of the tunnel. The model boundary is 50 m from the center point of the tunnel section to the bottom and left and right boundaries respectively, and the distance to the top surface is taken to the surface, with a tunnel depth of 200 m. The inner contour of the tunnel calculation model section adopts the form of three center circles. The vertical effective height of the tunnel structure is 9.85 m, and the structural width is 12.1 m. As the longitudinal structure of the tunnel is continuous and regular, the cross-sectional configuration remains unchanged and the tunnel longitudinal direction is taken to be 6 m. Only the seismic checking calculation in the cross-sectional direction of the tunnel was performed.

After the tunnel excavation, initial support and second lining have been completed, seismic waves were applied to the tunnel structure as a whole to assess the seismic performance of the tunnel support structure by analyzing the deformation, stress and plastic zone distribution of the tunnel second lining structure.

4.2 Unit Type and Parameter Setting

Taking into account the preliminary geological survey report, construction conditions, advanced geological forecasts and the graded determination parameters of the surrounding rocks on site, and referring to the physical and mechanical parameters of the surrounding rocks at all levels specified in the specification [14], the specific parameters selected are shown in Table 1. In the actual project, the test section of the tunnel has about 24 m thick gravel soil layer at the surface, and below the gravel soil layer are all grade V surrounding rocks.

According to the relevant literature [15], the elastic modulus of the surrounding rock in the anchor grouting area increases by 50%, the Poisson's ratio decreases by 9%, the cohesion increases by 65% and the angle of internal friction increases by 3%. The modulus of elasticity of the surrounding rock in the small conduit reinforcement circle of grade V surrounding rock is increased by 2.6 times, Poisson's ratio is unchanged, internal friction angle is increased by 1.2 times, cohesion is increased by 2 times and density is increased by 10%. The surrounding rock is simulated in solid units, using an elastoplastic principal structure model and satisfying the Mohr-Coulomb yielding criterion.

Table 1. Enclosure rock and structural parameters

Material	Modulus of elasticity (GPa)	Poisson's ratio	Cohesive force (MPa)	Angle of internal friction (°)
Gravelly soil	0.80	0.48	0.15	18
Surrounding rock	1.15	0.37	0.19	26
Anchor grouting area	1.73	0.34	0.31	27
Small conduit reinforced ring surround	3.00	0.37	0.38	31
Initial lining	28.6	0.20	–	–
Second lining	33.5	0.20	7.20	60

The initial support for all three support schemes consists of reinforcement mesh, I18 I-beam steel arches (75 cm longitudinal spacing) and 24 cm thick C25 shotcrete. The second lining is C35 molded reinforced concrete with a thickness of 45 cm. Both system anchors and locking foot anchors are hollow grouted anchors.

The initial support for all three support schemes was simulated using shell cells. The second lining was simulated using solid units. Overrunning small conduits within 120° of the arch crown were simulated with beam units. The anchors were simulated using cable units. The modulus of elasticity was calculated by equating the steel arch, reinforcement mesh and C25 shotcrete layer in the primary lining as a single support as shown in Eq. (1):

$$E = \frac{s_c \cdot E_c + s_g \cdot E_g}{s_c + s_g} \tag{1}$$

Where E is the converted modulus of elasticity of concrete (GPa). E_c is the modulus of elasticity of the original concrete (GPa). E_g is the modulus of elasticity of the steel (GPa). s_g is the cross-sectional area of the steel (m^2). s_c is the concrete cross-sectional area (m^2).

The code [16] states that the dynamic modulus of elasticity of a tunnel structure is 30% higher than the static modulus of elasticity, so the modulus of elasticity in Table 1 will be increased by 30% for the dynamic calculations.

4.3 Setting of Boundary Conditions and Mechanical Damping

The surrounding rock at the bottom of the tunnel is a rigid foundation with the bottom of the model horizontal and all sides straight. The dynamic boundary conditions are selected as free-field boundaries, which are imposed via the Apply ff command stream.

The form of damping chosen for this simulation is local damping, with a local damping factor of 0.05. The local damping calculations can reach convergence values as quickly as possible, the calculation time is reduced and better results can be obtained without having to determine the frequency.

4.4 Selection of Seismic Waves

In this paper, El-Centro wave, Chinese Taiwan Chi-Chi (ChiChi) wave and Japanese Hanshin wave were selected for computational analysis. The most significant parts of peak acceleration and amplitude frequency were in the selected periods.

The seismic importance factor of road tunnels was selected concerning domestic and international standards for shield, open cut and immersed tunnels. Therefore, the seismic importance factor is taken as 1.0 when the design ground shaking is a 50-year exceedance probability of 10% and the recurrence period is 475 years. In this paper, the design basic ground vibration peak acceleration of the tunnel is 0.20 g. The seismic category of the highway tunnel is B. Therefore, for E1 earthquake, the seismic importance factor is 0.43, the recurrence period is 75 years, and the peak acceleration is 0.086 g. For E2 earthquake, the seismic importance factor is 1.3, the recurrence period is 1000 years, and the peak acceleration is 0.26 g. The seismic wave adjustment equation is shown in Eq. (2):

$$a'(t) = \frac{a'_{max}}{a_{max}} a(t) \tag{2}$$

Where $a'(t)$ is the adjusted seismic acceleration time course of El-Centro wave. a'_{max} is the adjusted peak seismic acceleration of El-Centro wave. $\alpha(t)$ is the seismic acceleration time intervals recorded by El-Centro wave. α_{max} is the peak seismic acceleration recorded by El-Centro wave.

The seismic wave acceleration time course is adjusted and baseline corrected according to Eq. (2). The adjusted input seismic waves are shown in Fig. 2.

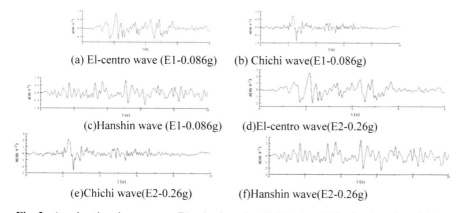

(a) El-centro wave (E1-0.086g)   (b) Chichi wave(E1-0.086g)

(c)Hanshin wave (E1-0.086g)   (d)El-centro wave(E2-0.26g)

(e)Chichi wave(E2-0.26g)   (f)Hanshin wave(E2-0.26g)

Fig. 2. Acceleration time curve at E1 seismic action (0.086 g) and E2 seismic action (0.26 g)

5   Analysis of Numerical Simulation Results

The code proposes a "multi-level defense, two-stage design". The tunnel structure should meet performance requirement 1 under E1 seismic action, i.e. be in an elastic state and structurally intact before and after the earthquake. The tunnel structure should meet performance requirement 2 under E2 seismic action, that is, in an elastic to elastoplastic transition state, the structure can be slightly damaged locally, after reinforcement treatment to ensure the safety of the tunnel.

Strength testing of the tunnel structure for E1 seismic action is required to meet performance requirement 1. The tunnel structure is subjected to deformation and strength tests for E2 seismic action and its seismic performance meets the performance requirement 2.

5.1   Deformation Analysis of the Tunnel Second Lining Structure

The maximum convergence values of the tunnel second lining structure for the three support schemes for E2 seismic action are shown in Table 2.

Table 2. Maximum convergence values for each part of the tunnel under E2 seismic action (mm)

Seismic wave	Support scheme	Convergence of the arch shoulder	Convergence of the arch waist	Convergence of the arch foot
EL wave	The original	18.795	8.995	2.097
EL wave	New scheme 1	19.722	9.612	2.024
EL wave	New scheme 2	19.749	9.606	2.023
Hanshin wave	The original	39.917	19.603	4.542
Hanshin wave	New scheme 1	42.401	20.994	4.312
Hanshin wave	New scheme 2	42.279	20.980	4.306
CC wave	The original	21.408	9.584	2.155
CC wave	New scheme 1	22.666	10.403	2.076
CC wave	New scheme 2	22.650	10.412	2.071

The specification requires that the maximum convergence value of the second lining structure in a drill and blast tunnel during E2 seismic action is 5‰ of the tunnel span. In this project, the tunnel span is 12.10 m and the maximum convergence value is calculated to be 60.5 mm.

As can be seen from Table 2, the maximum convergence values of the second lining structure for the original, new scheme 1, and new scheme 2 are 3.30‰, 3.50‰, and 3.49‰ of the span respectively for the E2 seismic action. The difference between the maximum horizontal convergence values of the three schemes is small and the difference in deformation is not significant. The maximum convergence values for each support scheme are less than 60.5 mm for the shoulder, waist, and foot of the arch, and the tunnel structure is in a safe condition. Therefore, the convergence deformation of the tunnel second lining structure for all three support schemes in grade V rock meets the code's requirements.

5.2   Stress Analysis of Tunnel Second Lining Structure

Under E1 and E2 seismic action, the minimum peak principal stresses at each monitoring point of the tunnel second lining structure for the three support schemes are shown in Tables 3 and 4, and the maximum peak principal stresses are shown in Tables 5 and 6.

As indicated in Tables 3 and 4, the minimum peak principal stresses at each monitoring point of the second lining structure in the three support schemes are compressive, and they are all less than the standard value of dynamic compressive strength of C35 concrete of 28.08 MPa. Therefore, under E1 and E2 seismic action, the tunnel second lining structure did not produce compression damage.

As indicated in Tables 5 and 6, under E1 and E2 seismic action, the maximum value in the maximum principal stress peak at each monitoring point of the tunnel second lining structure is tensile stress, and all of them are less than the standard value dynamic tensile strength of C35 concrete of 2.64 MPa. Under E2 seismic action, the maximum peak principal stress at the arch shoulder and foot of the tunnel second lining structure is larger compared to other parts.

Since the tunnel structure is in a state of elastic to elastoplastic transition under E2 seismic action, the second lining principal structure model is the Mohr-Coulomb principal structure ideal elastic-plastic model. However, under the ideal elastic-plastic model, the maximum tensile stress value will not exceed the standard value of dynamic tensile strength of C35 concrete of 2.64 MPa. Therefore, whether the second lining structure is damaged in tension cannot be discerned only by whether the peak tensile stress exceeds 2.64 MPa. At this point, the distribution of the post-earthquake plastic zone of the second lining structure should be further analyzed to determine the part of the second lining structure that has been damaged.

Table 3. Peak value of minimum principal stress at monitoring point under E1 earthquake (MPa)

Seismic wave	Support scheme	Arch crown	Right arch shoulder	Right arch waist	Right arch foot	Arch base	Left arch foot	Left arch waist	Left arch shoulder
EL wave	The original	−2.059	−3.636	−5.749	−6.684	−0.093	−6.273	−5.880	−3.502
EL wave	New scheme 1	−2.097	−3.810	−6.744	−6.789	−0.100	−6.349	−6.557	−3.765
EL wave	New scheme 2	−2.103	−3.788	−6.788	−6.830	−0.086	−6.401	−6.703	−3.712
Hanshin wave	The original	−2.492	−6.276	−5.753	−9.291	−0.884	−8.905	−5.896	−6.073
Hanshin wave	New scheme 1	−2.503	−6.317	−6.798	−9.394	−0.747	−9.091	−6.671	−6.224
Hanshin wave	New scheme 2	−2.503	−6.271	−6.851	−9.451	−0.759	−9.138	−6.832	−6.145
CC wave	The original	−2.045	−3.541	−5.721	−6.818	−0.094	−6.218	−5.887	−3.806
CC wave	New scheme 1	−2.081	−3.716	−6.725	−6.932	−0.124	−6.326	−6.566	−4.056
CC wave	New scheme 2	−2.087	−3.693	−6.773	−6.974	−0.111	−6.376	−6.711	−4.002

Table 4. Peak value of minimum principal stress at monitoring point under E2 earthquake (MPa)

Seismic wave	Support scheme	Arch crown	Right arch shoulder	Right arch waist	Right arch foot	Arch base	Left arch foot	Left arch waist	Left arch shoulder
EL wave	The original	−2.956	−8.970	−5.809	−11.865	−2.026	−11.401	−6.051	−9.560
EL wave	New scheme 1	−2.887	−8.814	−7.009	−12.155	−1.789	−11.568	−6.881	−9.403
EL wave	New scheme 2	−2.891	−8.757	−7.064	−12.225	−1.808	−11.618	−7.052	−9.332
Hanshin wave	The original	−5.277	−20.041	−8.184	−20.435	−5.846	−20.855	−8.081	−20.311
Hanshin wave	New scheme 1	−5.942	−18.793	−8.439	−20.959	−5.656	−21.371	−8.216	−18.615
Hanshin wave	New scheme 2	−5.938	−18.841	−8.439	−21.003	−5.668	−21.407	−8.329	−18.624
CC wave	The original	−3.050	−10.182	−5.737	−11.338	−2.214	−11.988	−6.249	−8.309
CC wave	New scheme 1	−2.936	−9.907	−6.767	−11.527	−2.092	−12.378	−7.146	−8.290
CC wave	New scheme 2	−2.942	−9.847	−6.811	−11.597	−2.108	−12.437	−7.315	−8.221

Table 5. Peak value of maximum principal stress at monitoring point under E1 earthquake (MPa)

Seismic wave	Support scheme	Arch crown	Right arch shoulder	Right arch waist	Right arch foot	Arch base	Left arch foot	Left arch waist	Left arch shoulder
EL wave	The original	0.041	0.253	−0.280	−0.893	0.872	−0.916	−0.278	0.399
EL wave	New scheme 1	0.051	0.090	−0.390	−1.057	0.826	−0.930	−0.364	0.102
EL wave	New scheme 2	0.041	0.050	−0.380	−1.079	0.825	−0.972	−0.363	0.103
Hanshin wave	The original	0.164	1.889	−0.280	0.841	0.873	0.893	−0.279	1.878
Hanshin wave	New scheme 1	0.185	1.607	−0.400	0.690	0.817	0.903	−0.373	1.562
Hanshin wave	New scheme 2	0.176	1.626	−0.390	0.602	0.820	0.888	−0.371	1.612
CC wave	The original	0.054	0.537	-0.278	−0.906	0.860	−0.914	−0.278	0.320
CC wave	New scheme 1	0.066	0.272	−0.388	−1.064	0.813	−0.957	−0.368	0.078
CC wave	New scheme 2	0.056	0.255	−0.378	−1.086	0.812	-0.997	−0.366	0.080

Table 6. Peak value of minimum principal stress at monitoring point under E2 earthquake (MPa)

Seismic wave	Support scheme	Arch crown	Right arch shoulder	Right arch waist	Right arch foot	Arch base	Left arch foot	Left arch waist	Left arch shoulder
EL wave	The original	0.283	2.164	−0.282	1.508	0.853	1.483	−0.305	2.173
EL wave	New scheme 1	0.323	1.994	−0.389	1.257	0.798	1.315	−0.376	1.959
EL wave	New scheme 2	0.312	2.015	−0.378	1.189	0.802	1.314	−0.374	1.994
Hanshin wave	The original	0.751	2.482	−0.476	1.636	0.891	1.905	−0.471	2.486
Hanshin wave	New scheme 1	0.564	2.465	−0.515	1.368	0.838	1.546	−0.441	2.444
Hanshin wave	New scheme 2	0.541	2.458	−0.504	1.339	0.841	1.567	−0.449	2.437
CC wave	The original	0.336	2.276	−0.281	1.464	0.852	1.601	−0.323	2.145
CC wave	New scheme 1	0.380	2.140	−0.384	1.229	0.797	1.379	−0.378	1.940
CC wave	New scheme 2	0.363	2.151	−0.374	1.157	0.801	1.379	−0.376	1.979

5.3   Plastic Zone Analysis of Tunnel Second Lining Structure

Under E1 seismic action (El-Centro wave, Hanshin wave and CC wave), the post-earthquake plastic zone distribution of tunnel second lining structure of three support schemes is shown in Fig. 3 and Fig. 4.

Under the action of El-Centro wave or Hanshin wave, the post-earthquake plastic zone distribution of the second lining structure of each support scheme is the same, as shown in Fig. 3(a) and Fig. 3(b) respectively.

Figure 3 and Fig. 4 show that the tensile failure area of the tunnel second lining structure of the three support schemes is at the arch bottom. The tensile failure volume is small, and the volume of plastic zone per linear meter is 1.096 m^3–1.436 m^3. Therefore, the tunnel is in a safe and stable state as a whole.

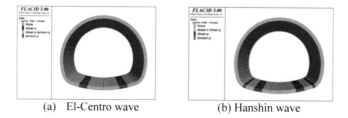

(a)   El-Centro wave   (b) Hanshin wave

Fig. 3. Plastic zone of tunnel second lining structure during E1 seismic action

(a) The original   (b) New scheme 1   (c) New scheme 2

Fig. 4. Plastic zone of tunnel second lining structure during E1 seismic action (CC wave)

Under E2 seismic action (El-Centro wave, Hanshin wave and CC wave), the post-earthquake plastic zone distribution of tunnel second lining structure of three support schemes is shown in Fig. 5, Fig. 6 and Fig. 7.

(a)The original   (b)New scheme 1   (c)New scheme 2

Fig. 5. Plastic zone of tunnel second lining structure during E2 seismic action (El-Centro wave)

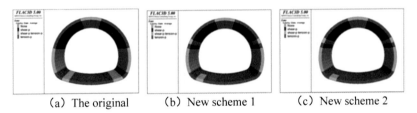

(a) The original   (b) New scheme 1   (c) New scheme 2

Fig. 6. Plastic zone of tunnel second lining structure during E2 seismic action (Hanshin wave)

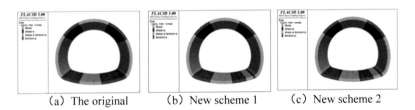

(a) The original   (b) New scheme 1   (c) New scheme 2

Fig. 7. Plastic zone of tunnel second lining structure during E2 seismic action (CC wave)

Figure 5, Fig. 6 and Fig. 7 show that under the action of El-Centro wave and CC wave, the failure area of tunnel second lining structure of the three support schemes is conjugate 45° with the horizontal direction, and the distribution area of plastic zone is mainly concentrated in the spandrel, arch foot and arch bottom, all of which are mainly tensile failure.

Under the action of Hanshin wave, the second lining structure of the tunnel with three support schemes has tensile failure at the arch shoulder, arch foot and arch bottom. Through analysis, it is considered that the natural vibration period of the tunnel structure may be very close to the predominant period of seismic wave, so the seismic response of Hanshin wave is amplified. At the same time, the tunnel second lining structure of the three support schemes produces a small range of shear failure, but the difference of shear failure volume is small. In the original scheme, new New scheme 1 and new scheme 2, the volume of shear failure per linear meter of tunnel second lining structure is 1.266 m³, 0.780 m³ and 0.820 m³ respectively.

Under E2 seismic action (El Centro wave, Hanshin wave and CC wave), the tensile damage volume statistics per linear metre in the plastic zone for the three support schemes after the earthquake are shown in Fig. 8.

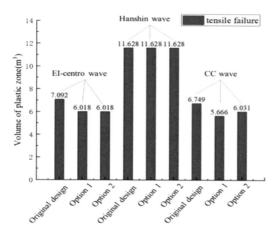

Fig. 8. Volume statistics of the plastic zone of the tunnel second lining structure

As shown in Fig. 8, the amount of tensile damage of the tunnel second lining structure is the same for the three support schemes when Hanshin wave is acting. However, the volume of tensile damage in the plastic zone of the tunnel second lining structure for the new New scheme 1 and 2 is reduced during the action of other seismic waves. The maximum reduction in the tensile damage volume of the tunnel second lining structure for the new New scheme 1 is 16.04% compared to the original scenario for the CC wave action, and the maximum reduction in the tensile damage volume of the tunnel second lining structure for the new scheme 2 is 15.14% compared to the original scenario for the El-Centro wave action. Overall the tensile damage areas produced by the three scenarios are not very different.

Comparing the deformation, stress and plastic zone distribution of the second lining of the three support schemes, it can be seen that the seismic performance of all three is basically the same. Comparing the support measures, it can be seen that the new support structure reduces the number of anchors used compared to the original support structure. This reduces disturbance to the surrounding rock and has obvious advantages in terms of optimising construction costs, making the new support structure worth promoting.

6 Conclusions

Through the numerical simulation of Xinglinpu tunnel of Yanchong expressway by FLAC3D, this paper analyzed the seismic dynamic response of the tunnel under grade V surrounding rock when the original support structure scheme and the new support structure scheme 1 and 2 were adopted respectively. The following conclusions are drawn:

(1) Under E1 seismic action, the compressive stress of tunnel second lining structure in the three support schemes is less than the standard value of dynamic compressive strength of C35 concrete, and the tensile stress is less than the standard value of dynamic tensile strength of C35 concrete. There is less plastic damage at the bottom of the plastic zone of the second lining structure after the earthquake. The tunnel structure meets the requirements of "no damage in small earthquake" and is safe and stable as a whole.

(2) Under E2 seismic action, the maximum convergence value of tunnel second lining structure in the three support schemes is less than the maximum convergence value (5‰ of tunnel span) required in the code. The plastic failure area of the second lining structure after the earthquake is conjugate 45° with the horizontal direction, which is mainly concentrated at the arch shoulder and arch foot. The plastic zone of the whole second lining structure is not penetrated, which is local failure and meets the "repairable level under moderate earthquake action".

(3) Compared with the seismic performance of the original support structure scheme and the new support structure scheme, the seismic performance of the two schemes is basically the same. The new support structure reduces the number of anchors, so it can reduce the disturbance to the surrounding rock. Because of its obvious advantages in construction cost optimization, the new support structure is worth popularizing.

Acknowledgement. This paper is obtained, based on the research project "Research Projects of the Hebei Provincial Department of Transport (YC-201906)".

References

1. Jian, M.A., et al.: Review on China's tunnel engineering research. China J. Highway Transp. **5**, 1–65 (2015)
2. Qian, Q.H., He, C., Yan, Q.X.: Investigation and study of seismic damage of Wenchuan earthquake. Chin. Soc. Rock Mech. Eng., 633–643 (2009)

3. Sun, Q.Q., Dias, D.: Seismic behavior of circular tunnels Influence of the initial stress state. Soil Dyn. Earthq. Eng. **126**, 1–17 (2019)
4. Yan, L.Q., Haider, A., Li, P., et al.: A numerical study on the transverse seismic response of lined circular tunnels under obliquely incident asynchronous P and SV waves. Tunn. Undergr. Space Technol. **97**, 103232 (2020)
5. Wang, M.N., Cui, G.Y.: Establishment of tunnel damping model and research on damping effect with model test in highly seismic area. Rock Soil Mech. **06**, 1884–1890 (2010)
6. Cui, G.Y., Wang, L.B., Wang, M.N., et al.: Model experimental study of rigid-flexible seismic mitigation measures in soft rock cavern sections of tunnels in strong earthquake zones. J. Vibr. Eng. **01**, 29–36 (2019)
7. Jiang, S.P., Wen, D.L., Zhang, S.B.: Large-scale shaking table test for seismic response in portal section of galongla tunnel. Chin. J. Rock Mech. Eng. **04**, 649–656 (2011)
8. Wang, Q.Y., Yang, K., Mao, J.L., et al.: Research on comprehensive seismic measures for secondary lining structures of highway tunnels in nine-degree seismic zones. Mod. Tunn. Technol. **05**, 42–49 (2019)
9. Liang, B., Zhao, F.B., Ren, Z.D., et al.: Analysis on seismic response of a super-large section tunnel in highway. Chin. J. Undergr. Space Eng. **01**, 243–248 (2020)
10. He, Z.G., Wang, X.W., Zuo, Z.B.: Numerical analysis of seismic response of lining structure for portal section in mountain tunnel. Sci. Technol. Eng. **17**, 7018–7024 (2020)
11. Liu, L.Y., Chen, Z.Y., Yuan, Y.: Impact of rock class on seismic responses of mountain tunnels under severe earthquakes. Chin. J. Undergr. Space Eng. **S1**, 1314–1318 (2011)
12. Xu, D.Q., Li, Y.Q., Zhang, N., et al:. Integrated support structure and construction process of highway tunnel primary lining steel arch with prestressed anchor CN206220978U (2017)
13. China Earthquake Administration, Seismic ground motion parameters zonation map of China (GB 18306–2015) Standards Press of China (2015)
14. Chongqing Communication Technology Research and Design Institute, Specifications for design of highway tunnels (JTG 3370.1–2018) China Communications Press (2018)
15. Zhang, L.Y., Zhang, Q.Y., Li, S.C., et al.: Analysis of impact of surrounding rock post-grouting for large oil cavern on its water seal ability based on fluid-solid coupling. Rock Soil Mech. **S2**, 474–480 (2014)
16. Chongqing Communication Technology Research and Design Institute, Specifications for Seismic Design of Highway Tunnels (JTG 2232–2019) China Communications Press (2019)

Research on Crack Propagation and Critical Water Pressure of Basalt Under Hydraulic Coupling

Daming Tang[1], Hai Quan[1], Xinkai Yu[1], Qizhong Xu[2,3], and Zeqi Zhu[3(✉)]

[1] Chengdu Survey, Design and Research Institute Co., Ltd., China Power Construction Group, Chengdu, Sichuan, China
[2] College of Architecture and Civil Engineering, Shenyang University of Technology, Shenyang, China
[3] State Key Laboratory of Geomechanics and Geotechnical Engineering, Institute of Rock and Soil Mechanics, Chinese Academy of Science, Wuhan, China
zqzhu@whrsm.ac.cn

Abstract. Hydraulic coupling triaxial test and acoustic emission test were carried out for Xiluodu basalt. The test results show that the peak strength of basalt increases with the increase of confining pressure, showing a typical hard brittle behavior. When the confining pressure remains unchanged, the peak strength decreases gradually with the increase of initial water pressure, and the hard brittleness decreases; Increasing the initial water pressure can promote the propagation of rock cracks, and the cumulative count of acoustic emission increases significantly with the increase of water pressure. Under the combined action of water pressure in the hole and external stress, the tensile failure occurs first and mainly in the internal cracks of basalt. Further, the functional relationship among crack initiation stress, confining pressure and water pressure is obtained through theoretical derivation and three-dimensional spatial data fitting. On this basis, the empirical relationship between critical water pressure, confining pressure and water pressure is established. The correlation is consistent with the test results, which has a good reference value for the study of critical water pressure of basalt crack propagation under hydraulic coupling.

Keywords: Hydraulic coupling · Acoustic emission · Crack propagation · Crack initiation stress · Critical water pressure

1 Introduction

In the hydropower projects under construction and built in China, the Underground Powerhouse Caverns, headrace tunnels or dam foundation slopes often face the combined action of high ground stress and strong osmotic pressure. Therefore, the research on the mechanical properties of rock mass or rock under hydraulic coupling has always been one of the hot issues in the field of geotechnical engineering. For example, Zhu Zhende et al. [1], Yu Jin et al. [2], Xu Jiang et al. [3] carried out stress seepage coupling indoor

© The Author(s) 2022
G. Feng (Ed.): ICCE 2021, LNCE 213, pp. 394–402, 2022.
https://doi.org/10.1007/978-981-19-1260-3_36

triaxial tests on limestone, sandstone and granite, studied the relationship between rock permeability, stress, strain and pore water pressure, and discussed the influence of pore water pressure on rock strength characteristics Deformation law and damage evolution. In addition, Li Junping [4], Zhao Xingdong [5] analyzed the acoustic emission characteristics in the whole process of rock fracture under hydraulic coupling by collecting the acoustic emission signals in the test process.

The above studies show that the pore water pressure in the micro crack in the rock has a great influence on the fracture process under the hydraulic coupling effect. When the pore water pressure reaches the critical value, it will lead to the initiation, propagation and penetration of internal cracks in rock (also known as hydraulic fracturing), which is also an important factor inducing a series of engineering accidents such as tunnel water gushing, rock slope landslide, reservoir earthquake and so on. K. S. min [6] and others successfully simulated the crack initiation and propagation under hydraulic fracturing by using vmib model, establishing equilibrium equation and introducing Weber Distribution Considering Rock heterogeneity; Tang Liansheng [7] derived a new fracture strength criterion of rock mass considering closed and open cracks under dynamic, hydrostatic pressure and hydrochemical damage based on the strength factor of crack tip; Bian Kang [8] combined the stress analysis of hydraulic tunnel with the crack propagation of surrounding rock from the macro and micro perspectives by using the propagation criterion of fracture mechanics, and deduced the calculation formula of critical water pressure for crack propagation of surrounding rock of hydraulic tunnel under tension shear and compression shear conditions; Sheng Jinchang [9] and others further studied the pressure shear crack propagation mechanism of the surrounding rock crack of the hydraulic tunnel on the basis of the mode I tensile failure of the crack; In addition, Cui Shaoying [10] also gave two methods to calculate the critical pore water pressure of hydraulic fracturing, and determined the critical pore water pressure in case of instability failure of surrounding rock. However, most of the above studies on critical pore water pressure are based on the analytical solution of fracture mechanics theory, which has insufficient adaptability to the popularization and application of engineering. In addition, more in-depth research is needed on the relationship between critical pore water pressure and external hydraulic conditions.

Critical pore water pressure is one of the important breakthroughs in the study of rock fracture mechanism under hydraulic coupling. Taking Xiluodu basalt as the research object, through hydraulic coupling indoor triaxial test and acoustic emission test, this paper studies the relationship between critical pore water pressure, confining pressure and initial water pressure of basalt, and puts forward an empirical calculation formula of critical pore water pressure during hydraulic fracture of basalt, The related research has important reference value and significance for rock fracture behavior and mechanism under hydraulic coupling and related engineering practice.

2 Test Principle and Method

The riverbed bedrock in the dam site area of Xiluodu hydropower station is Emeishan basalt of Upper Permian system. The color of basalt specimen is dark cyan, its particle composition is fine and uniform, and there are no obvious macro cracks and holes in the appearance. The cylindrical standard specimen with diameter of 50mm and height of 100mm is uniformly made by drilling, cutting and grinding. The test was carried out on mts815.04 electro-hydraulic servo stiffness testing machine of Wuhan Institute of geotechnical mechanics, Chinese Academy of Sciences. Acoustic emission acquisition system is a PC-II acoustic emission (AE) three-dimensional positioning real-time monitoring and display system developed by physical acoustics company. The sampling frequency is 5 MHz and the data recording threshold is 45 dB.

In the triaxial compression test of basalt, the confining pressure is selected as three different confining pressure values of 20 MPa, 25 MPa and 30 MPa. Under different confining pressures, the initial water pressure P increases according to the gradient, which are 5 MPa, 10 MPa and 15 MPa respectively. During the test, the initial water pressure remains unchanged. In order to carry out the test smoothly and keep the basalt sample in a closed state under water force coupling, it is necessary to ensure that the initial water pressure P is less than the confining pressure. Before the test, according to the test requirements, the sample shall be soaked freely to reach the saturation state, and the soaking time shall not be less than 48h. The overall test method is as follows:

(1) In the initial stage, the axial pressure and confining pressure are applied by controlling the load rate. Apply axial pressure and confining pressure at the same time until the predetermined confining pressure value, and keep the confining pressure value unchanged in the subsequent test process.

(2) Apply pressurized water. Apply pressurized water at the port at the rate of 1 MPa/min and gradually reach the predetermined water pressure. The water outlet at the top of the sample is connected with the atmosphere to ensure that the water pressure at both ends of the sample is the same. Throughout the test, the end water pressure of the sample remains unchanged.

(3) Keep the confining pressure and water pressure unchanged, and load the axial pressure at a constant loading rate. When the sample enters the unstable expansion stage, the axial pressure control mode is changed to circumferential strain rate control until the sample is damaged.

3 Test Results and Analysis

3.1 Typical Stress-Strain Curve

Figure 1 shows the whole process curves of stress-strain under different water pressures when the confining pressure is 20 MPa. The analysis shows that the phenomenon of stress drop behind the peak of basaltic rock is very obvious, showing typical hard brittle behaviour; With the increase of water pressure, the peak strength of rock gradually decreases and the stress drop gradually weakens, indicating that water pressure will weaken the compressive strength of rock and weaken its hard brittleness to a certain extent.

3.2 Analysis of Acoustic Emission Results

3.2.1 Acoustic Emission Event Rate Characteristics

According to the characteristics of the rock failure process, the rock failure process can generally be divided into four stages [11]. Since basalt is a hard brittle rock, the compaction section of the stress-strain curve in Fig. 2 is not obvious or even visible. It can be considered that basalt mainly displays three stages in the compression failure process, namely elastic stage (I) and crack stable propagation stage (II) And the unstable crack growth stage (III).

Figure 2 shows the whole process curve of basalt stress time and the change curve of acoustic emission event rate under 5 MPa water pressure and 20 MPa confining pressure. The analysis shows that in the elastic stage, the rock sample is mainly elastic deformation, and there is almost no acoustic emission activity.When entering the stable crack growth stage, acoustic emission events occur, indicating that the pressurized seepage has entered the rock and led to the crack to begin to crack. Therefore, this paper takes the time when the initial acoustic emission event occurs as the boundary between stage I and stage II, and considers this time as the time point of crack initiation; In the unstable crack growth stage, the AE event rate increases sharply and reaches the peak.

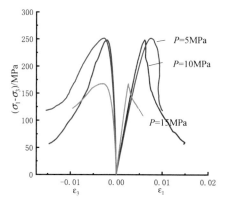

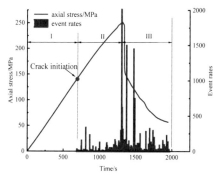

Fig. 1. Typical stress-strain whole process curves

Fig. 2. Changes of basalt axial stress and acoustic emission event rate with time under the conditions of 20 MPa confining pressure and 5 MPa water pressure

3.2.2 Ra-Af Value Characteristics

The Japanese concrete AE monitoring code introduces a fracture type discrimination method based on AE parameters. In this method, the ratio of ring count to duration is defined as average frequency AF (unit kHz), and the ratio of rise time to maximum amplitude is defined as RA (unit MS/V). RA value and AF value are usually used as reference methods to judge rock failure types. High RA value and low AF value indicate shear failure of rock, and low RA value and high AF value indicate tensile failure of rock [12].

In order to study the characteristics of rock fracture behaviour in different time domains, taking basalt samples under 20 MPa confining pressure and 15 MPa initial water pressure as an example, Fig. 3 shows the distribution of RA-AF values of basalt at different stages. It can be seen from the figure that the basalt is dominated by crack tension failure in stage II, the AF value of acoustic emission is high and increases rapidly, and the RA value is small and basically unchanged; In the third stage, there are two failure modes of shear and tension in the internal cracks of basalt. The acoustic emission signals with high RA value and low AF value are significantly increased, and the acoustic emission signals with low RA value and high AF value are more dense, indicating that the crack failure at this stage is mainly shear failure.

In general, it can be considered that there is pressurized seepage in the internal cracks of the rock. When the pore pressure in the crack reaches a critical state, the crack will firstly undergo tension failure, and then gradually expand and extend, due to the asymmetry and bending characteristics of the crack When entering the unsteady growth stage of cracks, under the continuous hydraulic-mechanical coupling action, the crack gap gradually becomes larger, and frictional slippage occurs at the upper and lower interfaces of the crack, and then the shear failure is the main form and the tension failure is supplemented. The fracture surface is partially sheared and finally penetrated to form a macro fracture surface, which is similar to the conclusion of T. Backers [13].

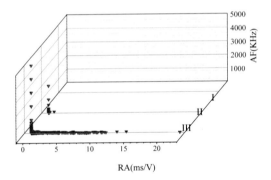

Fig. 3. Distribution map of RA-AF at different stages

3.3 Crack Initiation Stress and Critical Water Pressure

According to the characteristic values of RA-AF, under the condition of hydraulic coupling, the internal crack of basalt mainly occurs tensile failure under the combined action of internal water pressure and external stress. This failure mode shows that when the initial water pressure fills the pores in the rock and reaches a certain threshold, the tensile failure will occur around the pores first. In order to facilitate the research, the vertical load borne by the rock at this time is called crack initiation stress, and the corresponding pore water pressure is called critical water pressure.

Figure 4 shows the relationship curve between crack initiation stress, confining pressure and initial water pressure. The analysis shows that when the initial water pressure is constant, the crack initiation stress increases with the increase of confining pressure; When the confining pressure is constant, the crack initiation stress decreases with the increase of initial water pressure. At the same time, the variation law of crack initiation stress with initial water pressure and confining pressure is approximately linear.

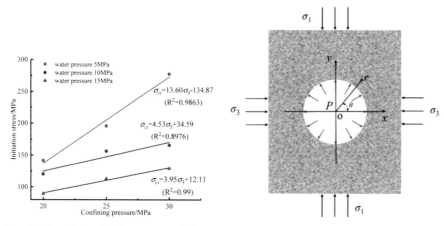

Fig. 4. The relationship between confining pressure, water pressure and crack initiation stress

Fig. 5. Crack model filled with water inside

4 Calculation Formula of Critical Pore Water Pressure

At present, the research on critical water pressure of rock under hydraulic coupling conditions is mainly based on theoretical analytical methods. The relevant analytical formulas are difficult to verify through experimental test data, which restricts the promotion and application of these theoretical formulas. This article assumes that the original internal cracks in basalt are circular pores. For cracks filled with water, the mechanical model of the crack under pseudo triaxial test conditions is shown in Fig. 5. The infinite body with circular cracks in Fig. 5 bears bidirectional uniform compressive stress and the initial water pressure is completely filled inside the pores, that is the pore water pressure is equal to the initial water pressure.

According to the relevant theories of elasticity and considering the initial water pressure P and critical water pressure P_i, the stress around the circular crack can be expressed as [14]:

$$
\begin{cases}
\sigma_r = \frac{\sigma_1+\sigma_3}{2}\left(1 - \frac{R^2}{r^2}\right) + \frac{\sigma_3-\sigma_1}{2}\left(1 - \frac{4R^2}{r^2} + \frac{3R^4}{r^4}\right)\cos(2\theta) + \frac{PR^2}{r^2} + \frac{P_iR^2}{r^2} \\
\sigma_\theta = \frac{\sigma_1+\sigma_3}{2}\left(1 + \frac{R^2}{r^2}\right) - \frac{\sigma_3-\sigma_1}{2}\left(1 + \frac{3R^4}{r^4}\right)\cos(2\theta) - \frac{PR^2}{r^2} - \frac{P_iR^2}{r^2} \\
\tau_{r\theta} = -\frac{\sigma_3-\sigma_1}{2}\left(1 + \frac{2R^2}{r^2} - \frac{3R^4}{r^4}\right)\sin(2\theta)
\end{cases}
\tag{1}
$$

In formula (1): σ_1 and σ_3 represent the maximum and minimum compressive stress respectively. σ_r and σ_θ represent the radial normal stress and the tangential normal stress at the distance r from the hole centre and the angle θ with the σ_3 direction. R is the radius of the circular crack.

Let $r = R$ to get the stress state on the circular hole wall:

$$\begin{cases} \sigma_r = P + P_i \\ \sigma_\theta = (\sigma_3 + \sigma_1) - 2(\sigma_3 - \sigma_1)\cos(2\theta) - P - P_i \\ \tau_{r\theta} = 0 \end{cases} \qquad (2)$$

The surface of a circular crack will generate tangential stress under the action of pore water pressure. It can be seen from Eq. (2) that σ_θ is a function of θ, that is $\sigma_\theta = f(\theta)$. Taking the derivative to get $\theta = 0°$ and σ_θ has a minimum. It means that the hole wall tension failure will occur in the direction parallel to σ_1. At this time, σ_θ reaches the tensile strength T of basalt:

$$T = 3\sigma_1 - \sigma_3 - P - P_i \qquad (3)$$

In this paper, the compressive stress is positive, the tensile stress is negative, and the tensile strength T is changed. When the circular crack hole wall undergoes tensile failure, the corresponding σ_1 is the initiation stress σ_{ci}, then the formula (3) can be expressed as:

$$\sigma_{ci} = \frac{1}{3}(T + \sigma_3 + P + P_i) \qquad (4)$$

Equation (4) gives the functional relationship between σ_{ci}, σ_3 and P. However, it is inconsistent with the linear relationship with σ_{ci}, σ_3 and P in Fig. 4. Author believes that this inconsistency is mainly related to P_i. Since σ_{ci} is a physical quantity related to σ_3 and P. P_i is the pore water pressure inside the crack when the axial stress reaches the initiation stress σ_{ci}. Therefore, it can be considered that P_i in Eq. (4) is also a variable related to σ_3 and P. To suppose $P_i = f(\sigma_3, P)$, at the same time, in order to simplify the research, the paper assumes that the tensile strength T is a constant.

The relationship between σ_{ci}, σ_3 and P in Fig. 4 is plane fitted in three-dimensional space. The plane equation is obtained as shown in Eq. (5):

$$242\sigma_3 - 310.21P - 28.44\sigma_{ci} + 1437.38 = 0 \qquad (5)$$

Combining Eq. (4), the relationship between critical water pressure P_i, σ_3 and P is obtained.

$$P_i = 24.81\sigma_3 - 34.06P + 153.15 - T \qquad (6)$$

It can be seen from the empirical relationship of Eq. (6), Under the same initial water pressure, the critical water pressure P_i increases with the increase of the confining pressure. Under the same confining pressure, the greater the initial water pressure, the critical water pressure P_i decreases, which shows this empirical relationship is consistent with the experimental test results, indicating that formula (6) describes the relationship between P_i, σ_3 and P is accurate.

5   Conclusion

Taking Xiluodu basalt as the research object, the hydraulic coupling triaxial test and acoustic emission test are carried out, and the following conclusions are obtained.

(1)   The hydraulic coupling triaxial test shows that when the initial water pressure is constant, the post peak stress drop phenomenon of basaltic rock is obvious, and the peak strength increases with the increase of confining pressure, showing a typical hard brittle behavior; When the confining pressure remains unchanged, its peak strength decreases gradually with the increase of initial water pressure, and the stress drop decreases gradually, indicating that the water pressure will weaken the compressive strength of rock and weaken its hard brittleness to a certain extent.

(2)   Acoustic emission test results show that water pressure has a great impact on acoustic emission activities, mainly in the post peak stage of rock. The cumulative acoustic emission count increases with the increase of water pressure, indicating that water pressure will stimulate the generation and propagation of rock cracks to a certain extent. The distribution of RA-AF value of basalt at different stages shows that the tensile failure occurs first and mainly under the combined action of pore water pressure and external stress.

(3)   Based on the theoretical derivation of single circular hole and the fitting of three-dimensional spatial data, the functional relationship between σ_{ci}, σ_3 and P is obtained respectively. Through comparative analysis, the empirical relationship between critical water pressure P_i, σ_3 and P is established. The correlation is consistent with the test results, which shows that the description of the relationship between P_i, σ_3 and P in this paper is accurate.

Acknowledgement.   The authors acknowledge the strong support and help by China Three Gorges Corporation.

References

1. Zhu, Z.D., Zhang, A.J., Xu, W.Y.: Experimental study on seepage characteristics of brittle rock in full stress-strain process. Rock Soil Mech. (05), 555–558+563 (2002)
2. Yu, J., Li, H., Chen, X., et al.: Triaxial test study on the correlation between permeability and deformation of sandstone under osmotic pressure-stress coupling. Chin. J. Rock Mech. Eng. **32**(06), 1203–1213 (2013)
3. Xu, J., Yang, H.W., Peng, S.J., et al.: Experimental study on the mechanical properties of sandstone under the action of pore water pressure and confining pressure. Chin. J. Rock Mech. Eng. **29**(08), 1618–1623 (2010)
4. Li, J.P., Yu, Z.X., Zhou, C.B., et al.: Experimental study on acoustic emission characteristics of rock under hydraulic coupling. Chin. J. Rock Mech. Eng. **03**, 492–498 (2006)
5. Zhao, X.D., Tang, C.A., Li, Y.H., et al.: Study on acoustic emission characteristics of granite fracture process. Chin. J. Rock Mech. Eng. **S2**, 3673–3678 (2003)

6. Min, K.S., Zhang, Z.N., Ghassemi, A.: Hydraulic fracturing propagation in heterogeneous rock using the VMIB method. Geotherm. Reservoir Eng. **35**, 1–10 (2010)
7. Tang, L.S., Zhang, P.C., Wang, Y.: Discussion on the fracture strength of rock mass under water. Chin. J. Rock Mech. Eng. **19**, 3337–3341 (2004)
8. Bian, K., Xiao, M., Hu, T.Q.: Analytical solution of critical water pressure for crack propagation in surrounding rock of hydraulic tunnel. Rock Soil Mech. **33**(08), 2429–2436 (2012)
9. Sheng, J.C., Zhao, J., Su, B.Y.: Hydraulic fracturing analysis of hydraulic pressure tunnel under high water head. Chin. J. Rock Mech. Eng. **7**, 1226–1230 (2005)
10. Cui, S.Y., Bao, T.F., Cui, J.J.: Calculation method of critical pore water pressure for hydraulic fracturing in hydraulic tunnels. Hydropower Energy Sci. **30**(05), 62–64+170 (2012)
11. Eberhardt, E.D.: Brittle rock fracture and progressive damage in uniaxial compression, Ph. D. Thesis. University of Saskatchewan, Saskatoon (1998)
12. Dimitrios, G.A.: Classification of cracking mode in concrete by acoustic emission parameters. Mech. Res. Commun. **38**(3), 153–157 (2011)
13. Backers, T., Stanchits, S., Dresen, G.: Tensile fracture propagation and acoustic emission activity in sandstone: the effect of loading rate. Int. J. Rock Mech. Min. Sci. **42**(7–8), 1094–1101 (2005)
14. Deng, G.Z., Huang, B.X., Wang, G.D., et al.: Theoretical analysis of pressure parameters of cracks in the wall of a circular hole with hydraulic pressure expansion. J. Xi'an Univ. Sci. Technol. (04), 361–364 (2003)

Theory and Technology of Digital Twin Model for Geotechnical Engineering

Jiaming Wu[1,2,3(✉)], Linfabao Dai[1,2], Guangqiao Xue[1,2], and Jian Chen[2,4]

[1] Tunnel Design and Research Institute, China Railway Siyuan Survey
and Design Group Co., Ltd., Wuhan 430063, China
1345233582@qq.com

[2] National and Local Joint Engineering Research Center of Underwater Tunnelling Technology,
Wuhan 430063, China

[3] School of Civil and Hydraulic Engineering, Huazhong University of Science and Technology,
Wuhan 430074, China

[4] State Key Laboratory of Geomechanics and Geotechnical Engineering, Institute of Rock and
Soil Mechanics, Chinese Academy of Sciences, Wuhan 430071, China

Abstract. As an innovative information technology, Digital Twin has greatly promoted the development of intelligent manufacturing in the industry. However, in the field of geotechnical engineering, there are still few researches on this aspect, which is still a "new territory" and "no man's land". The concept of digital twin coincides with the needs of geotechnical engineering informatization, so the introduction of digital twin technology into the field of geotechnical engineering will help to promote the development process of geotechnical engineering informatization and digitization. This paper puts forward and defines the digital twin model of geotechnical engineering, describes the connotation of the digital twin model, and studies the architecture of the digital twin model of geotechnical engineering. On this basis, the integration and sharing mechanism of geotechnical engineering digital twin data based on BIM technology is proposed. In order to break through the defect that the geotechnical engineering information model has not fully played its role for a long time, an integrated model of geological body and structural body is constructed based on the construction and integration module of geotechnical digital twin model. Furthermore, the geotechnical engineering digital twin simulation analysis module is developed to initially form the geotechnical engineering digital twin model, so as to realize the geotechnical engineering digital design, collaborative construction, visual decision-making and transparent management.

Keywords: Digital twin model of geotechnical engineering · Sharing mechanism · Integration · BIM · Simulation analysis

1 Introduction

Around the world, with manufacturing development strategies put forward by many countries such as Made in China 2025, German Industry 4.0 and American Industrial

G. Feng (Ed.): ICCE 2021, LNCE 213, pp. 403–411, 2022.
https://doi.org/10.1007/978-981-19-1260-3_37

Internet, intelligent manufacturing has become a common trend and goal of global manufacturing development [1–3]. How to realize the mapping and integration of physical world and virtual world is one of the crucial obstacles in the field of intelligent manufacturing at home and abroad. As a key technology to solve the problem of the integration of physical world and virtual world in the process of intelligent manufacturing, digital twin has been widely concerned and studied in recent years, and has been gradually applied in some fields.

The idea of digital twin was first proposed by Professor Michael Grieves [4]. For the concept of virtual digital expression, it was named as "mirror space model" and "information mirror model" at first, and then evolved into the term "digital twin" in 2011. NASA was the first to propose the use of digital twin technology. By building a virtual aircraft model that is the same as the real aircraft, the flight state of the real aircraft can be accurately simulated to assist the pilot to make correct decisions [5, 6]. Since then, the concept of digital twinning has been taken seriously and popularized in the aviation industry.

During the continuous development of the concept of digital twin, Thomas introduced the theoretical research of real-time data acquisition based on digital twin [7]. Wang defined the man-machine cooperation framework system based on digital twin technology, and developed the man-machine co assembly system [8]. Liu and others proposed the research on digital twin modeling methods in their respective fields based on digital twin technology [9]. David studied the use of digital twin technology to enhance the level of disaster management [10]. During the COVID-19 pandemic, the Central South Architectural Design Institute used digital twin technology to plan and design the world-famous Wuhan Raytheon hospital, which help the hospital quickly put into operation.

In terms of digital twin city construction, it is in the early stage of construction and has made a certain contribution to urban decision-making, but many applications are relatively macro. Specifically in the field of civil engineering, a lot of research on digital twin is basically still in the theoretical research stage, most of them only solve local details, and there is a big gap compared with satellite, automobile, aircraft and other fields. Subdivided into the field of geotechnical engineering, the research is still a "new territory", "no man's land".

By introducing digital twin technology into the field of geotechnical engineering, we will focus on the construction of geotechnical engineering digital twin model, promote the professional application of geotechnical digital twin, and explore a new geotechnical engineering digital construction path and practice mode. However, there are still a lot of careful work to be carried out, but also need theoretical innovation and technical breakthrough. Therefore, this paper puts forward and defines the digital twin model of geotechnical engineering, describes the connotation of the digital twin model, studies the system structure of the digital twin model of geotechnical engineering, further builds the integrated model of geotechnical engineering, develops the simulation and analysis module of geotechnical engineering digital twin, and preliminarily forms the digital twin model of geotechnical engineering.

2 Definition and Connotation of Digital Twin Model in Geotechnical Engineering

Please follow these instructions as carefully as possible so all articles within a conference have the same style to the title page. This paragraph follows a section title so it should not be indented.

At present, the research on geotechnical engineering digital twin model is in the conceptual stage, lack of systematic definition and description. Considering the characteristics of geotechnical engineering, based on the concept and evolution process of digital twin, this paper gives the definition of geotechnical engineering digital twin model: use digital technology to describe and express the object characteristics, construction process, and real-time status of geotechnical engineering physical entities, so as to form a multi-scale, interactive and dynamic simulation virtual model (Digital twin body) accurately mapped with the objective geotechnical physical entities (completely corresponding) in the real world, which will be used to simulate and predict the state in the whole life cycle of geotechnical engineering.

Through the definition of geotechnical digital twin model, we can further analyze the connotation of digital twin model: (1) geotechnical engineering digital twin model is a virtual model of the physical entity of geotechnical engineering project in the real world, which can store all kinds of data and information in the whole life cycle of geotechnical engineering project. (2) Considering that the engineering geological body is naturally formed and has certain concealment, the geological information will be updated frequently with the progress of geotechnical engineering projects, so the geotechnical digital twin model needs to be constantly updated and interacted with the physical entities of geotechnical engineering projects. (3) There are multi-scale problems in the process of simulating the physical entity of geotechnical engineering projects. In the process of simulating geotechnical engineering for different purposes (such as visualization, computability, etc.), we need to extract different information from the digital twin model. (4) We can simulate and predict the possible status of geotechnical engineering projects in the specific construction process, and take measures in advance to ensure the progress of the projects.

The digital twin model of geotechnical engineering is formed in the stage of geotechnical engineering investigation and design. The three-dimensional information model constructed based on the actual survey data and design data forms the basis of the digital twin model of geotechnical engineering, which is further applied in the whole life cycle stage of geotechnical engineering projects. Digital twin model is not only the expression of geometric shape of geotechnical engineering model, but also can contain various semantic information of geotechnical engineering object and reflect some mechanical properties of geotechnical engineering object. Through the real-time interaction with the data and information of the construction site, the accuracy of the digital twin model of geotechnical engineering can be continuously improved. By the dynamic display of geotechnical engineering project, the real-time change of state in the life cycle process is simulated, and the possible state of actual geotechnical engineering is predicted.

3   Architecture of Digital Twin Model in Geotechnical Engineering

In order to construct the digital twin model of geotechnical engineering, it is necessary to build the underlying architecture based on the digital twin model to store and express the digital twin model. Then, the multi-source data of geotechnical engineering need to be standardized and systematized. For engineering geological body and engineering structure object, parameterization and digitalization techniques are used to build 3D information model respectively. On the basis of 3D information model and unified data standard, the integration of multiple models is realized. It is necessary to use the IOT technology to map the model to the geotechnical engineering physical entity, so as to ensure the high simulation of the digital twin model and physical entity. Further, in combination with the needs of specific geotechnical engineering projects, we can combine numerical simulation, big data, augmented reality, cloud computing, lightweight and other technologies to carry out model expression, simulation, intelligent prediction and decision-making assistance for digital twin models.

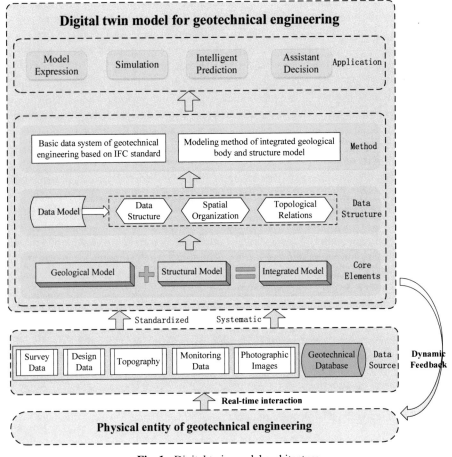

Fig. 1. Digital twin model architecture

The architecture of digital twin model in geotechnical engineering is described from three aspects: underlying architecture, method system and application requirements (Fig. 1).

3.1 Underlying Architecture

In order to realize the top-level design of geotechnical engineering digital twin model theory, it is necessary to further define the internal data structure of the model object, organize the internal spatial data relations of the model object, form a clear hierarchy structure inside the model object, and complete the description of geotechnical engineering digital twin model. The data model for digital twin model of geotechnical engineering should abide by the following principles:

- Applicability: the model adopts a unified data structure, which can not only describe the data characteristics of geological body and structural body, but also describe the geometric network relationship within the entity.
- Universality: it can not only build geological body and structure model, but also meet the needs of geotechnical mechanical analysis.
- Consistency: the data model supports Boolean operation analysis of engineering geological body and engineering structure model, and ensures the geometric topology consistency of the two models at the location of the common contact surface after operation.
- Compatibility: data structure can be easily converted with other data structures.

3.2 Method System

- Model construction: develop 3D geological modeling method for engineering geological body. Aiming at the engineering structure, the existing BIM software is used to develop the engineering structure component library and parametric component modeling method.
- Data fusion: based on the unified data standard, the integration of engineering geological body model and engineering structure model is realized.
- Simulation analysis: Dynamic loading of construction progress and time information, the simulation method of dynamic construction is established. Combined with mechanical parameters, the information model can be calculated, and the simulation analysis technology for geotechnical engineering digital twin model is formed.

4 Data Integrating and Sharing Mechanism of Digital Twin Model Based on Bim

Considering that IFC (Industry Foundation Classes) are extensible and unified data formats published by buildingSMART International for defining BIM models, so we exchange and share BIM model information and data based on IFC standard. By analyzing the existing IFC Standard Framework system, the IFC Standard expansion mechanism is studied.

According to the geotechnical engineering characteristics, aiming at the engineering structure and engineering geological object, the fusion mechanism of geotechnical information model based on IFC Standard is developed, and the basic data system of geotechnical digital twin model based on IFC Standard is formed.

5 Modeling and Integration of Digital Twin Model in Geotechnical Engineering

According to the borehole information, the interface information of each borehole stratum can be obtained. The DEM method is used to interpolate each formation interface, and the depth of each layer of borehole is used to calculate the formation layer. By intersecting each layer, topological relationship is re-established, and the upper and lower layers are formed. The intersecting strata are intersected to form a 3D geological model.

Based on BIM technology, considering the structural component characteristics of different types of engineering, the structural component families of various types of geotechnical engineering are generated through parametric modeling method. On this basis, the structural component library of geotechnical engineering is formed. Using geotechnical structural components, according to geotechnical engineering construction technology and construction method, the assembly technology of geotechnical structural components is developed, and a set of parameterized modeling method of geotechnical structural components based on BIM technology is established. Based on the above methods, various BIM models of geotechnical engineering constructed are shown in the Fig. 2.

Fig. 2. BIM model of geotechnical engineering

6 Simulation Analysis Technology for Geotechnical Engineering Digital Twin

In order to break through the defect that the information model of geotechnical engineering has not played a full role for a long time, we developed the simulation and analysis module of geotechnical engineering digital twin based on the construction and integration module of geotechnical engineering digital twin model.

Based on the integrated model of geological body and structural body, the grid generation technology of BIM model for numerical calculation is developed, and the implementation method of numerical calculation function is formed. The geological model cutting program for construction process is developed, and the construction dynamic simulation method based on BIM Technology is constructed. From the two aspects of numerical analysis and construction simulation, the application research of geotechnical engineering digital twin model is promoted, and the simulation analysis technology for geotechnical engineering digital twin is established to make the geotechnical engineering digital twin model "both beautiful and useful".

Based on the digital twin model theory of geotechnical engineering, combined with data standards, modeling methods and simulation analysis, the preliminary construction of digital twin model of geotechnical engineering is shown in Fig. 3.

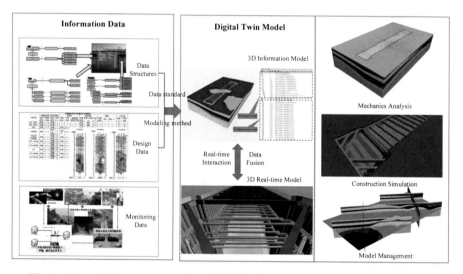

Fig. 3. The preliminary construction of digital twin model of geotechnical engineering

7 Conclusion

In this paper, digital twin technology is introduced into geotechnical engineering, and we have preliminarily established the geotechnical engineering digital twin model. By describing the definition, connotation and system structure of geotechnical digital twin model, the theoretical basis of geotechnical digital twin model is formed. The basic data system of geotechnical engineering digital twin model based on IFC standard is formed, which promotes the data fusion of information model built on BIM. Supported by BIM technology, we have formed the key modeling technology of digital twin. Driven by mechanical calculation and analysis, the data conversion and information exchange between BIM software and geotechnical engineering general numerical analysis software are realized. In order to realize the synchronization of virtual and real, the dynamic construction simulation of geotechnical excavation and structural application is realized, and the key technology of digital twin modeling and simulation in geotechnical engineering is formed. The research in this paper lays a solid foundation for the preliminary construction of digital twin model in geotechnical engineering, which promotes the popularization of digital twin technology in geotechnical engineering and helps the development of geotechnical engineering informatization and digitization.

Acknowledgments. The research was supported by National Key R&D Program of China (Grant No. 2021YFB2600400), Software Development Project (Grant No. 2020D049), National Natural Science Foundation of China (Grant No. 52079135).

References

1. Tao, F., et al.: Digital twin-driven product design framework. Int. J. Prod. Res. **57**(12), 3935–3953 (2019)
2. Quan, Y., Park, S.: Review on the application of Industry 4.0 digital twin technology to the quality management. J. Korean Soc. Qual. Manag. **45**(4), 601–610 (2017)
3. Tao, F., et al.: Digital twin driven prognostics and health management for complex equipment. CIRP Ann. Manuf. Technol. **67**(1), 169–172 (2018)
4. Grieves, M., Vickers, J.: Digital twin: mitigating unpredictable, undesirable emergent behavior in complex systems. In: Kahlen, F.-J., Flumerfelt, S., Alves, A. (eds.) Transdisciplinary Perspectives on Complex Systems, pp. 85–113. Springer, Cham (2017). https://doi.org/10.1007/978-3-319-38756-7_4
5. Tuegel, E.J., et al.: Reengineering aircraft structural life prediction using a digital twin. Int. J. Aerosp. Eng. **2011** (2011)
6. Glaessgen, E.H., Stargel, D.S.: The digital twin paradigm for future NASA and U.S. air force vehicles. In: Proceedings of the 53rd AIAA/ASME/ASCE/AHS/ASC Structures Structural Dynamics and Materials Conference, pp. 7274–7260 (2012)
7. Uhlemann, T.H.J., et al.: The digital twin: demonstrating the potential of real time data acquisition in production systems. Procedia Manuf. **9**, 113–120 (2017)
8. Wang, X.V., Kemény, Z., Váncza, J., Wang, L.: Human–robot collaborative assembly in cyber-physical production: classification framework and implementation. CIRP Ann. **66**(1), 5–8 (2017)
9. Liu, S.M., et al.: Digital twin modeling method based on biomimicry for machining aerospace components. J. Manuf. Syst. **58**, 180–195 (2020)
10. Ford, D.N., Wolf, C.M.: Smart cities with digital twin systems for disaster management. J. Manag. Eng. **36**(4): 4020027.1–4020027 (2020)

Study on Optimization of Row Spacing Between Steel Arches in Deep Buried Fault Cave Sections

G. Y. Zhang[1], Y. X. Xiao[1], J. H. Zhang[1(✉)], and J. Y. Luo[2]

[1] College of Water Resources and Hydropower, Sichuan University, Chengdu 610065, China
zhangjianhai@scu.edu.cn
[2] Guangxi Water Resources and Hydropower Survey Design and Research Institute, Nanning 530023, China

Abstract. As a widely used support form in tunnel support, the support effect of steel arch is influenced by the row distance between steel arches. In F8 fault fracture zone of the North Main Canal of Letan Reservoir in Guangxi, the support system of "steel arch + shotcrete" in this faulted cavern section was equalized with elastic modulus and yield stress by using theoretical analysis and numerical simulation, and the characteristic curves of rock support of deeply buried circular cavern under modified axisymmetric loading were obtained. The sensitivity analysis and optimization study of the spacing between steel arches were conducted by using FLAC^{3D}. The results show that with the increase of steel arch spacing, the cavity wall displacement increases, the support reaction force decreases nonlinearly, and the radial displacement and plastic zone around the cavity continue to increase. When the distance between steel arches >600 mm, the deformation of cavern perimeter changes abruptly and the plastic zone increases significantly. Based on comprehensive analysis, the optimization suggestions of steel arch are proposed.

Keywords: Cross-sectional equivalence · Numerical modelling · Steel arch · Supporting spacing · Sensitivity analysis

1 Introduction

The construction of the supporting structure can effectively restrain the deformation of the surrounding rock and improve the overall bearing capacity of the surrounding rock. For the weak surrounding rock section, a reasonable combined support system such as grid mesh or steel arch frame is often selected according to the degree of rock fragmentation. In actual projects, to ensure the safety of the project, a tighter steel arch spacing is often used, but the reduction of the steel arch spacing will increase the engineering cost. How to use economical and reasonable steel arch spacing, which can not only limit the deformation of the surrounding rock, but also be economical and reasonable, is a technical issue that the project is concerned about.

Many experts and scholars have studied the influence of steel arches on surrounding rock support. Gao et al. [1] conducted numerical simulations on the mechanical properties of three-composite steel arches and sprayed concrete initial support systems,

G. Feng (Ed.): ICCE 2021, LNCE 213, pp. 412–429, 2022.
https://doi.org/10.1007/978-981-19-1260-3_38

and gave recommended value of steel arch spacing under class IV surrounding rock. Mei Hua et al. [2] used the theory of common deformation to calculate the composite support of profiled steel arches, suspended steel mesh, and shotcrete. The deformation and stress of the three complied with the theory of common deformation. The effect of steel arch spacing on the surrounding rock of class V was analyzed, and corresponding suggestions were put forward for the actual situation of Caigu tunnel project. Li Xuefeng et al. [3] used the method of equivalent elastic modulus to calculate the supporting system formed by the steel arch and the primary shotcrete. By comparing the calculation results of different steel arch spacing, in view of reducing cave deformation, it is not so effective to reduce the steel arch spacing than by changing the thickness of the initial lining support. Liao Wei et al. [4] used numerical simulation to calculate the tensile damage of the surrounding rock, and then selected the safe arch spacing for class V surrounding rock according to the tensile damage area, and verified it by actual monitoring data. Zuo Qiankun et al. [5] used beam elements to simulate steel arches and proposed the optimal steel arch spacing of class IV surrounding rock when there is no inverted arch. Li et al. [6] used beam elements to simulate steel arches and proposed a support arch yielding criterion based on the arch section compression-bending bearing capacity equation, and embedded the modified beam elements through FISH language programming. The modified numerical simulation method is more reliable for large deformation tunnels using combined arch-bolt support, especially in supporting the bearing and damage behaviour of the arch and anchor rods. Song et al. [7] used a beam element to simulate steel arches, and the arrangement of the steel arches had a significant effect on controlling the sinking of the surrounding rock. As the spacing of the steel arches increased, the vertical displacement of the dome and floor slab increased, and the areas where larger displacements occurred tended to expand in the direction of the arches. Wang et al. [8] proposed a steel-concrete composite support system for loess tunnels, which consisted of a steel arch layer, a reinforced skeleton layer and a concrete filling layer, and found that the steel-concrete composite support system was superior to the traditional support system in terms of structural safety and load-bearing capacity. Liu et al. [9] conducted numerical analysis and field monitoring tests for shallow buried tunnels under soft surrounding rocks, proposing that under shallow buried tunnel conditions, the tunnel vault is the most unfavourable location and pre-supporting measures such as pre-emptive small tube grouting should be taken to ensure the safety of tunnel construction.

In this paper, the theoretical analysis method is applied to equate the elastic modulus and yield stress of the "steel arch + shotcrete" support system of the Letan Reservoir TBM diversion tunnel, establishing an equivalent calculation model of the joint support considering the steel arch. The influence of the steel arch on the surrounding rock displacement and surrounding rock reaction force of the deeply buried circular cavern is quantitatively evaluated under axisymmetric loading by using modified Fenner formula and elastic thin shell theory.

The finite difference software "FLAC^{3D}" is used to establish a three-dimensional model to simulate the stress-strain characteristics of the surrounding rock and the force characteristics of the support structure, and to carry out parametric sensitivity analysis on different steel arch spacing under three-dimensional conditions. The comparison verifies the results of the theoretical analysis and provides reference for engineering support design and construction.

2 Project Overview

The first phase of the Guangxi Guizhong Drought Control Letan Reservoir Diversion and Irrigation District Project consists of two parts: the main trunk canal and the northern trunk canal. The former section of the North Main Canal starts from the village of Nengrong and goes eastwards for about 1.5 km to the village of Yaowa. From the village of Yaowa, it passes through Chencun and Beisi to Liulang, with a total length of about 25.47 km, mainly in the form of tunnels.

The TBM in the Yaowa-Lulang Tunnel (5.94 m diameter) in the North Main Canal encountered the F8 fault zone near $17 + 833.2$ m (Fig. 1). The F8 fault has the following attitude: N11°E, SE∠85°, tilting downstream and intersecting the axis of the tunnel at a large angle, with the fault zone partially twisted. The parent rock is mainly muddy siltstone with a small amount of muddy tuff. The nature of the filling is not homogeneous, and the upper part of the filling has collapsed many times under self-weight and mechanical vibration, forming pits of different sizes, so it is necessary to take reasonable support measures to ensure construction safety and overall stability of the surrounding rock.

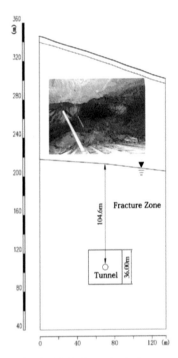

Fig. 1. F8 fault geological map.

3   Basic Theory

3.1   Equivalent Model of the Support System

Due to the intermittent arrangement of the steel arches in the actual project, it is difficult to obtain the analytical solution for each steel arch. To simplify the calculation, for the support system of "steel arch + shotcrete" shown in Fig. 2(a), the modulus of elasticity and yield strength of the steel arch are converted into concrete according to the principle of equal modulus of elasticity and yield stress. And the concrete parameters are modified to replace the steel arch for the equivalent of the support system.

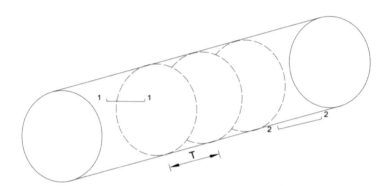

(a) Schematic diagram of the "steel arch + shotcrete" support for the tunnel

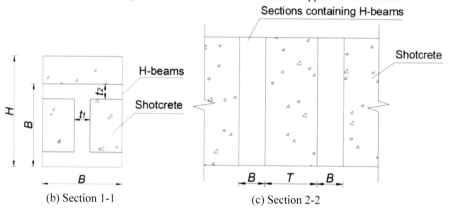

(b) Section 1-1   (c) Section 2-2

Fig. 2. Schematic diagram of equivalent model of "steel arch + shotcrete".

As shown in Fig. 2(b), the steel arch is intercepted in section 1-1, the modulus of elasticity of the H-shaped steel arch is E_g, the yield stress is σ_g, the height and width of the steel arch are B. The flange width is t_1, the web thickness is t_2, the area is A_1, the modulus of elasticity of plain concrete is E_c, the yield stress is σ_c, the area is A_2, the total area of section 1-1 is A, and the lining thickness is H. The equivalent modulus of elasticity E_h and the equivalent yield stress σ_{sh} of the 1-1 profile are:

$$E_h = E_g \cdot \frac{A_1}{A} + E_c \cdot \frac{A_2}{A} \tag{1}$$

$$\sigma_{sh} = \sigma_g \cdot \frac{A_1}{A} + \sigma_c \cdot \frac{A_2}{A} \tag{2}$$

As shown in Fig. 2(c), the equivalent elastic modulus E_1 and equivalent yield stress σ_{s1} of the surrounding rock support system for profile 2-2 with steel arch spacing T are:

$$E_1 = E_h \cdot \frac{B}{B+T} + E_c \cdot \frac{T}{B+T} \tag{3}$$

$$\sigma_{s1} = \sigma_{sh} \cdot \frac{B}{B+T} + \sigma_c \cdot \frac{T}{B+T} \tag{4}$$

3.2 Enclosure Support Characteristic Curves

Previous studies have shown [10] that when a tunnel is in a deep burial situation (burial depth $Z \geq 20R$, R being the excavated hole diameter), a deeply buried tunnel can be simplified to an axisymmetric plane strain problem [11].

3.2.1 Surrounding Rock Characteristic Curves

(1) Elastic state of the surrounding rock. The excavation of a circular cavern chamber of diameter R in the case of deep burial can be equated to a thick-walled cylinder, where the external pressure is p_0 (mountain rock pressure) and the internal pressure is p_1 (support reaction force).

According to the axial symmetry problem in elastic mechanics and the constitutive equation of plane strain, when there is a support reaction force p_1 acting on the surrounding rock, the elastic radial displacement around the cavern chamber is:

$$u_o = \frac{1+\mu}{E} \frac{R^2}{r} (p_0 - p_1) \tag{5}$$

E -- modulus of deformation of the surrounding rock.
μ -- poisson's ratio of the surrounding rock.

The inverse relationship between the radial displacement and the support reaction force is obtained from Eq. (5), and the characteristic curve of the surrounding rock in the fully elastic state is:

$$p_1 = p_0 - \frac{u_o E}{(1+\mu)R} \tag{6}$$

(2) Plasticity state of the surrounding rock

When the surrounding rock enters the plastic zone, assuming that the surrounding rock is an ideal elastic-plastic body, the radius of the plastic zone is R_p, the support reaction force provided by the lining is p_1, and the mountain rock pressure is p_0, it is known from the modified Fenner equation [11] that:

The radius of the plastic zone of the enclosing rock is:

$$R_p = R\left[\frac{(p_0 + c\cot\varphi)(1 - \sin\varphi)}{(p_1 + c\cot\varphi)}\right]^{\frac{1-\sin\varphi}{2\sin\varphi}} \tag{7}$$

c -- the cohesive force of the surrounding rock.
φ -- the angle of internal friction of the surrounding rock.

The support reaction forces are:

$$p_1 = (p_0 + c\cot\varphi)(1 - \sin\varphi)\left(\frac{R}{R_p}\right)^{\frac{2\sin\varphi}{1-\sin\varphi}} - c\cot\varphi \tag{8}$$

Radial displacement in the plastic zone:

$$u_o = \frac{\sin\varphi}{2G}(p_0 + c\cot\varphi)R\left[\frac{(p_0 + c\cot\varphi)(1 - \sin\varphi)}{p_1 + c\cot\varphi}\right]^{\frac{1-\sin\varphi}{\sin\varphi}} \tag{9}$$

G -- the shear deformation modulus of the surrounding rock.

The inverse relationship between the radial displacement and the support reaction force is obtained from Eq. (9), and the characteristic curve of the surrounding rock in the plastic state is:

$$p_1 = (p_0 + c\cot\varphi)(1 - \sin\varphi)\left[\frac{\sin\varphi(p_0 + c\cot\varphi)R}{2Gu_o}\right]^{\frac{\sin\varphi}{1-\sin\varphi}} - c\cot\varphi \tag{10}$$

3.2.2 Support Characteristic Curves

(1) Initial displacement of the surrounding rock

In the relationship of the surrounding rock-support structure, u_{oc} is defined as the radial displacement of the cavern wall before the support is set, which is the main unknown quantity and can be calculated by actual field measurements or by the empirical formula proposed by Hoek [12].

$$\frac{u_{oc}}{u_m} = [1 + \exp(\frac{-x/R}{1.10})]^{-1.7} \tag{11}$$

x -- the distance of the applied support from the palm face.
u_m -- the maximum value of the radial displacement of the cave wall.

(2) Maximum support force

The maximum support force can be calculated according to the theory of a circular tube under external pressure, and since lining thickness $H > 0.04R$, the maximum support resistance p_{max} that can be provided by the "steel arch + shotcrete" support system is [13]:

$$p_{max} = \frac{1}{2}\sigma_{s1}[1 - \frac{R^2}{(R + H)^2}] \tag{12}$$

(3) Radial displacement

Before the support structure reaches the maximum support force, it can be regarded as a thick-walled cylinder subjected to uniform external pressure p_0. According to the thick-walled cylinder formula, the radial displacement of the support structure is:

$$u_o = \frac{p_0 R^3 (1 + \mu_1)}{E_1 [R^2 - (R-H)^2]} [1 - 2\mu_1 + \frac{(R-H)^2}{R^2}] \tag{13}$$

μ_1 -- poisson's ratio of the support structure.

From Eqs. (12) and (13), the minimum displacement of the support structure u_{omin} can obtained when support structure reach the maximum support force, at this time the cave displacement is u_{om}, $u_{om} = u_{oc} + u_{omin}$.

Then the support curve considering the support timing consists of two sections: When $u_{oc} < u < u_{om}$:

$$p_1 = \frac{(u - u_{oc}) E_1 [R^2 - (R-H)^2]}{[1 - 2\mu_1 + \frac{(R-H)^2}{R^2}] R^3 (1 + \mu_1)} \tag{14}$$

When $u > u_{om}$:

$$p_1 = p_{max} = \frac{1}{2} \sigma_{s1} [1 - \frac{R^2}{(R+H)^2}] \tag{15}$$

3.3 Study on the Influence of F8 Fault Support Parameters on the Stability of the Surrounding Rock

3.3.1 F8 Fault Calculation Parameters

F8 fault fracture zone is encountered during tunnel boring of Letan TBM construction. The tunnel was excavated at a depth of 215 m, with a radius of $R = 2.97$ m. The tunnel was supported by a full section of HW150 × 150 steel, with a wet sprayed C25 concrete thickness of $H = 200$ mm and a steel arch spacing of $T = 300$ mm.

The surrounding rock parameters of the F8 fault zone tunnel are shown in Table 1 and the support parameters in Table 2.

Table 1. Surrounding rock parameters of tunnel in F8 section.

E/GPa	μ	$\rho/(t/m^3)$	$\varphi(°)$	c/MPa
0.3	0.38	2.3	26.56	0.05

Table 2. Supporting parameters of "section steel + concrete".

Support structures	E/GPa	μ	σ_s/MPa	B/mm	t_2/mm	t_1/mm
C25concrete	28	0.2	28	/	/	/
Steel	206	0.2	215	150	7	10

3.3.2 Analysis of Support Reaction Forces and Deformation Characteristics of the Surrounding Rock

Take the burial depth of F8 section $h = 215$ m, the pressure of surrounding rock is calculated by self-weight stress:

$$p_0 = \rho g h = 4.85 \text{ MPa}$$

In calculating the shotcrete characteristic curve, a 3-day strength can generally be used [13]:

$$E_{c3} = 0.67E_c = 18.76 \text{ GPa}$$

$$\sigma_{c3} = 0.5\sigma_c = 14 \text{ MPa}$$

The equivalent modulus of elasticity and equivalent yield stress of the steel section + concrete can be calculated as:

$$E_1 = \left(E_g \cdot \frac{A_1}{A} + E_{c3} \cdot \frac{A_2}{A}\right) \cdot \frac{B}{B+T} + E_{c3} \cdot \frac{T}{B+T} = 25.87 \text{ GPa}$$

$$\sigma_{s1} = \left(\sigma_g \cdot \frac{A_1}{A} + \sigma_{c3}\frac{A_2}{A}\right) \cdot \frac{B}{B+T} + \sigma_{c3} \cdot \frac{T}{B+T} = 21.73 \text{ MPa}$$

As the F8 cave section is a fault zone, the pre-emptive strengthening measure of grouting is adopted, taking $x = -2.94$ m, the displacement before support is:

$$u_{oc} = u_m[1 + \exp(\frac{-x/R}{1.10})]^{-1.7} = 43.19 \text{ mm}$$

Maximum support force is:

$$p_{max} = \frac{1}{2}\sigma_{s1}[1 - \frac{R^2}{(R+H)^2}] = 1.33 \text{ MPa}$$

From the Eq. (12), (13) joint solution can obtain u_{omin}, and $u_{oc} = 43.19$ mm, so the cave wall displacement u_{om} when the support structure reaches the maximum support force is 45.26 mm.

According to the parameters of the surrounding rock and the above calculation results, the characteristic curve of the support structure and the characteristic curve of the surrounding rock are shown in Fig. 3.

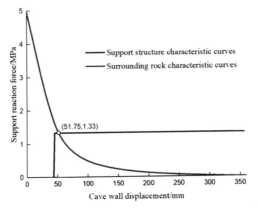

Fig. 3. Supporting structure and surrounding rock characteristic curve with T = 300 mm, E = 0.3 GPa.

From Fig. 3: $u_0 = 51.75$ mm, $p_1 = 1.33$ MPa.
The radius of the plastic zone is:

$$R_p = R \left[\frac{(p_0 + c \cot \varphi)(1 - \sin \varphi)}{(p_1 + c \cot \varphi)} \right]^{\frac{1 - \sin \varphi}{2 \sin \varphi}} = 3.43 \text{ m}$$

It can be seen that when the deformation of the support structure and the deformation of the surrounding rock are equal, the displacement of the cave wall is 51.75 mm, at which time the surrounding rock and the support structure reach a state of coordinated deformation.

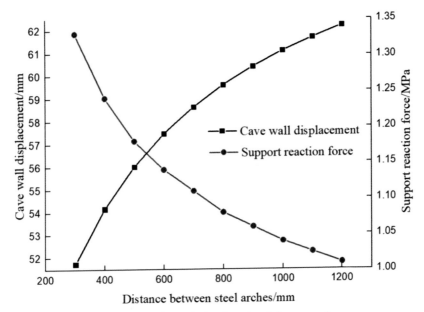

Fig. 4. Sensitivity analysis of steel arch frame spacing.

3.4 Sensitivity Analysis of Row Spacing Between Steel Arches

When the spacing between steel arches is between 300 and 1200 mm, the cave wall displacement and support reaction force changes as shown in Fig. 4:

Figure 4 shows the variation curves of steel arch spacing and cavern wall displacement u_o and support reaction force p_1 at $E = 0.3$ GPa. It can be seen that:

(1) With the increase of steel arch spacing, the cave wall displacement increases non-linearly, when $T < 600$ mm, the cave wall displacement grows faster, when $T > 600$ mm, the cave wall displacement grows gradually, and tends to level off.
(2) With the increase of steel arch spacing, the support reaction force decreases non-linearly, when $T < 600$ mm, the support reaction force decreases rapidly, and when $T > 600$ mm, the support reaction force decreases slowly and finally converges.

4 Numerical Simulation Studies

4.1 Computational Models and Research Scheme

4.1.1 3D Calculation Range for the F8 Fault

The geological map of the F8 fault is shown in the Fig. 1. As seen in Table 3, for 17+757 m of fault F8, the horizontal direction of the section is taken as the X-axis direction, the interception length is 36 m. The tunnel axis is the Y-axis direction, the interception length is 20 m. The vertical direction Z is taken from 92 m elevation to 128 m elevation.

Table 3. F8 fault section 3D model calculation range table.

Profile stake number	Elevation of the top of the cave/m	Depth of burial of the rock at the top of the cave/m	Waterline Elevation/m	Length in horizontal direction X/m	Elevation Z range/m	Length in the Y direction of the axis/m	Nodes	Elements
17+757	108.217	215.0	204.718	36	92–128 m	20	185623	181440

Self-weight stress is taken as the initial geo-stress. The self-weight of the mountain above an elevation of 128 m is applied to the top surface of the calculation model in the form of a top surface force.

4.1.2 Reinforcement Parameters for Hole Section F8

In order to facilitate the sensitivity analysis of the row spacing between steel arches, it is advisable to ignore the sprayed layer of concrete and design a support scheme that only considers the effect of steel arch support: top arch with drainage holes (top arch centre angle 120° range) $\Phi50@3 \times 3$ m, L = 3 m arranged in intervals, top arch 120° backfill grouting, full section HW150 × 150 section @300.

4.1.3 Calculated Working Conditions and Parameters

Table 4 shows the four working conditions in the sensitivity analysis of the steel arch spacing. Varying the spacing of the steel arches (300 mm, 600 mm, 900 mm and 1200 mm) to analyse the support effect of the steel arches provides a direct response to the effect of the spacing between the steel arches on the support effect.

Table 4. Sensitivity analysis condition of row spacing between steel arches.

Distance between steel arches/mm	Working conditions	Calculation content
300	GK0.3	Excavation + Support
600	GK0.6	Excavation + Support
900	GK0.9	Excavation + Support
1200	GK1.2	Excavation + Support

4.1.4 F8 Holesection Gridding

Figure 5 shows the grid diagram of the computational model.

The rock interface, steel arch and grouting area were simulated in detail in the 3D model.

1. Tunnel surrounding rock: the boundary range of the model section is about 6 times the diameter of the tunnel. $36 \times 36 \times 20$ m was chosen for the tunnel surrounding rock, with a reserved excavation diameter of 5.94 m. The model was set up using Extrusion in FLAC^{3D} 6.0.
2. Steel arch: the steel arch model was drawn by 'Rhino3D', then imported into '.dxf' using FLAC^{3D}6.0 import command, 'beam' structural unit is used to simulate the steel arch.
3. Grouting zone: the solid unit was used to simulate a grouting zone of 2.94 m in depth within 120° above the top arch.

The rock mass parameters are provided by test result of the design institute, and the support structure parameters are selected from Hydraulic Tunnel Design Code SL279-2016, shown as Table 5.

4.2 Geo-Stress Regression Calculation Results and Discussion

Self-weight load was applied to calculate the three-dimensional geo-stress field in the tunnel surrounding rock. Due to the depth of the tunnel, the top of the calculation area is not taken to the free surface of the ground. The gravity load is applied with an equivalent surface force load of the overlying rock pressure at the top boundary of the model. The geo-stress relief factor around the hole is taken to be 0.5. The initial geo-stress at the centre of the tunnel in the F8 fault was obtained as follows: $\sigma_{xx} = 3.3417$ MPa, $\sigma_{yy} =$

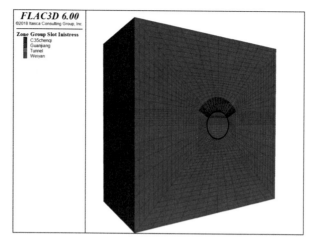

Fig. 5. Computational model grid.

Table 5. Model parameters.

Category	E/GPa	μ	$\varphi(°)$	c/MPa	$\rho/(t/m^3)$
Surrounding rocks	0.3	0.38	26.56	0.05	2.3
Grouting area	20.55	0.4	35	0.075	2.54
Steel arch	206	0.2	/	/	7.9

3.3411 MPa, $\sigma_{zz} = 4.8091$ MPa, $\tau_{xy} = 1.81 \times 10^{-4}$ MPa, $\tau_{yz} = -0.2692$ MPa, $\tau_{zx} = 1.52 \times 10^{-5}$ MPa, the principal stresses are known to be: $\sigma_1 = 4.8096$ MPa, $\sigma_2 = 3.3413$ MPa, $\sigma_3 = 3.3410$ MPa.

According to the results of the geo-stress regression calculation, it can be concluded that.

(1) For the F8 fault, the calculated burial depth at the left boundary is 226 m and at the right boundary is 209 m. Therefore, under the action of self-weight, the initial geo-stress is larger and there is a weak bias pressure phenomenon.

(2) Under self-weight action, the regional stress field is to some extent influenced by the structure, and the vertical stress is slightly larger than the horizontal stress, and the value of the shear stress component τ_{yz} is small.

(3) The major and minor principal stresses are roughly distributed along the vertical and horizontal directions respectively, and the stress magnitude is directly related to the depth of burial.

4.3 Sensitivity Analysis of Row Spacing Between Steel Arches

4.3.1 Sensitivity Analysis of the Inter-Row Spacing Between Steel Arches on the Deformation of the Surrounding Rock

The deformation of the surrounding rock after excavation shows significant symmetry. Due to the limited space, Fig. 6 gives the cloud diagram of the vertical displacement along the Z direction in the Y-Z section under GK0.6 and GK1.2 working conditions. Fig. 7 shows the cavern perimeter displacements for each working condition, Fig. 8 shows the maximum displacements for each working condition, and Table 6 lists the maximum displacements for each working condition.

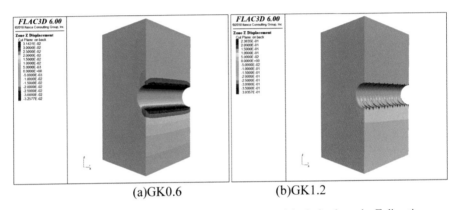

(a)GK0.6   (b)GK1.2

Fig. 6. Uz/m contour map of the displacement around the hole along the Z direction.

It can be seen that the excavation deformation has the following characteristics:

(1) Under all working conditions, the cavern shows an overall inward deformation trend, and the horizontal and vertical displacements are basically symmetrical. When the row distance between steel arches is 300 mm, the top arch sinks by 15.00 mm, when the spacing between steel arches is 600 mm, the top arch sinks by 37.41 mm, and when the spacing is 900 mm and 1200 mm, the top arch sinks by 185.11 mm and 320.00 mm respectively, resulting in danger of cave collapse.

(2) With the increase of the spacing between steel arches, the displacement of the cavern perimeter gradually increases. When the spacing is greater than 600 mm, the maximum displacement of the steel arches increases significantly.

4.3.2 Sensitivity Analysis of Inter-Row Spacing Between Steel Arches on Surrounding Rock Stresses

Due to space limitations, the main stress cloud diagramin the Y-Z profile of GK0.6 is given in Fig. 9.

Figure 10 plots the relationship between the variation of the Maximum principal stresses at each characteristic node under different working conditions. It can be seen

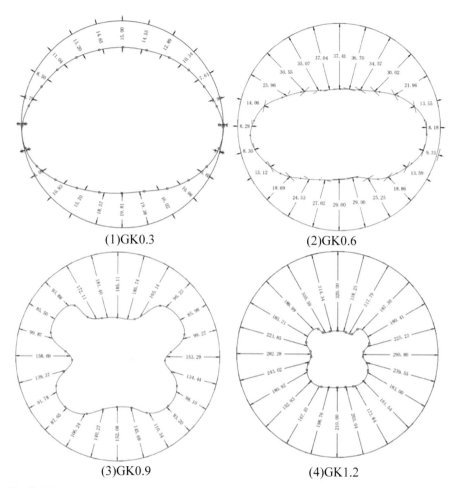

(1)GK0.3   (2)GK0.6

(3)GK0.9   (4)GK1.2

Fig. 7. Displacement contour map of the center section around the hole with different steel arch spacing/mm.

that the distribution of the principal stresses in the F8 cavern section has the following characteristics.

(1) The stress distribution pattern is consistent under different working conditions, with the left and right waists slightly larger than the top and bottom of the arch.

(2) As the spacing of steel arch increases, the surrounding rock enters the plastic state from the elastic state, and the stress around the hole gradually increases. When the surrounding rock enters the failure state, the stress around the hole rapidly decreases. Study shows that when the spacing of steel arch is 900 mm and 1200 mm, the rock around the hole failed, and the stress around the hole is close to the level of complete unloading.

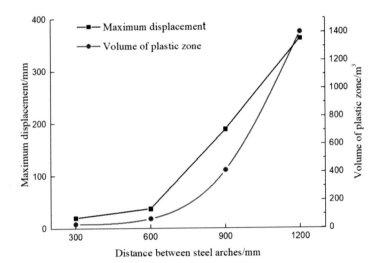

Fig. 8. Trend chart of maximum displacement and total plastic zone volume change under various working conditions.

Table 6. Maximum displacement and plastic zone volume under various working conditions.

Distance between steel arches/mm	300	600	900	1200
Maximum displacement/mm	19.98	37.49	189.21	362.75
Volume of plastic zone/m^3	29.00	66.95	415.86	1404.21

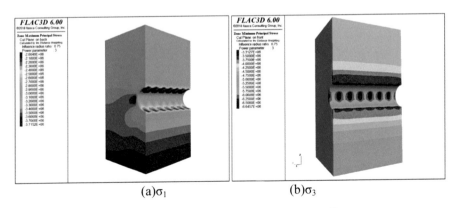

(a)σ_1 (b)σ_3

Fig. 9. GK0.6 (steel arch spacing 600 mm) stress diagram.

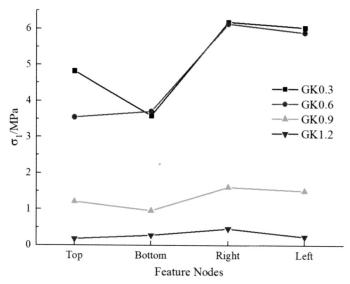

Fig. 10. Maximum principal stress σ_1 diagram of characteristic nodes of surrounding rock under various working conditions/MPa.

4.3.3 Sensitivity Analysis of Inter-Row Spacing Between Steel Arches on the Distribution of Damage Zones of The Surrounding Rock

Table 6 lists the plastic zone volumes for each working condition and Fig. 8 demonstrates the trend in plastic zone volume for each working condition. It can be seen:

With the increase of steel arch spacing, the plastic zone around the cavern increases non-linearly. When the spacing between steel arches is small, the volume of plastic zone grows slowly. But when the spacing is large, the volume of plastic zone increases rapidly, till the surrounding rock enters the failure state, indicating that the steel arches can no longer sustain the stability of the rock around the cave. For example, when the spacing is 600 mm, the volume of the plastic zone is 66.95 m³, but when the spacing is 1200 mm, the volume of the plastic zone reaches 1404.21 m³, at which point the cavern is already in danger of collapse.

5   Conclusion

Sensitivity analysis of the stability of the surrounding rock with different steel arch spacing in the F8 fault zone was carried out by means of theoretical analysis and numerical simulation, and the main results are as follows.

(1) With the increase of steel arch spacing, the cave wall displacement increases and the support reaction force decreases non-linearly.
(2) Under all working conditions, the cavern shows an overall inward deformation trend, the horizontal and vertical displacements are basically symmetrical, and the uplift of the bottom of the arch is slightly larger than the sinking of the top of the

arch. With the increase of the spacing between steel arches, the deformation around the cavern gradually increases, and the non-linearity of the plastic zone around the cavern increases.

(3) The stress distribution pattern of the surrounding rock under different working conditions is consistent, with the left and right waists slightly larger than the top and bottom of the arch. With the increase of steel arch spacing, the surrounding rock enters the plastic state from the elastic state, and the stress around the cavern gradually increases. When the spacing of steel arch is 900 mm and 1200 mm, the rock around the cave entered into failure state and the stress around the cave is close to the level of complete unloading.

(4) From the analysis results of the support characteristic curve of the surrounding rock and the sensitivity analysis of the steel arch spacing, it can be obtained that for the F8 fault, the steel arch spacing should be less than 600 mm, and the spacing of 300 mm used in the actual construction is safe and feasible.

References

1. Gao, X., et al.: Study on bearing mechanism and coupling mechanism of steel arch-concrete composite structure of initial support system of large section tunnel. Geotech. Geolog. Eng. **37**(6), 4877–4887 (2019). https://doi.org/10.1007/s10706-019-00948-4
2. Mei, H., Qu, Z., Liu, W.X.: The numerical simulation analysis of different spacing between steel arch in the case of shallowly buried soft rock. Low Temp. Archit. Technol. **40**(10), 77–81 (2018)
3. Li, X.F., Shang, Y.C., Gu, X.X., Yu, J., Yang, W.B., Jiang, Y.J.: Research on the optimization of initial support parameters for tunnel passing through red clay contact zone. J. Railw. Eng. Soc. **9**, 49–53 (2019)
4. Liao, W., He, P., Yan, D.M., Chen, Z., Gao, H.J., Wang, X.Y.: Study on stress distribution law of steel arch of initial tunnel support. J. China Railw. Soc. **39**(09), 140–147 (2017)
5. Zuo, Q.K., Li, T.B., Meng, L.B., Zheng, Y.X.: Numerical simulation analysis of tunnel steel arch supporting structure stress characteristics. J. China Foreign Highw. **31**(04), 196–199 (2011)
6. Li, W.T., et al.: An improved numerical simulation approach for arch-bolt supported tunnels with large deformation. Tunn. Undergr. Space Technol. **77**, 1–12 (2018)
7. Song, X., Zhu, J.X.: IOP Conference Series: Earth Environmental Science, vol. 455, p. 012159 (2020)
8. Wang, Z.C., Su, X.L., Lai, H.P., Xie, Y.L., Qin, Y.W., Liu, T.: Conception and evaluation of a novel type of support in loess tunnels. J. Perform. Constr. Facil. **35**(1), 04020144 (2021)
9. Liu, J., Liu, X., Zhang, Y., Xiao, T.: Numerical analysis and field monitoring tests on shallow tunnels under weak surrounding rock. J. Central South Univ. **22**(10), 4056–4063 (2015). https://doi.org/10.1007/s11771-015-2950-7
10. Jiang, Y.C., Zhang, J.H., Li, Z.Z.: Elasticity and Finite Element Method, pp. 117–119. Science Press, Beijing (2006)
11. Xu, Z.Y.: Rock Mechanics, 3rd edn., pp. 132–135. China Water Power Press, Beijing (2007)
12. Qi, M.S.: Study on rheological properties of soft rock with large deformation and its application in tunnel engineering, Shanghai Tongji University (2006)
13. Guan, B.S.: Introduction to Tunnel Mechanics, pp. 76–77. Southwest Jiaotong University Press, Chengdu (1993)

Seismic Performance of a Precast Hollow Insulated Shear Wall

Zhexian Chen[1], Wenfu He[1], Sen Yang[1(✉)], Cheng Chang[2], and Min Ji[1]

[1] Department of Civil Engineering, School of Mechanics and Engineering Sciences, Shanghai University, Shanghai 200444, China
{chenzhexian,kongxiaosen}@shu.edu.cn
[2] Department of Chemical Engineering, The School of Engineering and Natural Sciences, Heriot-Watt University, Edinburgh EH14 4AS, UK

Abstract. A new precast hollow insulation shear wall (PHISW) is proposed in this paper. To study the seismic behaviors of the new PHISW, two cast-in-place solid shear wall (CSW) specimens, two precast monolithic hollow insulated shear wall (PMW) specimens, and two precast hollow insulated shear wall (PSW) specimens with vertical seams were produced and subjected to low-cyclic reversed loadings. The seismic indices obtained from low-cyclic reversed tests include the failure pattern, hysteretic curves and energy dissipation. The experiment results indicate that flexural failure is the main failure mode of the specimens, but a noticeable difference is detected in the cracking distribution between the three types of shear walls. The bearing capacity of each characteristic point of PMW and PSW is comparable to that of CSW. The ductility coefficient of the newly proposed precast shear wall is slightly lower than that of CSW.

Keywords: Precast hollow insulated shear wall · Low cyclic loading · Bearing capacity · Displacement ductility · Energy dissipation

1 Introduction

In recent years, precast shear walls have been intensively studied and rapidly developed owing to the excellent superiority of a short construction period, less polluting, and good construction quality [1, 2]. However, their relatively large volume and weight limit the wider usage of precast shear walls. Embedding thermal insulation materials (TIMs) in precast shear walls is a reasonable alternative solution, as the TIM can reduce the structural weight and simultaneously enhance the thermal insulation and fire resistance performance [3]. The achieved structure is generally known as the precast hollow insulated shear wall (PHISW). In 1997, Salmon et al. studied the shear resistance performance of a prototype sandwich panel by transverse loading tests [4]. In 1998, Bush et al. proposed a precast sandwich panel with diagonal connectors and explored the flexural performance experimentally and numerically [5]. In 2014, Palermo et al. carried out a shaking table model test on a 3-story building employing precast sandwich shear walls and a corresponding nonlinear numerical simulation on the shear wall components, which proved that structures using the precast sandwich insulation shear wall

© The Author(s) 2022
G. Feng (Ed.): ICCE 2021, LNCE 213, pp. 430–439, 2022.
https://doi.org/10.1007/978-981-19-1260-3_39

could meet the seismic requirements of the current code [6]. Ricci et al. conducted a low cyclic loading test on precast sandwich insulation shear wall specimens, which showed that the precast sandwich insulation shear wall had a larger bearing capacity and better ductility than the cast-in-place shear wall [7].

Current studies of PHISWs are mainly focused on precast sandwich panel walls (PSPWs), which are built up by two outward concrete layers and a middle insulation layer. However, the weakness caused by connectors between the insulation layer and concrete layers and the large requirement of on-site wet work suppresses the applications of PSPWs. For overcoming the above problems, this paper presents a new precast hollow insulated shear wall, which consists of a hollow concrete shear wall and filled-in-hollow polyurethane insulation materials, as shown in Fig. 1. The insulation materials can be directly embedded in the concrete volume at fabrication, which avoids using weak connectors between the thermal insulation material and the concrete.

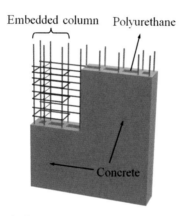

Fig. 1. Precast hollow insulation shear wall.

To analyze the seismic performances of the newly proposed PHISW, 6 shear wall specimens of 3 different configurations were produced and subjected to low-cyclic reversed loading tests. The obtained failure pattern, strength, displacement ductility, and energy dissipation capacity of the specimens were meticulously studied to assess the seismic performance of the new structure.

2 Experimental Investigation

2.1 Test Walls

In this paper, according to the different axial compression ratios, two groups of 6 full-sized specimens were designed, including two cast-in-place solid shear wall (CSW) specimens, two precast monolithic hollow insulated shear wall (PMW) specimens, and two precast hollow insulated shear wall (PSW) specimens with vertical seams, which were designated CSW30, PMW30, PSW30, CSW50, PMW50, and PSW50 (30 and 50 indicate the axial compression ratio by percentage).

The CSW specimens were designed to be 2900 mm high with a cross-section of 1300 mm × 200 mm, as shown in Fig. 2(a). The heights of PMW and PSW were the same as those of CSW, while the cross-section size was 1300 mm × 250 mm, as shown in Fig. 2(b) and (c), because of the insulation fillers. The horizontal reinforcement of the PSW was connected by a straight thread sleeve. The associated parameters of the specimens are presented in Table 1.

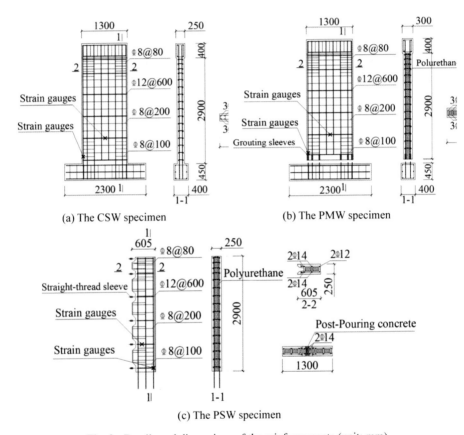

(a) The CSW specimen

(b) The PMW specimen

(c) The PSW specimen

Fig. 2. Details and dimensions of the reinforcements (unit: mm).

2.2 Material Properties

After the test, several concrete core samples with a diameter of 75 mm were picked from the undamaged area of each specimen by drilling. The core drilling method for the compression tests was employed based on the Chinese standard [8]. Table 2 listed the relevant properties of concrete. The strength of reinforcement was measured by a tensile test. The grade of reinforcement used in the shear wall is HRB400, and the diameters are 6 mm, 8 mm, 10 mm, 12 mm, and 14 mm. The achieved yielding and ultimate strengths are summarized in Table 3.

Table 1. Specimen description.

Specimens	Axial compression ratio	Construction forms	Section size (mm)
CSW30	0.3	Cast-in-place solid wall	200 × 1300
PMW30	0.3	Precast monolithic hollow insulated wall	250 × 1300
PSW30	0.3	Precast hollow insulated wall with vertical seams	250 × 1300
CSW50	0.5	Cast-in-place solid wall	200 × 1300
PMW50	0.5	Precast monolithic hollow insulated wall	250 × 1300
PSW50	0.5	Precast hollow insulated wall with vertical seams	250 × 1300

Table 2. Properties of concrete.

Category	Class	f_{cu} (MPa)
CSW30	C35	32.6
PMW30	C50	52.7
PSW30	C50	52.8
CSW50	C35	51.7
PMW50	C50	52.6
PSW50	C50	47.3

Table 3. Properties of steel reinforcements.

Category	f_y (MPa)	f_u (MPa)
6	–	612.5
8	–	646.1
10	–	633.7
12	453	616.3
14	436	641.5

Note: After cold treatment, the yield strength of steel bars with diameters of 6, 8, and 10 was not measured.

2.3 Experimental Setup and Measuring System

The schematic of the loading device is shown in Fig. 3. Considering that the PMW and PSW were hollow, their bearing capacity was uncertain, so the axial pressure of the pre-cast specimen was determined according to the CSW. The corresponding vertical loads under axial compression ratios of 0.3 and 0.5 are 1340 kN and 2240 kN, respectively.

As shown in Fig. 3, the applied loadings generated by the hydraulic actuator were measured by transducers built inside. The displacement of the top of the wall was gauged using LVDT-1. The transducer LVDT-2 for monitoring the displacement of the bottom beam ensured that the bottom beam had no slippage displacement. In addition, high-precision sensors were installed to the longitudinal reinforcements and stirrups at the reference points, as represented in Fig. 2, monitoring the rebar strain state. A hybrid loading protocol alternately controlled by force and displacement was employed in horizontal loading, as illustrated in Fig. 4.

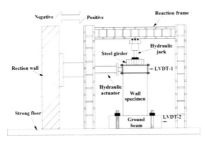

Fig. 3. Test setup.

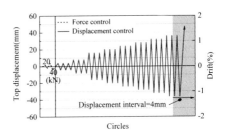

Fig. 4. Loading protocol.

3 Experimental Results

3.1 Cracks and Failure Patterns

The six specimens exhibited similar breakage patterns dominated by flexural failure, including concrete spalling in the corners, and the appearance of bend and shear cracks at both sides of the wall. Typically, the seismic capacity of the reinforced concrete shear wall can be evaluated with crack distribution [9, 10]. Figure 5 reveals the crack distribution of the 6 specimens.

However, the evolution of the cracks differed markedly. The number of cracks of specimens under an axial compression ratio of 0.5 was significantly less than those of 0.3. The cracks of the CSW and PMW were evenly distributed with abundant development compared with the PSW specimens. This could be attributed to the gap growth of the vertical assembling seam in the PSW, which led to insufficient crack propagation. Because of the existence of a vertical assembling seam, PSW specimens had fewer shear cracks, which can make full use of the strength of the material [11].

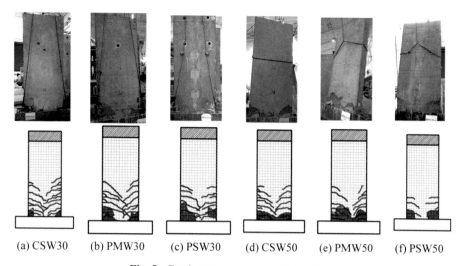

(a) CSW30   (b) PMW30   (c) PSW30   (d) CSW50   (e) PMW50   (f) PSW50

Fig. 5. Crack patterns under cyclic loading.

3.2 Strain Distributions of the Reinforcement Rebars

The strain distributions of the vertical reinforcement bars measured by the arranged gauges are plotted in Fig. 6. Before the specimen cracked, the strain of the longitudinal reinforcements in the most lateral of the wall appeared linearly. In the plastic phase, with the neutral axis of the specimen moving to the compression side, the strain of the vertical tension reinforcement was far larger than that of the compressed reinforcement. The longitudinal bars of the PMW specimens yielded earlier than those of the CSW and PSW.

Figure 7 demonstrates the strain evolution ruler of the middle stirrup of the wall at a height of 650 mm from the ground beam. The strain developed slowly before the horizontal displacement reached 20 mm. However, due to the extension of flexural-shear cracks, when the horizontal displacement was larger than 20 mm, the stirrup strain increased significantly, and the ultimate strain of the PMW specimens was larger than that of the other two types of specimens.

3.3 Hysteresis Behavior and Envelope Curves

The property of hysteretic loops of the specimens is illustrated in Fig. 8. At the beginning of loading, all specimens worked elastically, and the hysteresis loop area, i.e., the dissipated energy, was rather small. With increasing loading displacement, the specimens gradually participated in the elastic-plastic phase, and the area of the hysteresis curve increased gradually. The stiffness of the specimen decreased obviously after reaching the peak load, and the pinching phenomenon began to appear in the hysteresis curves. In the end, the hysteretic area of the PSW specimens was smaller than that of the other two types of specimens according to the slight slip of the straight thread sleeves under high-stress repeated loading.

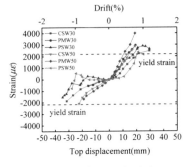

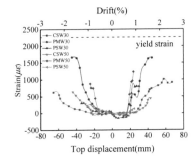

Fig. 6. The strain of vertical reinforcement.   **Fig. 7.** The strain of stirrup.

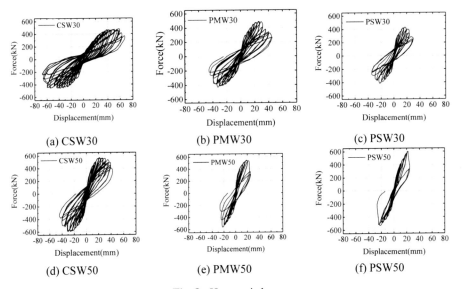

(a) CSW30   (b) PMW30   (c) PSW30

(d) CSW50   (e) PMW50   (f) PSW50

Fig. 8. Hysteresis loops.

Table 4 listed the cracking load F_{cr}, yielding load F_y and peak load F_P of the specimens. The mean values of F_y of PMW30 and PSW30 were approximately 3.6% and 13.1% lower than those of CSW30, respectively. Under the same axial pressure, the average F_P values of PMW30 and PSW30 were 2.1% and 9.8% lower than those of CSW30, respectively. The bearing capacity at feature points of wall specimens with an axial load ratio of 0.5 is greater than those with small axial pressure. Significantly, the F_{cr} values of PSW30 and PSW50 were 7.9% and 8.4% greater than those of CSW30 and CSW50, respectively. Because of the existence of the vertical assembling seam, PSW specimens can convert the overall shear failure of the ordinary shear wall into the bending failure of each wall limb element and can capitalize the strength of the material [11].

Table 4. The bearing capacity of the characteristic points (unit: kN).

Specimens	F_{cr}			F_y			F_P		
	Pos	Neg	Ave	Pos	Neg	Ave	Pos	Neg	Ave
CSW30	287.15	281.08	284.12	410.22	380.31	395.26	477.32	460.98	469.15
PMW30	297.45	238.38	267.92	398.10	363.95	381.03	485.98	432.33	459.16
PSW30	329.81	283.58	306.70	353.40	333.25	343.32	428.84	417.64	423.24
CSW50	387.80	412.10	399.95	468.59	486.17	477.38	565.91	593.53	579.72
PMW50	312.51	416.35	364.43	448.59	448.47	448.53	543.20	556.86	550.03
PSW50	442.91	424.50	433.71	508.35	427.35	467.85	602.80	525.60	564.20

Table 5. Displacement at feature point (mm) and ductility coefficient.

Specimens	Δ_{cr}		Δ_y		Δ_μ		μ		Ave
	Pos	Neg	Pos	Neg	Pos	Neg	Pos	Neg	
CSW30	16.04	12.10	29.54	19.64	62.58	59.47	2.12	3.03	2.48
PMW30	8.09	8.66	15.05	18.97	31.96	36.82	2.12	1.94	2.03
PSW30	8.70	8.09	10.22	12.64	21.26	28.03	2.08	2.22	2.15
CSW50	8.00	8.12	13.09	13.39	37.99	37.99	2.90	2.84	2.87
PMW50	4.34	12.55	10.42	13.50	23.45	23.93	2.25	1.77	2.01
PSW50	12.14	12.45	16.49	13.79	23.08	24.81	1.40	1.80	1.60

3.4 Ductility

As the axial load ratio increased, the ductility of PMW and PSW was reduced. The ductility coefficients of PMW30 and PSW30 were 18.1% and 13.4% lower than those of CSW30, respectively. However, the ultimate drifts were 1.21% and 0.85%, respectively, which were larger than the criterion of 1/120 (0.83%) specified by Chinese codes [12]. The ductility coefficients of PMW50 and PSW50 with a designed axial compression ratio of 0.50 were 29.9% and 44.3% lower than those of CSW50, and their ultimate drifts were 0.817% and 0.828%, respectively, which were slightly smaller than the limit value of 1/120 (0.83%). Therefore, the four precast hollow insulated shear wall specimens exhibited good ductility. It's worth noting that the ductility of PSW was better under an axial compression ratio of 0.3, because larger axial pressure would weaken the structural integrity (Table 5).

3.5 Energy Dissipation

Energy-absorbing ability can also be used to evaluate seismic performance [13]. The cumulative hysteretic energy E and equivalent viscous damping coefficient h_e are normally used in seismic analysis of structures [14]. Figure 9 shows the cumulative energy dissipation of specimens E. And the equivalent viscous damping coefficient h_e is exhibited in Fig. 10. From the diagram, the trends of the E and h_e of the three types of 6 specimens were almost the same. The equivalent viscous damping coefficient h_e of the PMW and CSW increased faster along with enhanced axial pressure, but that of the PSW did not change significantly. This means that due to the existence of a vertical seam, PSW was less vulnerable to the influence of the axial force.

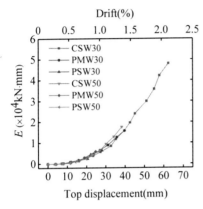 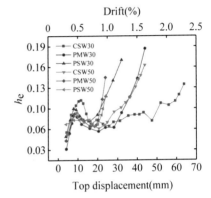

Fig. 9. Cumulative energy dissipation.   **Fig. 10.** Equivalent viscous damping coefficient.

4 Conclusions

In this paper, a new precast shear wall filled with polyurethane materials for thermal insulation is put forward and investigated. A low cyclic loading test was carried out about six full-scale specimens, which include two cast-in-place solid shear wall specimens, two precast monolithic hollow insulated shear wall specimens, and two precast hollow insulated shear wall specimens with vertical seams. Based on the test results, some meaningful conclusions can be achieved:

(1) Bending failure was the primary destructive mode of cast-in-place and precast walls, although the crack distributions were different. The cracks of the cast-in-place wall are denser than those of the precast wall. In addition, the cracks of the PSW continuously ran through the vertical seams of the wallboard, which manifested that the PSW had a good cooperative performance.

(2) The peak bearing capacity of the precast shear wall newly brought forward was close to that of the cast-in-place wall. And the hysteretic area of the PMW was larger than that of the PSW.

(3) The ductility of the two kinds of precast walls was slightly worse than that of the cast-in-place walls under the two axial compression ratios. However, the ductility coefficient of the precast wall came close to the value of Chinese specification, which indicated good ductility performance of the precast wall.

References

1. Wang, Z., Pan, W., Zhang, Z.Q.: High-rise modular buildings with innovative precast concrete shear walls as a lateral force resisting system. Structures **26**, 39–53 (2020)
2. Smith, B.J., Kurama, Y.C., McGinnis, M.J.: Behavior of precast concrete shear walls for seismic regions: comparison of hybrid and emulative specimens. J. Struct. Eng. **139**, 1917–1927 (2013)
3. Michele, P., Tomaso, T.: Experimentally-validated modeling of thin RC sandwich walls subjected to seismic loads. Eng. Struct. **119**, 95–109 (2016)
4. Salmon, D.C., Eiena, A., Tadros, M.K., Culp, T.D.: Full scale testing of precast concrete sandwich panels. ACI Struct. J. **94**, 354–362 (1997)
5. Bush, T.D., Wu, Z.: Flexural analysis of prestressed concrete sandwich panels with truss connectors. PCI J. **43**, 76–86 (1998)
6. Palermo, M., Ricci, I., Silvestri, S., et al.: Preliminary interpretation of shaking-table response of a full-scale 3-storey building composed of thin reinforced concrete sandwich walls. Eng. Struct. **76**, 75–89 (2014)
7. Ricci, I., Palermo, M., Gasparini, G., Silvestri, S., Trombetti, T.: Results of pseudo-static tests with cyclic horizontal load on cast in situ sandwich squat concrete walls. Eng. Struct. **54**, 131–149 (2013)
8. JGJ/T384-2016: Technical Specification for Testing Concrete Strength with Drilled Core Method. China Architecture Industry Press (2016)
9. Ebrahimkhanlou, A., Farhidzadeh, A., Salamone, S.: Multifractal analysis of crack patterns in reinforced concrete shear walls. Struct. Health Monitor.-An Int. J. **15**, 81–92 (2016)
10. Zhi, Q., Guo, Z.X., Xiao, Q.D., et al.: Quasi-static test and strut-and-tie modeling of precast concrete shear walls with grouted lap-spliced connections. Constr. Build. Mater. **150**, 190–203 (2017)
11. Cui, Z.: An experimental research on seismic behavior of reinforced concrete hollow shear wall with vertical seams. Xi'an University of Architecture and Technology (2004)
12. GB 50011-2010: Code for seismic design of buildings. China Architecture Industry Press (2010)
13. Rao, G.A., Poluraju, P.: Cyclic behaviour of precast reinforced concrete sandwich slender walls. Structures **28**, 80–92 (2020)
14. JGJ/T101-2015: Specification for Seismic Test of Buildings. China Architecture Industry Press (2015)

Study on Surface Deformation Model Induced by Shield Tunneling Based on Random Field Theory

Jianbin Li[1,2(✉)], Junguang Huang[1], Huawei Tong[2], and Shankai Zhang[3]

[1] Guangzhou MTR Design Academy Co., Ltd., 3 TIYU Dongheng Street, TIYU East Road, Tianhe District, Guangzhou, China
lijianbin1992@foxmail.com
[2] School of Civil Engineering, Guangzhou University, 230 Waihuan West Road, University Town, Guangzhou, China
[3] School of Urban and Environmental Sciences, Huaiyin Normal University, 111 Changjiang West Road, Huai'an, China

Abstract. Based on the shield tunnel engineering in weathered granite stratum in Xiamen, Stochastic calculations, by combining the random field theory and the finite difference analysis together with Monte Carlo simulation, are used to carry out the change law of the characteristics of surface deformation curve and surface deformation model. Results show that with the increase of the vertical scales of fluctuation, the decrease of the transverse scales of fluctuation or the increase of the coefficient of variation, the low peak distribution characteristics of the location of the maximum surface settlement induced by shield tunneling become more obvious, and the randomness and chaos of the shape of surface deformation curve gradually increase. The diversity of surface deformation model is affected by parameter correlation and randomness. Under the condition of small transverse scales of fluctuation and large vertical scales of fluctuation, the sensitivity of coefficient of variation to surface deformation mode is limited.

Keywords: Shield tunnel engineering · Surface deformation model · Random field theory · Monte Carlo simulation · Distribution characteristics

1 Introduction

Rail transit shield tunnel with shallow buried depth often passes through the urban core area. The site stratum is seriously weathered, the underground pipe network is dense, and there are many ground buildings. When the soil disturbance caused by tunnel construction is transmitted to the surface, it will lead to surface deformation. And if the stratum disturbance is large, it will bring risks to the adjacent existing rail transit lines and buildings (structures). Therefore, mastering the law of surface deformation caused by shield tunnel construction is the key link to realize project safety risk control. For the problem of stratum deformation caused by tunnel construction, Many researches have been carried out by domestic and abroad researchers, but they mostly ignore the influence of spatial

© The Author(s) 2022
G. Feng (Ed.): ICCE 2021, LNCE 213, pp. 440–454, 2022.
https://doi.org/10.1007/978-981-19-1260-3_40

variability of geotechnical parameters. Random field theory [1] provides an effective means to describe the spatial variation characteristics of geotechnical parameters. In recent years, it is widely used in the reliability analysis of tunnel engineering. Based on the random field theory, Cheng et al. [2] studied the effects of elastic modulus variation coefficient and scales of fluctuation on surface settlement. Considering the spatial variability of parameters, Wen et al. [3] analyzed the mechanical response of surrounding rock after large section tunnel excavation. Li et al. [4] studied the influence of scales of fluctuation and variation coefficient of soil elastic modulus on stratum deformation during twin shield tunnel construction. Miro et al. [5] studied the influence of parameter distribution type on surface deformation and pointed out that when the variability is small, the influence of parameter probability distribution type is very small. Xiao et al. [6] studied the influence of low stiffness random field location on surface deformation during tunnel construction.

At present, the analysis on the surface deformation response of tunnel construction considering the spatial variability of parameters mostly focuses on the influence of the spatial variability of parameters on the digital characteristics of deformation and the dispersion degree of deformation curve. There are relatively few systematic studies on the surface deformation mode of shield tunnel due to the spatial variability of geotechnical parameters, It is very important to clarify the meso characteristics and mode of surface deformation curve for the protection of shallow old houses and underground pipelines. In view of this, taking the spatial variability of soil elastic modulus as the starting point, this paper systematically studies the influence of two basic spatial variability characteristics of parameter spatial autocorrelation (vertical and transverse scales of fluctuation) and randomness (variation coefficient) on surface deformation during shield tunnel construction by using the combination of random field theory, finite difference method and Monte Carlo strategy. The change law of the shape and characteristics of surface deformation curve is discussed, and the surface deformation model is summarized and refined.

2 Random Analysis Method of Surface Deformation in Shield Construction

2.1 Parameter Spatial Variability

The geotechnical parameters have the dual characteristics of local randomness and overall structure. The random field theory regards the geotechnical parameters at any point as a random variable that approximately obeys a certain probability distribution. The spatial structure of the parameters is characterized by spatial concepts such as fluctuation range and autocorrelation structure.

2.2 Random Analysis Process of Surface Deformation

The random analysis method of surface deformation induced by shield tunneling based on random field theory is constructed. The analysis process is shown in Fig. 1: (1) The spatial variability characteristics of geotechnical parameters is statistically analysed, including the probability distribution characteristics of parameters (mean value, standard

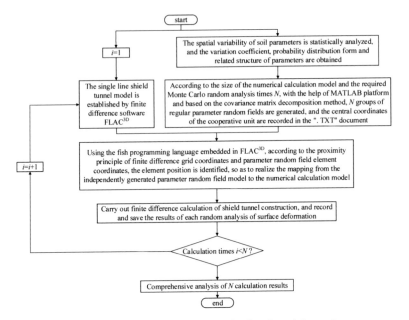

Fig. 1. Flow chart of random analysis of surface deformation.

deviation and distribution type) and spatial correlation characteristics (related structure and scales of fluctuation). The finite difference software FLAC^{3D} is used to construct the numerical model of shield tunnel, divide the grid and record the model size. (2) The parametric random field model is generated with the help of MATLAB platform. (3) Realize the mapping from the independently generated parameter random field model to the numerical calculation model. (4) With the help of Monte Carlo strategy, repeat steps (2)–(4) to realize multiple random analysis of surface deformation caused by shield tunnel construction. (5) With the help of probability statistics method, the surface deformation results obtained by Monte Carlo random calculation are analyzed.

3   Characteristics and Mode Analysis of Surface Deformation Curve

3.1   Numerical Calculation Model

Based on the shield tunnel project in weathered granite stratum in Xiamen, this paper simplifies the tunnel excavation problem into a two-dimensional plane strain model to carry out the random analysis of the surface deformation response of shield construction. Model size is 76 m × 34 m (width × Height), tunnel diameter is $D = 6.2$ m, axis buried depth is $H = 15.4$ m, and the maximum size of grid is about 0.75 m. Except that the surface is a free boundary, other boundaries are subject to normal constraints. The elastic-perfectly plastic body of M-C yield criterion is adopted for the soil, and the shell element is adopted for the simulation of lining structure. The values of physical and mechanical parameters of materials are shown in Table 1. The model is divided into upper and lower

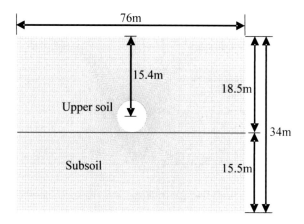

Fig. 2. Schematic diagram of model

Table 1. Physical and mechanical parameters of soil and segments in the model.

Medium	Material name	Lining thickness d (m)	Gravity γ (kN•m^{-3})	Elastic modulus E (Mpa)	Poisson's ratio μ	Friction angle φ (°)	Cohesion c (kPa)
Soil mass	Elastic-perfectly Plastic material	–	1850	25.0 (mean value)	0.35	25.0	25.0
Lining structure	Linear elastic material	0.35	2450	24.44e^3	0.20	–	–

layers. When assigning the elastic modulus of subsoil, taking 3 times of the original modulus [7] to simply consider the loading and unloading characteristics of soil. The numerical calculation model is shown in Fig. 2.

This paper focuses on the influence of the spatial variability of soil elastic modulus on the surface deformation mode of shield tunnel construction, and other physical and mechanical parameters are constant. Considering the calculation scale of Monte Carlo random simulation, and the surface deformation mainly occurs in the stress release stage [7], the calculation result analysis is only carried out for the surface deformation in the stress release stage.

3.2 Deterministic Analysis

Figure 3 shows the surface settlement curve under different stress release coefficients λ. According to the figure, it can be seen that the stress release coefficient increases, the surface settlement value increases accordingly, and the fitting results of peck formula are in good agreement with the numerical calculation results, which basically obeys the Gaussian distribution. It should be pointed out that the actual engineering situation and

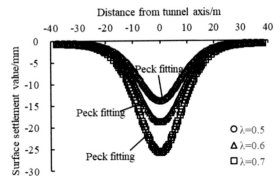

Fig. 3. Relationship between stress release coefficient and surface settlement deformation curve.

site environment are complex and there are many influencing factors, so it is generally difficult to accurately balance the corresponding stress release coefficient.

3.3 Random Analysis

Combined with random field theory, finite difference method and Monte Carlo strategy, the random analysis of surface deformation during shield tunnel construction is carried out. Without losing generality, the stress release coefficient is taken $\lambda = 0.5$ [7], the effects of vertical and transverse scales of fluctuation (θ_z, θ_x) and variation coefficient (*COV*) of soil elastic modulus on surface settlement curve model are systematically studied.

Referring to the suggestions on the value of scales of fluctuation of geotechnical parameters [8] (the transverse scales of fluctuation is generally 10.0–80.0 m and the vertical scales of fluctuation is 1.0–3.0 m), the basic value of soil elastic modulus scales of fluctuation is selected as $\theta_z = 0.3D = 1.86$ m and $\theta_x = 6.0D = 37.2$ m. On this basis, the random analysis condition is designed and divided into three types of random calculation condition groups, including MCS-z*-x (variable θ_z, there are 20 simulated working conditions, see Table 2, MCS-z-x* (variable θ_x, A total of 20 simulated working conditions) and MCS-E*-θ (variable *COV*, a total of 15 simulation conditions, see Table 3. The log normal distribution [9] is used to describe the uncertainty of soil elastic modulus. And the log modulus field satisfies the anisotropic exponential autocorrelation function [10], which can be expressed as

$$\rho_{\ln E}(\tau_x, \tau_z) = \exp\left(-\frac{2\tau_x}{\theta_x} - \frac{2\tau_z}{\theta_z}\right) \tag{1}$$

Where $\rho_{\ln E}(\tau_x, \tau_z)$ is the autocorrelation coefficient of two points in the logarithmic modulus field, and $0 \leq \rho_{\ln E}(\tau_x, \tau_z) \leq 1$, τ_x, τ_z are the horizontal and vertical distances respectively, θ_x, θ_z are transverse and vertical scales of fluctuation respectively.

Considering the calculation accuracy and efficiency, 1000 times are selected as the calculation times of random analysis under each working condition.

Table 2. Random analysis conditions of vertical scales of fluctuation.

Case name	Variable	Coefficient of variation COV	Scales of fluctuation		Case name	Variable	Coefficient of variation COV	Scales of fluctuation	
			Vertical θ_z	Transverse θ_x				Vertical θ_z	Transverse θ_x
MCS-z1-x1	θ_z	0.3	0.2D	1.5D	MCS-z1-x4	θ_z	0.3	0.2D	9.0D
MCS-z2-x1			0.3D		MCS-z2-x4			0.3D	
MCS-z3-x1			0.4D		MCS-z3-x4			0.4D	
MCS-z4-x1			0.5D		MCS-z4-x4			0.5D	
MCS-z1-x2	θ_z	0.3	0.2D	3.0D	MCS-z1-x5	θ_z	0.3	0.2D	12.0D
MCS-z2-x2			0.3D		MCS-z2-x5			0.3D	
MCS-z3-x2			0.4D		MCS-z3-x5			0.4D	
MCS-z4-x2			0.5D		MCS-z4-x5			0.5D	
MCS-z1-x3	θ_z	0.3	0.2D	6.0D					
MCS-z2-x3			0.3D						
MCS-z3-x3			0.4D						
MCS-z4-x3			0.5D						

Noting that tunnel diameter is $D = 6.2$ m, the same below.

Table 3. Coefficient of variation random analysis condition.

Case name	Variable	Coefficient of variation COV	Scales of fluctuation		Case name	Variable	Coefficient of variation COV	Scales of fluctuation	
			Vertical θ_z	Transverse θ_x				Vertical θ_z	Transverse θ_x
MCS-E1-θ1	COV	0.1	0.3D	1.5D	MCS-E1-θ3	COV	0.1	0.3D	6.0D
MCS-E2-θ1		0.2			MCS-E2-θ3		0.2		
MCS-E3-θ1		0.3			MCS-E3-θ3		0.3		
MCS-E4-θ1		0.4			MCS-E4-θ3		0.4		
MCS-E5-θ1		0.5			MCS-E5-θ3		0.5		
MCS-E1-θ2	COV	0.1	0.3D	3.0D					
MCS-E2-θ2		0.2							
MCS-E3-θ2		0.3							
MCS-E4-θ2		0.4							
MCS-E5-θ2		0.5							

3.4 Analysis of Surface Deformation Curve Results

Surface deformation is an important index to reflect the impact of shield tunnel construction on the surrounding environment. Based on the random calculation results of surface deformation of three working conditions, the characteristics and types of settlement curve are discussed from the distribution of the location of maximum surface settlement, and the surface deformation mode is summarized and refined.

3.4.1 Shape Analysis of Surface Deformation Curve

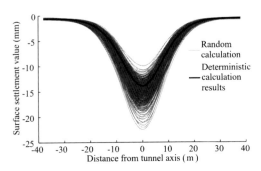

Fig. 4. Surface deformation curve obtained by random and deterministic calculation.

Figure 4 shows the ground settlement results of tunnel construction obtained from 1000 random calculations (gray curve in the figure) and deterministic calculations (modulus is the mean value used for random analysis) (black curve in the figure) under the working condition which the scales of fluctuation is $\theta_x = 6.0D, \theta_z = 0.5D$ and coefficient of variation is $COV = 0.3$. For each realization of the parameter random field, the soil elastic modulus is spatially heterogeneous, and the calculation results are also different, which is shown as a cluster of discrete curves. Relevant studies [2, 4] have also shown that with the increase of spatial correlation and randomness of parameters, the dispersion degree of settlement curve distribution caused by tunnel construction increases accordingly, and the influence of parameter variation coefficient is more significant than scales of fluctuation.

3.4.2 Analysis of Surface Deformation Model

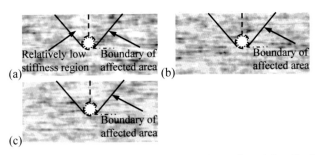

Fig. 5. Relationship between maximum settlement location and modulus distribution. (a) The maximum settlement position is left. (b) The maximum settlement position is right. (c) The maximum settlement is located on the axis.

Comparing the settlement curve characteristics of deterministic analysis and stochastic analysis in Fig. 4, it can be seen that the surface settlement curve obtained by deterministic analysis is single peak, and the maximum value is located above the tunnel axis.

Although the settlement curve obtained by random analysis still presents a single peak distribution, it is different from the deterministic results. The maximum settlement is no longer above a single tunnel axis. The specific location of the maximum settlement is closely related to the random distribution of soil modulus within the influence range of tunnel excavation [11]. When a relatively high (low) stiffness area appears on one side of the tunnel axis, the maximum value of the surface deformation curve obtained by random analysis will also deviate (deviate) from that side, which is more appropriate to the on-site monitoring [12], as shown in Fig. 5. Obviously, this influence will gradually weaken with the increase of the distance between the high (low) stiffness area and the excavation surface. When the high (low) stiffness area is outside the influence range of tunnel excavation, the influence on surface deformation can be ignored.

The location of the maximum surface settlement is counted, and the influence of the spatial variability of soil parameters on the surface deformation model is discussed.

(1) Location of maximum surface settlement

1) Influence analysis of vertical scales of fluctuation

Figure 6 shows the distribution of the location of the maximum surface settlement caused by tunnel excavation under the conditions of different vertical scales of fluctuation in MCS-z^*-x working condition group (4 groups of different transverse scales of fluctuation). It can be seen that considering the influence of spatial variability of modulus, the location of maximum surface settlement caused by tunnel excavation is a probability distribution interval, in which Under the condition of $\theta z = 0.5D$ and $\theta x = 6.0D$, the distribution range of maximum surface settlement is $(-1.1625, 1.1625)$.

As can be seen from Fig. 6(a)–(d), with the increase of vertical scales of fluctuation, the interval where the maximum surface settlement occurs expands, the dispersion degree also increases, and its low peak distribution characteristics become more obvious. Among which under the condition of $\theta z = 0.5D$ and $\theta x = 1.5D$, the probability that the maximum settlement point is directly above the tunnel is only 29.7%. The main reason for this phenomenon is that the size of the element concentration area in the high (low) stiffness area in the parameter random model is affected by the scales of fluctuation, and the correlation of the horizontal parameters remains unchanged. The larger the vertical scales of fluctuation, the probability of a large range of high (low) stiffness area increases accordingly, and the probability of asymmetric distribution of parameters on both sides of the tunnel also increases.

2) Influence analysis of transverse scales of fluctuation

Figure 7 shows the distribution of the location of the maximum surface settlement caused by tunnel excavation under the conditions of different transverse scales of fluctuation in MCS-x^*-z working condition group (4 groups with different vertical scales of fluctuation). It can be seen that under different transverse scales of fluctuation conditions, the distribution law of the location where the maximum surface settlement occurs is opposite to the vertical scales of fluctuation condition. With the increase of transverse scales of fluctuation, the peak distribution characteristics of the location where the maximum surface

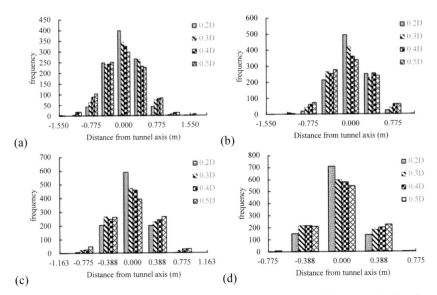

Fig. 6. Location statistics of maximum surface settlement under different vertical scales of fluctuation conditions. (a) $\theta_x = 1.5D$. (b) $\theta_x = 3.0D$. (c) $\theta_x = 6.0D$. (d) $\theta_x = 12.0D$.

settlement occurs are prominent. Among which Under the condition of $\theta_x = 12.0D$ and $\theta_z = 0.2D$, the probability that the maximum settlement point is directly above the tunnel is as high as 70.9%. This is mainly because in the anisotropic random field, with the increase of transverse scales of fluctuation, the correlation of horizontal parameters strengthens and gradually tends to the horizontal mean value, that is, the overall zonal distribution. The distribution of geotechnical parameters on both sides of the tunnel axis is more symmetrical and uniform, and the probability of relatively high (low) stiffness area on one side is small. Therefore, the location of maximum settlement value is mostly directly above the tunnel.

3) Influence analysis of coefficient of variation

Figure 8 shows the distribution of the location of the maximum surface settlement caused by tunnel excavation under the conditions of different variation coefficients in MCS-E^*-θ working condition group (two groups with different scales of fluctuation). It can be seen that with the increase of modulus variation coefficient, the dispersion degree of location distribution of maximum surface settlement also increases, where under the condition of $COV = 0.5$, $\theta_z = 0.3D$ and $\theta_x = 1.5D$, the probability that the maximum settlement point is directly above the tunnel is only 23.5%. This is mainly because the larger the coefficient of variation in the parameter random field model, the probability of asymmetric distribution of parameters on both sides of the tunnel will be significantly improved.

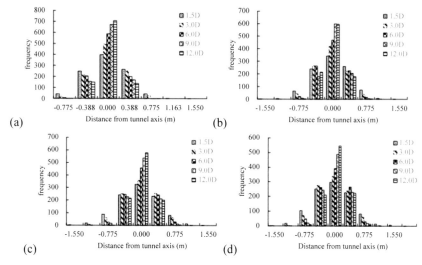

Fig. 7. Location statistics of maximum surface settlement under different transverse scales of fluctuation conditions. (a) $\theta_z = 0.2D$. (b) $\theta_z = 0.3D$. (c) $\theta_z = 0.4D$. (d) $\theta_z = 0.5D$

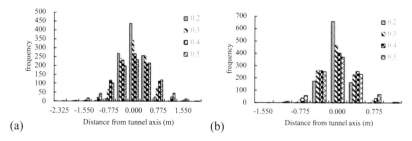

Fig. 8. Location statistics of maximum surface settlement under different coefficient of variation conditions. (a) $\theta_x = 1.5D$. (b) $\theta_x = 6.0D$.

(2) Settlement curve type

According to the different positions of the maximum surface settlement, the surface settlement curve is divided into three types. I –III successively represent the single peak type which the wave peak is on the left, the single peak which type wave peak is above the tunnel axis, and the single peak type which wave peak is on the right. The distribution of various surface deformation modes under various random working conditions is studied.

1) Influence analysis of vertical scales of fluctuation

Table 4 shows the variation of the type and number of surface deformation curves with the vertical scales of fluctuation. It can be roughly seen that in the MCS-z^*-x random working condition group, with the increase of the vertical scales of fluctuation, the probability of the same surface model as the deterministic analysis decreases, that is, the random characteristics of the surface settlement form gradually increase. Under different transverse scales of fluctuation conditions, the vertical scales of fluctuation increases from 0.2D to 0.5D, and the probability of type II settlement mode decreases by 10%–20%. Under the condition of $\theta_x = 1.5D$, the number of three settlement curves is roughly the same.

Table 4. Statistics of surface settlement curve types under different vertical scales of fluctuation conditions.

MCS-z*-x	θ_z	Type and number of surface deformation curves			MCS-z-x*	θ_x	Type and number of surface deformation curves		
		I	II	III			I	II	III
$\theta_x = 1.5D$	0.2D	290	398	312	$\theta_x = 9.0D$	0.2D	157	673	170
	0.3D	312	343	345		0.3D	189	599	212
	0.4D	350	325	325		0.4D	243	536	221
	0.5D	373	297	330		0.5D	264	489	247
$\theta_x = 3.0D$	0.2D	232	494	274	$\theta_x = 12.0D$	0.2D	151	709	140
	0.3D	313	422	265		0.3D	218	597	185
	0.4D	321	358	321		0.4D	217	578	205
	0.5D	354	338	308		0.5D	218	546	236
$\theta_x = 6.0D$	0.2D	209	589	202					
	0.3D	285	470	245					
	0.4D	271	458	271					
	0.5D	310	392	298					

Selecting a combination of scales of fluctuation (e.g. $\theta_x = 3.0D, \theta_z = 0.3D$), only one scales of fluctuation value is changed each time. Taking the number of class II curves as the research object, the influence of the change of vertical and transverse scales of fluctuation on the number of curves is analyzed. It is found that the vertical scales of fluctuation increases by only 0.1D, resulting in the change of the number of class II curves, which is roughly the same as that when the transverse scales of fluctuation increases by one time. This also shows that the vertical scales of fluctuation has a more significant impact on surface deformation.

2) Influence analysis of transverse scales of fluctuation

Similarly, Table 5 shows the relationship between the type and number of surface deformation curves and the transverse scales of fluctuation. It can be seen that with the increase of transverse scales of fluctuation, the probability of the same surface model as the deterministic analysis increases significantly, that is, the random characteristics of surface settlement form gradually weaken. Under different vertical scales of fluctuation conditions, the transverse scales

Table 5. Statistics of surface settlement curve types under different transverse scales of fluctuation conditions.

MCS-z-x*	θ_x	Type and number of surface deformation curves			MCS-z-x*	θ_x	Type and number of surface deformation curves		
		I	II	III			I	II	III
$\theta_z = 0.2D$	1.5D	290	398	312	$\theta_z = 0.4D$	1.5D	350	325	325
	3.0D	232	494	274		3.0D	321	358	321
	6.0D	209	589	202		6.0D	271	458	271
	9.0D	157	673	170		9.0D	243	536	221
	12.0D	151	709	140		12.0D	217	578	205
$\theta_z = 0.3D$	1.5D	312	343	345	$\theta_z = 0.5D$	1.5D	373	297	330
	3.0D	313	422	265		3.0D	354	338	308
	6.0D	285	470	245		6.0D	310	392	298
	9.0D	189	599	212		9.0D	264	489	247
	12.0D	218	597	185		12.0D	218	546	236

of fluctuation increases from 1.5D to 12.0D, the probability of type II settlement mode increases by 25%–30%, and the sensitivity of transverse scales of fluctuation to surface deformation mode is significantly weakened when the transverse scales of fluctuation increases to 9.0D. Under the condition of $\theta_x = 12.0D$, the probability of type II settlement mode is more than 50%.

3) Influence analysis of coefficient of variation

Table 6 shows the variation of the type and number of surface deformation curves with the modulus variation coefficient. With the increase of modulus variation coefficient, the random characteristics of surface settlement form gradually increase. Under different wave distance conditions, the coefficient of variation increases from 0.2 to 0.5, and the probability of type II settlement mode decreases by 20% –30%. When $\theta_x = 1.5D$ and $\theta_z = 0.3D$, the number of three settlement curves is roughly the same under different coefficient of variation conditions. Combined with the analysis conclusion of scales of fluctuation condition, it can be seen that the diversity of surface deformation modes is affected by parameter correlation and randomness.

Table 6. Statistics of surface settlement curve types under different coefficient of variation conditions.

MCS-z-x*	COV	Type and number of surface deformation curves			MCS-z-x*	θ_x	Type and number of surface deformation curves		
		I	II	III			I	II	III
$\theta_x = 1.5D, \theta_z = 0.3D$	0.2	282	438	280	$\theta_x = 6.0D, \theta_z = 0.3D$	0.2	176	659	165
	0.3	312	343	345		0.3	285	470	245
	0.4	144	231	625		0.4	300	404	296
	0.5	368	235	397		0.5	322	373	305
$\theta_x = 3.0D, \theta_z = 0.3D$	0.2	207	591	202					
	0.3	313	422	265					
	0.4	319	355	326					
	0.5	368	286	346					

4 Conclusion

Aiming at the random response of surface deformation caused by shield tunnel construction, combined with random field theory, finite difference method and Monte Carlo strategy, this paper constructs a random analysis method of surface deformation of shield tunnel construction based on random field theory, and systematically studies the influence of autocorrelation and randomness of soil elastic modulus on surface deformation mode of tunnel construction. The main conclusions are as follows:

(1) The location of the maximum surface settlement is closely related to the random distribution of parameters in the influence area above the tunnel. According to the location of the maximum surface settlement, three surface deformation modes are summarized.

(2) The size of the element concentration area in the high (low) stiffness area and the probability of asymmetric distribution of parameters on both sides of the tunnel are closely related to the scales of fluctuation and variation coefficient.

(3) With the increase of the vertical scales of fluctuation, the decrease of the transverse scales of fluctuation or the increase of the variation coefficient, the low peak distribution characteristics of the location of the maximum surface settlement caused by tunnel construction become more and more obvious. The influence of vertical scales of fluctuation is more significant than transverse scales of fluctuation, and the influence of parameter variation coefficient is significantly stronger than scales of fluctuation.

(4) With the increase of the vertical scales of fluctuation, the decrease of the transverse scales of fluctuation or the increase of the coefficient of variation, the probability of

the same surface model as the deterministic analysis decreases, that is, the randomness and chaos of the surface settlement curve gradually increase. When the vertical scales of fluctuation increases by $0.1D$, the change of the number of type II curves is roughly the same as that when the transverse scales of fluctuation increases by one time.

(5) The diversity of surface deformation modes is affected by parameter correlation and randomness, and there is an obvious superposition effect.

The existence of spatial variability of geotechnical parameters will have a significant impact on the deformation response characteristics of the ground surface during shield tunnel construction. Accurately estimating and characterizing the spatial variability of formation geotechnical parameters should be one of the main contents of routine design of shield tunnel engineering.

Acknowledgments. This work is supported by National Natural Science Foundation of China (No. 52008122 and No. 41807512).

References

1. Vanmarcke, E.H.: Random Fields: Analysis and Synthesis. MIT Press, Cambridge (1983)
2. Cheng, H.Z., Chen, J., Li, J.B., et al.: Study on surface deformation induced by shield tunneling based on random field theory. Chin. J. Rock Mech. Eng. **35**(Supp. 2), 4256–4264 (2016)
3. Wen, M., Zhang, D.L., Fang, Q.: Stochastic analysis of surrounding rock behavior of high speed railway tunnel considering spatial variation of rock parameters. Chin. J. Rock Mech. Eng. **36**(7), 1697–1709 (2017)
4. Li, J.B., Chen, J., Luo, H.X., et al.: Study on surrounding soil deformation induced by twin shield tunneling based on random field theory. Chin. J. Rock Mech. Eng. **37**(7), 1748–1765 (2018)
5. Miro, S., König, M., Hartmann, D., et al.: A probabilistic analysis of subsoil parameters uncertainty impacts on tunnel-induced ground movements with a back-analysis study. Comput. Geotech. **68**, 38–53 (2015)
6. Xiao, L., Huang, H.W., Zhang, J.: Effect of soil spatial variability on ground settlement induced by shield tunnelling, Geo-Risk [S. l.] [s. n.], pp. 330–339 (2017)
7. Möller, S.C.: Tunnel induced settlements and structural forces in linings, Ph. D. thesis. Stuttgart Germany Universität Stuttgart (2006)
8. El-Ramly, H., Morgenstern, N.R., Cruden, D.M.: Probabilistic stability analysis of a tailings dyke on presheared clay-shale. Can. Geotech. J. **40**(1), 192–208 (2003)
9. Fenton, G.A., Griffiths, D.V.: Three-dimensional probabilistic foundation settlement. J. Geotech. Geoenviron. Eng. **131**(2), 232–239 (2005)
10. Fenton, G.A., Griffiths, D.V.: Probabilistic foundation settlement on spatially random soil. J. Geotechn. Geoenviron. Eng. **128**(5), 381–390 (2002)
11. Wei, G.: Study on calculation for width parameter of surface settlement trough induced by shield tunnel. Ind. Constr. **39**(12), 74–79 (2009)
12. Wang, Z.C., Wang, H.T., Zhu, X.G., et al.: Analysis of stratum deformation rules induced by the construction of double-tube parallel shield tunnels for metro. China Railw. Sci. **34**(3), 53–58 (2013)

Analysis of Deformation Characteristics of Large Diameter Shield Tunnel with Construction Load

Weidong Zhu[1], Xiwen Zhang[1,2(✉)], and Liangliang Zhang[3]

[1] School of Civil Engineering and Architecture, University of Jinan, Jinan, Shandong, China
cea_zhangxw@ujn.edu.cn
[2] The Engineering Technology Research Center for Urban Underground Engineering Supporting and Risk Monitoring of Shandong Province, Jinan 250022, Shandong, China
[3] China Railway Siyuan Survey and Design Group Co., Ltd., Wuhan, Hubei, China

Abstract. With the rapid development of the method of shield tunnel in our country, the shield in the tunnel and the internal structure exist a lot of prefabricated and assembled internal structure can be roughly divided into full cast-in-place, full precast, and precast and cast-in-place structure. However, it has the disadvantages of more internal structural joints, poor waterproof performance, and complex structural stress. Therefore, combined with the actual engineering case, the construction process is simulated by using the finite element analysis software, and the influence of the segmental and internal structure cooperative deformation and joint mechanical performance is considered to further analyze the deformation characteristics of shield tunnel assembly internal structure, which can provide a reference for the construction and design of similar projects.

Keywords: Shield tunnel · Internal structure · Structure deformation · Construction load · Assembly structure

1 Introduction

For the past few years, large-diameter shield tunnels have been widely used in the construction of underwater tunnels and tunnels combining highway and metro, such as Nanjing Yangtze River tunnel, Wuhan Sanyang road Yangtze River tunnel, Jinan Yellow River tunnel, etc., because of its efficient utilization of space, high cost-effectiveness and small disturbance to surrounding soil [1, 2]. The internal space layout of shield tunnels presents a variety of ways, but the internal structure of shield tunnels is mainly semi-prefabricated in China [3, 4]. However, few studies on the stress analysis of the fabricated internal structure of shield tunnel at home and abroad, and there is no mature calculation theoretical basis; The construction and service conditions of the prefabricated assembled structure of the fabricated internal structure of large-diameter shield tunnel are also different from those of aboveground buildings [5–7]. Its construction technology and environment are more complex, and the requirements for the waterproof and seismic performance of the structure are strict.

© The Author(s) 2022
G. Feng (Ed.): ICCE 2021, LNCE 213, pp. 455–461, 2022.
https://doi.org/10.1007/978-981-19-1260-3_41

Therefore, it is necessary to further study the fabricated internal structure of large-diameter shield tunnels. At present, domestic scholars have studied the assembled underground structure. Tao [8] used ABAQUS software to calculate and study the overall stress and joint deformation of Yuanjiadian subway station of Changchun under different support modes; Taking Guangzhou Metro as an example, Zhang [9] established a three-dimensional coupling model by ABAQUS to study the dynamic response of shield tunnel lining and bolts under train load; Song [10] establishes load structure of multiple subway stations which is used to calculate and analyze the overall deformation, local joint deformation, structural stress and internal force of the fabricated underground station structure; Li [11] used Midas software to analyze and calculate the mechanical performance of arch and rectangular fabricated subway stations respectively, and the results show that the mechanical performance of arch structure stations is better; Li [12] establishes the three-dimensional discontinuous medium of shield segments and bending bolts Qualitative model and reasonable segment joint contact model are set to study the stress and deformation of shield segment under actual load; Liu [13] established joint test model, conducted joint test, and established joint bending stiffness model. The difference between the test value and the calculated value is compared, and the main reasons for this difference are analyzed. In the above research results, the mechanical properties of the prefabricated subway station and shield tunnel lining are analyzed through numerical simulation and full-scale experiment, but the internal structure of shield tunnels is rarely analyzed, and the mechanical characteristics of its internal structure joints are not clear.

Based on the finite element software ABAQUS, a three-dimensional finite element analysis model of soil layer-tunnel-internal structure is established. To provide a reference for the construction and design of similar projects, the deformation rule and internal force characteristics of shield tunnel internal structure under different load grades are systematically compared and studied.

2   Finite Element Model and Parameters

2.1   Model Overview

The segment of a river-crossing tunnel in Shandong Province, a tunnel combing highway and metro, is 2519.2 m long. The segment is made of C60 reinforced concrete, with an outer diameter of 15.2 m.

The ABAQUS finite element software is used to establish the calculation model of soil-tunnel segment-assembled internal structure. The Mohr-Coulomb constitutive model is used for the soil layer and the elastic constitutive model is used for concrete segments and assembled internal structure. Five ring segments are set in the model and five box culverts are set accordingly. The model size is $120 \times 75 \times 12$ m. The soil layer parameters are shown in the following Table 1. The finite element model is shown in Fig. 1. C3D8 element is used for the segment, internal structure, and soil layer. The box culverts are connected by longitudinal connecting bolts which grade is 6.8 with the yield strength of 480 MPa. Assuming that no relative slip and separation occurs between the asphalt concrete pavement and the internal structure, the "Tie" constraint is used between asphalt concrete pavement and internal structures. Static step analysis is used,

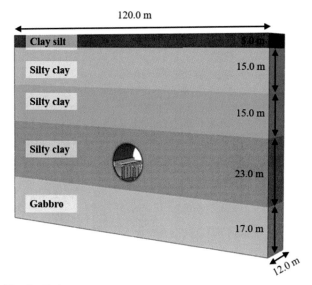

Fig. 1. Finite element model of tunnel segment-internal structure.

with fixed constraints at the bottom boundary of the model and horizontal and vertical constraints at both sides of the model.

2.2 Model Parameters

To improve the calculation efficiency, the elastic constitutive model is used for the reinforced concrete segment and the assembled internal structure. The segment is C60 concrete and the internal structure is C40 concrete. The Mohr-Coulomb constitutive model is used for the rock and soil. The basic physical and mechanical parameters of the model are shown in Table 1.

Table 1. Parameters of DEM simulations.

	Material	Elastic modulus E (MPa)	Poisson ratio μ	Density ρ (kg/m^3)	Cohesion c (kPa)	Friction angle φ (°)
1	Clay silt	28.56	0.30	1846.2	20.5	15.6
2	Silty clay	24.90	0.30	1897.2	18.7	10.4
3	Silty clay	30.18	0.30	2009.4	35.4	16.0
4	Silty clay	34.65	0.30	2019.6	40.5	24.6
5	Gabbro	8333.33	0.25	2244.0	40.0	38.0
Shield tunnel segment		36000.0	0.2	2500.0	–	–
Internal structures		32500.0	0.2	2400.0	–	–

2.3   Cases for Analysis

The construction of internal structures of the tunnel completed is regarded as the initial state of the calculation model after completion of construction. According to the driving positions of the transporting vehicles in the construction stage, the load is simplified into two equal concentrating forces which are applied to the span of the box culvert slab. The load level is divided into 5 levels, which are 100 kN, 150 kN, 200 kN, 250 kN, and 300 kN respectively. The cross section of the loading mode is shown in the following Fig. 2.

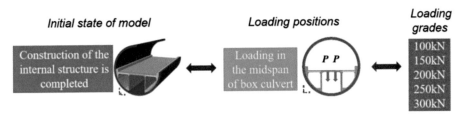

Fig. 2. Working condition diagram.

3   Result Analysis

3.1   Vertical Displacement Analysis

The displacement curves obtained under the monitoring path are shown in the following Fig. 3(a). Figure 3(b) extracted the vertical displacement values of each key position of the internal structure under different levels of load.

It can be seen from the following drawings that when construction load is applied to the span of the box culvert, the vertical deformation of internal structures mainly concentrates on the span of intermediate box culvert and decreases along the direction of box culvert support, which has an effect on the pre-fabricated lane slab after assembly on both sides. A certain amount of deformation also occurs at the support of the lane slab, and the greater the load level, the more obvious the effect is. With the increase of load grade, the vertical displacement of the key position of the internal structure increases linearly, and the growth rate is the largest at the middle position of the box culvert span. This is caused by the loading position close to the middle of the box culvert span. However, the growth rate of W1 and W2 positions far from the middle position of the box span is small and close.

3.2   Structural Internal Force Analysis

In addition, the response of rebar and connecting bolt of intermediate box culvert is analyzed. Figure 4 shows the rebar and connection bolts maximum principal stress clouds diagram under P = 300 kN. All rebar does not yield. Rebar of the top plate of the box culvert and the upper plate of land are subject to large tensile stress. The maximum principal stress is much lower than the yield strength, which is observed at the bottom of the roof in the loading box culvert with the largest deformation, and the maximum stress is 4.05 Mpa.

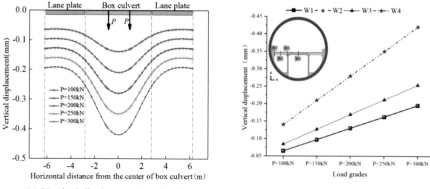

(a) Vertical displacement under the monitoring path.

(b) Vertical displacement at key positions.

Fig. 3. Vertical displacement.

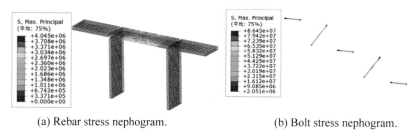

(a) Rebar stress nephogram.

(b) Bolt stress nephogram.

Fig. 4. Maximum principal stress of rebar and bolt.

3.3 Synergetic Deformation Analysis of Soil-Tunnel Segment-Assembled Internal Structures

Taking the calculation result of load P of 300 kN as an example, the cloud diagram of the calculation result when the loading position is in the middle of the box culvert span is shown as follows. It can be seen that the vertical deformation of the intermediate box culvert mainly concentrates on the box culvert at two coupling points where the load is applied. The maximum vertical displacement of the box culvert span reaches 0.42 mm; The soil-segment-internal structure has a cooperative deformation, and the soil around the segmental also has a certain settlement with the load applied (Figs. 5 and 6).

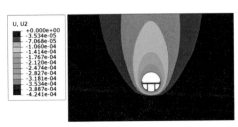 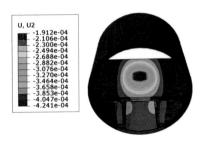

Fig. 5. Vertical deformation of the whole model.　**Fig. 6.** Vertical deformation of internal structure.

4　Conclusion

Based on ABAQUS finite element simulation, the simulated stress state is analyzed according to different load sizes. The following conclusions are drawn:

(1) When the inner structure of the shield tunnel is under construction load and the inner structure undergoes non-uniform longitudinal deformation, the box culvert concrete mainly bears tensile stress and the reinforcement and connecting bolts do not yield. When the load is applied, the soil-segment-internal structure synergistically deforms. With the load applied, the soil around the segmental also has a certain amount of settlement.

(2) When the construction load is applied to the middle of the box culvert span, the vertical deformation of the inner structure is mainly concentrated in the box culvert span. With the further construction of the inner structure, the maximum vertical displacement of the box culvert span is 0.4 mm, which indicates that the ability of the assembled inner structure to resist deformation is further improved.

(3) Under different load levels, the vertical displacement in the span of the box culvert increases with the increase of load.

References

1. Yu, C., Zhou, A.N., Chen, J., et al.: Analysis of a tunnel failure caused by leakage of the shield tail seal system. Undergr. Space 5(2), 105–114 (2020)
2. Liu, J.W., Shi, C.H., Lei, M.F., et al.: A study on damage mechanism modelling of shield tunnel under unloading based on damage-plasticity model of concrete. Eng. Fail. Anal. **123**, 105261 (2021)
3. Tang, F., He, Y.D.: Research on the internal structures of road - metro shield tunnel. J. Railw. Eng. Soc. **30**(12), 57–63 (2013)
4. Liu, N.: Analysis of and countermeasures against the assembly deviations of prefabri-cated and assembled lane structure in double-deck shield-driven tunnels. Mod. Tunn. Technol. **58**(04), 210–217 (2021)
5. Geng, P., Wang, Q., Guo, X.Y., et al.: Pull-out test of longitudinal joints of shield tunnel. China J. Highw. Transp. **33**(07), 124–134 (2020)

6. Feng, K.: Research on mechanical behavior of segmental lining structure for underwater shield tunnel with large cross-section. Southwest Jiaotong University (2012)
7. Li, C.L., Wang, G.Q., Zhao, K.J., et al.: Vertical mechanical behavior on shield tunnel under loads on ground surface. J. Jilin Univ. (Eng. Technol. Ed.) **41**(S2), 180–184 (2011)
8. Tao, L.J., Li, Z.Y., Yang, X.R., et al.: Research of the mechanical behaviors of subway station structure assembled with prefabricated elements based on abaqus. Mod. Tunn. Technol. **55**(05), 115–123 (2018)
9. Zhang, B.W.: Dynamic response characteristics analysis of shallow buried shield tunnel segment under-crossing the high speed railway. Railw. Eng. **58**(09), 50–54 (2018)
10. Song, R., Zhang, J.Q., Cui, T., et al.: Analysis on mechanical characteristics of double-span column-free fabricated subway station. Railw. Stan. Des. **65**(02), 123–7+78 (2021)
11. Li, X.H., Liu, C.Y., Liu, B., et al.: Study on bending stiffness of tenon groove joints in a assembled subway station. Constr. Technol. **48**(16), 5–7 (2019)
12. Li, Y.J., He, P., Qin, D.P.: Stress analysis of metro shield tunnel segment. Chin. Civil Eng. J. **44**(S2), 131–134 (2011)
13. Liu, H.M.: Study on the mechanics characters of the fabricated linings in the subway constructed by open-cut method. Southwest Jiaotong University (2003)

Study on Deterioration Rule of Water-Binder Ratio on Mechanical Properties and Frost Resistance of Concrete

Xu Gong[1], Hongfa Yu[2(✉)], and Chengyou Wu[1]

[1] School of Civil Engineering, Qinghai University, Xining 810016, China
[2] Department of Civil Engineering, Nanjing University of Aeronautics and Astronautics, Nanjing 210016, China
yuhongfa@nuaa.edu.cn

Abstract. Based on the rapid freeze-thaw (F-T) cycle test, the changes of relevant indexes (mass, relative dynamic elastic modulus, mechanical properties and thickness of concrete F-T damage layer) of concrete and mortar specimens with three water-binder ratios under F-T cycle was systematically studied. The results show that the quality, relative dynamic elastic modulus and mechanical properties of concrete and mortar decrease with the increase of F-T cycles, the thickness of F-T damaged layer of concrete increases in different degrees. The degree of F-T damage of mortar specimen is obviously lower than that of concrete specimen. Through regression analysis, it is find that the relative dynamic elastic modulus of concrete were significantly related to the relative dynamic elastic modulus of mortar, the thickness of concrete damage layer and the relative dynamic elastic modulus of concrete, it shows that one of the factors causing the F-T damage of concrete is the damage of mortar.

Keywords: Concrete · Freeze-Thaw (F-T) cycle · Thickness of F-T damaged layer · Mechanical properties · Quantitative analysis

1 Introduction

Concrete is an anisotropic, heterogeneous, artificial material that is made of a cementitious binder, water, coarse aggregate, fine aggregate and possibly one or more admixtures to facilitate mixing, molding, pouring, and curing processes, through a complex series of physical changes and chemical reactions [1, 2]. The solid components of hardened concrete consist primarily of hardened cement paste, aggregates, and interfacial transition zone (ITZ) [3]. Presently, the hydrostatic pressure hypothesis [4] and the osmotic pressure hypothesis [5] had generally accepted as the mechanisms behind the F-T damage of concrete. There are two conditions for F-T failure of concrete: one is that the concrete saturated with water, and the other is the alternation of temperature. At present, most of the research on the frost resistance of concrete focuses on qualitative analysis [6–9], but there are few studies on the quantitative analysis of concrete related performance indicators under the action of F-T.

© The Author(s) 2022
G. Feng (Ed.): ICCE 2021, LNCE 213, pp. 462–473, 2022.
https://doi.org/10.1007/978-981-19-1260-3_42

In this study, three kinds of concrete and mortar specimens with different water-binder ratios (0.35, 0.42 and 0.53) were prepared. Based on the rapid F-T cycle test, the mass loss rate, relative dynamic modulus of elasticity, change of mechanical properties and thickness of F-T damaged layer of concrete had measured. The internal damage and mechanical properties of specimens with different water-binder ratio under F-T action had studied systematically. The quantitative relationships between the relative dynamic elastic modulus of concrete and the relative dynamic elastic modulus of mortar, the thickness of F-T damage layer and the relative dynamic elastic modulus of concrete are established.

2 Experimental Details

2.1 Materials

P·II 52.5-grade Portland cement, Grade I fly ash, limestone graded gravel, river sand, and drinking water produced by Qinghai Qilian Mountain Cement Co., Ltd. were used in this study. The physical and mechanical properties of P·II 52.5 Portland cement are listed in Table 1, and the chemical compositions of the cement and fly ash are listed in Table 2. The limestone is produced by Jiaheng Magnesium Industry Co. Ltd. in Qinghai Province. The physical and mechanical properties are shown in Table 3, and the gradation curve of graded gravel is shown in Fig. 1. The river sand used was obtained from the Huangshui River in Xining. Its fineness modulus of 2.70 indicates that it is medium grain sand. The basic performance indicators shown in Table 4, the gradation curve of sand is shown in Fig. 2.

Table 1. Physical and mechanical properties of P·II 52.5 Portland cement.

Fineness (Percentage of sieve residue/%)	Specific surface area/ (m2/kg)	Ignition loss (%)	Stability	Setting time/min		Compressive strength/MPa		Flexural strength/MPa	
0.3	376	0.96	Qualified	Initial	Final	3d	28d	3d	28d
				171	216	25.8	58.3	5.3	8.0

Table 2. Chemical composition of cementitious materials/%.

Element	CaO	SiO_2	Al_2O_3	Fe_2O_3	MgO	SO_3	K_2O	TiO_2	Na_2O	MnO_2	SrO
Cement	62.65	23.21	4.98	2.89	2.44	2.41	0.79	0.31	0.06	0.05	0.17
Fly ash	11.38	51.56	18.65	9.92	0.99	2.67	1.76	0.88	1.91	0.20	–

Table 3. Basic performance index of limestone graded gravel.

Stacking density/Kg·m^{-3}		Apparent density/Kg·m^{-3}	Elongated flaky particle/%	Crushing value/%	Water absorption/%
Compact packing density	Loosely packing density				
1565	1425	2710	9.05	10.43	1.15

Table 4. Basic performance index of sand.

Stacking density/Kg·m^{-3}		Apparent density/Kg·m^{-3}
Compact packing density	Loosely packed density	
1840	1720	2670

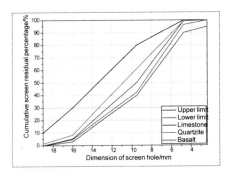

Fig. 1. Grading curve of coarse aggregate.   **Fig. 2.** Grading curve of sand.

2.2 Sample Preparation

The concrete and mortar specimens with water–binder ratios of 0.35, 0.42 and 0.53 were used (Table 5).

2.2.1 Concrete Specimens

Firstly, the cement, fly ash, aggregate and sand was mixed into the horizontal ready-mix concrete for 1 min, after which the water and admixture were mixed for 3–4 min. Finally, the mixture was poured into 100 × 100 × 100 mm molds and placed on a shaking table for vibration and compaction, then placed in an environment of 20 ± 2 °C and a relative humidity of >50%, and sealed and hardened whilst wrapped in a plastic film. After curing for 24 h, the mold was removed and then placed in the standard curing chamber with a temperature 20 ± 2 °C and a humidity of ≥95% for 28 days.

Table 5. Mixture ratio of concrete specimen.

Serial number	Mixed proportion/Kg·m^{-3}		Aggregate/Kg·m^{-3}			Water/Kg·m^{-3}	Water-cement ratio
	Cement	FA	Coarse aggregate	Coarse aggregate	Sand		
CFL35	425	75	1035	Limestone	690	175	0.35
CFL42	382.5	67.5	1056.6	Limestone	704.4	189	0.42
CFL53	340	60	1019.16	Limestone	768.84	212	0.53

2.2.2 Mortar Specimens

By using the method of "wet screen mortar", the concrete mixture with the corresponding proportion was poured into the square screen with the aperture of 4.75 mm. It was put on the shaking screen machine in order to get rid of the aggregate whose particle size is more than 5mm, and then the mortar with corresponding proportion can be prepared. The mortar was poured into a cubic mold of 70.7 mm, and placed on the shaking table for compaction. The mortar specimens were placed in the same environment as the concrete specimens to harden, after whom the molds were removed, and the samples were left to harden in the same curing chamber as the concrete samples. The serial numbers of mortar specimens prepared from CFL35, CFL42 and CFL53 concrete are MF35, MF42 and MF53.

2.3 Sample Testing and Analysis

2.3.1 Compressive Strength Test of Concrete and Mortar Cube Specimen

The compressive strength of concrete and mortar cube specimens was measured by microcomputer controlled electro-hydraulic servo pressure tester (YAW4306) in accordance with the *Standard for Test Methods of Mechanical Properties of Ordinary Concrete* (GB/T 50081-2019). The calculation result is accurate to 0.01 MPa.

2.3.2 Rapid F–T Cycle Test

The rapid F-T test machine (NJW-HDK-9) is used, and the F-T cycle method and test method refer to the relevant methods of the *Standard for Long-term Performance and Durability Test Methods of Ordinary Concrete* (GB/T 50082-2009) [11]. The concrete prism specimens were put into the rapid F-T testing machine, and the mass loss rate and relative dynamic elastic modulus (Fig. 3) were measured after every 25 cycles of F-T (NM-4B). When the mass loss reaches 5% of the initial mass or the relative dynamic modulus decreases to 60% of the initial value, the experiment is finished. The mass loss rate and relative dynamic modulus of elasticity of the specimen are calculated according to Eqs. (1) and (2):

$$\Delta W_n = [(W_0 - W_n)/W_0] \times 100 \qquad (1)$$

where ΔW_n is the mass loss rate of specimens after F-T cycles, W_n is the mass of the specimen after several F-T cycles and W_0 is the Initial mass of the specimen before the F-T cycle test.

$$RDEM = (E_t/E_0) \times 100 = (V_t/V_0)^2 \times 100 = (T_0/T_t)^2 \times 100 \qquad (2)$$

where $RDEM$ is the relative dynamic elastic modulus of concrete and wet screen mortar, E_0, V_0, T_0 are the initial dynamic modulus, ultrasonic wave velocity and ultrasonic acoustic time respectively. E_t, V_t, T_t are the dynamic modulus, ultrasonic wave velocity and ultrasonic sound time at the end of every 25 F-T cycles respectively.

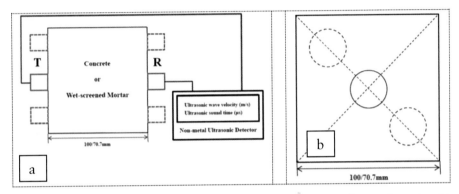

Fig. 3. Schematic diagram of ultrasonic detection (a) and arrangement of measuring points (b).

At the same time, at the end of every 25 F-T cycles, the thickness of F-T damage layer of concrete prismatic specimens was measured according to technical specification for detecting concrete defects by ultrasonic method (CECS21-2000) [12–14]. The test method is shown in Fig. 4. Firstly, the "Time-Interval" diagram of corroded concrete is drawn by using the acoustic time value and the corresponding distance value of each measuring point (Fig. 5). The thickness of concrete damage layer is calculated by the ultrasonic time value data and the inflection point of the curve in the "Time-Interval" diagram of each specimen, and then the regression linear equations of damaged and undamaged concrete l and t are obtained by regression analysis method. Finally, the thickness of the damaged layer of the concrete specimen is calculated according to Eqs. (3) and (4):

$$l_0 = (a_1 b_2 - a_2 b_1)/(b_2 - b_1) \qquad (3)$$

$$h_f = (l_0/2) \cdot [(b_2 - b_1)/(b_2 + b_1)]^{1/2} \qquad (4)$$

where h_f is thickness of F-T damaged layer of concrete, a_1, b_1, a_2, b_2 is the regression coefficient of the line being the intercept and slope of the damaged and undamaged concrete line in Fig. 5.

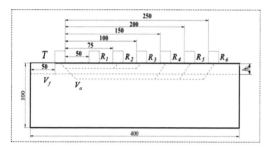

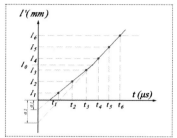

Fig. 4. Schematic diagram of thickness detection of concrete damage layer.

Fig. 5. Time-distance diagram of damage layer detection.

3   Results and Discussion

3.1   Surface Spalling of Concrete Under F-T

Figure 6 shows the surface spalling of concrete specimens with different water-binder ratios under F-T. It can be seen from the figure that as the number of F-T cycles increases, the peeling off the concrete specimen surface becomes more and more serious. When the specimen is damaged, the degree of F-T damage increases with the increase of water-binder ratio. In the early stage of F-T, the surface of the specimen was partially rough, a small amount of cement mortar peeled off the attached coarse aggregate, and the damage was unobvious. With the number of F-T cycles increased, the amount of cement mortar spalling also increased, and pits appeared on part of the surface. When the concrete is about to fail, the surface of the specimen shows serious peeling, the coarse aggregate is exposed, and the corners of the local corners are dropped.

3.2   Variation of Mass and Relative Dynamic Elastic Modulus of Specimens Under F-T

Figure 7 shows the mass loss and relative dynamic elastic modulus change curves of concrete and mortar specimens with different water-binder ratios under F-T. It can be seen that with the increase of F-T times, the mass and relative dynamic elastic modulus of the specimen decrease to some extent. After 250, 75 and 50 cycles of F-T cycles, the relative dynamic modulus of elasticity of CFL35, CFL42 and CFL53 decreased to less than 60% (49.84%, 54.28% and 56.01%), the mass loss rate was 1.135%, 0.456% and 0.523% respectively. After 250, 75 and 50 cycles of F-T cycles of MF35, MF42 and MF53, the relative dynamic elastic modulus decreased to less than 60% (52.9%, 58.18%, 59.24%), the mass loss rate was 0.836%, 0.184% and 0.417% respectively. The basic reason for this phenomenon lays in the difference of pore structure parameters in concrete. With the increase of water-binder ratio, the internal structure of concrete also tends to be loose, resulting in a decline in its frost resistance.

At the same time, it is found that the relative dynamic elastic modulus of concrete is lower than that of mortar under the same F-T cycles. Through the regression analysis of the relative dynamic elastic modulus of concrete and the relative dynamic elastic modulus

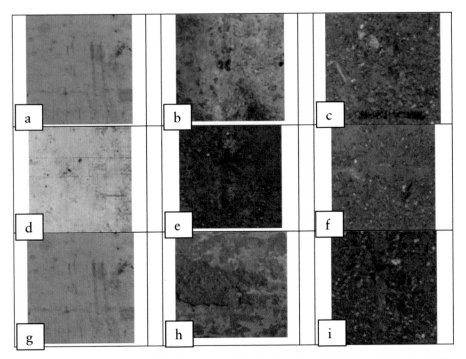

Fig. 6. Destruction mode of CFL35 (a, b, c), CFL42 (d, e, f) and CFL53 (g, h, i) under F-T.

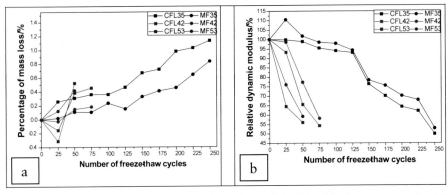

Fig. 7. Mass loss rate (a) and relative dynamic modulus of elasticity (b) of specimens under F-T.

of mortar under F-T (Fig. 8), it is found that there is a significant linear relationship (Eq. (5)) between them:

$$RDEM_{,c} = 1.033RDEM_{,m} - 7.224 \qquad (5)$$

where $RDEM_{,c}$ is the relative dynamic elastic modulus of concrete, $RDEM_{,m}$ is the relative dynamic elastic modulus of mortar. In this fitting relationship, the number of samples is 15, and it can be seen from the figure that $R = 0.9798 > R_{0.001}(13) = 0.760$.

This shows that the relative dynamic elastic modulus of concrete and the relative dynamic elastic modulus of mortar in the F-T are more than 99.9% significant. It is preliminarily shown that one of the factors of F-T damage of concrete is the damage of mortar.

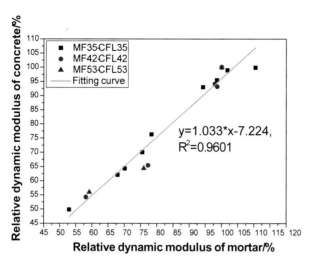

Fig. 8. Relation between relative dynamic modulus of concrete and relative dynamic modulus of mortar.

3.3 Comparative Analysis of Basic Mechanical Indexes of Specimens Before and Aater F-T

Figure 9 shows the change rule of the basic mechanical indexes of concrete and mortar specimens with different water-binder ratios before and after F-T. It can be seen that the mechanical properties of the concrete and mortar specimens have different degrees of loss under F-T. When the concrete and mortar specimens with water-binder ratios of 0.35, 0.42 and 0.53 have undergone 250, 75 and 50 F-T cycles in sequence, they reached the condition of F-T failure. It is found from Fig. 9a that the compressive strength loss rates of CFL35, CFL42 and CFL53 specimens are 61.24%, 26.93% and 37.50% respectively after F-T failure. The compressive strength loss rates of corresponding mortar specimens (MF35, MF42 and MF53) were 53.11%, 21.27% and 29.02% respectively. It is found from Fig. 9b that the splitting loss rates of CFL35, CFL42 and CFL53 specimens are 41.54%, 51.0% and 66.02% respectively, and the splitting loss rates of corresponding mortar specimens (MF35, MF42 and MF53) are 40.81%, 24.32% and 57.92% respectively. It is found from Fig. 9c that the loss rates of flexural strength of CFL35, CFL42 and CFL53 specimens are 73.12%, 76.15% and 83.71% respectively. It can be seen from the above data that the loss of mechanical properties of the mortar is less than that of the concrete specimens. The reason for this phenomenon is that the larger the porosity of the inner structure and the looser the structure, the worse the frost resistance of the specimen with large water-binder ratio. Compared with the concrete specimen, the mortar specimen is more homogeneous, which leads to a lower loss rate of strength.

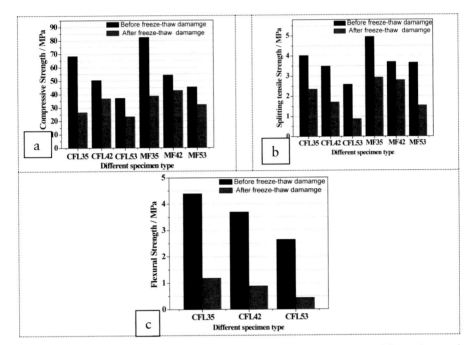

Fig. 9. Variation of compressive strength (a), splitting tensile strength (b) and flexural strength (c) of concrete and mortar specimens before and after F-T.

3.4 Variation Law of Thickness of F-T Damaged Layer of Concrete Under F-T

Figure 10 is the change curve of thickness of F-T damaged layer of concrete with different water-binder ratios under the action of F-T. The results show that the thickness of damage layer increases with the increase of F-T cycles, and the growth rate increases with the increase of water-binder ratio. When the specimen reached the failure condition, the thickness of the F-T damage layer from high to low was CFL35 (47.52 mm), CFL53 (43.59 mm) and CFL42 (31.49 mm). This is due to the expansion of the pores caused by the water expansion in the pore structure of concrete under the action of F-T, which leads to the increase of the thickness of the damaged layer. When the specimen reaches the condition of F-T damage, the thickness of the F-T damage layer is between 31–48 mm, which indicates that the specimen is in the state of thoroughly frozen.

Through further analysis of the thickness of F-T damage layer of concrete, it is found that the variation law is basically consistent with the relative dynamic elastic modulus of concrete. Regression analysis of the two (Fig. 11) revealed the following nonlinear relationship (Eq. (6)):

$$h_f = -0.0079(RDEM_{,c})^2 + 0.5306(RDEM_{,c}) + 36.456 \tag{6}$$

where h_f is the thickness of F-T damaged layer of concrete, $RDEM_{,c}$ is the relative dynamic elastic modulus of concrete. In this fitting relationship, the number of samples is 19, and it can be seen from the figure that $R = 0.860 > R_{0.001}(17) = 0.693$. The results show that the prediction model of the thickness of the F-T damage layer of concrete and

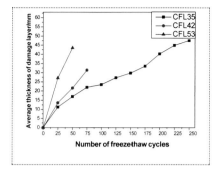

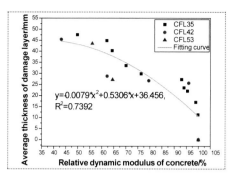

Fig. 10. Thickness variation of F-T damage layer of concrete specimen under F-T.

Fig. 11. Relationship between the thickness of F-T damaged layer and the relative dynamic elastic modulus of concrete.

relative dynamic elastic modulus of concrete is more than 99.9% significant. It shows that as the relative dynamic elastic modulus of concrete decreases, the thickness of the F-T damage layer is also increasing. At the same time, it can be seen that the F-T damage of concrete is a process from the outside to the inside, which not only occurs inside the concrete but also occurs on the surface of the concrete. Through the above analysis, it is found that one of the reasons for the F-T damage of concrete is caused by partial F-T damage of the mortar.

4 Conclusions

This study systematically explored the changes of relevant indexes of concrete and mortar specimens with different water-binder ratios under F-T. Based on the experimental results, the following conclusions can be drawn:

1. With the increase of F-T times, the quality and relative dynamic elastic modulus of the specimens both decreased to varying degrees. This is caused by the difference in the parameters of the pore structure inside the concrete. Through the regression analysis of the relative dynamic elastic modulus of concrete and the relative dynamic elastic modulus of mortar, it is found that there is a very significant linear relationship between them. It is preliminarily shown that one of the factors leading to concrete F-T damage is mortar damage.
2. After the concrete and mortar specimens experienced F-T effects, their mechanical properties showed varying degrees of loss. The loss rate of strength of mortar is relatively low. This is caused by the relatively homogeneous interior of the mortar specimen.
3. When the concrete reaches F-T failure, the degree of exfoliation on the surface of the test piece increases with the increase of the water-binder ratio, and the concrete is basically in a state of being frozen through. Through regression analysis of the thickness of the concrete F-T damage layer and the relative dynamic elastic modulus of concrete, it is found that the two have a very significant non-linear relationship. It is further confirmed that one of the causes of concrete F-T damage is caused by

mortar damage. At the same time, it shows that the F-T damage of concrete is a process from surface to interior.

Acknowledgments. This study was supported by the National Natural Science Foundations of China (Grant No. 52002202) and the applied fundamental research project of Qinghai Province (Grant No. 2019-ZJ-7005) and the Science and Technology Basic Conditions Platform of Qinghai Province (2018-ZJ-T01) and the West Light Project of Chinese Academy of Sciences.

References

1. Yang, H.Q., Li, W.W.: Research and application on durability of hydraulic concrete. M. China Electric Power Press, Beijing (2004)
2. ACI committee 209Prediction of creep, shrinkage and temperature effects in concrete structures. Manual of concrete practice.Part1 (1992)
3. Deng, Y.F., Leng, Z., Wang, J.Q.: Effect of coarse aggregate types on interface transition zone of fly ash polymer concrete. Commercial Concrete **01**, 31–36 (2019)
4. Powers, T.C.: Air requirement of frost-resistant concrete.In: Highway Research Board Proceedings, vol. 29 (1950)
5. Powers, T.C.: Freezing effects in concrete, durability of concrete. American Concrete Institute, Detroit (1975)
6. Shang, J.L., Xing, L.L.: Study on interface transition zone of steel slag coarse aggregate concrete. J. Build. Mater. **16**, 217–220 (2013)
7. Qiu, C., Zhang, Y.M.: Influence of aggregate size and water-cement ratio on microstructure of interfacial transition zone of concrete. C. Abstracts from the 3rd annual meeting of Cement Branch of China Silicate Society and the 12th national meeting of cement and concrete chemistry and application technology (2011)
8. Wang, Z.Q., Li, J.Z., Zhou, S.H., Shi, Y.: Evolution of internal microstructure of concrete during freezing-thawing cycle. Concrete **01**, 13–14 (2012)
9. Petersen, L.L., Polak, M.: Influence of freezing and thawing damage on behavior of reinforced concrete elements. ACI Mater. J. **104**, 369 (2007)
10. GB/T 50081-2019 Standard for experimental method of mechanical properties of ordinary concrete. Standards Press of China, Beijing (2019)
11. GB/T 50082-2009 Standard test method for long-term performance and durability of ordinary concrete. Standards Press of China, Beijing (2009)
12. CECS 21: 2000 Technical specification for ultrasonic detection of concrete defects. S. Beijing: China Association for Standardization of Engineering Construction (2000)
13. Akhras, N.M.: Detecting freezing and thawing damage in concrete using signal energy. Cem. Concr. Res. **28**, 1275–1280 (1998)
14. Mehta, P.K., Monteiro, P.J.M.: Concrete Microstructure, Properties and Materials, 3rd edn. McGraw-Hil, New York (2006)

Analysis of Pile-Soil Interaction of Precast Pile Driven in Coastal Strata

Hongping Xie[1], Chao Han[1], Changqing Du[1], Bo Wang[2(✉)], Yuchi Zhang[2], and Pinqiang Mo[2]

[1] State Grid Jiangsu Electric Power Engineering Consulting Co., Ltd., Nanjing, China
[2] State Key Laboratory for Geomechanics and Deep Underground Engineering, China University of Mining and Technology, Xuzhou, Jiangsu, China
wangbocumt@163.com

Abstract. In order to further reveal the pile-soil interaction mechanism during precast pile driving in saturated soft soil in coastal areas, the compaction effect and excess pore pressure response of a single pile and adjacent pile penetration under hammer driven pile construction are analyzed by using the cavity expansion and model test method. The results show that pile driving in saturated soil layer will cause large soil compaction and accumulation of excess pore water pressure. Under the model test conditions, the variation range of soil pressure and excess pore pressure is about 0.7–3.0 times and 0.5–1.5 times of soil mass weight stress. As the driving of adjacent pile, soil pressure at the constructed pile-soil interface increases gradually and fluctuates at the same time, and multi peak phenomenon appears under the influence of different soil layers. At the initial stage of driving, the pile driving force is mainly borne by the pile side friction, and the pile tip resistance will actions as the increase of penetration depth, and the relationship between them is basically linear. These results have certain guiding and reference value for the construction of precast pile driving in saturated soft soil in coastal areas.

Keywords: Precast pile · Coastal strata · Pile-soil interaction · Cavity expansion theory · Model test

1 Introduction

Due to the special geological causes and forming environment, there are a large number of muddy soft soil and silt foundations in coastal areas, which have low bearing capacity, large deformation or liquefaction characteristics [1, 2]. In practice, it is often necessary to adopt the form of pile foundation to meet the bearing requirements of the superstructure. Precast pile is a common pile type in China's coastal areas because of its low cost per unit of bearing capacity and relatively suitable geological conditions in coastal areas. However, no matter how to construct precast piles by hammer sinking, vibration sinking or static pressing, it will inevitably disturb the soil around the piles, and there will be obvious soil squeezing effect and excess pore water pressure accumulation in the soil, etc. It is not uncommon for pile foundation construction quality problems to be caused

© The Author(s) 2022
G. Feng (Ed.): ICCE 2021, LNCE 213, pp. 474–486, 2022.
https://doi.org/10.1007/978-981-19-1260-3_43

by insufficient understanding of pile-soil interaction mechanism during pile sinking [3–5]. In this regard, scholars and engineers at home and abroad have carried out a lot of research work from theoretical analysis, experimental research and field measurement [6–8]. In the aspect of theoretical research, Liu et al. [8] derived an analytical solution to the problem of pore expansion of saturated soil in K_0 consolidation state, and the anisotropy induced by natural soil consolidation had a great influence on the soil stress and excess pore water pressure around the pile after pore expansion. Mo et al. [9] derived the analytical solution of pore expansion under completely undrained condition, and analyzed the displacement-controlled pile driving process in clay. Zhou et al. [10] used the unified strength theory to study the pore expansion in unsaturated soil under different drainage conditions. Zheng et al. [11] adopted the critical constitutive model of sand considering the characteristics of particle breakage and dilatancy, and established a semi-analytical solution to the problem of pore expansion in sand. It can be seen that in order to better reflect the mechanical properties and actual engineering conditions of various soils, scholars have used the ideal elastic-plastic constitutive model to the more complex critical state model to analyze the problem of column hole expansion considering actual complicated situations such as drainage, particle breakage and stress path. In terms of experimental and measured research, Xu et al. [7] analyzed the development law of pore water pressure and lateral deformation of soil under the action of ground motion through shaking table test of liquefaction site-pile group foundation interaction. Wang et al. [12] carried out the field test of pile sinking of anti-liquefaction drainage rigid pile and ordinary rigid pile. Su et al. [13] established numerical simulation of shaking table of pile group in liquefied lateral expansion site by means of OpenSees numerical method, and the interaction between pile group, pile cap and soil has great influence on lateral deformation of the system. Zhou et al. [14] carried out the model test of transparent soil driven by rectangular piles, and derived the modified reaming theory of rectangular pile section according to the test results. Wang et al. [15] analyzed the variation law of soil pressure and excess pore water pressure of open pile and closed pile in saturated clay during static pressure through laboratory tests. Many scholars used shaking table, centrifuge and numerical simulation methods to carry out the simulation test and field measurement research of pile driving process at different scales. Combined with transparent soil and sensing test technology, the development and distribution law of soil deformation, earth pressure and excess pore water pressure around pile driving process were analyzed.

At present, with the development and utilization of marine wind power resources in the eastern coastal areas of Jiangsu Province, the number and scale of pile foundation construction in coastal soft and liquefiable soil layers are constantly expanding, but the construction quality control still faces many problems. In this paper, based on the construction of precast pile driving in Jiangsu coastal stratum, theoretical analysis of pile driving and soil squeezing effect and indoor model test research are carried out, and the interaction mechanism between pile and soil, the response law of soil pressure and excess pore water pressure in the process of precast pile driving are discussed. The results can provide support for improving pile driving efficiency and construction quality control of precast pile driving in coastal soil stratum.

2 Theoretical Analysis of Pile Squeezing Effect

2.1 Cavity Expansion Theory

The pore expansion theory is a theory to study the changes of stress, pore water pressure and displacement caused by the expansion of cylindrical or spherical holes. In this paper, the analytical solution of undrained pore expansion derived by Mo et al. [9] is used to analyze the pile driving process.

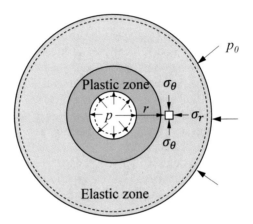

Fig. 1. Diagram of cavity expansion theory

Referring to Fig. 1, the equilibrium equation expression of pore expansion is:

$$\sigma_\theta - \sigma_r = \frac{r}{m} \frac{\partial \sigma_r}{\partial r} \tag{1}$$

And satisfy two boundary conditions:

$$\sigma_r|_{r=a} = p \tag{2}$$

$$\sigma_r|_{r=b} = p_0 \tag{3}$$

Among them, σ_r and σ_θ are the radial and circumferential stresses in the soil, r is the distance from the calculation point to the center of the hole, m is the hole shape coefficient, m = 1 is the cylindrical hole, m = 2 is the spherical hole, a and b are the inner diameter and outer diameter of the hole after the expansion of the soil, p is the inner wall pore pressure of the expanded soil and p_0 is the outer wall pore pressure of the soil. In this paper, it is assumed that the soil is an infinite medium, that is, $b = \infty$. In order to consider the large deformation of soil caused by pore expansion, it is necessary to use the large strain analysis method to derive the analytical solution, and the logarithmic strain expression is as follows:

$$\varepsilon_r = -\ln\left(\frac{dr}{dr_0}\right) \tag{4}$$

$$\varepsilon_\theta = -\ln\left(\frac{r}{r_0}\right) \tag{5}$$

Among them, ε_r and ε_θ are the radial and circumferential strains in the soil, respectively, and r_0 is the initial distance from the calculation point to the center of the hole. In this paper, the stress and strain are positive in compression. For the actual pile driving process, an approximate analysis is made by a series of spherical small hole expansions moving down the pile body, and the soil at the pile bottom is controlled by the small hole expansions at the pile end. The process of hole expansion is assumed to be an initial hole (about 0.1 mm in diameter) that is expanded to the pile diameter (that is, r = D/2, D is the equivalent diameter of precast pile), so as to simulate the soil squeezing effect of precast pile driving.

The unified sand-clay constitutive model CASM is adopted to analyze the constitutive model of soil, which can better simulate the stress-strain characteristics of sand and clay under different loading paths and drainage conditions [16, 17]. The yield surface equation expression of CASM model is:

$$\left(\frac{\eta}{M}\right)^n = -\frac{\ln\left(p'/p_y'\right)}{\ln r^*} \tag{6}$$

Where η is the effective stress ratio $= q/p'$, q is deviatoric stress, p' is the effective average stress, and p_y' is the pre-consolidation pressure; n is the stress state coefficient and r^* is the spacing ratio, both of which belong to the newly introduced material parameters of CASM. The model uses an unrelated flow criterion to control the plastic strain of materials, and the expression is:

$$\frac{\dot{\delta}^p}{\dot{\gamma}^p} = \frac{9(M - \eta)}{9 + 3M - 2M\eta} \times \frac{m}{m + 1} \tag{7}$$

In which $\dot{\delta}^p$ is the plastic volumetric strain rate, $\delta = \varepsilon_r + m\varepsilon_\theta$; $\dot{\gamma}^p$ is the plastic shear strain rate, $\gamma = \varepsilon_r - \varepsilon_\theta$; M is the slope of the critical state line in $p' - q$ space.

The specific values of soil parameters are shown in Table 1, and the saturation degree γ_{sat} of soil is 18 kN/m³. The friction angle δ_f of pile-soil interface is $\phi/2$, where ϕ is the internal friction angle of soil, which is calculated by $M = \frac{6\sin\phi}{3-\sin\phi}$. The maximum penetration depth of pile foundation is 1 m.

Table 1. Soil parameters used in calculation

Elastic parameter	$\kappa = 0.025$	$\mu = 0.3$	
Critical state parameter	$M = 0.9$	$\lambda = 0.093$	$\Gamma = 2.06$
CASM parameter	$r^* = 2.714$	$n = 4.5$	

2.2 Calculation Method of Pile Sinking Resistance

The pile driving force Q when the penetration depth z is controlled according to the displacement consists of the pile end force Q_{tip} and the pile side force Q_{shaft}, and the calculation formulas are as follows:

$$Q_{tip} = q_t A = q_t \pi D^2 / 4 \tag{8}$$

$$Q_{tip} = \int_0^z q_{s,z} \pi D dz \tag{9}$$

Among them: q_t is the pile end resistance, and $q_{s,z}$ is the pile side resistance at depth z. The specific expression is:

$$q_t = \left(\sigma_r' + \Delta u\right) \times \left(1 + \sqrt{3}\tan\phi\right) \tag{10}$$

$$q_{s,z} = \left(\sigma_{r,z}' + \Delta u_Z\right) \times \tan\delta_f \tag{11}$$

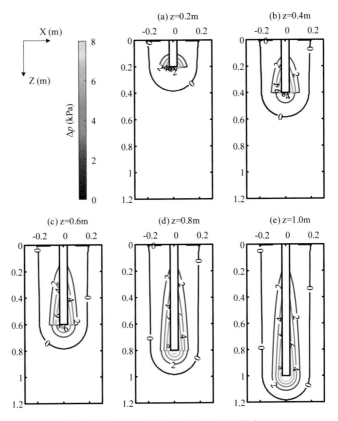

Fig. 2. Stress variation during pile driving

Among them: σ_r' is the effective radial stress at the pile tip (depth z) caused by pile sinking, and Δu is the excess pore water pressure at the pile tip caused by pile sinking; $\sigma_{r,z}'$ and Δu_z are the effective radial stress and excess pore water pressure at the pile side at the depth z, respectively, which can be calculated by the expansion of spherical holes at the corresponding depth.

2.3 Calculation Result Analysis

Soil squeezing effect will occur in the process of pile sinking, which will cause the stress change of the soil around the pile. Figure 2 shows the calculated average stress increment Δp caused by pile driving, where the average stress $p = \sigma_r + 2\sigma_\theta/3$. The pile sinking causes the average stress around the pile to increase, and the influence range is within 3 times of the pile diameter around the pile, and the change of soil stress within 3 times of the pile diameter depth on the surface can be neglected. Figure 3 shows the distribution of excess pore water pressure Δu caused by pile driving, and the influence range is similar to the average stress increment, with the value slightly larger by about 20%.

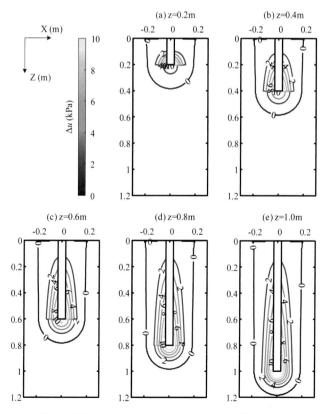

Fig. 3. Excess pore water pressure during pile driving

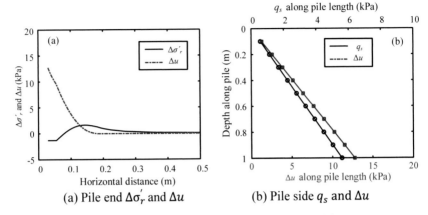

(a) Pile end $\Delta\sigma'_r$ and Δu (b) Pile side q_s and Δu

Fig. 4. Stress state of pile tip and side after pile driving

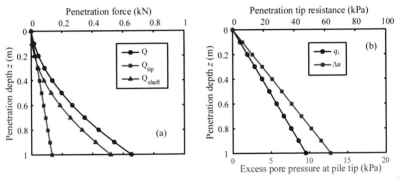

(a) Pile sinking force Q, pile tip force Q_{tip} and pile side force Q_{shaft} (b) Pile end q_t and Δu

Fig. 5. Stress state of pile tip and side during pile driving

The distribution of effective radial stress and excess pore water pressure at the pile tip along the horizontal direction after the pile is sunk by 1 m is shown in Fig. 4(a). The effective radial stress around the pile is distributed horizontally, which indicates that the soil after soil compaction disturbance is in a critical state, and the negative value indicates that there is excessive excess pore water pressure in this area. With the increase of horizontal distance, the effective radial stress away from the soil first increases and then decreases. It can be seen that the size of plastic zone is about 2.5D, and the overall influence range of pile sinking is about 4D. The excess pore water pressure decreases rapidly with the horizontal distance, and the influence range is less than the effective radial stress, which is about 3D. Figure 4(b) shows the distribution of pile body friction (black line) and excess pore water pressure (red line) along the depth of pile body after pile sinking for 1m, and its linear distribution may be due to the influence of near-surface and too small in-situ stress level.

The evolution of the total penetration force Q, tip force Q_{tip} and side force Q_{shaft} of the displacement pile in the process of pile sinking is shown in Fig. 5(a). When the penetration depth is greater than 0.5 m, the pile driving force is mainly borne by the pile side friction. The variation of pile end resistance q_t and excess pore water pressure at pile end with penetration depth is shown in Fig. 5(b).

3   Model Test on Pile Sinking Process

3.1   Test Device and Method

As shown in Fig. 6, steel drum model box (inner diameter 1000 mm) is used, and earth pressure box and pore pressure meter are respectively arranged on both sides of concrete model pile body to measure the earth pressure and pore water pressure at the pile-soil interface. The size of the concrete model pile is 60 mm * 60 mm * 1000 mm, and the length-diameter ratio of the concrete model pile is 16.7. The wires of the earth pressure box and pore pressure meter arranged inside the concrete model pile are led out at a distance of 200 mm from the top of the model pile. The concrete model pile is poured with C30 concrete. Due to the size limitation and the need of embedded sensors, the coarse aggregate of poured concrete is screened with a 5 mm screen to better fill the inner space of the model pile mold, and at the same time, the contact between large aggregate and sensors can be avoided to affect the measurement results.

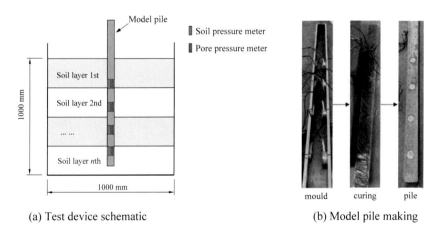

(a) Test device schematic   (b) Model pile making

Fig. 6.   Diagram of model test

3.2   Soil Sample and Sensor Layout

The soil samples used in the test are silty clay from the construction site of a substation in Yancheng. According to the test requirements, soil samples with 15% and 25% water contents are prepared, sealed and stored for 24 h, and then packed into the model box in layers in turn. The packing density of each layer of soil samples is strictly controlled according to the test design. Density of the top, middle and bottom soil layers are about

1.8, 1.6 and 2.1 g/cm³. Water contents of the top, middle and bottom soil layers are 15, 25 and 25%.

In order to measure the changes of soil pressure and pore water pressure in the soil around the pile caused by pile sinking, sensors are arranged at positions 3 times and 6 times away from the model pile body by using positioning rods. Among them, two positioning rods are respectively fixed with three pore pressure meters, and the other two positioning rods are respectively fixed with three earth pressure boxes, and each sensor is arranged in turn with an upward spacing of 200 mm from the bottom of the positioning rod 300 mm. Sensors of the pile body are upward from the pile bottom 100 mm in turn, with a spacing of 150 mm. See Fig. 7 for specific sensor arrangement.

Considering the effect of hammering into piles, the impact force on precast piles is about $1 \times 10^6 N$, and the force similarity constant in model test is about $S_F = 0.001$. Therefore, in the test, a weight plate weighing 10 kg is selected, and it is allowed to fall freely along the guide rod by 150 mm, and the simulated impact force is about $1 \times 10^3 N$, with hammering interval of 2 s. In the process of hammering, laser levels is used to proofread the verticality of the model pile to ensure the effective application of hammering force and the stability of the pile body in the model test.

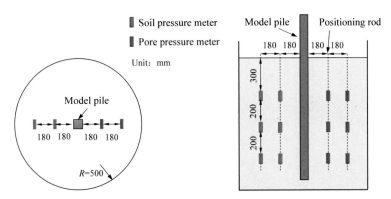

Fig. 7. Location of measure sensors in model test

3.3 Test Results and Analysis

In this model test, the soil squeezing effect of pile sinking and the response of excess pore pressure in soil are analyzed under two working conditions of single pile and adjacent pile penetration. In order to facilitate the post-processing of data, the pile body is marked every 50 mm During the process of pile sinking, the data collector continuously collects data every 20 s. In the later stage, five groups of data from the sinking position to the vicinity of the pile body mark are selected, and the average value is taken.

3.3.1 Analysis of Soil Squeezing Effect

The change of earth pressure of the soil around the pile in the process of pile sinking is shown in Fig. 8, and the data of the change of earth pressure in the figure is normalized by the self-weight stress at the corresponding position. It can be seen that when the pile driving depth is within the range of 400 mm, although there is a certain upward trend in the change of the earth pressure around the pile with 3 times and 6 times the pile diameter, the magnitude is small. The reason is that the consolidation stress level of the shallow soil layer is low, and the deformation limit is small. This phenomenon is basically consistent with the theoretical analysis results of small hole expansion. When the pile depth is in the range of 400–600 mm, the model pile body passes through the relatively soft saturated soil layer in the middle, and the data of earth pressure at 3 times of pile diameter and 6 times of pile diameter are greatly increased, with the largest increase at the depth of about 700 mm, and then there is a downward trend. On the whole, the variation range of earth pressure is 0.7–3.0 times of self-weight stress level. When the pile depth is 600–800 mm, when the model pile enters the fourth layer of soil, there are some differences in the earth pressure changes of 3 times the pile diameter and 6 times the pile diameter, and the law is not obvious. The main reason for the analysis is that the sensor placement position is displaced to the upper side of the sinking depth, while the lateral soil squeezing during the model pile penetration mainly affects the soil within the sinking depth range. The test results also confirm that it is reasonable to assume that the soil around the pile is only affected by the expansion of small holes with the same depth in theoretical analysis.

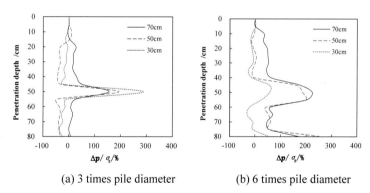

(a) 3 times pile diameter   (b) 6 times pile diameter

Fig. 8. Soil pressure response during pile driving

3.3.2 Response of Excess Pore Pressure in Surrounding Soil

The variation of soil pore pressure around the pile during pile sinking is shown in Fig. 9. Similarly, the pore pressure data are normalized according to the self-weight stress at the corresponding position. At 3 times of pile diameter, with the increase of pile sinking depth, the pore pressure meter data at different depths of soil around the pile all show an obvious upward trend, with the range of 0.5–1.5 times of self-weight stress level. 6 times of pile diameter and different depths, the change of pore pressure count value is obviously less than 3 times of pile diameter, and the maximum increase of excess pore pressure is about 0.5 times of deadweight stress level. When the pile depth is shallow at 150 mm, the response of the excess pore pressure with the embedded depth of 300 mm is the most obvious. With the increase of the embedded depth, the excess pore pressure with the embedded depth of 700 mm increases greatly, and then keeps a high level. The reason may be that the depth of 200–600 mm in the model box is saturated weak soil layer, and the outside of the model box is an undrained boundary, so the pore pressure meter with the embedded depth of 700 mm is more obviously affected by the accumulation of excess pore pressure in saturated soil layer.

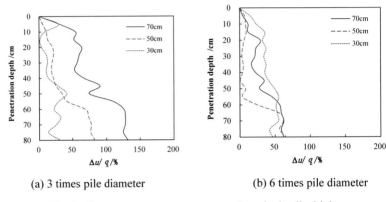

(a) 3 times pile diameter   (b) 6 times pile diameter

Fig. 9. Excess pore pressure response when single pile driving

4   Conclusion

In this paper, the squeezing effect of precast pile driving in saturated soil and the response of excess pore water pressure in soil are analyzed by theoretical calculation and model test, and the variation law of soil pressure and excess pore water pressure in the pile-soil interface and the soil around the pile during driving is analyzed. The main conclusions are as follows:

(1) Based on the theory of spherical pore expansion, considering the complex stress-strain relationship of soil, the large deformation during pore expansion and the development of plastic zone around pile caused by pile penetration, the evolution law of soil stress, excess pore water pressure and pile penetration resistance in the process of precast pile driving is analyzed.

(2) The phenomenon of soil squeezing and excess pore water pressure accumulation caused by precast pile driving in saturated soil is studied by model test. Under the condition of model test, the increase range of soil pressure is 0.7–3.0 times of deadweight stress level, and the variation range of excess pore water pressure is 0.5–1.5 times of deadweight stress level. When adjacent piles are sunk, the earth pressure at the pile-soil interface shows a fluctuating upward trend, and multi-peak phenomenon occurs when affected by different soil properties, and the earth pressure increases by 1.0–3.0 times of the self-weight stress level.

(3) The model test results are in good agreement with the theoretical analysis in the variation law of soil stress and excess pore water pressure, but the model test data fluctuates obviously, mainly due to the influence of the filling of test soil samples and the response of sensors. In the later stage, necessary field measurement work will be carried out to enhance the understanding and grasp of pile-soil interaction mechanism in the process of deep pile, and provide more valuable guidance for actual construction.

Acknowledgments. This research was supported by the National Science Foundation of China (Grant No. 51408595), the Scientific Research Fund of Institute of Engineering Mechanics, China Earthquake Administration (Grant No. 2020D16) and the 2020 Science and Technology Project of State Grid Jiangsu Electric Power Engineering Consulting Co., Ltd. (Grant No. J202006). These supports are gratefully acknowledged.

References

1. Lin, Z.Q.: Mechanism of rigid pile composite foundation influenced by marine soft soil. Geotech. Eng. Tech. **33**(6), 328–333, 371 (2019)
2. Lv, G.R., Ge, J.D., Xiao, H.T.: Treatment of coastal soft foundation with cement-soil mixing pile. J. Shandong Univ. (Eng. Sci.) **50**(3), 73–81 (2020)
3. Li, S.Y., Liang, S., Wang, B., et al.: Research and application of prestressed pipe pile construction quality control under ultra-deep soft soil foundation. Constr. Technol. **46**(1), 18–24 (2017)
4. Zhong, J.M.: Quality problem and treatment of pile foundation caused by extrusion of precast pile. Build. Struct. **47**(s2), 458–463 (2017)
5. Zhang, Z.Z., Gui, Z.P., Zhang, R.J.: Horizontal additional response analysis of adjacent existing loaded pile foundation caused by embankment filling on deep soft soil. J. China Foreign Highway **38**(6), 7–13 (2018)
6. Fu, Y., Huang, F.Y., Chen, B.C., et al.: Shaking table test on structure-soil-pile of PHC in coastal soft soil area. China J. Highw. Transp. **30**(10), 81–92 (2017)
7. Xu, C.S., Dou, P.F., Du, X.L., et al.: Dynamic response analysis of liquefied site-pile group foundation-structure system—large-scale shaking table model test. Chin. J. Geotech. Eng. **41**(12), 2173–2181 (2019)
8. Liu, S.P., Shi, J.Y., Lei, G.H., et al.: Elastoplastic analysis of cylindrical cavity expansion in K0 consolidated saturated soil. Rock Soil Mech. **34**(2), 389–403 (2013)
9. Mo, P.Q., Yu, H.S.: Undrained cavity expansion analysis with a unified state parameter model for clay and sand. Geotechnique **67**(6), 503–515 (2017)
10. Zhou, F.X., Mou, Z.L., Yang, R.X., et al.: Analytical analysis on the expansion of cylindrical in unsaturated soils under different drainage conditions. Chin. J. Theor. Appl. Mech. **53**(5), 1496–1509 (2021)
11. Zheng, J.H., Qi, C.G., Wang, X.Q., et al.: Elasto-plastic analysis of cylindrical cavity expansion considering particle breakage of sand. Chin. J. Geotech. Eng. **41**(11), 186–194 (2019)
12. Wang, X.Y., Liu, H.L., Jiang, Q., et al.: Field tests on response of excess pore water pressures of liquefaction resistant rigid-drainage pile. Chin. J. Geotech. Eng. **39**(4), 645–651 (2017)

13. Su, L., Tang, L., Ling, X.C., et al.: Numerical simulation of shake-table experiment on dynamic response of pile foundation in liquefaction-induced lateral spreading ground. J. Disaster Prev. Mitig. Eng. **39**(2), 227–235 (2019)
14. Zhou, H., Yuan, J.R., Liu, H.L., et al.: Model test of rectangular pile penetration effect in transparent soil. Rock Soil Mech. **40**(11), 307–316 (2019)
15. Wang, Y.H., Zhang, M.Y., Liu, X.Y., et al.: Laboratory test comparative study on open-close piles during jacked pile-jacking based on pile-soil interface stress measurement. J. Central South Univ. (Sci. Technol.) **52**(2), 599–606 (2021)
16. Yu, H.S.: CASM: A unified state parameter model for clay and sand. Int. J. Numer. Anal. Meth. Geomech. **22**(8), 621–653 (1998)
17. Mo, P.Q., Gao, X.W., Huang, Z.F., et al.: Analytical method for settlement control of displacement pile induced by undercrossing tunnel excavation. Rock Soil Mech. **40**(10), 121–130, 141 (2019)

Reliability Analysis of Gravity Retaining Wall

Baohua Zhang[1]([✉]), Zhenkang Zhang[2], and Peng Zheng[3]

[1] School of Civil and Architectural Engineering, Nanchang Institute of Technology, Nanchang, China
2351459436@qq.com
[2] Nanchang Road Bridge Engineering Corporation, Nanchang, China
[3] Nanchang Institute of Technology, Nanchang, China

Abstract. In this article, When considering effect of earthquake and under ground water. The failure mode of gravity earth-retaining wall and the way of settling the limit of reliability are analyzed. Based on project improved JC Method is applied to calculate unitary reliability, the pertinence with the invalid mode being considered farther, the limit of reliability is calculated.

Keywords: Earthquake function · Under ground water function · Gravity earth-retaining wall · Failure mode · Limit analysis of reliability

1 Foreword

Gravity retaining wall is widely used in water conservancy, highway, construction, port, railway, mine and other projects because of its advantages of local materials, convenient construction and good economic benefits. It is of great significance to conduct a more comprehensive and accurate reliability analysis. At present, the fixed value analysis method is mostly used to analyze the retaining wall in engineering, Although this method has been proved to be an effective method by long-term engineering practice, it has obvious shortcomings: firstly, the load, shear strength index of soil, unit weight of soil, groundwater level and material strength are not considered; Secondly, the failure mode correlation of retaining wall overturning failure, horizontal sliding failure, insufficient foundation bearing capacity failure and overall sliding failure is not considered. Therefore, some retaining walls calculated by the fixed value method are sufficient, but they are damaged in practical application, which has been confirmed by many failure examples at home and abroad.

During an earthquake, the earth pressure often increases due to the earthquake, resulting in the destruction of the retaining wall. Therefore, the impact of the earthquake on the earth pressure should be considered when building the retaining wall in the earthquake area. In the area with sufficient rainfall, the soil seepage and wall drainage speed are limited, resulting in the rise of the water level behind the wall and the increase of the pressure behind the wall, The retaining wall is often damaged due to the increase of water pressure during or after rain. For low retaining wall, the tensile bending tensile stress is low due to the heavy wall. In addition, the compressive stress generated by the self weight of the wall can offset part of the tensile stress, so the tensile stress of the wall body is very small; For the slightly higher retaining wall, in addition to the heavy wall, structural measures such as reinforcement can also be adopted. Therefore, the damage caused by insufficient material strength of the wall body will not be considered temporarily.

2 Failure Mode of Retaining Wall Considering Earthquake and Groundwater

2.1 The Function and Selection of Constant and Random Variable

In order to simplify the calculation, the amount with small variation is regarded as a constant and the amount with large variation is regarded as a random variable. Taking the retaining wall with homogeneous fill behind the wall and horizontal bottom as an example (the load acting on the retaining wall is shown in Fig. 1, for the retaining wall with layered fill behind the wall and inclined bottom, the failure mode can be analogized, and the reliability limit analysis method is the same

1) Functional function corresponding to overturning failure of retaining wall

$$g_1 = G_{x0} + E_{az}'' x_f + E_p z_p - E_{ax}'' z_f - F_t z_k \tag{1}$$

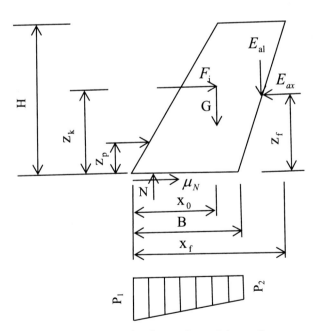

Fig. 1. The load acting on the retaining wall

In function Where G is the weight of retaining wall with length of 1 m. x_0 is the horizontal distance between the center of gravity of the retaining wall and the wall toe; E''_{az} is the vertical component of static active earth pressure and the vertical inertial force caused by earthquake on the sliding wedge acting on the wall back. Generally, the vertical component of the vertical inertial force caused by earthquake on the sliding wedge acting on the wall back is not considered. Therefore, E''_{az} is the vertical component of static active earth pressure E_{az}; x_f is the horizontal distance from the static active earth pressure action point to the wall toe; E_p is the static and passive earth pressure on the wall chest; z_p is the height from the action point of static and passive earth pressure on the wall chest to the wall toe. E''_{ax} is the horizontal component of total active earth pressure under horizontal seismic action [1], $E''_{ax} = E_{ae} = (1 + K_h C_z C_e \tan \varphi) E_a$, Where K_h is the horizontal seismic coefficient(When Design crack is 7, the $K_h = 0.1$,When Design crack is 8, the $K_h = 0.2$, When Design crack is 9, the $K_h = 0.4$), C_z is the comprehensive influence coefficient, taken as 0.25, C_e is the seismic earth pressure coefficient (see Table 1 below), φ is the friction angle of fill at the back of the wall, E_a is the static active earth pressure. z_f is the height from the static active earth pressure action point to the wall toe; F_i is the horizontal inertial force caused by earthquake on the sliding wedge.; z_k is the height from the action line of horizontal inertial force caused by earthquake on the sliding wedge to the wall toe.

Table 1. Seismic dynamic earth pressure coefficient

Dynamic earth pressure	Fill slope (°)	Internal friction angle φ (°)				
		21–35	26–30	31–35	36–40	41–45
Active earth pressure	0	4.0	3.5	3.0	2.5	2.0
	10	5.0	4.0	3.5	3.0	2.5
	20	–	5.0	4.0	3.5	3.0
	30	–	–	–	4.0	3.5
Passive earth pressure	0–20	3.0	2.5	2.0	1.5	1.0

2) The function corresponding to horizontal sliding failure of retaining wall

$$g_2 = \mu G + \mu E''_{ax} + E_p - E''_{ax} - F_i \tag{2}$$

μ is the friction coefficient between soil and wall bottom. G, E''_{ax}, E''_{az} E_p and F_i same as function (1)

3) The function of Failure of retaining wall foundation due to insufficient bearing capacity.

For the soil foundation with the design value of foundation bearing capacity $f \leq 200\,kPa$

$$g_3 = \frac{\sum N''}{b}(1 + \frac{6e}{b}) - f_{SE} \quad e \leq \frac{b}{6}$$

$$g_3 = \frac{2\sum N''}{3c} - f_{SE} \quad e \leq \frac{b}{6} \tag{3}$$

Where $\sum N''$ is the component of the force acting on the back of the wall perpendicular to the sliding surface at the bottom of the wall caused by the self weight of the retaining wall, earth pressure and inertia force caused by earthquake on the sliding wedge. If the sliding surface at the bottom of the wall is horizontal, then $\sum N'' = G + E''_{az}$ is the bottom width of retaining wall; e is the eccentricity of $\sum N''$.

4) Function corresponding to integral sliding failure of retaining wall

$$g_4 = \sum_{i=1}^{n} c_i l_i + \sum_{i=1}^{n} (G_i + q_i b_i) \tan \varphi \cos \beta_i - \sum_{i=1}^{n} (G_i + q_i b_i) \sin \beta_i$$

When there is homogeneous fill behind the wall [2, 3]:

$$g_4 = c \sum_{i=1}^{n} l_i + \tan \varphi \sum_{i=1}^{n} (G_i + q_i b_i) \cos \beta_i - \sum_{i=1}^{n} (G_i + q_i b_i) \sin \beta_i \tag{4}$$

Where c_i is the cohesion of the soil at the sliding arc surface of the ith soil strip, When there is homogeneous fill behind the wall $c_i = c$; l_i, b_i, ρ_i is ith slip arc length of soil strip, Width of soil strip and horizontal inclination of tangent at the midpoint of slip arc, φ_i is the friction angle of the soil at the sliding arc surface of the ith soil strip, When there is homogeneous fill behind the wall $\varphi_i = \varphi$; G_i is ith Weight of soil strip; q_i is the overload concentration on the top surface of the ith soil strip. When the filling surface is overloaded, it is uniformly distributed load, $q_i = q$.

2.2 Limit State Equation

In structural reliability analysis, the structural limit state is generally the case that the functional function is equal to 0, that is

$$Z = g(x_1, x_2, \cdots, x_n) = 0 \tag{5}$$

For retaining walls considering earthquake and groundwater, the limit state equations are as follows: formula (1)–(4) is equal to 0.

3 Reliability Limit Analysis of Retaining Wall Considering Earthquake and Groundwater

3.1 General Limit of Reliability of Retaining Wall Considering Earthquake and Groundwater

Considering the effect of earthquake and groundwater, the four failure modes of retaining wall are connected through random variables, so its reliability must fall within a range, and its failure probability limit can be solved by equation [4, 5] (6)

$$\max_{i} P_{f_i} \leq P_f \leq 1 - \prod_{i=1}^{4} (1 - P_{f_i}) \tag{6}$$

3.2 Narrow Limit of Reliability of Retaining Wall Considering Earthquake and Groundwater

Considering the narrow limit of reliability of retaining wall under earthquake and groundwater, when the failure probability calculated by Eq. (6) is wide, the narrow limit formula of O. Ditleven [6] on structural failure probability can also be used to calculate the narrow limit of reliability of retaining wall under earthquake and groundwater. The steps are as follows:

1) The failure probability and design check point of single reliability index under various failure modes are calculated by improved C method
2) Find the correlation coefficient of each functional $\rho_{g_r g_j}$

$$\rho_{g_r g_j} = \sum_{i=1}^{n} \alpha_{ik}^* \, \alpha_{ik}^* \tag{7}$$

Type: $\alpha_{ik}^* = \left(\dfrac{\partial g_i}{\partial x_k} \right)_{x^*} \Bigg/ \sqrt{\sum_{k} \left(\dfrac{\partial g_i}{\partial x_k} \right)_{x^*}^{2}}$; $\alpha_{jk}^* = \left(\dfrac{\partial g_j}{\partial x_k} \right) \Bigg/ \sqrt{\sum_{k} \left(\dfrac{\partial g_j}{\partial xk} \right)_{x^*}^{2}}$ and $x_k' = \dfrac{x_k - \mu_{x_k}}{\sigma_{x_k}}$; $\dfrac{\partial g_i}{\partial x_k} = \sigma_x \dfrac{\partial g_i}{\partial x_k}$.

3) Probability of finding $p_f(E_i E_j)$

When all random variables are normally distributed and the correlation coefficient $\rho_{g_r g_j} \geq 0$, by Reliability index of E_i, E_j, Determined by

$$S_i + S_j \geq P_f(E_i E_j) = P_f(g_i < 0 \cap g_j < 0) \geq \max[S_i, S_j]$$

4) The narrow limit of failure probability of retaining wall considering earthquake and groundwater is obtained from the following formula

$$P_f(E_1) + \max \left[\sum_{i=2}^{n} \left\{ P_f(E_i) - \sum_{j=1}^{i-1} P_f(E_i E_j) \right\}, 0 \right] \leq P_f \leq \sum_{i=1}^{n} P_f(E_i) - \sum_{i=2}^{n} \max_{j<i} P_f(E_i E_j) \tag{8}$$

Then the narrow limit of reliability is obtained from $P_r = 1 - P_f$.

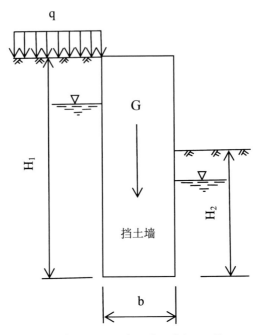

Fig. 2. Cross section of retaining wall

4 Case Analysis

Gravity retaining wall is adopted in a project, The weight of wall material concrete is, Coefficient of variation is 0.05, The section is rectangular, as shown in Fig. 2. The buried depth is 3.2 M and the foundation pit excavation is 5.0 M. The foundation soil is assumed to be a single soil layer. The friction angle between the retaining wall bottom and the soil is 31°. The average height of groundwater level is 1.0 m below the surface. Coefficient of variation is 0.42. During the flood period, the groundwater does not overflow the wall top, the probability characteristics of each soil index are listed in Table 2, and the geometric dimensions of the retaining wall are regarded as fixed values.

Table 2. Random variables and their statistical characteristics [7]

Random variable	Random variable	Coefficient of variation	Distribution type
Unit weight of soil γ	18(KN/m)	0.1	Normal distribution
Unit weight of wall γ_0	24(KN/m^3)	0.05	Normal distribution
Cohesion c	8.65(kPa)	0.22	Lognormal distribution
Internal friction angle φ	11.33°	0.13	Lognormal distribution
High water level H_0	3.5(m)	0.42	Extreme Value I
Ground overload q	25(kPa)	0.2	Extreme Value I

Note: Correlation between random variables: Correlation coefficient between c and is −0.3; coefficient between c and H_0 is −0.4; between H_0 and is −0.3, other variables are independent of each other.

Among the four failure modes of gravity retaining wall listed above, 1–3 is common way. The improved JC method calculation program is adopted. The calculated reliability indexes corresponding to various failure modes are shown in Table 3.

Table 3. Reliability index corresponding to various failure modes

Failure mode of gravity retaining wall	Reliability index	Reliability (%)	Failure probability
Anti overturning failure of retaining wall	2.387	99.150 64	0.008 493 6
Anti horizontal sliding failure of retaining wall	2.692	99.64 82	0.003 551 8
Bearing capacity of retaining wall foundation	2.416	99.215 36	0.007 846 4

Considering that the three failure modes are connected by random variables and are interrelated, their reliability must fall within a range, and the general limit of failure probability can be solved by Eq. (6)

$$0.0084936 \leq P_f \leq 1 - (1 - 0.0084936)(1 - 0.0035518)(1 - 0.0078464)$$

Then the limit of reliability is obtained from $P_r = 1 - P_f$

$$99.15064\% \geq P_r \geq 98.02326\%$$

The reliability range calculated above is narrow, so the narrow limit of reliability of retaining wall can not be calculated.

5 Conclusion

Four failure modes corresponding to earthquake, overturning failure of retaining wall under groundwater, horizontal sliding failure, insufficient foundation bearing capacity failure and overall sliding failure, as well as the method of calculating the reliability limit of retaining wall are analyzed. Combined with engineering practice, the single reliability of a gravity retaining wall is calculated by improved JC method, and the correlation of failure modes is further considered, The boundary of the reliability of gravity retaining wall is obtained, and a more objective, comprehensive and accurate evaluation of the reliability of gravity retaining wall is made, which provides a preliminary and feasible way for further studying the reliability of gravity retaining wall under the condition of inclined bottom surface and layered filling, and considering the action of horizontal and vertical seismic inertia force and groundwater, It can be used as a reference for studying the reliability of other forms of retaining wall, slope stability and foundation pit support.

References

1. Wang, X.L.: Foundation Engineering. Chongqing University Press, Chongqing (2001)
2. Zhao, M.H.: Soil Mechanics Foundation Engineering. Wuhan University of Technology Press, Wuhan (2003)
3. Han, Y.F., Liu, D.F., Dong, S.: Reliability analysis method of slope stability. Coast. Eng. **4**, 9–13 (2001)
4. Xu, C., Wang, Z., Liu, Q.H.: Application Optimization of Dendriform Structure Considering Stability. Acta Constructivi Tectonica (2018)
5. Li, Z.Y., Xiao, S.G.: Stability analysis of embankment with multi-stage cantilever retaining wall. Railway Construction (2020)
6. Dlitlevsen, O.: Narrow reliability bounds for structural system. J. Strut. Mech. **4**, 45–48 (1979)
7. Liang, J., Yang, Y.W.: Safety analysis of reinforced earth-rock dams with internal walls. Eng. Sci. Technol. (2020)

Research on Key Technology of Anti-sliding Treatment of Resin Thin Layer of Bascule Bridge Steel Panel

Di Wu, Yafang Han[✉], and Xiayu Jin

Guangdong Jianke Traffic Engineering Quality Testing Center Co. Ltd., Guangzhou, China
826540401@qq.com

Abstract. There is relatively little research on steel deck pavement of bascule bridge at home and abroad, which usually adopts the same paving materials and structures as ordinary steel deck, but it can't meet the normal application conditions. In order to improve the anti-sliding performance of steel deck pavement of bascule bridge, this paper provides a set of steel deck pavement technology of bascule bridge through theoretical analysis of mechanics of bascule bridge and research of pavement materials. The engineering application research shows that the anti-sliding performance of the pavement back layer is obviously improved. After 7 months of application, the pavement layer has a good overall structure, and the bonding strength is still stable at 3.57 MPa, with good durability and good social benefits.

Keywords: Bascule bridge · Steel bridge deck · Resin thin layer · Anti-sliding performance · Bonding property

1 Introduction

With the rapid development of bridge construction, steel deck pavement has become one of the most complicated and critical technical difficulties in road engineering. At present, the main pavement modes of steel bridge deck are asphalt concrete, epoxy asphalt concrete, asphalt concrete and epoxy asphalt concrete double-layer pavement, etc. [1]. With the increase of service life, various diseases, such as cracks, ruts, slipping, etc., have appeared, and the maintenance problem has gradually attracted attention.

The bridge studied in this paper is Zhongshan bascule bridge, and its structural type is bascule bridge. Bascule bridge mainly refers to the bridge whose span structure can move or rotate [2]. According to the mechanical research and analysis report of related literature, the mechanical characteristics of large cantilever box girder structure of bascule bridge are obviously different from those of long-span suspension bridge or cable-stayed bridge. With the change of bridge opening angle, the surface layer is subjected to complex transverse tensile stress and shear stress [3, 4]. The main requirements of bascule bridge for bridge deck pavement are adaptability to deformation, stability of pavement structure and excellent durability, which are reflected in the performance of pavement

© The Author(s) 2022
G. Feng (Ed.): ICCE 2021, LNCE 213, pp. 495–507, 2022.
https://doi.org/10.1007/978-981-19-1260-3_45

materials such as good flexibility, thermal stability, waterproof, rutting resistance and high bonding strength with steel bridge deck, and the performance of pavement surface is good service performance. At present, there is relatively little research on steel deck pavement of bascule bridge at home and abroad, which usually adopts the same paving materials and structures as ordinary steel deck, and there is little research on targeted application.

Therefore, this paper aims to provide a steel deck paving technology of bascule bridge through theoretical analysis of mechanics of bascule bridge and research of paving materials. In this paper, taking an bascule bridge in Zhongshan as the research object, through theoretical analysis, combined with the use environment and other factors, the maintenance and reconstruction scheme of steel bridge deck is comprehensively determined, and the performance of paving materials is targeted for research. Finally, the application performance of this paving technology is verified by tracking.

The project relies on bascule bridge in Zhongshan city, which was built in the mid-1990s. The bridge is 95.78 m long and 24 m wide. The main span is an open steel box girder with a span length of 25.9 m. The approach bridge is composed of 5-span reinforced concrete simply supported plate beam, 69.88 m in length, the substructure adopts φ 150 cm bored pile foundation, the pier and the foundation diameter is the same, except the 4# pier, the design load is steam - super 20, the checking load is hanging − 100, the design speed is 50 km/h, seismic grade 7°. The opening time of the bridge is from 2 am to 4 am every day, and the maximum opening Angle is 60°.

Midas Civil, a THREE-DIMENSIONAL finite element analysis software, was used to establish the open span model of the open bridge and simulate the stress state when it was opened to 60°, as shown in Fig. 1. According to the mechanical analysis, the stress of the bridge surface varies greatly during the opening process of the bridge, which requires higher mechanical properties of pavement materials.

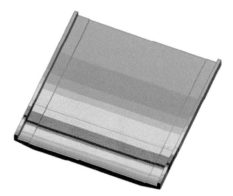

Fig. 1. Schematic diagram of stress of opening span model of opening bridge to 60°

2   The Choice of Bridge Deck Pavement Scheme

The opening bridge is located in the central business district, with a large traffic flow, and the deck layer is made of patterned steel plate, which is relatively smooth. The bridge is located in subtropical monsoon climate zone, where the temperature is mild in winter, hot and rainy in summer, and the annual average precipitation is higher by about 1700 mm. In rainy days, the anti-sliding performance of the surface layer drops sharply, and there is a great potential safety hazard in driving [5]. In addition, the design life of this bascule bridge is long, and the load of steel deck pavement is not specified in the design, so the pavement layer cannot be overloaded and should not be too thick.

Based on the comprehensive analysis of factors such as steel deck structure, traffic flow and climate environment of bascule bridge, the main engineering characteristics of the steel deck pavement scheme are as follows: the anti-sliding performance of steel deck surface is poor; Heavy traffic, short maintenance time; South China is located in high temperature and rainy area, which requires high weather resistance of materials [6]. In order to meet the performance of steel bridge deck pavement, materials should have excellent anti-sliding performance, aging resistance, deformation coordination, waterproof performance and bonding strength.

According to the comparison and selection of the recommended paving schemes for opening bridges in Table 1, and considering the cost and the requirements of paving schemes comprehensively, the resin anti-sliding thin-layer paving scheme is finally selected. The total design thickness of bridge deck paving is 5 mm, which can greatly

Table 1. Recommended paving scheme for bascule bridge

Project	Option 1	Option 2	Option 3
	Resin thin layer	Epoxy concrete	Asphalt + epoxy asphalt concrete
Thickness	5 mm	50 mm	80–90 mm
Technology maturity	Short time and mature technology	Long time, mature technology	Long time, mature technology
Construction difficulty	Simpler	Complex, epoxy asphalt needs heating	Complex, asphalt needs heating
Road performance	Good skid resistance and good performance under high temperature and heavy traffic conditions	Good performance under high temperature and heavy traffic conditions	
endurance	Meet 5–8 years of application	Local rutting and shifting are easy to occur	Local rutting and shifting are easy to occur
Repair and maintenance	Little difficulty	Be extremely difficult	Be extremely difficult
Cost	Option 3 > Option 2 > Option 1		

reduce the load increase caused by surface paving. The concrete paving structure is shown in Fig. 2, which consists of steel panel + resin adhesive layer (1.5 kg/m²) + anti-sliding aggregate (8 kg/m², particle size range of 1.18–1).

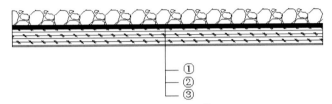

① Steel panel; ② resin adhesive layer; ③ Anti-sliding black aggregate;

Fig. 2. Schematic diagram of bridge deck structure section

3   Research on Pavement Material Performance

3.1   High Temperature Stability

The opening bridge is located in the south humid area, and the main problem is that the steel deck absorbs heat and transfers heat quickly. In high temperature summer, the service temperature of steel deck is higher than that of asphalt pavement or cement concrete bridge surface. Therefore, the steel bridge deck pavement is required to have better thermal stability to prevent rutting, pushing, bulging and other diseases [7]. In this paper, the dynamic stability of resin thin layer paving materials is tested, and the data are shown in Table 2. As can be seen from Table 2, the resin mixture prepared from resin pavement binder and aggregate has good high temperature stability. When the test conditions are 60 °C and 0.7 MPa, the dynamic stability of resin mixture reaches 29,929 times/mm, and with the increase of test temperature and pressure, the dynamic stability of resin mixture decreases to some extent but still reaches 26,662 times/mm. In addition, in the immersion test, the dynamic stability of the resin mixture is obviously lower than that without water under the same conditions. There are still 18,828 times/mm [8] when immersed in water at 70 °C and 0.9 MPa. It can be seen that the resin mixture for steel bridge deck has excellent high temperature stability. This is because resin adhesive is a highly crosslinked three-dimensional network structure after resin and curing agent react, and it still maintains a certain mechanical strength at high temperature. Pure asphalt materials have poor bonding strength at high temperature and are easy to loose, so it is not suitable for steel bridge deck pavement.

3.2   Bonding Performance Between Resin Thin Layer and Steel Base Surface

The bonding performance between resin paving material and steel bridge deck is the key factor to ensure the service performance of thin layer and steel bridge deck after bonding, which mainly includes the function of meeting the vehicle load and the stress function during the cycle of opening and closing the bridge.

Table 2. Dynamic stability of resin mixture

Serial number	Material	Stability (times/mm)			
		60 °C, 0.7 MPa	60 °C, 0.7 MPa immersion	70 °C, 0.9 MPa	70 °C, 0.9 MPa immersion
1	Resin mixture	29045	25010	25200	19030
2	Resin mixture	29205	26890	25245	18700
3	Resin mixture	30145	27090	25245	18635
4	Resin mixture	30205	27530	25245	18775
5	Resin mixture	31045	26790	25245	19000
	Average value	29929	26662	25236	18828

Testing the bonding strength between the resin thin layer and the steel plate surface layer indoors, the main steps are as follows: firstly, pretreating the steel surface, namely, no treatment of the original steel surface, sand blasting treatment and deep grinding treatment; Then, coating the resin anti-sliding layer on the surface of the steel plate, and curing at 40 °C for 2 days; The bonding strength of the resin thin layer material was tested by pulling-out. The test results are as shown in Figs. 3 and 4,and the failure locations are all at the anti-sliding surface layer of the planting bar adhesive and resin,As shown in Table 3, the test results of the bonding strength of the three steel plates are 2.85 MPa, 4.46 MPa and 4.47 MPa respectively. From the results, it can be seen that the bonding strength of the surface layer is obviously improved after sand blasting and grinding rough treatment, which shows that the resin anti-sliding layer has a high interfacial bonding strength with the steel base surface.

Table 3. Bonding strength of steel plate surface with different treatment methods

Steel plate treatment method	Test item	
	Adhesive strength (MPa)[a]	Destructional forms
Steel plate is not treated	2.85	Damage at the bonding surface
Sand blasting treatment of steel plate surface	4.46	Damage at the bonding surface
Deep grinding of steel plate surface	4.47	Damage at the bonding surface

Note a: The bonding strength was tested at 40 °C for 2 days

Fig. 3. Failure modes of adhesive strength test

Fig. 4. Adhesion test between mixture and steel base

4   Construction Technology Research

In this paper, the construction technology of steel deck pavement is confirmed through the paving of thin resin pavement on steel deck of bascule bridge. The main construction technology includes the following aspects: base surface treatment-glue mixing-glue spreading-aggregate spreading-maintenance-aggregate cleaning-open traffic.

(1) Base surface treatment

Through modeling and analysis, it can be seen that the different stress in each area of the bridge deck pavement during opening has higher requirements for the bonding strength of pavement materials. Combined with the indoor steel plate treatment methods and test results (Table 3), the steel bridge deck is treated by shot blasting machine in this application project. After shot blasting, the roughness of the surface layer is generally uniform, with an average roughness of 115 μm and a cleanliness of Sa2.5, as shown in Fig. 5. For the anchor bolt of steel plate surface layer, seam leaving treatment can not only facilitate the bolt loosening and welding treatment, but also prevent the surface layer damage caused by the loosening of raised bolts, as shown in Fig. 6.

(2) Resin thin layer bridge deck paving technology

After shot blasting treatment of steel bridge deck, glue mixing, surface paving, curing, aggregate recovery and other steps are carried out, as shown in Fig. 7. In this process, it is necessary to strictly control the proportion of components and the amount of glue used for glue mixing. The glue used for paving the steel bridge surface is controlled to be 1.4–1.5 kg/m^2, and the amount of aggregate is 6–8 kg/m^2. The curing should be

Fig. 5. Shows the steel bridge deck before and after shot blasting on the left

Fig. 6. Treatment of seam leaving at anchor bolt

carried out according to the ambient temperature and the recommended time, as shown in Table 4. Finally, the excess aggregate is cleaned and recycled.

Table 4. Curing time of resin thin layer at different temperatures

Serial number	Temperature/°C	Operation time/min	Health time/h
1	15	43	12
2	20	40	8
3	25	38	5
4	30	35	4
5	35	30	2
6	40	25	1.5

| a) Shot blasting treatment of steel bridge deck | b) Glue paving |
| c) Aggregate recovery | d) The steel bridge deck pavement is completed. |

Fig. 7. Steel deck pavement of bascule bridge

(3) Quality control

The technical requirements and indoor test results of the resin adhesive are shown in Table 5. The tensile strength of the resin adhesive reaches 20.3 MPa, the steel-steel bonding strength reaches 19.5 MPa, and the elongation at break reaches 53.5%, which shows that the material has good mechanical properties and ductility, and can adapt to the deformability of the steel bridge deck while ensuring firm bonding with the steel bridge deck.

In addition, on-site pull-out test was carried out on the quality of the layer behind the resin anti-sliding layer pavement, and the test results are shown in Table 6. The results of on-site raw material inspection and pull-out test after on-site paving show that the average bonding strength reaches 4.08 MPa, which fully meets the technical requirements ($\geq$3.0 MPa), indicating that the resin paving thin layer after paving has good mechanical.

Table 5. Main technical indexes of resin thin-layer binder

Test item	Technical requirement	Test result	Reference standard
Mass ratio of components A/B	1:0.75	1:0.75	–
Tensile strength/MPa (50 mm/min)	≥15	20.3	GB/T 1040.3-2006
Adhesive strength/MPa (2.0 mm/min)	≥1.5	≥2.5(Concrete damage)	GB/T 16777-2008
Elongation at break/%(2mm thick, 23°C)	≥35	53.5	GB/T 1040.3-2006
Steel-steel bonding strength/MPa (2.0 mm/min)	≥15	19.5	GB/T 6329-1996
Steel-steel tensile shear strength/MPa (2.0 mm/min)	≥10	17.8	GB/T 7124-2008

Table 6. Drawing test results of resin thin steel bridge deck pavement

Serial number	Post-paving bonding strength (25 °C)/MPa	Damage location	Technical requirement
1	3.98	Modification of resin binder	≥3.0 MPa
2	3.95	Modification of resin binder	
3	4.15	Modification of resin binder	
4	4.12	Modification of resin binder	
5	4.23	Modification of resin binder	
6	4.07	Modification of resin binder	

Table 7. Friction coefficient of steel bridge surface layer

Project		Original steel bridge deck	After resin anti-sliding thin layer treatment	Specification requirements
Friction factor	0M	36	86	≥54
	1M		83	
	2M		80	
	3M		80	
	4M		79	
	5M		78	
	6M		75	
	7M		74	

Note: M stands for month, which is the time after opening traffic

5 Application Performance of Resin Thin Layer After Paving

5.1 Sliding Resistance

Before the bridge deck is paved, the anti-sliding performance of patterned steel plates on the surface is poor, especially in wet weather after rain, and the anti-sliding performance drops rapidly. For specific data, please refer to Table 7. The anti-sliding performance was improved from the friction pendulum value BPN36 to 86 by 139% after the resin anti-sliding thin layer was added on the surface layer. Even after 7 months of use, the friction coefficient remains above 74. In addition, the tracking of the pavement performance of the bridge deck shows that the structural depth of the bridge deck is stable at about 1.4 mm (as shown in Fig. 8) after the anti-sliding reconstruction, and the water permeability of the resin thin layer is 0 ml/min, which plays a good role in protecting the steel plate at the bottom. It can be seen that the resin thin layer added not only obviously improves the anti-sliding performance of the original bridge deck, but also plays a role of waterproof and anti-corrosion.

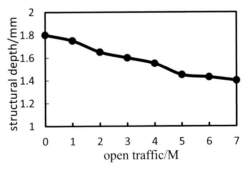

Fig. 8. Changes of structural depth with open traffic time

5.2 Durability

After the completion of the steel deck pavement (January-July), the application performance was tracked and tested, and the overall application effect of the deck pavement of the bascule bridge was good, without falling off and other diseases, as shown in Fig. 9. In addition, the adhesive strength between the resin thin layer and the base surface was tracked and tested, and after 7 months of open traffic, the adhesive strength of the surface layer remained stable above 3.71 MPa, as shown in Fig. 10.

a) Overall application of surface layer   b) Surface local structure

c)On-site drawing test   d)On-site drawing test

Fig. 9. Application effect tracking return visit

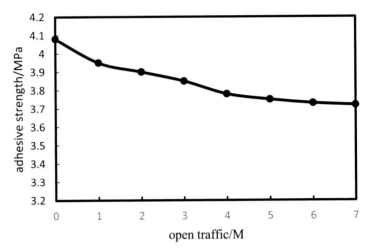

Fig. 10. The change of adhesive strength between resin anti-sliding thin layer and bridge deck with open traffic time

6   Conclusion

(1) Through the stress analysis of the opening bridge and the comprehensive analysis of the use environment factors, the resin anti-sliding thin layer pavement scheme is selected for the steel deck pavement of the opening bridge, with a total thickness of about 5 mm, which has little influence on the original bridge deck load.

(2) In order to ensure the good durability of the resin anti-sliding thin layer, the performance of pavement materials was studied in laboratory. The results show that the resin anti-sliding thin layer has excellent high temperature resistance and mechanical properties. At 60 °C and 0.7 MPa, the dynamic stability of the mixture reaches 29,929 times/mm, and the bonding strength between the resin thin layer and the steel plate surface reaches 4.08 MPa.

(3) Through the research on the construction technology of steel deck pavement of bascule bridge, shot blasting of steel deck is beneficial to improve the bonding strength between anti-sliding layer and base surface, thus improving the durability of thin layer.

(3) The tracking application test shows that the anti-sliding performance of the resin thin layer pavement is obviously improved. After 7 months of application, the pavement surface is in good condition as a whole, the structural depth is more than 1.4 mm, and the bonding strength is still stable at 3.71 MPa, with good durability and good social benefits.

References

1. Wang, C.H., Fu, Y., Chen, Q., Chen, B., Zhou, Q.W.: Research and application progress of epoxy asphalt concrete bridge deck pavement materials. Mater. Guide **17**, 2992–3009 (2018)

2. Shen, C., Wei, J.F., Guo, H.Z., Jiang, K.Q., et al.: Key technologies and difficulties in the research of bridge deck pavement with vertical turning and opening. Highw. Eng. 132–137 (2018)
3. Yu, L.Q., Guo, H.Z., Sui, H.R., Liu, T.G., et al.: Mechanical response analysis of steel deck pavement of vertical-turn-bascule bridge. Road Traffic Technol. 29–12 (2009)
4. Wei, P., Liu, Y., Kuang, Y.: Dynamic response analysis of bascule bridge deck pavement considering cantilever structure rotation. J. East China Jiaotong Univ. 15–21 (2020)
5. Rong, B.S.: Analysis of anti-sliding performance and noise reduction technology of cement concrete pavement in highway tunnel. Shanxi Archit. 106–107 (2019)
6. Xu, R.H., Zeng, G.D., Huang, H.L.: Research on the whole maintenance scheme of wide epoxy asphalt pavement of Pingsheng Bridge. Chin. Foreign Highw. 62–66 (2020)
7. Zheng, Y.: Study on the performance of reactive resin mixture and its application in steel bridge deck pavement. Chongqing Jiaodong University (2010)
8. Yang, Y., Chen, S.T., Yin, C.Y.: Application of resin anti-sliding layer in anti-sliding reconstruction of steel deck of Guangming Bridge in Zhongshan. Highw. Autom. Transp. 160–163 (2014)

Influence of Fiber on Properties of Graphite Tailings Foam Concrete

Xiaowei Sun[1], Miao Gao[1], Honghong Zhou[1(✉)], Jing Lv[2], and Zhaoyang Ding[1,3]

[1] School of Material Science and Engineering, Shenyang Jianzhu University, Shenyang, China
SXW@sjzu.edu.cn
[2] School of Science, Shenyang Jianzhu University, Shenyang, China
[3] School of Civil Engineering, Shenyang Jianzhu University, Shenyang, China

Abstract. The project used graphite tailings as a filler to prepare graphite tailings foamed concrete. Mainly studied the physical properties, mechanical properties and thermal properties of the foam concrete by graphite tailings, also studied the combination of polypropylene fiber and glass fiber influence of foam concrete compressive strength and cracking strength. The experimental results show that in the case of the same dry density grade, adding 20% graphite tailings can make the foam concrete strength reach its peak. When the water-binder ratio is 0.65 and the self-made chemical foaming agent content is 7%, the optimal total fiber volume blending rate is 0.18%, and the blending ratio of polypropylene fiber and glass fiber is 2:1. The compounding of polypropylene fiber and glass fiber can improve the flexural performance of foam concrete, which is not conducive to the thermal insulation performance of foam concrete, but the test results are still better than industry standards.

Keywords: Graphite tailings · Foam concrete · Strength · Polypropylene fiber · Glass fiber

1 Introduction

Foam concrete is a lightweight, thermal insulation, fire resistant, sound insulation and anti freezing concrete material. In recent years, many people have done a lot of research work in the field of foam concrete from different angles [1–8]. However, there are few reports on the preparation of foam concrete with graphite tailings as filler. Graphite tailings, as industrial wastes discharged from mines after beneficiation, have a very negative impact on the ecological environment. In addition to effective cover and reclamation, it is more important to find ways to turn graphite mine tailings into treasure and reuse them. Using graphite tailings to replace part of the cementitious materials to prepare foamed concrete is another new way to turn graphite tailings into treasure. Therefore, the research on graphite tailings modified foamed concrete can not only improve its crack resistance and thermal insulation performance, but also solve the problem of industrial waste pollution. On the one hand, it solves the problems of storage and land occupation of graphite tailings in graphite production sites; on the other hand, it can open up a

© The Author(s) 2022
G. Feng (Ed.): ICCE 2021, LNCE 213, pp. 508–515, 2022.
https://doi.org/10.1007/978-981-19-1260-3_46

way for mines to turn waste into treasure. It has extremely important economic and social significance for the rational use of energy and the improvement of the ecological environment. Graphite tailings are used as filler to prepare foamed concrete and the effect of fiber on properties of foam concrete are studied in this paper.

2 Text

2.1 Raw Materials

The graphite tailings produced in Heilongjiang Province are selected for this experiment, with a particle size range of 0–0.4 mm and an apparent density of 2.85 g/cm^3. The particle morphology and the mineral composition of graphite tailings are analyzed by X-ray diffraction, and the diffraction pattern is shown in Fig. 1.

   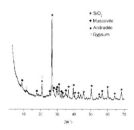

(a) Particle morphology of graphite tailings   (b) XRD pattern of graphite tailings

Fig. 1. Microscopic morphology of graphite tailings

The chemical composition of graphite tailings is analyzed in Table 1.

Table 1. Chemical composition of graphite tailings (%)

SiO$_2$	Al$_2$O$_3$	CaO	Fe$_2$O$_3$	SO$_3$	K$_2$O	MgO
51.56	14.55	10.03	7.64	6.47	3.77	3.60

The cement is P·O 42.5 grade cement produced by Liaoning Jidong Cement Co., Ltd. The foaming agent used in this experiment is a self-made chemical foaming agent with a foaming multiple of 25 times, a settlement distance of less than 5 mm in 1 h, and a bleeding volume in 1 h of less than 18 ml. The quality of polycarboxylate superplasticizer provided by Jiangsu Subote New Material Co., Ltd. accounts for 0.1% of the total quality of cement. Technical indicators of glass fiber and polypropylene fiber are shown in Table 2.

Table 2. Technical indicators of glass fiber and polypropylene fiber

Fiber	Length specification (mm)	Tensile strength (MPa)	Elastic modulus (GPa)	Equivalent diameter (μm)	Density (g/m^3)
Glass fiber	24	2800	86	13	2.63
Polypropylene fibers	20	420	3.58	34	0.91

2.2 Influence of Graphite Tailings Fineness on the Strength of Foam Concrete.

Aiming at the graphite tailings foamed concrete with a dry density of 900 kg/m^3, under the condition that the mixing ratio of graphite tailings is unchanged at 20%, graphite tailings of different particle sizes are screened to study the effect of fineness on the strength of foam concrete and the line chart is drawn as shown in Fig. 2.

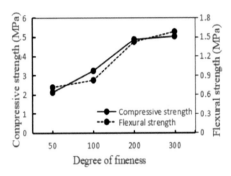

Fig. 2. The influence of graphite tailings fineness on the strength of foam concrete

It can be seen from Fig. 2 that the mechanical properties of foamed concrete have a great relationship with the fineness of graphite tailings. Both the flexural and compressive strengths gradually increase with the increase in fineness. When the fineness of graphite tailings is 50 mesh (0.325 mm), the flexural strength of the foamed concrete is 0.71 MPa and the compressive strength is 2.10 MPa. When the fineness of graphite tailings reached 300 mesh (0.045 mm), flexural strength and compressive strength reaches 1.6 MPa and 5.0 MPa respectively.

2.3 The Influence of Dry Density on the Performance of Graphite Tailings Foamed Concrete

As the dry density of foam concrete changes, its strength, water absorption and thermal conductivity have been greatly affected. Under the condition that the water-to-material ratio was 0.50 and the mixing ratio of graphite tailings is 20% unchanged, by adjusting the foam content, 7 kinds of foam concrete with different dry density grades are prepared. The corresponding line graphs is Fig. 3 respectively.

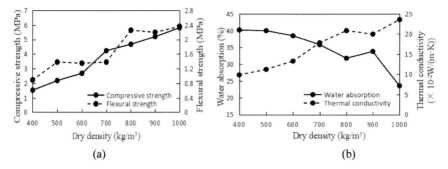

Fig. 3. The effect of dry density on the performance of foam concrete

It can be clearly observed from Fig. 3 (a) that the strength and dry density of foam concrete change almost linearly. The greater the dry density, the higher the flexural and compressive strength of foam concrete. When the dry density was 400 kg/m³, the flexural and compressive strengths are only 0.9 MPa and 1.5 MPa. When the dry density increases to 1000 kg/m³, the flexural and compressive strengths increas by 1.5 MPa and 4.3 MPa respectively.

It can be seen from Fig. 3 (b) that there are two opposite trends in the effect of dry density of foam concrete on water absorption and thermal conductivity. The water absorption of foam concrete decrease with the increase of dry density, the thermal conductivity increases with the increase of dry density. To express the pore structure of graphite tailings foam concrete more intuitively, this project uses the image processing and analysis system of ImageJ software to select three groups of tests of 700 kg/m³, 800 kg/m³, 900 kg/m³ dry density to measure the pore structure characteristics of foam concrete. Analyzed and researched factors such as porosity, pore size and roundness.

It can be seen from Fig. 4 that the dry density is related to the slurry content per unit volume. The larger the dry density, the more the corresponding slurry quantity, the stronger the cementing ability in the hardened foam concrete, the thicker the hole wall on the foam wall, and the stronger the ability to resist external forces. Generally speaking, the smaller the pore size will improve the mechanical properties of the foam concrete. The porosity decreased with the increase of the dry density. It can also be seen that the strength of the foam concrete increased with the increase of the dry density. Foam concrete belongs to porous material. When the dry density is large, the porosity is relatively small, and the number of connected pores decrease, which eventually leads to foam concrete water absorption decrease. While the dry density increases and the porosity decrease, the content of solid matter in unit volume increases. According to theoretical knowledge, the thermal conductivity of gas is only about one-tenth of that of ordinary solid materials, so in this process, the thermal conductivity of foamed concrete is also increasing.

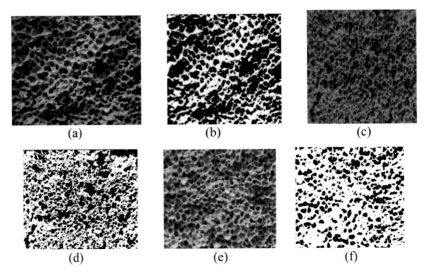

Fig. 4. (a) (c) (e) the gray level processing image of the original image of the hole structure under three dry densities. (b) (d) (f) Binary processed image of the original pore structure at three dry densities

2.4 Effect of Glass Fiber and Polypropylene Fiber on Properties of Foam Concrete

The glass fiber and polypropylene fiber are mixed into the reference foam concrete to test the flexural strength, compressive strength, water absorption and thermal conductivity of the foam concrete. Among them, the total volume doping ratio of the two fibers is 0%, 0.06%, 0.12%, 0.18%, 0.24%, 0.30%, and the volume doping ratio is 1:2, 1:1 and 2:1 respectively.

2.4.1 Effect of Glass Fiber and Polypropylene Fiber on Mechanical Properties of Foam Concrete

Figure 5 shows the effect of fiber blending on the compressive strength of foam concrete. When the total fiber content is 0.18%, the ratio of polypropylene fiber to glass fiber is 1:2, compressive strength of foam concrete up to 4.47 MPa. When the ratio of polypropylene fiber to glass fiber is 1:1, the fiber content is 0.06%, Maximum compressive strength is 4.48 MPa. When the ratio is 2:1, the total doping rate is 0.18%, compressive strength up to 4.55 MPa maximum. Thus it can be seen that fiber blending has little effect on the compressive strength of foam concrete, and the fiber mainly plays the role of toughening, crack resistance and folding resistance in foam concrete.

The composite fiber have obvious influence on the flexural strength of foam concrete. When the total fiber volume ratio increases, the flexural strength of the foam concrete mixed with three proportions of fibers all show a trend of first increasing and then decreasing. The flexural strength of foam concrete can be improved by fiber blending. When the ratio of polypropylene fiber to glass fiber is 2:1 and the total volume ratio is 0.18%, the synergy of the two reaches the maximum, and the increase rate of foam

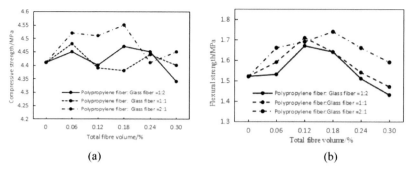

Fig. 5. Effect of fiber remixing on the compressive strength and bending strength of foam concrete

concrete flexural strength was the highest. The results showed that the glass fiber ratio can strengthen the strength of foam concrete at 0.06%, while the polypropylene fiber ratio can play a better role in toughening and cracking resistance under 0.12% condition. When the two cooperate, polypropylene fiber forms a dense network structure in foam concrete, in which glass fiber acts as a skeleton to resist the deformation of foam concrete under the action of external force. The two cooperate to play a good anti-folding and toughening effect. When the ratio of polypropylene fiber to glass fiber was 1:2 or 1:1, the glass fiber occupied a large proportion. The excessive glass fiber content made the crack and sudden stress change easily in the foam concrete. The corresponding decreased in the fiber content reduced the crack resistance and toughness of the foam concrete, which ultimately led to a decrease in the flexural strength of the foam concrete.

2.4.2 Effect of Fiber Blending on Water Absorption and Thermal Conductivity of Foam Concrete

Figure 6 (a) reveals the influence of fiber blending on the water absorption of foam concrete. It can be seen that the water absorption of foam concrete increases gradually with the increase of total fiber volume. When the fiber content is 0.30% and the fiber content ratio is 1:2, the mass water absorption of foam concrete reached the maximum value, which was 43.3% and 42.6% respectively. When the fiber content ratio is 2:1, the water absorption of foam concrete reached the maximum value of 45.4% when the volume content is 0.24%, respectively. Increasing the total fiber content can destroy the pore structure of concrete to a certain extent, resulting in the increase of porosity, and then increase the mass water absorption of foam concrete.

As shown in Fig. 6(b), fiber blending can enhance the thermal conductivity of foam concrete and reduce its thermal insulation performance. The ratio of the three fibers to the thermal conductivity of foam concrete was different. When the ratio of polypropylene fiber to glass fiber is 1:2,1:1 and 2:1, the maximum growth rate of thermal conductivity is 2.6%, 6.4% and 7.8%, respectively. With the increase of glass fiber content, the change rate of thermal conductivity decreased. It can be seen from the Fig. 6 that the ratio of mixing rate was 1:2, the thermal conductivity of foam concrete increases slightly and tends to be stable. The phenomenon is mainly due to the fact that glass fiber had less influence on the thermal conductivity of foam concrete than polypropylene fiber.

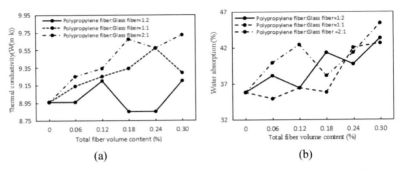

Fig. 6. Effect of fiber remixing on the water absorption rate and thermal conductivity of foam concrete

3 Conclusion

It is concluded that the optimal amount of graphite tailings is 20%. The pore structure distribution of foam concrete under different dry densities is analyzed by using the roundness. Fiber compounding has a significant effect on increasing the flexural strength of foam concrete. The optimal total fiber volume blending rate is 0.18%, and the blending ratio of polypropylene fiber and glass fiber is 2:1; the two fibers have achieved complementary advantages and significant improve the flexural performance of foam concrete. Fiber blending can improve the water absorption and thermal conductivity of foam concrete, thus reducing the insulation performance of foam concrete, but the test results are still better than the industry standard.

Acknowledgments. This research was funded by the Basic research project of Liaoning Provincial Department of Education, grant number lnjc201916.

References

1. Sun, W.B., Li, J.H., Zhang, Z.C.: A study on strength and influencing factors of potamatory foam concrete. J. Harbin Constr. Univ. **3**, 79–83 (2002)
2. Gai, G.Q.: A study on the pore structure of ceramsite foam concrete and its effect on its properties. Silic. Build. Prod. **5**, 13–15 (1995)
3. Li, X.Z., Zhang, R.R., Gao, D.W.: A study on rapid restoration and reconstruction of graphite tailings vegetation. Heilongjiang Water Conserv. Sci. Technol. **3**, 67–68 (2007)
4. Mao, H.L., Liu, W.M.: A study on cement-stabilized graphite tailings used as subbase of highway. Shandong Traffic Technol. **3**, 21–23 (2003)
5. Bai, Z.M., Liao, L.B.: Preparation technology and properties of graphite tailings sinterbrick. Study Mineral Rock Geochem. Bull. **4**, 221–225 (1999)
6. Kearsleya, E.P., Wainwright, P.J.: Ash content for optimum strength of foamed concrete. Cem. Concr. Res. **2**, 241–246 (2002)

7. Jones, M.R., Mccarthy, A.: Utilising unprocessed low-lime coal fly ash in foamed concrete. Fuel **84**(11), 1398–1409 (2005)
8. Nambuar, E.K., Ramamurthy, K.: Models relating mixture composition to the density and strength of foam concrete using response surface methodology. Cem. Concr. Compos. **9**, 752–760 (2006)

Influence of Oxide Molar Ratio on Size Effect of Geopolymer Recycled Aggregate Concrete

Xiaowei Sun[1], Hongguang Bian[1], Zhaoyang Ding[1,2(✉)], and Lin Qi[3]

[1] School of Material Science and Engineering, Shenyang Jianzhu University, Shenyang, China
SXW@sjzu.edu.cn
[2] School of Civil Engineering, Shenyang Jianzhu University, Shenyang, China
[3] Shenyang Urban Construction University, Shenyang, China

Abstract. Four different size of concrete cube (70 mm^3, 100 mm^3, 150 mm^3, and 200 mm^3) of geopolymer recycled aggregate concrete (GRAC) were prepared by replacing cement with geopolymer and natural aggregate with wast concrete. The effect of oxide molar ratio of raw material on compressive strength and its size effect of GRAC was studied.The results show the size conversion coefficient of GRAC cannot adopt the values from the current national standard GB/T50081. The relationship of size conversion coefficient α and oxide molar ratio ε of GRAC was worked out. It was found that the compressive strength of GRAC of all sizes were in line with the Bazant's size theory. Oxide molar ratio on critical size and critical value of GRAC were calculated.

Keywords: Geopolymer recycled aggregate concrete · Oxide molar ratio · Compressive strength · Size conversion coefficient · Size effect

1 Introduction

The production process of concrete consumes abundant resources, it is known that 1700–2000 kg of sand and stone and 350–450 kg of cement will be consumed for 1 m^3 of concrete [1]. According to the National Bureau of Statistics, the consumption of sand and stone and cement in 2019 for China is 3.58 billion tons 810 million tons. Sand, stone, and calcium carbonate mineral (main material of cement) are natural resources. The sustained consumption of those natural resources will damage the environment and exhausted the resource, even although China is a rich resource country.

Geopolymer recycled aggregate concrete (GRAC) was prepared by substituting cement with geopolymer and coarse aggregate with recycled aggregate concrete (RAC). GRAC is a green and environmentally friendly building material [2], not only reducing the consumption of natural resources, such as stone and calcium carbonate mineral [3], but also recycling industrial waste such as slag and fly ash [4]. However, due to the difference of the nature of cement to geopolymer and RAC to natural aggregate, some performances of GRAC, such as its size conversion coefficient and size effect, are uncertain. Moreover, because the raw material of geopolymer are different by each researcher, it is hard to give a uniform law to describe the effect of raw material to the properties of

G. Feng (Ed.): ICCE 2021, LNCE 213, pp. 516–524, 2022.
https://doi.org/10.1007/978-981-19-1260-3_47

GRAC. Therefor, in this study, oxide molar ratio of raw material is selected as the main factor, and its effect on the performance of GRAC is studied.

2 Experimental Program

2.1 Raw Materials

Blast furnace slag used in the test is produced by Anshan Iron and Steel Co.Ltd and a first grade fly ash of Benxi is selected, the main chemical components of which are shown in Table 1. Water glass used is produced by Shandong Yousuo Chemical Technology Co.Ltd. The original modulus is 3.3, the chemical composition of which is shown in Table 2. Recycled aggregate with original strength grade of C40 is used as coarse aggregate, which is artificially broke into a maximum particle size of 25 mm and continuous gradation. Natural medium-fine river sand is used as fine aggregate.

Table 1. Chemical composition of mineral slag and fly ash/wt%

	CaO	SiO_2	Al_2O_3	Fe_2O_3	MgO	Na_2O	K_2O	loss on ignition
Slag	43.10	32.26	14.69	2.06	6.19	–	–	0.97
Fly ash	5.51	48.54	28.35	6.37	2.42	3.01	3.90	0.96

Table 2. Chemical composition of water glass

Chemical composition (Oxide percentage)			$n(SiO_2)/n(Na_2O)$	pH	$\rho/g \cdot cm^{-3}$
$Na_2O/\%$	$SiO_2/\%$	$H_2O/\%$			
7.96	26.1	66	3.3	13.1	1.47

2.2 Experimental Method and Instrument

According to the previous research [5], the optimal alkali-activator solution is used, of which the content of water glass in alkaline activator solution is 40% to the total mass of liquid, and the concentration of NaOH is 9mol/L, liquid binder ratio is 0.5. Sand coarse aggregate ratio is 0.44. Oxide molar ratio ($n(CaO):n(SiO_2 + Al_2O_3)$) is the main research variable in this research, the value is selected as 0.7, 0.75, 08, 0.85, 0.9. The mixing ratio of each group are shown in Table 3.

GRAC was prepared into cube samples with side lengths of 70 mm, 100 mm, 150 mm and 200 mm. A total of 6 samples of each size and proportion were made, and a total of 120 samples were tested. The compressive strength value and compressive strength test instrument use the RGM-100A microcomputer-controlled universal testing machine produced by Shenzhen Regal Instrument.

Table 3. Mix ratio of GRAC/kg·m^{-3}

Oxide molar ratio		Slag	Fly ash	NaOH	Water-glass	Water	Sand	Recycled -aggregate
n(CaO): n(SiO$_2$ + Al$_2$O$_3$)	n(SiO$_2$): n(Al$_2$O$_3$)							
0.7	3.36	393	172	88	113	170	847	1050
0.75	3.40	416	149	88	113	170	847	1050
0.8	3.44	441	124	88	113	170	847	1050
0.85	3.48	461	104	88	113	170	847	1050
0.9	3.53	482	83	88	113	170	847	1050

3   Results and Discussion

3.1   The Influence of Oxide Molar Ratio on Compressive Strength

Figure 1 presents compressive strength under different oxide molar ratio and size.It can be observed that with the increasing of n(CaO):n(SiO$_2$ + Al$_2$O$_3$), the compressive strength of GRAC shows a trend of increases at beginning and then decreases. According to the research of J. L. Provis [6], geopolymers are divided into tow system: N-A-S-H (low or no calcium, three-dimensional network structure) and C-A-S-H (high-calcium, layered structure). The diagrammatic drawings of low-CaO and high-CaO structure are shown in Fig. 2.

As n(CaO): n(SiO$_2$ + Al$_2$O$_3$) increased from 0.7 to 0.8, the structure of geopolymer binder transfers from low-calcium system to high-calcium system. Moreover, according to the research of Wang Qing [7], when the molar ratio exceeds 0.8, the CaO in the system is saturated, then the strength is controlled by the molar ratio of n(SiO$_2$):n(Al$_2$O$_3$). For C-A-S-H structure, as n(SiO$_2$):n(Al$_2$O$_3$) increases from 3 to 4, the strength decreases. Becaease this process is converted from PSS type ([-Si-Al-Si-]) geopolymer to PSSS type ([-Si-Al-Si-Si-]) geopolymer, and the structure of PSS-geopolymer is more dense [8].

The law of the size of the test block on the compressive strength of GRAC is: f_{100} > f_{150} > f_{200} > f_{70}. The general rule is that the larger the side length of the cube test block, the greater the compressive strength. However, the strength of GRAC cube specimen with a side length of 70 mm does not conform to this law, and its compressive strength is the lowest. This is because there are a great deal of original cracks in the recycled aggregate, and as the size of the test block is close to the size of the aggregate, the strength of the concrete will be affected by the cracks more significantly [9].

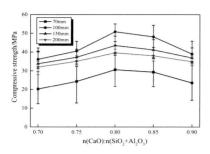

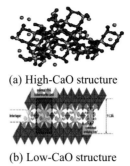

(a) High-CaO structure

(b) Low-CaO structure

Fig. 1. Compressive strength under different oxide molar ratio and dimensions

Fig. 2. Low-CaO and High-CaO structure of geopolymer

3.2 The Influence of Oxide Molar Ratio on Size Conversion Coefficient

Through the size conversion factor (α), compressive strength of non-standard test block can be calculated by the standard specimen. The cube specimen with a side length of 150 mm is a standard specimen, and the conversion coefficients of other non-standard samples are shown in Eqs. (1)–(3):

$$\alpha_{70} = f_{cu,70}/f_{cu,150} \tag{1}$$

$$\alpha_{100} = f_{cu,100}/f_{cu,150} \tag{2}$$

$$\alpha_{200} = f_{cu,200}/f_{cu,150} \tag{3}$$

Figure 3 shows effect of the oxide molar ratio on the size conversion coefficient. It can be seen from Fig. 3 that there are only 2 test data is in the region of 0.95–1.05. That indicates that the size conversion coefficient of GRAC cannot current adopt the value form the national standard "Standard for Test Methods for Mechanical Properties of Ordinary Concrete" GB/T50081, in which the value is 0.95 and 1.05 for samples that size are 100 mm^3 and 200 mm^3. In this paper, the mathematical equations of oxide molar ratio (ε) and size conversion coefficient (α) are established by the linear fitting. As shown in Fig. 4 and the results are as: $a_{70} = 0.463 + 0.28\varepsilon$, $a_{100} = 1.016 + 0.12\varepsilon$, $a_{200} = 0.966 - 0.04\varepsilon$.

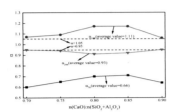

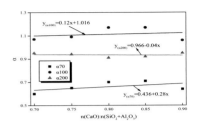

Fig. 3. Effect of the oxide molar ratio on size conversion coefficient

Fig. 4. Fitting curve of the oxide mole ratio on size conversion coefficient

3.3 Bazant Size Effect Fitting

GRAC is a kind of quasi-brittle material. The strain energy released by crack propagation under load causes the existence of size effect. According to the size effect theory of Bazant [10], the relationship between the nominal compressive strength of concrete and the size D is shown in Eq. (4)

$$f_N = f_\infty(1 + \frac{D_b}{D})$$ (4)

In the equation, f_∞ is the nominal compressive strength of the infinite size of GRAC, and D_b is the effective thickness of the boundary layer cracking. After decomposition:

$$f_N = f_\infty + f_\infty \times \frac{D_b}{D}$$ (5)

As X=1/D, $Y = f_N$, $C = f_\infty$, $A = f_\infty \times D_b$, then Eq. (5) can become a linear equation as shown in Eq. (6)

$$Y = AX + C$$ (6)

Table 4. Parameter calculation of the theoretical formula of the dimension effect

Oxide molar ratio	0.70	0.75	0.80	0.85	0.90
A	28.1285	29.9343	28.5871	27.8429	31.2721
C	913.428	1069.714	2226.857	2027.1429	789.857
R^2	0.9722	0.9969	0.9994	0.9989	0.93372

X and Y can be directly calculated by the strength and the size of the specimen. Table 4 shows parameter calculation of the theoretical equation of the size effect. Through the study of compressive strength and size effect degree (Δa), it is found that the data of the 70 mm cube specimen does not conform to this law, so its data is not used in the calculation.

Figure 5 presents the comparison diagram of measured strength values of cube specimen with different side lengths and the Bazant theoretical strength. It can be seen that under different oxide molar ratio conditions, the measured strength values of compressive strength are all on the theoretical curve, so Bazant theory can be used for GRAC.

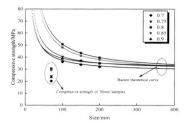
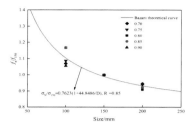

Fig. 5. Comparison diagram of the measured strength value and the Bazant theoretical strength

Fig. 6. Comparison diagram of dimensionless strength and Bazant theoretical strength

3.4 Critical Size and Critical Strength

None-dimensionalized the value of size effect of GRAC [10], shown in Eq. (7), in which f_{150} is the measured strength of the GRAC in a cube specimen with a side length of 150 mm, and b is the undetermined coefficient of the equation. Then the undetermined coefficients of none-dimensionalized of size effect of GRAC were be worked out by regression analysis of Eq. (6), shown in Eq. (8). The comparison diagram of none-dimensionalized strength and Bazant theoretical strength is shown in Fig. 6,

$$\frac{f_N}{f_{150}} = \frac{f_\infty}{f_{150}}(1 + \frac{b}{D}) \tag{7}$$

$$\frac{f_N}{f_{150}} = 0.7623 \times (1 + \frac{44.8486}{D}) \quad R^2 = 0.85 \tag{8}$$

It can be seen from Fig. 6 that as the side length of the cube specimen is 200 mm, the data of each oxide molar ratio is close to the theoretical curve. The oxide molar ratio of 0.8 is the demarcation point of the size effect change, and its compressive strength is the highest, and the strength will be reduced if it is greater or less than the strength. Therefore, a piece-wise function is used to fit the relationship between oxide molar ratio (ε) and f_∞/f_{150} and D_b. Figure 7 shows schematic diagram of the relationship between the oxide molar ratio and f_∞/f_{150} and D_b. Equations (9)–(12) are the fitting curve equation.

When $0.7 \leq \varepsilon \leq 0.8$:

$$\frac{f_\infty}{f_{150}} = 2.24027 - 1.97934 \times \varepsilon \tag{9}$$

$$D_b = -347.09643 + 531.1485 \times \varepsilon \tag{10}$$

When $0.8 \leq \varepsilon \leq 0.9$:

$$\frac{f_\infty}{f_{150}} = -0.62015 + 1.59552 \times \varepsilon \tag{11}$$

$$D_b = 431.98236 - 442.4970 \times \varepsilon \tag{12}$$

According to Eqs. (9) to Eq. (12), piece-wise function has higher applicability to oxide mole ratio, and substitute it into Eq. (8). In summary, the predictive equation of the nominal compressive strength of the size effect rate of the compressive strength of GRAC can be obtained by the coupling effect of the size effect and the molar ratio of oxides. As shown in Eq. (13) and Eq. (14).

When $0 \leq \varepsilon \leq 50\%$:

$$\frac{f_N}{f_{150}} = 2.24027 - 1.97934 \times \varepsilon \times (1 + \frac{-347.09643 + 531.1485 \times \varepsilon}{D}) \qquad (13)$$

$$\frac{f_N}{f_{150}} = -0.62015 + 1.59552 \times \varepsilon \times (1 + \frac{431.98236 - 442.4970 \times \varepsilon}{D}) \qquad (14)$$

According to Eq. (13) and Eq. (14), the critical strength characteristic value (f_{cr}) can be calculated under the condition of different oxide molar ratios and the side length of the specimen is infinite. Consider the scope of application of engineering size effect, when the difference between the nominal compressive strength and the characteristic value of the critical dimension is within 5%, the size of the specimen corresponding to the nominal compressive strength can be considered as the critical dimension (D_{cr}). Figure 8 presents relationship of the oxide molar ratio to the critical strength and the critical size. It can be seen that as the molar ratio of oxide increases, its critical size gradually increases. However, the critical strength reached its maximum value when the oxide molar ratio was 0.8. The prediction equations proposed by Eq. (13) and Eq. (14) have wider applicability mainly reflected in two aspects: Considering effect of the amount of recycled aggregate and the size effect coupling, so it has higher applicability. The use of dimensionless methods has certain reference significance for predicting the compressive strength of GRAC with other strength grades and oxide molar ratios.

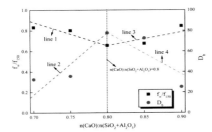

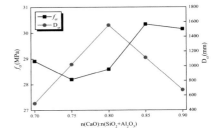

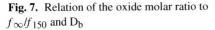

Fig. 7. Relation of the oxide molar ratio to f_∞/f_{150} and D_b

Fig. 8. Relationship of the oxide molar ratio to the critical strength and the critical size

4 Conclusions

The strength of GRAC increases first and then decreases with the increase in the molar ratio of oxides, and reaches the maximum when $n(CaO):n(SiO_2 + Al2O3) = 0.8$. The law of compressive strength under specimen of different sizes is as follows: $f_{100} > f_{150} > f_{200} > f_{70}$.

The size conversion coefficient α can be obtained by linear fitting to the relationship between the oxide molar ratio ε.

The compressive strength of the cube specimen with side lengths of 200 mm, 150 mm and 100 mm are in line with Bazant's theoretical curve of size effect. The non-dimensional method can be used to obtain the prediction equation of the combined effect of the compressive strength of GRAC with the oxide molar ratio and the size effect. Finally, the critical size and critical strength values of GRAC under different oxide molar ratios are obtained.

Acknowledgments. This research was funded by the National Natural Science Foundation of China, grant number 51678374; the Science and Technology Program of the Ministry of Housing and Urban-Rural Development of China, grant number 2019-K-48; State Key Laboratory of Silicate Materials for Architectures (Wuhan University of Technology), grant number SYSJJ2021-13.

References

1. Li, Q.Y., Quan, H.Z.: Recycled Aggregate for Concrete M. China Building Materials Industry Press, Beijing (2010)
2. Davidovits, J.: 30 years of successes and failures in geopolymer applications. In: Geopolymer 2002 Conference, Melbourne, Australia, 1–16 October 2002
3. Davidovits, J.: Geopolymers: inorganic polymeric new materials. J. Therm. Anal. **37**, 1633–1656 (1991)
4. Hai, R., Li, D.D., Hui, C., et al.: Experimental study on workability and mechanical performance of liquid high strength recycled concrete. J. Sci. Technol. Eng. **18**(19), 256–261 (2018)
5. Ding, Z.Y., Zhou, J.H., Su, Q., et al.: Study on mechanical properties of geopolymer recycled aggregate concrete. J. J. Shenyang Jianzhu Univ. (Nat. Sci.) **37**(1), 138–146 (2021)
6. Provis, J.L., Myers, R.J., White, C.E.: X-ray microtomography shows pore structure and tortuosity in alkali-activated binders. J. Cement Concr. Res. **42**(12), 855–864 (2012)
7. Wang, Q., Kang, S.R., Wu, L.M.: Molecular simulation of N-A-S-H and C-A-S-H in geopolymer cementitious system. J. J. Build. Mater. **023**(001), 184–191 (2020)
8. Xu, H.., Van Deventer, J.S.J.: Ab initio calculations on the five-membered alumino-silicate framework rinks model: implications for dissolution in alkaline solutions. J. Comput. Chem. **24**(5), 391–404 (2000)
9. Guo, Z.H.: Strength and Deformation of Concrete M. Tsinghua University Press, Beijing (1997)
10. Bazant, Z.P.: Size effect. J. Int. J. Solids Struct. **37**(4), 69–80 (2000)

Fast Algorithm for Completion State Calculating of Wire Rope Cable Bridge

Yang Feng, Dan Song$^{(\boxtimes)}$, and Minghui Li

Power China Northwest Engineering Corporation Limited, Xi'an 710065, China
346201135@qq.com

Abstract. When determining the completion state of the main cable of a cableway bridge, although the catenary theory can accurately consider the nonlinear mechanical effect of the cable, the iterative calculation is cumbersome and not convenient for engineering applications. Although the calculation based on the parabola theory is simple, the calculation accuracy for long cables is low. In this paper, based on catenary theory and considering the calculation accuracy and avoiding iterative calculation, a fast algorithm for the completion state of the main cable of the wire rope cableway bridge is proposed. The results show that the bridge state can be quickly and accurately determined by approximating the horizontal component of the cable tension and avoiding multiple iterative calculations. The proposed algorithm can be used in engineering design and construction.

Keywords: Fast algorithm · Wire rope · Cable Bridge · Catenary theory · Parabola theory

1 Introduction

Cable bridges are essential transport channels that play an important role in water conservation, hydropower, emergency relief, and military fields. The main cable is the main load-bearing component of the cable bridge, which is crucial for the stress state of the entire structure. The main cable produces significant deformation under the stress state, which directly affects the bridge shape and stress state of the structure. The completion state of the main cable bridge directly determines the ideal completion state of the cable bridge, which affects the safety of the cable bridge under dead load and live load. This directly leads to the calculation of the bridge state of the main cable multiplied; therefore, it is urgent to find a fast algorithm for the bridge state of the main cable of the wire rope track bridge.

The construction control theory of conventional suspension bridges has been widely used, and many studies have reported results on the calculations of the main cable [1–3]. Given the structural differences, these results cannot be directly applied when calculating the main cables of cable bridges, and only a few studies have specifically investigated cable bridges. The catenary theory curve was first studied in [4], and a solution method was established. However, the catenary theory curve is a transcendental function that is difficult to calculate. Accordingly, a parabola theory was proposed [5]. Using this

G. Feng (Ed.): ICCE 2021, LNCE 213, pp. 525–531, 2022.
https://doi.org/10.1007/978-981-19-1260-3_48

theory, the first two terms of the catenary curve expansion series were approximately calculated, and the application scope of each theoretical method was described. In [6], a hybrid finite element model combining a three-node quadratic curve element and a straight beam element was proposed using the finite element method combined with the structural characteristics of the cable bridge. This method ignores the influence of high-order displacement and needs to be improved. In [7], a five-node curve element was used to simulate the initial shape of the cable, and a high calculation accuracy was achieved. However, for a suspension cable structure with a small sag, several degrees of freedom need to be considered in the calculation, which increases the calculation cost. In [8], the spline element method was used for the nonlinear analysis of cable bridges. In [9], the algorithm of the main cable erection parameters of a cable bridge based on an approximate analytical method was used, and the cable alignment of the main span was parabolic. In [10], the initial cable shape of the cable bridge was considered to be catenary. The catenary theory in this case it can be simulated well only by establishing the model of the mid-span cable, and a large calculation cost is required to determine the cable shape that meets various requirements through iterative trial calculation. Therefore, based on the catenary theory and the basic principle of the stress-free state control method, as well as considering the calculation accuracy and avoiding iterative calculation, this study proposes a fast algorithm for calculating the completion state of wire rope cable bridges.

2   Analytic Algorithm Theory

2.1   Catenary Cable Shape Theory

Because the main cable of the cableway bridge is fixed at both ends and suspended in the middle span, the main cable conforms to the catenary clue shape without requiring any external force. Therefore, the catenary clue shape theory can be used for analysis.

As shown in Fig. 1, assuming that the main cable is a completely flexible cable that can only bear tension and cannot be bent, the balanced analysis of any micro-segment can be performed as follows.

$$\frac{y''}{\sqrt{1 + (y')^2}} = -\frac{q}{H}, \tag{1}$$

where q is the weight per unit cable length, and H is the horizontal component of cable tension, which is determined by the cable tension T.

$$H = \frac{T}{\sqrt{1 + (y')^2}} \tag{2}$$

After obtaining the integral solution of formula (1) and considering the boundary conditions (x = 0, y = 0) and (x = l, y = h), the catenary cable shape can be obtained as follows.

$$y = \frac{H}{q}\left[cha - ch\left(a - \frac{q}{H}x\right)\right] \tag{3}$$

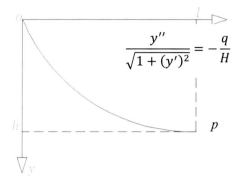

$$\frac{y''}{\sqrt{1 + (y')^2}} = -\frac{q}{H}$$

Fig. 1. Schematic diagram of main cable

In the formula, the parameter $a = arsh\left[\dfrac{qh}{2H \cdot \backslash sh\left(\frac{ql}{2H}\right)}\right] + \dfrac{ql}{2H}$ can be obtained by integrating the length s of the catenary cable from Eq. (3) as follows.

$$S = \int_0^1 \sqrt{1 + (y')^2}\,dx$$

$$= -\frac{H}{q}\left[sh\left(\frac{ql}{H} - a\right) + cha\right] \tag{4}$$

The elastic elongation ΔS of the catenary due to cable tension T is given as follows.

$$\Delta S = \int_0^S \frac{T}{EA}\,ds = \frac{H}{EA}\int_0^l\left[1 + (y')^2\right]dx$$

$$= \frac{H}{2EA}\left\{l + \frac{H}{2q}\left[sh\left(\frac{2gl}{H} - 2a\right) + sh(2a)\right]\right\} \tag{5}$$

The unstressed cable length S_0 can be calculated as follows.

$$S_0 = S - \Delta S \tag{6}$$

From Eqs. (2) to (6), it can be seen that given the tension T of one end cable, the horizontal component H and the cable shape y are coupled with each other. Thus, the unstressed cable length S_0 can be determined by iterative calculation. In the calculation, the iterative parameter can be selected as the horizontal component force H, and its initial iterative value H_0 is often taken as the component force of the tower end cable tension T along the chord line.

$$H_0 = T \cdot \frac{l}{\sqrt{l^2 + h^2}} \tag{7}$$

2.2 Parabolic Cable Shape Theory

The force and load of the cable bridge can be applied according to the span and bridge width. Under the action of an external force, the external load of the main cable is equal per linear meter in the span direction, which agrees with the parabola theory. Therefore, the parabola thread shape theory can be used for the cable shape approximate analysis.

The approximate analytical algorithm assumes that the load and deformation of several main cables of the cable bridge are uniform, and it is equivalent to a single cable plane model, as shown in Fig. 1. It is assumed that the cable shape of the main span is parabolic, and both ends are hinged at the theoretical intersection points A and B of the main cable axis at the saddle. The influence of the horizontal dip angle and the sag of the main cable of the anchor span was ignored, and a horizontal cable force was used to replace the axial cable force of all sections of the anchor cable span and main span. Based on the relationship between the mechanical balance and the physical properties of the materials, according to the geometric conditions of equal cable suspension speed of the whole cable for the same cable bridge under any two load conditions, a cubic algebraic equation of horizontal cable force under the calculated load state is obtained. Finally, the main cable shape is obtained from the obtained balance conditions of the horizontal cable force and moment [11].

At present, after the cable bridge is completed, the deck alignment can only be adjusted by removing the deck system components. The economical and ideal construction technology for main cable erection is as follows: initial suspension of empty cable, linear measurement accuracy meets the requirements, the end of the steel cable is permanently anchored to the anchorage, S_0 of the cable of the entire bridge is fixed immediately and is unchanged in the subsequent construction process [12]. The actual blanking length S_c includes the blanking length S_0 of the steel cable and a certain operating length.

2.3 Fast Algorithm

According to the catenary cable shape given in Eq. (3), the cable slope of the tower end can be obtained as follows.

$$y'(0) = sha = \frac{qh}{2H \cdot sh\left(\frac{ql}{2H}\right)} \cdot ch\left(\frac{ql}{2H}\right) - \sqrt{1 + \left[\frac{gh}{2H \cdot h\left(\frac{ql}{2H}\right)}\right]^2} \cdot sh\left(\frac{ql}{2H}\right) \quad (8)$$

By Eq. (8),

$$shx = \sum_{n=0}^{\infty} \frac{x^{2n+1}}{(2n+1)!} \qquad chx = \sum_{n=0}^{\infty} \frac{x^{2n}}{2n!}. \quad (9)$$

It can be seen that, when $(ql)/(2H) < 1$, it is desirable.

$$h\left(\frac{ql}{2H}\right) \approx \frac{ql}{2H} \qquad ch\left(\frac{ql}{2H}\right) \approx 1 \quad (10)$$

Equation (8) can be simplified as follows.

$$y'(0) = \frac{h}{l} + \sqrt{1 + \frac{h^2}{l^2}} \cdot \left(\frac{ql}{2H} \right) \tag{11}$$

Y in the tower ends are determined using

$$y'(0) = \frac{\sqrt{T^2 - H^2}}{H}. \tag{12}$$

By combining Eqs. (11) and (12), the horizontal component H can be decomposed as follows.

$$H = \frac{l}{l_0} \cdot T \cdot \left[\sqrt{1 - \left(\frac{ql}{2T} \right)^2} - \frac{qh}{2T} \right] \tag{13}$$

Equation (13) shows that, for a given cable tension T, the horizontal component force h can be approximately solved, and the unstressed cable length S_0 can be quickly determined using Eq. (6) without iteration. According to the actual engineering situation, in the approximate solution condition, $(ql)/(2H) < 1$, which can be met under most cable tension levels. Therefore, the accuracy of this method can be easily guaranteed.

3   Real Bridge Verification

A wire rope cable bridge was used for comparative calculation and analysis. The geometric and material characteristics are as follows. The sag was 2.275 m, the span was 95 m, the area of the wire rope was 1568 mm^2, the density was 0.171 kN/m, the elastic modulus was 1.10×10^5 MPa, and the external load was 3.5 kN/m^2.

By using the above three methods, the unstressed length of the main cable of the wire rope cable bridge under different tension levels was calculated. The calculation results are summarized in Table 1. For comparison, the errors in the latter two methods are taken as the deviation values relative to the catenary theory.

According to the results in Table 1, the following conclusions can be drawn. 1) The horizontal force H and cable length S obtained by the fast algorithm have certain errors for the theoretical calculation results of the catenary; however, the horizontal force errors are approximately 1‰, and the cable length errors are approximately 1%. 2) The error of the fast algorithm results is less than that of the parabola theory. 3) Compared with the parabolic theoretical error, the error of the fast algorithm decreased with an increase in the cable force. 4) The absolute value of the H error calculated by the fast algorithm was within 1 kN, and the error was within the acceptable range. 5) The recommended methods for calculating the main cable alignment of cable bridges are the catenary and parabola theories, and the performance of the fast algorithm is between them; therefore, the calculation results are reliable. 6) The calculation results of the fast algorithms H and S are in the range of the calculation results of the catenary and parabola theories under different cable force conditions. The fast algorithm has high stability.

Table 1. Calculation results

Working condition	Cable force /kN	Results of catenary theory		Results of parabolic theoretical				Results of fast algorithm			
		H /kN	S /m	H /kN	Err or	S /m	Error	H /kN	Err or	S /m	Error
	①	②	③	③/①	④	④/②	⑤	⑤/①	⑥	⑥/②	
Empty cable condition	610.4	603.79	94.987	604.7	1.52 ‰	95.47	5.04 ‰	604.46	1.11 ‰	95.36	3.88 ‰
Completed state	628.37	621.72	94.443	622.5	1.26 ‰	95.48	11.0 ‰	622.25	0.85 ‰	95.37	9.77 ‰
Full load condition	752.96	744.77	94.466	745.8	1.38 ‰	95.49	10.87‰	745.63	1.16 ‰	95.43	10.24‰

4 Conclusion

In this study, based on the catenary theory, the completion state of the main cable of the cable bridge was quickly obtained through reasonable simplification, avoiding the tedious process of the accurate catenary theoretical calculations. Compared with the parabola theory, the calculation results were more accurate, and the error results were within the acceptable range. Compared with the catenary theory, the fast algorithm had a higher calculation speed and fewer iterations. Therefore, by approximately solving the horizontal component of cable tension and avoiding multiple iterative calculations, the fast algorithm can quickly determine the completed state of the main cable of the cable bridge with high precision, which can be useful in engineering design and construction.

References

1. Wang, L.: Discussion on load acceptance test method of pedestrian cable and bridge. Sichuan Cement **000**(011), 86 (2017)
2. Liu, J.H.: Form-finding analysis of wind cable with stable cable based on subsection catenary theory. Sichuan Build. Mater. **244**(12), 46 (2020)
3. Wang, L., et al.: Calculation of main cable of pedestrian ropeway bridge and analysis of completed state. Steel Struct. **034**(002), 34 (2019)
4. Shan, S.D.: Engineering Cableway, p. 36. China Forestry Press, Beijing (2001)
5. Chen, Z.L., et al.: Study on practical dynamic characteristic table of suspension cable engineering. Highw. Automob. Transp. (003), 03 (2007)
6. Qu, B.M., et al.: Cable-beam hybrid finite element model and its application in cable-bridge analysis. Comput. Struct. Mech. Appl. **7**(4), 8 (1990)
7. Dong, M., et al.: Nonlinear finite element analysis of tension structures. J. Comput. Mech. **14**(3) (1997)
8. Yang, Y.P.: Spline calculation theory of cable bridge. Eng. Mech. **1**(a01) (1999)
9. Zhou, Y.J., et al.: Load resistance analysis and test of cable bridge. J. PLA Univ. Sci. Technol. (Nat. Sci. Edit.) **11**(6) (2010)

10. Tian, Z.C., et al.: Determination of completed state and load test of temporary cable bridge. J. Changsha Jiaotong Univ. (2005)
11. Zhou, X.N.: Engineering Cableway and Suspension Bridge. People's Communications Press, Waveland (2013)
12. Guo, X., et al.: Analysis method of Literature cable of self-anchored suspension bridge based on segmental unstressed cable length of main cable. Highway **7** (2019)

Effect of Grouting Defect Sleeve on Seismic Performance of Concrete Column

Changjun Wang[1], Zhijian Zhao[2], Xiaonan Xu[2], Sen Pang[1], and Hongguang Zhu[2(✉)]

[1] Beijing Building Research Institute Co. Ltd of CSCEC, Beijing 100076, China
[2] School of Mechanics and Civil Engineering, China University of Mining & Technology, Beijing, China
Zhg@cumtb.edu.cn

Abstract. The development of assembly construction technology has become an inevitable way for the transformation and upgrading of China's construction industry. However, in the development process, the grouting sleeve connection quality problem is the most prominent. In this paper, ABAQUS model is used to establish a concrete column model with grouting defects, and the influence of grouting defects on the seismic performance of concrete columns is explored: Under the condition of the same defect location and quantity, the larger the defect size, the lower the bearing capacity, the worse the ductility, the weaker the energy dissipation capacity; In the case that the number and size of defects are the same, the positive and negative directions of the hysteresis curve are symmetric when the defects are located in the direction of stress, and the energy dissipation capacity is higher than that when all the defects are located on one side of the column section.

Keywords: Void defects · Mechanical properties · Concrete columns · Hysteresis loops · Abaqus

1 Introduction

The development of prefabricated building will become an inevitable way for the transformation and upgrading of the construction industry. However, in the process of development, assembly building also has its disadvantages, among which the grouting sleeve connection quality problem is the most prominent [1]. The primary problem of precast concrete structures is the connection between precast components such as beams, columns, plates and shear walls, which must be effectively integrated to ensure the safety, ease of use and durability of the structure under various load conditions [2] 53/5000.

Andrea Belleri et al. [3] (2012) showed through experimental research that grout steel sleeve is suitable for columnal-foundation connection in earthquake area. ZhengLu et al. [4] (2017) studied the seismic performance of prefabricated concrete columns with grouting sleeve joint. The test results show that the precast column has good energy dissipation capacity and stiffness degradation is slower than the cast-in-place column, and meets the requirements of interstorey displacement ratio during large earthquakes. GuoshanXu et al. [5] (2017) found that the assembled concrete shear wall sample was

© The Author(s) 2022
G. Feng (Ed.): ICCE 2021, LNCE 213, pp. 532–540, 2022.
https://doi.org/10.1007/978-981-19-1260-3_49

similar to the cast-in-place shear wall sample in terms of failure mode, interlayer displacement Angle, ultimate force, ductility and dissipated hysteretic energy. In this paper, ABAQUS is used to conduct numerical simulation of prefabricated concrete columns with defects under the action of low cyclic load, and the influence of size and location of grouting defects on the seismic performance of prefabricated concrete columns is analyzed.

2 Concrete Column Model

2.1 Column Size and Reinforcement

Figure 1 is the schematic diagram of this modeling model: The assembled concrete column consists of three parts: concrete base, concrete column body and loading column cap.

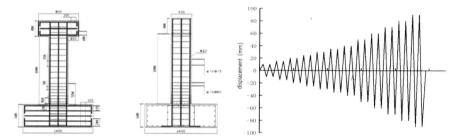

Fig. 1. Schematic diagram of size and reinforcement of prefabricated concrete column

Fig. 2. Control diagram of specimen loading system

2.2 Loading System

Two loads are applied during the experiment: axial pressure and horizontal thrust. The axial pressure belongs to the monotone static load and remains unchanged during the test after it is applied. In this experiment, the design value of axial compression ratio is 0.3, and the design value of compressive strength of C35 commercial concrete is 16.7 mpa. Therefore, the axial pressure of the specimen was 801.6 kN. The axial pressure of the specimen is loaded, and the axial force is slowly loaded to the predetermined design value. Thereafter, the axial pressure is kept unchanged during the test. The loading regime is shown in Fig. 2.

2.3 Specimen Number and Defect Setting

Four models are established this time. The defect Settings are shown in Table 1, and the naming rules are shown in Table 2.

Table 1. Defect settings

The first type (named 1)	The second type (named 2)

Note: black circle means full grouting, white circle means defects in grouting sleeve; Type 1 indicates that only one side of the force direction has grouting defect; Type 2 indicates that there are grouting defects on both sides of the section of the stressed directional column.

Table 2. Naming rules of specimens

Specimen	Defects
4d-D-1	The anchorage length of reinforcement is 4d, and the defect type is 1
5d-D-2	The anchorage length of reinforcement is 5d, and the defect type is 2
BM	No defects in grouting

3   Model

3.1   Reinforcing Steel Bar

The steel bar adopts an ideal elastic-plastic tripline model, in which the yield strength, ultimate strength, yield platform length and ultimate strain of the steel bar are taken from the Code for Design of Concrete Structures (GB50010-2010) [6].

3.2   Concrete

Concrete constitutive model adopts the concrete damage plastic model (CDP) of ABAQUS. See the detailed introduction. It should be noted that although stirrups are created in the finite element model in this paper, the Truss unit does not take into account the tangential action between stirrups and concrete. Therefore, in order to consider the restraint effect of stirrup on concrete in the core area of section, this paper adopts Mander constraint concrete constitutive model for concrete constrained by stirrup [7]. Peak strain and peak stress are calculated using the formula suggested by Dr. Hu Qi [8] of Tongji University.

3.3   Optimization of Grouting Sleeve in the Model

In this paper, the effect of grouting sleeve with end defects on seismic performance of assembled concrete column is studied. The key is how to deal with grouting sleeve and grouting defects. The former can not better simulate the mechanical behavior of the defective sleeve from tension to compression through the spring element, and the latter is easy to cause "over-constraint" and inconsistent with the actual force. Therefore, a more convenient and easy to operate method is adopted in this paper. The grouting sleeve connecting steel specimens are treated as equivalent steel bars, and the grouting defects are simulated by cutting off the corresponding steel bars.

4 Results Analysis

4.1 Seismic Performance of Concrete Columns with Different Anchorage Lengths

(1) Load-displacement curve

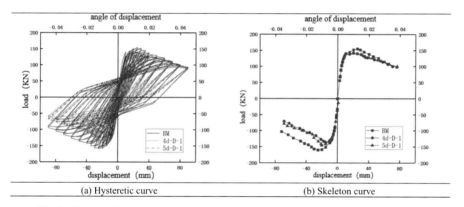

| (a) Hysteretic curve | (b) Skeleton curve |

Fig. 3. Hysteresis curve and skeleton curve of BM, 4D-D-1 and 5D-D-1 specimens

Figure 3 shows the hysteretic curves and skeleton curves of BM, 4D-D-1 and 5D-D-1 specimens. By comparison, it is found that the bearing capacity of the positive and negative directions decreases when the grouting defects are located only on the side of the section of the stressed direction column, and the bearing capacity of the side with grouting defects decreases more obviously than when the grouting is full. The larger the defect is, the smaller the anchorage length is, the lower the bearing capacity is and the more obvious the stiffness degradation is. Table 3 lists the bearing capacity of each specimen at the characteristic point.

(2) Ductility and deformation performance analysis

In this study, displacement ductility coefficient is expressed, which is defined as the ratio of the ultimate failure point displacement Δf of the specimen to the yield displacement Δy of the specimen.inflection point method is adopted for yield point and 85% peak load method is adopted for limit point.

It can be seen from Table 4 that the positive direction of each specimen is dense, and the yield point, limit point displacement and ductility coefficient are close. In the negative direction, the limit displacement and displacement ductility coefficient decrease due to grouting defects.

Table 3. Bearing capacity information of each specimen

Specimen	Peak load /kN		Ultimate load /kN	
	Containing no defects (Positive direction)	Containing defects (Negative direction)	Containing no defects (Positive direction)	Containing defects (Negative direction)
BM	154.90	−159.95	131.67	−135.96
4d-D-1	142.27	−137.11	120.93	−116.54
5d-D-1	141.70	−145.50	120.44	−123.68

Note: The ultimate load is 85% of peak load

Table 4. Ductility information of each specimen

Specimen	Yield point displacement/mm		Limit point displacement/mm		Displacement ductility coefficient	
	Positive direction	Negative direction	Positive direction	Negative direction	Positive direction	Negative direction
BM	5.94	−5.2	46	−44.31	7.74	8.52
4d-D-1	5.94	−5.04	47.27	−26.95	7.96	5.35
5d-D-1	6.2	−5.62	50.20	−29.52	8.1	5.25

(3) Energy dissipation capacity analysis

Figure 4 shows the relationship between energy dissipation and displacement of prefabricated concrete column specimens connected with grouting sleeve under the condition of end defects of different anchoring lengths. It can be seen from the Fig. 4 that the energy dissipation of concrete specimens increases gradually with the increase of displacement. In the horizontal displacement is small, when the graph is less than or equal to 40 mm, basic no grouting defects of energy dissipation of the specimens, even stronger than when grouting full energy dissipation ability, but when more than 40 mm, namely, with the increase of horizontal displacement concrete column is affected by grouting defects is more and more big, the filling defect, the greater the energy dissipation of the specimens.

4.2   Effect of Different Defect Locations on Seismic Performance of Concrete Columns

(1) Load-displacement curve

Figure 5 shows the hysteresis curves and skeleton curves of 5D-D-1 and 5D-D-2 specimens. It can be seen from the Fig. 5 that the location of grouting defects has a certain influence on the seismic performance of concrete columns when the number and

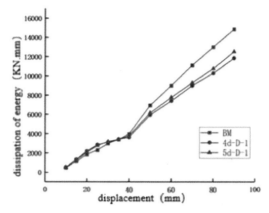

Fig. 4. Energy dissipation curves of 4D-D-1 and 5D-D-2 specimens

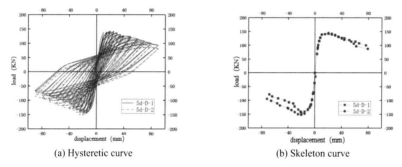

(a) Hysteretic curve   (b) Skeleton curve

Fig. 5. Hysteresis curve and skeleton curve of 5D-D-1 and 5D-D-2 specimens

size of defects are the same. The positive and negative directions of the hysteretic curves are symmetric when the grouting defects are found on both sides of the stress direction. Table 5 lists the bearing capacity of each specimen at the characteristic point.

Table 5. Bearing capacity information of each specimen

Specimen	Peak load /kN		Ultimate load /kN	
	(Positive direction)	(Negative direction)	(Positive direction)	(Negative direction)
5D-D-1	141.70	−145.50	120.44	−123.68
5D-D-1	144.69	−151.96	122.99	−129.17
Note: The ultimate load is 85% of peak load				

(1) Ductility and deformation performance analysis

Table 6. Ductility information of each specimen

Specimen	Yield point displacement /mm		Limit point displacement /mm		Displacement ductility coefficient	
	(Positive direction)	(Negative direction)	(Positive direction)	(Negative direction)	(Positive direction)	(Negative direction)
5d-D-1	6.2	−5.62	50.20	−29.52	8.1	5.25
5d-D-2	6.2	−6.22	41.73	−40.11	6.73	6.45

As can be seen from Table 6, when the number and size of defects are the same, for a certain side of the column section in the stress direction, when the number of grouting defects on this side is larger, the displacement ductility coefficient of this side is smaller and the ductility is worse.

(3) Energy dissipation capacity analysis

Figure 6 shows the relationship between energy dissipation and displacement of prefabricated concrete column specimens connected with grouting sleeve at different positions with the same defects. The Fig. 6 shows that when the horizontal displacement is less than or equal to 40 mm, for the same displacement, flaws in the column on both sides of the energy dissipation ability is slightlyless than a full energy dissipation along the side of the column, when more than 40 mm, with the increase of the horizontal displacement, the defects in the column on both sides of the energy dissipation capacity will gradually all energy dissipation along the side of the column above defects. It can be seen that all the defects on one side of the column are more unfavorable to the seismic performance of the prefabricated concrete column.

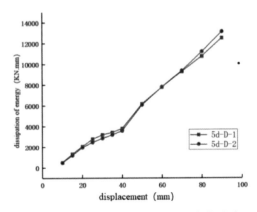

Fig. 6. Energy dissipation curves of 5D-D-1 and 5D-D-2 specimens

5 Conclusion

In this chapter, ABAQUS finite element software is used to simulate the influence of different defect sizes and positions on the seismic performance of concrete columns, and the main conclusions are as follows:

(1) In the case of the same defect location and number, the larger the defect size, the lower the bearing capacity, the more obvious the stiffness degradation, the smaller the displacement ductility coefficient, the worse the ductility, the weaker the energy dissipation capacity;

(2) In the case that the number and size of defects are the same, when the defect position is symmetric in the force direction, the hysteresis curve is symmetric in the positive and negative directions, and its energy dissipation capacity is higher than that of the defect located on one side of the column section Force.

Acknowledgments. This paper is supported by cSCEC-2019-Z-4, and we would like to express our sincere thanks.

References

1. Gao, R.D., Li, X.M., Xu, Q.F.: Existing problems and solutions of sleeve grouting in prefabricated monolithic concrete building. Constr. Technol. **47**(10), 5 (2018)
2. China Building Standards Design and Research Institute, China Academy of Building Research. Technical specification for fabricated concrete structures[S]. Beijing (2014)
3. Sayadi, A.A., Abd, A.B., Sayadi, A., et al.: Effective of elastic and inelastic zone on behavior of glass fifiber reinforced polymer splice sleeve. Constr. Build. Mater. **80**, 38–47 (2015)
4. Lu, Z., Wang, Z.X., Li, J.B., et al.: Studies on seismic performance of precast concrete columns with grouted splice sleeve. Appl. Sci. Basel **50**(9), 97–109 (2017)
5. Xu, G.S., Wang, Z., Wu, B., et al.: Seismic performance of precast shear wall with sleeves connection based on experimental and numerical studies. Eng. Struct. **45**(16), 346–358 (2017)
6. Code for design of concrete structures (GB50010-2010) [S]: Beijing: China Architecture and Architecture Press (2010)
7. Mander, J.B., Priestley, M.J.N., Park, R.: Theoretical stress-strain model for confined concrete. J. Struct. Eng. **114**(8), 1804–1826 (1998)
8. Qi, H., Li, Y.G., Lü, X.L.: Study of uniaxial mechanical behavior of concrete confined with hoops. Build. Struct. **1**, 79–82 (2011)

Developments of Microseismic Monitoring Technology in Deep Tunnels in China

Guangliang Feng[1(✉)], Qi Ma[1,2], Xun Zhang[3], Dingjun Qu[4], Guojun Wang[4],
Jian Liu[5], and Zongjun Zhu[5]

[1] State Key Laboratory of Geomechanics and Geotechnical Engineering, Institute of Rock and Soil Mechanics, Chinese Academy of Sciences, Wuhan 430071, China
glfeng@whrsm.ac.cn
[2] School of Engineering Science, University of Chinese Academy of Sciences, Beijing, China
[3] Central South Exploration & Foundation Engineering Co., Ltd., Wuhan 430081, China
[4] Hubei Yihua Group Mining Co., Ltd., Yichang 443000, China
[5] Hubei Shanshuya Mining Co., Ltd., Yichang 443100, China

Abstract. With the increasing demand for infrastructure construction as the global economy progresses, the need for exploration and utilization of deep underground space becomes more crucial. Microseismic (MS) monitoring technology has been widely used in deep underground tunnel projects for safety monitoring in China in recent years. In this paper, four aspects of MS monitoring technology developments, i.e. distribution of projects, environment and system characteristic, purpose, and effect of MS monitoring in deep tunnel projects in China were analyzed and summarized. The results show that the technology was mainly applied in the west of China with a wide range of project types. The maximum buried depth of the tunnels monitored reached 2525 m. The tunnel construction method was mainly drilling and blasting method. The lithologies of the tunnels were mainly marble, granite and basalt. The monitoring purpose was for disaster warning and mechanism understanding. In addition, the future development of MS monitoring technology in deep tunnels in China is prospected. The results will be helpful for a rapid development of MS monitoring technology in deep tunnels in China.

Keywords: Tunnel · Microseismic Monitoring · Deep · Development · China

1 Introduction

With the rapid development of underground engineering, various deep underground projects are planned, are under construction, and have been built. Microseismic (MS) monitoring technology has been widely used in deep tunnel engineering for safety monitoring as a geophysical and seismic-based method, and has achieved a series of helpful results. The MS wave signal associated with rockmass fracturing can be captured by MS sensors in spatial. Then, the MS wave signal can be systematic analyzed and information such as the time, location and energy of the fracture can be obtained. Therewith, the stability and its development trend of rockmass can be judged based on the information [1]. MS monitoring technology originated in the US rock explosion research of

© The Author(s) 2022
G. Feng (Ed.): ICCE 2021, LNCE 213, pp. 541–548, 2022.
https://doi.org/10.1007/978-981-19-1260-3_50

deep mines in the early 20th century. However, because the technology was not mature enough at that time, the research did not achieve satisfying results. The technology was first put into use officially by South Africa in the 1960s. South Africa successfully used the technology to monitor the microseismicity in deep gold mining, and the stability of rockmass was judged based on the microseismicity monitored. Later, the MS monitoring technology was widely used as an emerging rockmass stability monitoring method in the fields such as tunnels, slopes and mines in developed countries. In 2004, China introduced the ISS MS monitoring system from South Africa, and a MS monitoring system in Dongguashan copper mine was established [2]. In 2010, Chen et al. [3] introduced the MS monitoring technology to the deep TBM tunnel in hydropower engineering for rockburst monitoring in China. In 2015, A MS method for quantitative warning of rockburst development processes in deep tunnel was first proposed by Feng et al. [4]. And the MS method has been successfully applied to rockburst warning in deep tunnels at the Jinping II hydropower project.

In recent years, with the development of science and technology, MS monitoring technology realized informatization, automation and intelligence, which has been widely used in deep tunnels in China with a series of helpful results [5–19]. In order to understand the developments of MS monitoring technology in deep tunnels in China, four aspects, i.e. distribution of projects, environment and system characteristic, purpose, and effect of MS monitoring in deep tunnels in China were analyzed and summarized. In addition, the future development direction of MS technology in deep tunnel engineering is prospected. The results will be helpful for a rapid development of MS monitoring technology in deep tunnel in China.

2   Distribution Characteristic of Tunnel Projects with MS Monitoring

In order to understand the distribution of tunnel projects with MS monitoring, the literature in the past decade related to tunnel projects with MS monitoring was investigated, and then, the related information, i.e. project name, project region, application time and project type, were counted. Distribution of several tunnel projects with MS monitoring is shown in Table 1.

According to the statistics in Table 1, we can see that MS monitoring technology has been widely used in deep tunnel projects in China in the past decade, and the projects were mainly distributed in western China. The area was mainly concentrated in Sichuan Province and there were several projects in Tibet and Xinjiang. This is mainly because the terrain characteristics of China. The western region is with a lot of mountains, high altitude, and rich water conservancy resources. As an effective and mature mean of rockmass stability monitoring technology, MS monitoring technology was mainly applied in water conservancy and hydropower project, traffic project, and laboratory. The application types of projects were comprehensive.

Table 1. Distribution of several tunnel projects with MS monitoring

No.	Project	Area	Start time (year)	Project type
1#	Diversion tunnel of Baihetan hydropower station	Jinsha river	2013	Water conservancy and hydropower project
2#	Duoxiongla tunnel	Tibet autonomous region	2016	Traffic project
3#	Headrace and drainage tunnels of Jinping II hydropower station	Sichuan province	2010	Water conservancy and hydropower project
4#	Bayu tunnel	Tibet autonomous region	2017	Traffic project
5#	China Jinping underground laboratory phase II	Sichuan province	2015	Laboratory
6#	Micangshan road tunnel	Sichuan-Shaanxi junction	2017	Traffic project
7#	Qinling section of Han-Ji-Wei project	Shaanxi province	2016	Water conservancy and hydropower project
8#	Underground powerhouse of Shuangjiangkou hydropower station	Sichuan province	2018	Water conservancy and hydropower project
9#	Water tunnel of Xinjiang ABH project	Xinjiang autonomous region	2019	Water conservancy and hydropower project

3 Environment and System Characteristic of MS Monitoring

Furthermore, based on the analysis of the literature related to tunnel project with MS monitoring in the past decade, the maximum buried depth, lithology, construction method and monitoring system of the tunnel projects with MS monitoring were counted. And the environment and system characteristic of MS monitoring were obtained. The statistical results are shown in Table 2. The projects corresponding to the serial numbers in Table 2 are consistent with those described in Table 1.

According to the statistics of Table 2, the MS monitoring system applied in deep tunnel projects in China at present mainly includes ISS, IMS, ESG and SSS. The IMS is a new generation product of ISS. With the development of deep tunnel projects, the maximum buried depth with MS monitoring reached 2525 m. And most of the buried depth was deeper than 2000 m (Fig. 1). MS monitoring technology can be applied to a variety of lithologies, including basalt, marble, diorite, schist, gneiss, mudstone and

Table 2. Environment and system characteristic of MS monitoring

No.	Maximum buried depth (m)	Lithology	Construction method	Monitoring system
1#	395	Basalt	Drilling and blasting(D&B)	ISS
2#	832	Schist, gneiss	TBM	ESG
3#	2525	Marble	TBM; D&B	ISS; ESG
4#	2080	Granite	D&B	SSS; IMS
5#	2400	Marble	D&B	SSS; IMS
6#	1070	Quartz diorite	D&B	ESG
7#	2012	Quartzite, granite, diorite	TBM; D&B	ESG
8#	500	Granite	D&B	ESG
9#	2253	Siltstone, metamorphic mudstone	TBM; D&B	SSS; IMS; ESG

granite, in which marble and granite were the most ones. The tunnel construction methods were TBM and D&B methods, among which the D&B method was in the majority.

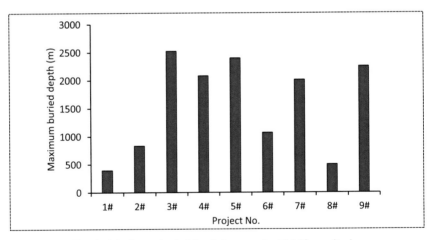

Fig. 1. Maximum buried depth for tunnel with MS monitoring

4   Purpose of MS Monitoring in Deep Tunnel

Furthermore, statistical analysis was made on the purposes of MS monitoring applied in the above tunnel projects. The statistical results are shown in Table 3. The projects corresponding to the serial numbers in Table 3 are consistent with those described in Table 1. As can be seen from Table 3, the purpose of MS monitoring technology applied in the deep tunnel projects in China was mainly for disaster warning, as well as the disaster mechanism understanding.

Table 3. Main purposes of MS monitoring in deep tunnel

No.	1#	2#	3#	4#	5#	6#	7#	8#	9#
Disaster early warning	√	√	√	√	√	√	√	√	√
Disaster mechanism understanding	√	√	√	√	√	√	√	√	√

5 Effect of MS Monitoring in Deep Tunnel: Case Study

5.1 Case 1: The Deep Tunnels in Jinping II Hydropower Project

Jinping II hydropower project was located on the Yalong River, Sichuan Province in southwestern China. The project and MS monitoring information have been described in detail by the references [4, 5] and [10]. A microseismicity-based method of rockburst quantitative warning in the deep tunnels was proposed to warn of the rockburst risk. The method was shown in Fig. 2. During the continuous MS monitoring activity, there were no serious rockburst casualties. The safety and construction schedule of the tunnels were ensured.

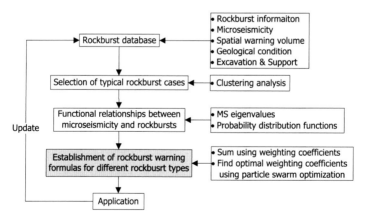

Fig. 2. Flowchart for quantitative warning of rockburst in deep tunnel using microseismicity [4]

5.2 Case 2: The Deep Tunnels in Baihetan Hydropower Station

The dam area of Baihetan hydropower station was a mountain canyon landform, and the rockmass along the diversion tunnels in this project was monoclinal rock with medium to high geostress. The columnar jointed basalt developed in the project was generally dark gray or gray, and the rock blocks were mostly columnar mosaic structure. MS monitoring technology was utilized during the excavation of columnar jointed basalt in the deep tunnels. The monitoring results provides a helpful understanding and suggestion for the application of MS monitoring technology, reasonable excavation and supporting design for the tunnels in the similar rockmass with columnar joint sets. Some results are shown in Fig. 3 [13].

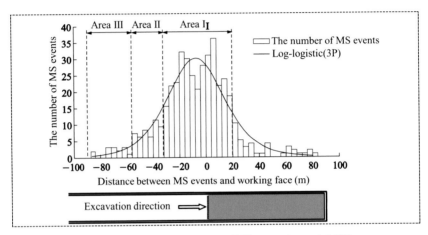

Fig. 3. Statistical distribution of MS events in spatial [13]

6 Conclusions

MS monitoring technology has achieved unprecedented development with a series of helpful results in deep tunnel projects in China. The technology was mainly applied in the western region of China, and the project distribution was mainly concentrated in Sichuan Province, and partly in Tibet and Xinjiang. It has been used in a wide range of project types, covering many fields such as water resources and hydropower engineering, transportation engineering and laboratory. The maximum buried depth of the projects reached a maximum of 2525 m and most of the buried depth was deeper than 2000 m. There were many types of MS monitoring systems used, i.e. IMS, ESG, ISS and SSS. The construction methods related were TBM and D&B methods, among which the D&B method was in the majority. And the lithologies of the projects were mainly marble and granite. The monitoring purposes were mainly for disaster warning and mechanism understanding. And the technology has been successfully used in a lot of deep tunnels in China.

The tunnel project is becoming deeper and deeper and with a more complex geological environment. Therewith, the type of disasters under high stress in deep tunnel and their mechanism are becoming complex. Therefore, a higher requirement for the performance and analysis method of the MS monitoring technology is required. The improvement of the performance of MS monitoring system, the optimization of the monitoring scheme, and the precision and efficiency of MS data analysis will be several key research directions in the future. The other one key research direction is the MS based disaster analysis technology, such as big data analysis on the large number of MS monitoring data, deeper understanding of disaster development process and establishment of a more accurate MS based early warning method for disaster.

Acknowledgments. The authors gratefully acknowledge the financial support received from the National Natural Science Foundation of China under Grant Nos 42177168 and 41972295, Key Research and Development Program Project of Hubei Province under Grant No 2020BCB078.

References

1. Mendecki, A.J.: Seismic Monitoring in Mines. Chapman & Hall, London (1997)
2. Tang, L.Z., Pan, C.L., Yang, C.X., et al.: Establishment and application of microseismicity monitoring system in Dongguashan copper mine. Metal Mine. **10**, 41–44, (2006). 86, (in Chinese)
3. Chen, B.R., Feng, X.T., Zeng, X.H., et al.: Realtime microseismic monitoring and its characteristic analysis during TBM tunneling in deep-buried tunnel. Rock Mech. Rock Eng. **30**(2), 275–283 (2011). (in Chinese)
4. Feng, G.L., Feng, X.T., Chen, B.R., et al.: A microseismic method for dynamic warning of rockburst development processes in tunnels. Rock Mech. Rock Eng. **48**(5), 2061–2076 (2015)
5. Feng, G.L., Feng, X.T., Chen, B.R., et al.: Microseismic sequences associated with rockbursts in the tunnels of the Jinping II hydropower station. Int. J. Rock Mech. Min. Sci. **80**, 89–100 (2015)
6. Feng, G.L., Feng, X.T., Chen, B.R., et al.: Sectional velocity model for microseismic source location in tunnels. Tunn. Undergr. Sp. Tech. **45**, 73–83 (2015)
7. Feng, G.L., Feng, X.T., Chen, B.R., et al.: Performance and feasibility analysis of two microseismic location methods used in tunnel engineering. Tunn. Undergr. Sp. Tech. **63**, 183–193 (2017)
8. Feng, G.L., Feng, X.T., Xiao, Y.X., et al.: Characteristic microseismicity during the development process of intermittent rockburst in a deep railway tunnel. Int. J. Rock Mech. Min. Sci. **124**, 104135 (2019)
9. Feng, G.L., Feng, X.T., Chen, B.R., et al.: Effects of structural planes on the microseismicity associated with rockburst development processes in deep tunnels of the Jinping-II Hydropower Station. China. Tunn. Undergr. Sp. Tech. **84**, 273–280 (2019)
10. Feng, G.L., Feng, X.T., Chen, B.R., et al.: Characteristics of microseismicity during breakthrough in deep tunnels: case study of Jinping-II hydropower station in China. Int. J. Geomech. **20**(2), 04019163 (2020)
11. Ma, C.C., Li, T.B., Zhang, H.: Microseismic and precursor analysis of high-stress hazards in tunnels: A case comparison of rockburst and fall of ground. Eng. Geol. **265**, 105435 (2020)
12. Feng, G.L., Chen, B.R., Jiang, Q., et al.: Excavation-induced microseismicity and rockburst occurrence: similarities and differences between deep parallel tunnels with alternating soft-hard strata. J. Central South Univ. **28**, 582–594 (2021)
13. Feng, G.L., Feng, X.T., Chen, B.R., et al.: Temporal-spatial evolution characteristics of microseismic activity for columnar jointed basalt tunnel at Baihetan hydropower station. Chin. J. Rock Mech. Eng. **34**(10), 1967–1975 (2015). (in Chinese)

14. Tang, Z.L., Liu, X.L., Li, Y.C., et al.: Microseismic characteristic analysis in deep TBM construction tunnels. J. Tsinghua Univ. (Sci. Technol.). **58**(5), 461–468 (2018). (in Chinese)
15. Du, L.J., Hong, K.R., Wang, J.X., et al.: Rockburst characteristics and prevention and control technologies for tunnel boring machine construction of deep-buried tunnels. Tunnel Constr. **41**(1), 1–15 (2021). (in Chinese)
16. Qian, B., Xu, N.W., Xiao, P.W., et al.: Damage analysis and deformation early warning of surrounding rock mass during top arch excavation of underground powerhouse of Shuangjiangkou hydropower station. Chin. J. Rock Mech. Eng. **38** (12), 2512–2524 (2019). (in Chinese)
17. Duan, W.S., Zhang, Z.Q., Tang, L.X., et al.: Microseismic response law of rock burst in No.5 branch of Qinling tunnel of water diversion project from Hanjiang River-to-Weihe River. Yangtze River **51**(03), 167–173 (2020). (in Chinese)
18. Li, S.J., Zheng, M.Z., Qiu, S.L., et al.: Characteristics of excavation disasters and long-term in-situ mechanical behavior of the tunnels in the China Jinping underground laboratory. J. Tsinghua Univ. (Sci. Technol.). **61**(8), 842–852 (2021). (in Chinese)
19. Du, X.Z.: Key technical issues in Qinling water diversion tunnel of Hanjiang-to-Weihe river project and research advancements. Yellow River **42**(11), 138–142 (2020). (in Chinese)

Author Index

Printed in the United States
by Baker & Taylor Publisher Services